Ferdinand Stoliczka

Scientific Results of the Second Yarkand Mission

Vol. 1

Ferdinand Stoliczka

Scientific Results of the Second Yarkand Mission
Vol. 1

ISBN/EAN: 9783337418557

Printed in Europe, USA, Canada, Australia, Japan

Cover: Foto ©berggeist007 / pixelio.de

More available books at **www.hansebooks.com**

SCIENTIFIC RESULTS

OF

THE SECOND YARKAND MISSION;

BASED UPON THE COLLECTIONS AND NOTES

OF THE LATE

FERDINAND STOLICZKA, Ph.D.

LEPIDOPTERA,

BY

FREDERIC MOORE, F.Z.S., ETC.,

ASSISTANT CURATOR, INDIAN MUSEUM, LONDON.

Published by order of the Government of India.

CALCUTTA:
OFFICE OF THE SUPERINTENDENT OF GOVERNMENT PRINTING.
1879.

CALCUTTA:
PRINTED BY THE SUPERINTENDENT OF GOVERNMENT PRINTING,
8, HASTINGS STREET.

SCIENTIFIC RESULTS

OF

THE SECOND YARKAND MISSION.

LEPIDOPTERA.

By FREDERIC MOORE, F.Z.S., ETC., *Assistant Curator, India Museum, London.*

Tribe—**PAPILIONES.**

Family—*NYMPHALIDÆ.*

Sub-Family—*SATYRINÆ.*

1. HIPPARCHIA LEHANA. Plate I, fig. 4, ♂.

Hipparchia lehana, Moore, Ann. and Mag. Nat. Hist. 1878, p. 227.

Allied to *H. baldiva,* Moore, from Upper Kunawur, the upperside being paler in colour, the discal transverse ochreous band broader on both wings, and its inner border, in the male, inwardly oblique. Both sexes above and beneath are without the small ocellus on the discal band above the anal angle. The underside is also very much paler, and the transverse sinuous lines wider apart.

Expanse ♂ 2, ♀ 2¼ inches.

Habitat.—Leh (September 6th, 1873), Kharbu, 13,000 feet, both in Ladák.

2. HIPPARCHIA CADESIA.

Hipparchia cadesia, Moore, Proc. Zool. Soc. 1874, p. 565, pl. 66, fig. 7.

Hab.—Leh, September 8th.

3. EPINEPHILE CHEENA.

Epinephile cheena, Moore, Proc. Zool. Soc. 1865, p. 501, pl. 30, fig. 6.

Hab.—Gaganghir, Kashmir.

4. AULOCERA SWAHA.

Satyrus swaha, Kollar, Hügel's Kaschmir, iv, p. 444, tab. 14, figs. 1, 2 (1844).
Satyrus brahminus (part), Blanch., Jacq. Voy. dans l'Inde, iv, Ins. p. 22, t. 2, figs. 5, 6, ♂.

Hab.—Gaganghir, Kashmir.

5. AULOCERA BRAHMINA.

Satyrus brahminus, Blanchard, Jacq. Voy. dans l'Inde, iv, Ins. p. 22, t. 2, fig. 4 (1814), ♂.
Aulocera weranga, Lang, Ent. Monthly Mag. iv, p. 247 (1868).

Hab.—Mataian, Dras Valley, 11,200 feet.

Sub-family—*NYMPHALINÆ*.

6. VANESSA LADAKENSIS. Plate I, fig. 2.

Vanessa ladakensis, Moore, Ann. and Mag. Nat. Hist. 1878, p. 227.

Nearest allied to *V. rizana*, Moore, from Cheeni, but is somewhat smaller, less angled below the apex of fore wing and at middle of the hind wing; the black markings on the upperside are much less prominent, the black oblique bands on forewing merging into the red and thus appearing somewhat confluent; the outer transverse discal yellow band on fore wing is also broader; other markings similar. On the underside the interspaces between the markings on fore wing are very much paler.
Expanse 1⅝ inch.
Hab.—Gogra, Changchenmo, 15,000 feet, October 1873; Karatágh Lake, on snow, midday temperature 33°, October 11th, 1873.

7. PYRAMEIS CARDUI.

Pyrameis cardui, Linn. Faun. Suec. p. 276 (1761).—Esper, Schmett. i, t. 10, fig. 3.—Eversmann, Ent. Imp. Ross. v, p. 107, t. 12, figs. 1, 2.—Erschoff, Lep. Turkestan, p. 15.

Hab.—Karghálik, November 11th, seen also south of Sánju and at Sánju, all in Eastern Turkestan.

8. ARGYNNIS JAINADEVA.

Argynnis jainadeva, Moore, Ent. Monthly Mag. i, p. 131 (1864); Proc. Zool. Soc. 1865, p. 495, pl. 30, fig. 1.

Hab.—Leh, September 6th.

LEPIDOPTERA.

Family—*PAPILIONIDÆ*.

Sub-family—*PIERINÆ*.

Genus BALTIA, Moore.

Baltia, Moore, Ann. and Mag. Nat. Hist. 1878, p. 228.

Fore wing very short; costa considerably arched from the base, apex and posterior angle rounded, exterior margin oblique, costal vein short, subcostal vein arched to end of the cell, six-branched, first and second branches arising at equal distances apart before the end of the cell and terminating on the costa before the apex, third branch bent near its base, middle, and immediately before its termination before the apex, the fourth, fifth, and sixth branches starting below from each of these angles, the fourth branch being very short; cell broad; discocellulars of nearly equal length, bent inwards; median vein three-branched, branches at equal distances apart; submedian vein curved: hind wing long, somewhat oval, slightly broader than fore wing, apex and exterior margin very convex, abdominal margin long; costal vein short; subcostal three-branched; cell broad; discocellulars oblique, upper the shortest; median vein three-branched; submedian nearly straight. Body small, abdomen short, thorax and front of head clothed with long lax hairs. Palpi very long, slender, densely hairy beneath. Legs short, femora fringed beneath with long lax hairs. Antennæ short, club large and spatulate.

Type. *Baltia shawii* (*Mesapia shawii*), Bates, in Henderson and Hume's Lahore to Yárkand, p. 305 (1873).

9. BALTIA SHAWII. Plate I, fig. 5, ♂.

Mesapia shawii, Bates, Henderson and Hume's Lahore to Yárkand, p. 305, ♀, 1873.
Baltia shawii, Moore, Ann. and Mag. Nat. Hist. 1878, p. 228.

Male. Upperside white; base of both wings densely black-speckled: fore wing with the costal edge ochreous and slightly black-speckled; a large black triangular oblique spot at end of the cell; a short discal transverse subapical black band, and a marginal row of black decreasing triangular spots: hind wing minutely and sparsely speckled with dark grey; a slight black streak at end of the cell, the speckles dense across the disc, and there forming a curved sinuous indistinct band. Body black. Palpi ochreous above and fringed with black beneath. Underside: fore wing with markings as above; costa and exterior margin tinged with ochreous: hind wing black-speckled, the speckles thickly disposed at the base, and also forming a narrow curved discal band; a slight black streak at end of the cell. Antennæ black, stem black-ringed. Abdomen beneath yellow. Legs black above, white beneath.

Female differs above in having the markings less prominently black, and the subapical band on fore wing continued across the wing on both upper and underside.

Expanse $1\frac{7}{8}$ inch.

Hab.—Aktâgh, north of the Karakoram Pass (15,500 feet), June 14th, 1874.

The male insect only was captured by Dr. Stoliczka; the female was taken on the Chang Lung Pass (18,000 feet) by Mr. R. B. Shaw during the expedition of 1870.

10. SYNCHLOE BRASSICÆ.

Pieris brassicæ, Linn., Faun. Suec. p. 269 (1761); Syst. Nat. i, p. 759.—Esper, Schmett. i, t. 3, fig. 1.—Erschoff, Lep. Turkestan, p. 4.
Pieris nipalensis, Gray, Lep. Ins. Nepal, pl. 6, fig. 1 (1846).

Hab.—Leh, September 6th.

11. SYNCHLOE RAPÆ.

Pieris rapæ, Linn., Faun. Suec. p. 270 (1761); Syst. Nat. i, p. 759.—Esper, Schmett. i, pl. 3, fig. 2.—Erschoff, Lep. Turkestan, p. 5.

Hab.—Yangihissár, April, Aktalla, May 17th, both in Eastern Turkestan.

12. SYNCHLOE DAPLIDICE.

Pieris daplidice, Linn., Syst. Nat. i, p. 760 (1767); Esper, Schmett. i, figs. 414, 415.—Erschoff, Lep. Turkestan, p. 5.

Hab.—Gond and Sonamarg, both in Kashmir.

13. SYNCHLOE CHLORIDICE.

Pieris chloridice, Hübner, Eur. Schmett. i, figs. 712, 713 (1803-1818).—Esper, Schmett. i, pl. 90, fig. 1.

Hab.—Sarikol, May 2nd and 8th.

14. COLIAS HYALE.

Colias hyale, Linn. Faun. Suec. p. 272 (1761); Syst. Nat. I, p. 764.—Esper, Schmett. I, pl. 4, fig. 2.

Hab.—Gaganghir, Kashmir; Sánju, October 30th; Sarikol, May 2nd; Yangihissár, April.

15. COLIAS FIELDII.

Colias fieldii, Menétries, Catal. Lep. Mus. Petrop. i, p. 79, t. 1, fig. 5 (1855).—Gray, Lep. Ins. of Nepal, pl. 5, fig. 2.

Hab.—Sonamarg, Kashmir, August 10th.

16. COLIAS STOLICZKANA. Plate I, fig. 1.

Colias stoliczkana, Moore, Ann. and Mag. Nat. Hist. 1878, p. 229.

Male. Upperside pale chrome-yellow, base of costal and abdominal borders greenish-yellow; base of wings speckled with blackish-brown; both wings with a broad yellowish-brown marginal band; a light narrow dusky-brown lunular streak at end of the cell in the fore wing. Underside: fore wing pale yellow; costal border and outer margin greenish-yellow; a dusky black-speckled lunular spot at end of the cell, and discal row of indistinct speckled spots: hind wing greenish-yellow, with darker green speckles; an ochreous-brown patch at end of cell, enclosing a white triangular mark and small spot; a discal series of dusky-brown dentate spots. Antennæ and legs reddish.

Expanse 1⅜ inch.

Hab.—North of Changla (17,000 feet), Ladák.

Differs from *C. eogene*, Feld. (Novara Reise, Lep. t. 27, fig. 7), in being smaller, and in having the wings, including the cilia, pale chrome-yellow instead of orange-yellow; the discocellular mark is less prominent and lunular, not oval; the broad marginal band is of a much yellower colour. On the underside, the discocellular mark on the fore wing is also lunular and is not pale-centred.

Sub-Family—*PAPILIONINÆ*.

17. PARNASSIUS CHARLTONIUS. Plate I, fig. 3, ♀.

P. charltonius, Gray, Catal. Lep. Ins. Brit. Mus. i, p. 77, pl. 12, fig. 7, ♂, (1852).

Hab.—Kharbu (13,000 feet), Ladák.

18. PARNASSIUS JACQUEMONTII.

P. jacquemontii, Boisd., Spéc. Gén. Lep. i, p. 400 (1836).—Blanchard, Jacq. Voy. dans l'Inde, iv, Ins. p. 16, t. 1, figs. 3, 4.—Gray, Catal. Lep. Ins. Brit. Mus. i, pl. 12, figs. 1, 2.

Hab.—North of Changla (17,000 feet), Ladák.

Sub-Family—*LYCÆNINÆ*.

19. POLYOMMATUS KASHGHARENSIS. Plate 1, fig. 7.

Polyommatus kashgharensis, Moore, Ann. and Mag. Nat. Hist. 1878, p. 230.

Male. Upperside pale blue, with narrow black exterior-marginal line; costal edge white. Cilia white, with dark inner border. Underside slightly pearly-grey, base of wings pale metallic green: fore wing with a white-bordered black spot in middle of the cell, and a curved discal series of five spots; a very indistinct spot at end of the cell, and a less distinct marginal series of spots: hind wing with three sub-basal and a curved discal series of six small white-circled black spots; an indistinct spot at end of the cell, and marginal row of spots with slightly ochreous interspaced upper dentated line.

Expanse 1¼ inch.

Hab.—Yangihissár, Eastern Turkestan, April 1874.

Allied to *P. semiargus*.

20. POLYOMMATUS LEHANUS. Plate I, fig. 6.

Polyommatus lehanus, Moore, Ann. and Mag. Nat. Hist. 1878, p. 230.

Male. Upperside violet-blue, somewhat brownish-blue at the margins. Cilia white. Underside leaden grey, palest at the apex and on hind wing: fore wing with a white-bordered black spot at end of the cell and a transverse discal oblique series of five spots: hind wing with a large triangular greyish-white spot at end of the cell, and a series of eight small round spots recurving from near base of costa across the disc to anal angle.

Expanse $\frac{7}{8}$ inch.
Hab.—Leh, 8th September 1873.
Allied to *P. pheretes*.

21. POLYOMMATUS YARKANDENSIS. Plate I, fig. 8.

Polyommatus yarkandensis, Moore, Ann. and Mag. Nat. Hist. 1878, p. 229.

Allied to *P. icarius*. Upperside dark blue, anterior and exterior borders dusky-brown: an indistinct streak at end of the cell on fore wing: hind wing with a marginal row of indistinct ochreous-bordered black spots. Cilia cinereous-white. Underside ochreous grey; fore wing with a white-circled black spot in middle of the cell, another below it, one at end of the cell, and a curved discal series of seven spots; a marginal row of indistinct spots bordered above by a dentated line with pale ochreous interspaces: hind wing with three white-circled black subbasal spots and a curved discal series of seven spots; a marginal row of prominent spots, bordered above by ochreous-interspaced dentated line.

Expanse 1⅜ inch.
Hab.—Yárkand, 23rd May 1873.

22. POLYOMMATUS ARIANA.

Polyommatus ariana, Moore, Proc. Zool. Soc. 1865, p. 504, pl. 31, fig. 2.

Hab.—Mataian, Drás valley (11,200 feet), Leh, September 6th and 8th.

23. POLYOMMATUS GALATHEA.

Polyommatus galathea, Blanchard, Jacq. Voy. dans l'Inde, iv, Ins. p. 21, pl. 1, figs. 5, 6, ♂, (1844).

Hab.—Sonamarg, Kashmir, 10th August.

24. DIPSAS ODATA.

Dipsas odata, Hewitson, Illustr. D. Lep. p. 66, pl. 30, fig. 13-4.—Moore, Proc. Zool. Soc. 1865, p. 507.

Hab.—Gaganghir, Kashmir.

LEPIDOPTERA.

Tribe—SPHINGES.

25. LEUCOPHLEBIA BICOLOR.

Leucophlebia bicolor, Butler, Proc. Zool. Soc. 1875, p. 16, pl. 2, fig. 5.

Hab.—Hatti, July 21st, Uri, July 23rd, both in Jhilam valley, on the road from Murree to Kashmir.

Tribe—BOMBYCES.

Family—ARCTIIDÆ.

26. HYPERCOMPA PRINCIPALIS.

Euprepia principalis, Kollar, in Hügel's Kaschmir, iv, p. 465, tab. 20, fig. 2 (1844).

Hab.—Gaganghir and Gond, in Kashmir.

27. ARCTIA ORIENTALIS.

Arctia orientalis, Moore, Ann. and Mag. Nat. Hist. 1878, p. 230.

Similar to *A. caja*, but differs, above, on the fore wing, in the general form of the bands, these being entire and transversely continuous, not broken longitudinally as in *A. caja*. On the hind wing the spot at the end of the cell is absent; this wing also has a yellowish-white narrow marginal line above, and brown cilia both above and beneath; the dorsal black band is on each segment and is moreover longer.

Expanse 2⅜ inches.
Hab.—Sonamarg, Kashmir, 8th August 1873.
This species has also been taken at Allahabad.

28. EUPROCTIS KARGHALIKA. Plate I, fig. 18.

Euproctis karghalika, Moore, Ann. and Mag. Nat. Hist. 1878, p. 231.

Male and female. Fore wing creamy-white, veins greyish-white; a large brown-speckled ochrey discocellular spot and submarginal row of spots: hind wing white. Thorax creamy-white; abdomen pale golden-yellow, of female grey slightly ringed with black, and tipped with large glossy golden-yellow tuft. Shaft of antennæ white, pectinations brown. Underside glossy white, costa of fore wing in male broadly suffused with brown.

Expanse, ♂ 1$\frac{7}{10}$, ♀ 1$\frac{7}{10}$ inch.
Hab.—Kárghalik Eastern Turkestan, May 29th and 30th.

29. EUPROCTIS LACTEA.

Euproctis lactea, Moore, Ann. and Mag. Nat. Hist. 1878, p. 231.

Uniform creamy-white, without markings. Abdomen tipped with pale yellow. Underside paler creamy-white; costal border of fore wing ochreous-brown. Palpi ochreous-brown. Antennæ pale ochreous-brown, shaft white. Fore tibiæ with ochreous-brown tuft.

Expanse 1⅜ inch.
Hab.—Kárghalik, May 29th, 1874.

Family—*NOTODONTIDÆ*.

30. PTILOPHORA KASHGHARA. Plate I, fig. 19.

Ptilophora kashghara, Moore, Ann. and Mag. Nat. Hist. 1878, p. 231.

Male. Fore wing dark grey, irrorated with brown scales, crossed by three indistinctly defined narrow zigzag brown bands, which are slightly dentated on the veins. Cilia alternately pale grey and brown: hind wing pale grey, sparsely sprinkled with brown scales. Thorax greyish-brown. Abdomen brown; three anterior segments with dorsal row of blackish tubercular scales; tip also black. Antennæ yellowish-testaceous. Underside grey, sparsely brown-speckled; long pubescence of abdomen brown and black. Legs pale brown.
Expanse $1\frac{7}{8}$ inch.
Hab.—Yangihissár, Eastern Turkestan, March 3rd, 1874.

31. OXICESTA MARMOREA. Plate I, fig. 17.

Oxicesta marmorea, Moore, Ann. and Mag. Nat. Hist. 1878, p. 231.

Male. Upperside greyish-brown: fore wing with a pale yellowish irregular streak along middle of cell to costa near apex, and a small spot beyond the cell, an indistinct pale streak below the cell; apical margin of costa and outer margin pale testaceous alternated with a short black streak, which extends through the cilia: hind wing uniform pale greyish-brown, slightly yellowish at base. Body and legs greyish-brown. Antennæ brown. Underside uniform greyish-brown; cilia of fore wing with black streaks.
Expanse $1\frac{1}{8}$ inch.
Hab.—Sasák Taka, Eastern Turkestan, May 16th, 1874.
Differs from *O. geographica* in being longer in the wings, of a different colour, and without the two transverse zigzag white bands on the fore wings.

Family—*SATURNIIDÆ*.

32. NEORIS SHAHIDULA.

Neoris shadulla, Moore, Proc. Zool. Soc. 1872, p. 577.

Hab.—Shahidula, Kuenlun (R. B. Shaw, 1870).
A distinct species from that figured by Felder (Nov. Reise, pl. 87, fig. 3), and named *Saturnia stoliczkai*, from Ladák.

Tribe—NOCTUES.
Family—*BOMBYCIDÆ*.

33. ACRONYCTA KARGHALIKA. Plate I, fig. 9.

Acronycta kargalika, Moore, Ann. and Mag. Nat. Hist. 1878, p. 232.

Female. Fore wing pale silvery brownish-grey; reniform and orbicular marks whitish, brown-bordered, and contiguous; a longitudinal streak from the base, a contiguous transverse subbasal recurved line, a discal transverse lunular line (crossed near posterior angle by a short streak), some short costal marks, and a streak on cilia between each vein, brown: hind wing glossy greyish-white, outer borders and veins pale greyish-brown. Thorax and abdomen dark grey. Antennæ grey. Underside greyish-white : fore wing with greyish-brown costal streaks and hinder margin : hind wing with brown basal costal streak and discocellular spot. Palpi brown at sides. Legs grey, femur tipped, tibia longitudinally streaked, and tarsi banded with black.

Expanse 1$\frac{4}{10}$ inch.

Hab.—Kárghalik, May 29th, 1874.

Nearest allied to *A. tridens*, but differs in being darker ; the markings are somewhat similar, but the basal longitudinal streak is shorter, which gives a wider interspace between the two transverse lines.

Family—*APAMIDÆ*.

34. HYDRÆCIA TIBETANA. Plate I, fig 21.

Hydræcia tibetana, Moore, Ann. and Mag. Nat. Hist. 1878, p. 232.

Male. Fore wing pale reddish-testaceous, crossed by two pale brown narrow lines with pale inner border, the first line subbasal and outwardly oblique, the other discal ; a submarginal row of blackish dots and pale marginal line ; orbicular and reniform marks indistinctly defined by a brown border : hind wing and abdomen paler. Underside palest on middle of wings, discal line on both wings and discocellular spot on hind wing slightly perceptible. Antennæ, palpi, and fore legs reddish-testaceous.

Expanse 1$\frac{1}{10}$ inch.

Hab.—Leh, September 1st, 1873.

35. MAMESTRA CANESCENS. Plate I, fig. 13.

Mamestra canescens, Moore, Ann. and Mag. Nat. Hist. 1878, p. 233.

Male. Fore wing brownish-grey : orbicular and reniform marks greyish-white with narrow black border ; a short double black streak below the base of the cell, and a quadrate mark below the orbicular spot ; an indistinct pale submarginal irregular fascia and black marginal lunular line with whitish inner border : hind wing pale greyish-brown. Antennæ brown. Underside glossy pale greyish-brown, each wing with indistinct short transverse discocellular streak.

Expanse 1$\frac{3}{8}$ inch.

Hab.—Kárghalik, Eastern Turkestan, May 30th, 1874.

36. MAMESTRA BRASSICÆ.

Phal. noct. brassicæ, Linn., Syst. Nat. i, p. 516.

Hab.—Srinagar, Kashmir, August 9th.

Family—*NOCTUIDÆ.*

37. AGROTIS SEGETUM.

Noctua segetum, Schiff., W. V. p. 252 (1776).—Eversm., Fauna Volgo-Ural, p. 196.
Agrotis segetum, Stoph., Haust. ii, p. 115.—Erschoff, Lep. Turkestan, p. 41.

Hab.—Tankse, Ladák; Kárghalik, Eastern Turkestan, May 29th.

38. AGROTIS AQUILINA.

Noctua aquilina, Schiff., W. V. p. 80 (1776).

Hab.—Tankse, 13,000 feet, Leh, August 29th, September 8th.

39. AGROTIS TIBETANA. Plate I, fig. 16.

Agrotis tibetana, Moore, Ann. and Mag. Nat. Hist. 1878, p. 233.

Upperside: fore wing greyish-brown, with indistinct dusky transverse subbasal double sinuous line, discal dentate lines, and pale outer-bordered wavy narrow submarginal band, speckled orbicular spot, and quadrate reniform mark. Cilia with narrow white marginal line: hind wing brownish-white, veins and outer margin brown; cilia white. Antennæ and body greyish-brown, tip of abdomen yellowish.
Underside: fore wing greyish-white, dusky-brown basally along the costa and hind margin, speckled on outer margin: hind wing whitish, an indistinct dusky spot at end of the cell, a spot medially on each vein, and narrow lunular marginal line. Legs greyish-brown, femora and tibiæ streaked, and tarsi banded, with black.
Expanse 1⅜ inch.
Hab.—Leh (August 8th, 1873).

40. SPÆLOTIS UNDULANS. Plate I, fig. 10.

Spælotis undulans, Moore, Ann. and Mag. Nat. Hist. 1878, p. 233.

Male and female. Fore wing grey-brown, irrorated with darker scales, crossed by sub-basal and ante and post-medial double pale-bordered undulated brown bands, each ending on the costa in a darker spot; a submarginal pale outer-bordered brown wavy fascia, and small black marginal lunules: hind wing glossy greyish-white with brownish-tinged borders, brown veins and lunular marginal line. Thorax grey-brown, abdomen greyish-white. Antennæ and palpi greyish-brown. Underside glossy greyish-white. Tibiæ streaked, and tarsi banded, with black.
Expanse 1₇/₁₆ inch.
Hab.—Ak Masjid, June 2nd, south-east of Chiklik, June 5th, 1874, both south of Yárkand.
Allied to *Spælotis pyrophila.*

LEPIDOPTERA. 11

Family—*ORTHOSIDÆ*.

41. TÆNIOCAMPA CHIKLIKA. Plate I, fig. 11.

Taeniocampa chiklika, Moore, Ann. and Mag. Nat. Hist. 1878, p. 234.

Male. Upperside grey : fore wing densely brown-speckled. Cilia with a brown-speckled line ; orbicular and reniform spots pale ; an indistinct transverse subbasal sinuous pale-bordered line : hind wing minutely brown-speckled, and with a pale brown cilial line. Underside paler ; both wings uniformly speckled, and with a very indistinct sinuous discal band. Antennæ blackish, shaft grey. Body, palpi, and legs brown-speckled.
Expanse 1⅜ inch.
Hab.—South-east of Chiklik, June 6th, 1874.

Family—*HADENIDÆ*.

42. HADENA STOLICZKANA. Plate I, fig. 12.

Hadena stoliczkana, Moore, Ann. and Mag. Nat. Hist. 1878, p. 234.

Male. Fore wing pale greyish-brown, crossed by three indistinct narrow brown zigzag double bands ; orbicular spot pale, reniform mark very indistinct ; two black spots below the apex ; a double narrow marginal blackish lunular line ; some short streaks on the costa : hind wing with the veins and a broad marginal band fuliginous-brown. Cilia white. Body pale greyish-brown. Antennæ brown. Underside greyish-white ; both wings crossed by a distinct curved discal brown band : fore wing with a discocellular brown lunule, and hind wing with a spot ; a marginal lunular dotted line. Legs grey ; tarsi banded with black.
Expanse 1⅜ inch.
Hab.—Kufelang (14,810 feet), June 6th, 1874.

Family—*HELIOTHIDÆ*.

43. HELIOTHIS SCUTOSA.

Heliothis scutosa, Schiff., Wien. Verz. p. 89 (1776).—Guén., Noct. ii, p. 182.

Hab.—Gaganghir, Kashmir.

44. HELIOTHIS DIPSACEA.

Heliothis dipsacea, Linn. Syst. Nat. ii, p. 856 (1776).—Guén., Noct. ii, p. 181.—Eversm., Fauna Volgo-Ural, p. 327.—Erschoff, Lep. Turkestan, p. 48.

Hab.—Posgám, near Yárkand, in lucerne-fields, May 28th. Yangihissár, April.

45. HELIOTHIS HYBLÆOIDES. Plate I, fig. 20.

Heliothis hyblæoides, Moore, Ann. and Mag. Nat. Hist. 1878, p. 234.

Upperside: fore wing grey, minutely brown-speckled; an indistinctly apparent brown curved streak at end of the cell, and a submarginal pale zigzag line: hind wing brownish-white, with a broad greyish-black medial transverse band (which is confluent with a curved discocellular black streak) and a large black oval spot on middle of outer margin; abdominal border tinged with brown; cilia white. Body grey, beneath whitish; legs greyish-white, brown-speckled.

Underside greyish-white: fore wing with a dusky-black transverse broad apical band and an outwardly-oblique medial band: hind wing with a dusky-black dentate streak at end of the cell, slight medial band, and oval marginal spot.

Expanse 1⅜ inch.

Hab.—Chiklik, south of Yárkand, June 3rd, 1874.

Family—*ACONTIIDÆ.*

46. AGROPHILA SULPHURALIS.

Agrophila sulphuralis, Bergstr., Ins. Suec. i, p. 16.—Guén., Noct., ii, p. 206.—Eversm., Fauna Volgo-Ural., p. 461.
Ph. trabealis, Scop., Ent. Carn. p. 40.
Agrophila trabealis, Erschoff, Lep. Turkestan, p. 52.

Hab.—Yárkand.

47. ACONTIA LUCTUOSA.

Acontia luctuosa, Schiff., Wien. Verz. p. 90 (1776).—Guén., Noct. ii, p. 223.—Eversm., Fauna Volgo-Ural, p. 331.—Erschoff, Lep. Turkestan, p. 50.

Hab.—Yangihissár, April.

Family—*ERASTRIDÆ.*

48. BANKIA ARGENTULA.

Bankia argentula, Hübn., Beit., i, p. 9, t. 2, fig. F. (1786).

Hab.—Ak Masjid, south of Yárkand.

Family—*CATOCALIDÆ.*

49. CATOCALA PUDICA, n. sp.

Allied to *C. puerpera*. Differs from Southern European specimens in the fore wing being prolonged at the apex and having its exterior margin more oblique; this wing is also much paler in colour, and has the two bands of the underside visible from above; the ante-and postmedial transverse sinuous lines and reniform mark are very indistinct, and the marginal row of black spots nearly obsolete; on the hind wing the inner black band is narrower and less irregularly angled in the middle.

Expanse 2¾ inches.

Hab.—Pashkyum, Ladák, 10,870 feet.

This species is described from a specimen taken by the late Mr. R. B. Shaw in 1870, and now in my own collection. A single wing only of a specimen of what appears to be this species, is preserved in the collection made by Dr. Stoliczka, having been taken at Sánju, 30th October.

Family—*TOXOCAMPIDÆ.*

50. APOPESTES PHANTASMA.

Noctua phantasma, Eversm., Bull. Mosc. 1843, p. 546.
Spintherops phantasma, Guén., Noct. ii, p. 422.—Erschoff, Lep. Turkestan, p. 58.

Hab.—Yárkand, 12th November.

Tribe—PYRALES.

Family—*BOTYDÆ.*

51. BOTYS FLAVALIS.

Pyralis flavalis, Schiff., W. V. p. 121 (1776).

Hab.—Ak Masjid, south of Yárkand, Sarikol, 2nd May. Yangihissár, April. Posgám, in lucerne-fields, 28th May.

Family—*ENNYCHIDÆ.*

52. PYRAUSTA CUPREALIS. Plate I, fig. 26.

Pyrausta cuprealis, Moore, Ann. and Mag. Nat. Hist. 1878, p. 235.

Upperside dark cupreous-brown: hind wing with a broad medial discal yellow band. Underside paler, basal two-thirds of both wings yellow, with brown-speckled subbasal patch. Antennæ black. Body beneath cupreous-black speckled with yellow. Palpi yellow beneath. Legs yellow, with cupreous speckles.

Expanse ⅝ inch.

Hab.—Gaganghir (near Sonamarg), Kashmir.

Family—*SCOPARIDÆ.*

53. EUDOREA GRANITALIS. Plate I, fig. 25.

Eudorea granitalis, Moore, Ann. and Mag. Nat. Hist. 1878, p. 235.

Upperside: fore wing pale brown, crossed by several irregular wavy grey-bordered black lines; cilia grey, alternated with black: hind wing greyish-white, traversed by numerous

D

short brown striæ somewhat regularly disposed between the veins, the wing being suffused with brown along exterior margin. Cilia grey, with dusky line. Body grey, brown-speckled. Palpi brown at apex, greyish at base. Legs grey, speckled with black. Underside as above; markings paler.

Expanse $\frac{6}{10}$ inch.

Hab.—South-east of Cliklik, hills south of Yárkand, 5th June 1874.

54. EUDOREA TRANSVERSALIS.

Eudorea transversalis, Moore, Ann. and Mag. Nat. Hist. 1878, p. 235.

Male. Upperside: fore wing grey, speckled with brown, crossed by an oblique subbasal and a recurved discal black speckled band; exterior margin black-spotted; some black speckles at end of the cell: hind wing pale brown, with darker marginal border. Cilia grey, with brown border. Body grey, brown-and black-speckled. Palpi speckled with black and white above. Antennæ dark brown. Underside pale ochrey-grey. Legs speckled with grey and black, fore and middle legs with black bands. Female paler, the bands across the wings broader and more distinct.

Expanse $\frac{6}{10}$ inch.

Hab.—Ighizyar (5,600 feet), 18th May 1874, Yangihissár (4,320 feet), April 1874, both in Eastern Turkestan.

Tribe—GEOMETRES.

Family—*BOARMIDÆ.*

55. HYPOCHROMA PSEUDOTERPNARIA.

Hypochroma pseudoterpnaria, Guén., Phal. i, p. 276.

Hab.—Uri, Jhilam valley, 23rd July.

56. GNOPHOS OBTECTARIA.

Gnophos obtectaria, Walker, Catal. Lep. Het. B. M. 35, p. 1597.

Hab.—Sonamarg, Kashmir.

57. GNOPHOS STOLICZKARIA. Plate I, fig. 22.

Gnophos stoliczkaria, Moore, Ann. and Mag. Nat. Hist. 1878, p. 235.

Upperside pale ochreous-grey, minutely brown-speckled, the speckles forming more or less numerous short transverse striæ; both wings with an indistinct oval brown spot at end of the cell, and marginal lunular dotted line: fore wing with a subbasal and discal, and hind wing with a discal, series of dentate brown points. Cilia white. Underside paler; speckles sparsely apparent; cell-spot less distinct.

Expanse 1⅜ inch.

Hab.—Ak Masjid, south of Yárkand, 2nd June 1874.

Family—*GEOMETRIDÆ.*

58. GEOMETRA DISPARTITA.

Geometra dispartita, Walker, Catal. Lep. Het. Brit. Mus. xxii, p. 520.

Hab.—Beshterek, south of Yárkand, 31st May.

Family—*LARENTIDÆ.*

59. EUPITHECIA SATURATA.

Eupithecia saturata, Guén., Phal. ii, p. 260.

Hab.—Chiklik, hills south of Yárkand, 3rd June.

60. THERA KASHGHARA. Plate I, fig. 23.

Thera kashghara, Moore, Ann. and Mag. Nat. Hist. 1878, p. 236.

Upperside pale brownish-cinereous: fore wing crossed by three equidistant pale-bordered blackish lines, the basal line nearly straight, the second slightly waved, the outer irregularly undulated, each darkest at costal end, the interspace between the two outer ones darker cinereous-brown; a slight short sinuous spot at apex; indistinct paler transverse undulating lines on outer margin; a distinct darker marginal narrow line. Underside paler; transverse lines very indistinctly visible. Legs dusky-brown above. Antennæ brownish.

Expanse 1⅜ in.

Hab.—Chiklik (3rd June 1874), 14,480 feet.

Tribe—CRAMBICES.

Family—*PHYCIDÆ.*

61. HOMŒOSOMA VENOSELLA. Plate I, fig. 24.

Homæosoma venosella, Moore, Ann. and Mag. Nat. Hist. 1878, p. 236.

Upperside: fore wing pale greyish-ochreous, minutely brown-speckled, the speckles sparsely disposed along the veins; having a transverse pale discal indented line and an indistinct space at end of the cell: hind wing cinereous-white with pale brown marginal line. Cilia white. Body and palpi above greyish-ochreous, paler beneath. Underside whitish-cinereous.

Expanse ⅞ inch.

Hab.—Ak Masjid, south of Yárkand (8,870 feet), June 2nd, 1874.

62. MYELOIS UNDULOSELLA. Plate I, fig. 27.

Myelois undulosella, Moore, Ann. and Mag. Nat. Hist. 1878, p. 236.

Male and female. Upperside ochreous-grey: fore wing speckled with brown, crossed by two medial oblique undulating pale-bordered blackish lines, both of which are sinuous at the costal end; a dark pale-centred streak at end of the cell; middle of hinder margin and the outer border grey, the latter with an indistinct pale sinuous line slightly black-speckled; cilia whitish, alternated with two dark marginal lines: hind wing pale brownish-cinereous externally; cilia white alternated with one dark marginal line, and having a dark patch situated at the middle of the margin. Body ochreous-grey. Underside pale cinereous.

Expanse 1⅖ inch.

Hab.—Ak Masjid, south of Yárkand (8,870 feet), June 2nd, 1874; Aktala, west of Yárkand (7,342 feet), May 17th, 1874.

63. MYELOIS GRISEELLA. Plate I, fig. 15.

Myelois griseella, Moore, Ann. and Mag. Nat. Hist. 1878, p. 236.

Upperside cinereous-grey: fore wing densely irrorated with brown, crossed by two medial undulating very indistinct speckled lines; an indistinct streak at end of the cell; both wings with an outer marginal narrow brown lunular line: hind wing whitish, with a very pale cinereous-brown marginal and an indistinct narrow submarginal band. Cilia whitish, with a narrow marginal dark line. Underside paler cinereous. Head and thorax brownish. Abdomen cinereous-brown.

Expanse 1½ inch.

Hab.—South-east of Chiklik, south of Yárkand (June 5th, 1874).

Tribe—TORTRICES.

64. CONCHYLIS STOLICZKANA. Plate I, fig. 14.

Conchylis stoliczkana, Moore, Ann. and Mag. Nat. Hist. 1878, p. 237.

Upperside: fore wing white, with three transverse outwardly oblique ochreous-brown bands, two inwardly oblique discal bands, and a spot at end of the cell; a brown-speckled marginal band: hind wing cinereous-white, with a narrow brown marginal band. Body white and black-speckled, with white segmental bands. Legs white. Palpi white, brown-speckled. Underside cinereous-white, outer bands on fore wing indistinctly visible.

Expanse ⅘ inch.

Hab.—South-east of Chiklik, (June 5th, 1874).

Tribe—TINEINES.

Family—*TINEIDÆ*.

65. ADELA SULZELLA.

Tinea sulzella, Schiff., W. V. 143 (1776).

Hab.—Gaganghir, Kashmir.

LEPIDOPTERA. 17

Family—*GELECHIDÆ*.

66. DEPRESSARIA STIGMELLA.

Depressaria stigmella, Moore, Ann. and Mag. Nat. Hist. 1878, p. 237.

Fore wing pale brownish-ochreous, greyish along the apical portion of the costa, interspersed with a few dusky speckles; a dusky-grey short straight streak at end of the cell, and a few speckles on outer margin. Legs pale ochreous. Hind wing pale ochreous-white. Underside of both wings paler.

Expanse $\frac{9}{8}$ inch.

Hab.—Yangihissár, Eastern Turkestan, (March 3rd, 1874).

This species is nearest allied to the European *D. subpropinquella*.

Tabular List showing geographical Distribution.

Kashmir.	Localities where captured.	Geographical Distribution.
Epinephile chena	Gaganghir	W. Himalayas (Pangi (Basahir); Kunawur).
Aulocera swaha	Gaganghir .	Ditto (Simla).
Synchloe daplidice	Gond, Sonamarg	W. Asia; Europe.
Colias hyale	Gaganghir . .	W. Himalayas (Masuri); W. Asia; S. and C. Europe.
Colias fieldii	Sonamarg . .	Ditto (Masuri); Punjab.
Polyommatus galathea .	Sonamarg.	
Dipsas odata	Gaganghir . .	Ditto (Upper Kunawur).
Leucophlebia bicolor	Hatti Uri . .	Ditto.
Hypercompa principalis	Gond, Gaganghir	Ditto.
Arctia orientalis, n. sp.	Sonamarg . .	Ditto N. W. Provinces of India (Allahabad).
Mamestra brassica	Srinagur . .	Ditto India; W. Asia; Europe.
Heliothis scutosa .	Gaganghir . .	Ditto W. Asia; Europe.
Pyrausta cuprealis, n. sp.	Gaganghir.	
Hypochroma pseudoterpnaria	Uri . .	Ditto Punjab.
Gnophos obtectaria	Sonamarg .	Ditto (Simla).
Adela sulzella	Gaganghir .	W . Asia; Europe.

LADÁK.

Hipparchia lehana, n. sp.	Leh; Kharbu.	
Hipparchia cadesia	Leh . . .	Kashmir.
Aulocera brahmina	Dras Valley .	Kashmir (Margan Pass).
Argynnis jainadeva	Leh . . .	Kashmir; Upper Kunawur.
Synchloe brassica	Leh . .	Kashmir; W. Himalayas; W. Asia; Europe.
Parnassius charltonius	Kharbu.	Ranang Pass, 13,000 feet.
Polyommatus lehanus, n. sp.	Leh.	
Polyommatus ariana	Dras Valley	Sanga (Puspa Valley); Kashmir; W. Himalayas; Pangi (Basahir).'
Hydræcia tibetana, n. sp.	Leh.	
Agrotis aquilina	Leh .	W. Asia; S. and C. Europe.
Agrotis tibetana, n. sp.	Leh. .	

E

SECOND YARKAND MISSION.

Tabular List showing geographical Distribution—continued.

MOUNTAIN RANGE BETWEEN LADÁK (LEH) AND PLAINS OF YÁRKAND.

Kashmir.	Localities where captured.	Geographical Distribution.
Vanessa ladakensis, n. sp.	Gogra, Karatágh Lake.	
Baltia shawii	Aktágh	Chang Lung Pass.
Colias stoliczkana, n. sp.	N. of Changla.	
Parnassius jacquemontii	N. of Changla.	Mountains of Ladák.
Parnassius acco	Lupsang or Lak Zung, 17,537	Ditto.
Neoris shahidula	Shahidúla.	
Hadena stoliczkana, n. sp.	Kufelang.	
Agrotis segetum	Tankse	N. W. India; W. Asia; Europe.
Agrotis aquilina	Tankse	W. Asia; Europe.

PLAINS OF YARKAND.

Pyrameis cardui	Sánju; Kárghalik	Asia; Africa; Europe; N. America.
Synchloe rapæ	Yangihissár	W. Asia; Europe.
Colias hyale	Sánju; Yangihissár	W. Asia; Europe.
Polyommatus kasgharensis, n. sp.	Yangihissár.	
P. yarkandensis, n. sp.	Yárkand.	
Euproctis karghalika, n. sp.	Kárghalik.	
Euproctis lactea, n. sp.	Kárghalik.	
Ptilophora kashghara, n. sp.	Yangihissár.	
Acronycta karghalika, n. sp.	Kárghalik.	
Mamestra canescens, n. sp.	Kárghalik.	
Agrotis segetum	Kárghalik	N. W. India; W. Asia; Europe.
Heliothis dipsacea	Posgám; Yangihissár	W. Asia; S. Europe.
Agrophila sulphuralis	Yárkand	W. Asia; S. and C. Europe.
Acontia luctuosa	Yangihissár	W. Asia; S. and C. Europe.
Catocala pudica, n. sp.	Sánju	Paskyum, Ladák, 10,870 feet (Shaw).
Apopestes phantasma	Yárkand; Bora (Shaw)	W. Asia.
Botys flavalis	Yangihissár; Posgám	W. Asia; S. and C. Europe.
Endorea transversalis, n. sp.	Yangihissár; Ighizyar.	
Geometra dispartita	Beshterek	N. W. India.
Depressaria stigmella, n. sp.	Yangihissár.	

HILLY COUNTRY WEST AND SOUTH-WEST OF THE PLAINS OF YARKAND.

Synchloe chloridice	Sarikol	W. Asia; S. Europe.
Colias hyale	Sarikol	W. Asia; Europe.
Oxicesta marmorea, n. sp.	Sasak Taka.	
Spælotis undulana, n. sp.	Ak Masjid; Chiklik.	
Tæniocampa chiklika, n. sp.	Chiklik.	
Heliothis hybleoides, n. sp.	Chiklik.	
Dankia argentula	Ak Masjid	W. Asia; Europe.
Botys flavalis	Ak Masjid; Sarikol	W. Asia; S. and C. Europe.
Eudorea granitalis, n. sp.	Chiklik.	
Gnophos stoliczkaria, n. sp.	Ak Masjid.	
Eupithecia satyrata	Chiklik	W. Asia; S. and C. Europe.
Thera khasgharia, n. sp.	Chiklik.	
Homœosoma venosella, n. sp.	Ak Masjid.	
Myelois undulosella, n. sp.	Ak Masjid; Aktala.	
Myelois grisella, n. sp.	Chiklik.	
Conchylis stoliczkana, n. sp.	Chiklik.	

ERRATUM.

In the names at foot of plate *for* "Myelois grisecla," *read* "Myelois grisella."

1. Colias Stoliczkana 2. Vanessa Ladakensis 3. Parnassius Charltonius 4. Hipparchia Lohana 5. Sultia Shawii 6. Polyommatus Lehanus 7. P. Kashgharensis 8. P. Yarkundensis 9. Acronycta Kargalica 10. Spilotis undulans 11. Tæniocampa Chikaka 12. Hadena Stoliczkana 13. Mamestra anescens 14. Conchyris Stoliczkana 15. Mamestra griseola 16. Agrotis Tibetana 17. Cxycesta marmorra 18. Euproctis Kargalica 19. Pelophora Kashgharia 20. Heliothis Hybbecides 21. Hydrœna Tibetana 22. Gnophos Stoliczkaria 23. Thera Kashghara 24. Homæosoma venosella 25. Eudorea granitalis 26. Pyrausta cupreaha 27. Myelois undulosella

CALIFORNIA

SCIENTIFIC RESULTS

OF

THE SECOND YARKAND MISSION;

BASED UPON THE COLLECTIONS AND NOTES

OF THE LATE

FERDINAND STOLICZKA, Ph.D.

COLEOPTERA,

GEODEPHAGA AND LONGICORNIA. BY H. W. BATES, F.R.S. (*Pp. 1—23, with one Plate.*)
PHYTOPHAGA. BY J. S. BALY, F.L.S. (*Pp. 25—36.*)
HALIPLIDÆ, DYTISCIDÆ, GYRINIDÆ, HYDROPTILIDÆ, STAPHYLINIDÆ, AND SCARABÆIDÆ (EXCEPT CETONIINI). BY D. SHARP, F.R.S. (*Pp. 37—53.*)
CETONIIDÆ. BY OLIVER JANSON. (*P. 54.*)
HETEROMERA. BY FREDERICK BATES. (*Pp. 55—79, with one Plate.*)

Published by order of the Government of India.

CALCUTTA:
OFFICE OF SUPERINTENDENT OF GOVERNMENT PRINTING, INDIA.
1890.

NOTE.—For the group CURCULIONIDÆ, see a paper by DR. FAUST in the STETTINER ENTOMOLOGISCHE ZEITUNG, Band XLVII., pp. 129-157, entitled *Verzeichniss auf einer Reise nach Kashgar gesammelter Curculioniden*.

SCIENTIFIC RESULTS

OF

THE SECOND YARKAND MISSION.

COLEOPTERA.

GEODEPHAGA AND LONGICORNIA.
By H. W. BATES, F.R.S., F.L.S.

INTRODUCTORY REMARKS.

THE Coleopterous insects of the two great tribes which form the subject of the present memoir were collected chiefly during the winter months. It is on this account, probably, that the collection contains so few species of Longicornia, which ought to be abundant in summer on flowers in the elevated valleys, as they are in Northern Europe, in Siberia, and in the Rocky Mountains. A similar remark may be made with regard to the *Cicindelidæ* family of Geodephaga, 4 species only of which were collected, three being Indian, taken in the Jhelam Valley, and one north of the Himalaya, which proves to be a new species, allied to a species of Palæarctic type found in the Altai. The *Carabidæ* are more numerous, the species of this family wintering generally in the imago state and being found readily in their usual haunts in the autumnal and early spring months. They afford occasion, however, for only one general remark, namely, that all the species without exception from the region north of the Himalaya are of European types, eight out of the 63 species collected being identical with European species, and the remainder either new species of European genera, or species of similar type previously described from the neighbourhood of the Caspian, or from Western and Northern Asia. The few that were found at Murree, in the Jhelam Valley, or in Ladak are either Indian and subtropical (*e.g., Colpodes ovaliceps, Pristomachærus chalcocephalus, Hypolithus perlucens*, &c.), or North Indian modifications of Palæarctic types (*e.g., Carabus caschmirensis et stoliczkanus, Hypsinephus ellipticus*), or well-marked and distinct species of, Palæarctic genera, *e.g., Bradytus compactus, Acinopus striolatus, Harpalus japonicus, Anchomenus politissimus, Molops piligerus.*

GEODEPHAGA.

1.—CICINDELA STOLICZKANA.
Bates, Proc. Zool. Soc. 1878, p. 713.

C. Burmeisteri (*Fisch.*) *affinis, sed minor, thorace breviori, etc. Nigra corpore subtus, pedibus, antennarumque basi chalybeo-violaceis, elytris lunula humerali et apicali (hac antice in maculam rotundatam dilatata) fasciaque mediana, lata, abbreviata, recta, flavo-albis; fronte inter oculos concava, subtiliter strigosa, albo-hirta; thorace brevi, lateribus fere rectis, supra subtilissime granulatim-strigoso: etytris minute, haud confertim grauulatis; palpis nigris, albo-setosis: labro albo, convexo: antice medio rotundatim producto, unidentato: corpore subtus pedibusque sparsim albo-pilosis.*

Long. 6—7½ lin.

In colour, sculpture, and form of labrum closely resembling *C. burmeisteri* (Fischer), but of shorter and less convex form; the thorax also being relatively smaller and the elytra more obtusely rounded at the apex. The white marks of the elytra are more numerous and much larger. They are variable in extent and sometimes all blended together along the lateral margin; but the characteristic feature of the non-flexuous, but broad and only slightl oblique, median belt remains constant. The apical lunule always forms a narrow border at the apex of the elytra, but expands into a large rounded spot at its anterior extremity.

Hab.—Without locality. Taken by Stoliczka shortly before his decease, probably on the northern slopes of the Kuen-lun. My specimens of *C. burmeisteri* came from the Tarbagatai Mountains.

2.—CICINDELA INTERMEDIA.

Chaudoir, Bull. Moscou, 1852, i. p. 6.

Hab.—Jhelam Valley.

3.—CICINDELA LIMBATA.

Wiedemann, Zool. Mag. ii, i. (1823), p. 64.

Hab.—Jhelam Valley. A single example.

4.—COLLYRIS ORTYGIA.

Buquet, Ann. Soc. Ent. France, 1835, p. 604.
Chaud., Monogr. Collyr. p. 502, t. 7, f. 6.

Hab.—Jhelam Valley. The single specimen of this species presents scarcely any points of difference from others with which I have compared it taken near Calcutta.

5.—NEBRIA PSAMMOPHILA.

Solsky, Fedchenko's Turkestan, Zool. tom. ii, v, Coleoptera i, p. 12.

Differs from Solsky's diagnosis only in the clearer-red head and thorax, these members according to him being "picescentibus."

Hab.—" Dras, Kargil, and Leh"; many examples. Fedchenko took it in Kokand, near the river Kizil-su.

COLEOPTERA.

6.—NEBRIA LIMBIGERA.

Solsky, Fedchenko's Turkestan, l.c. Col. i. p. 13.

Hab.—One example, same locality as the above. Differs from *N. psammophila* by its larger size and black abdomen. Fedchenko found it in Kokand, "near the Kizil-su and in the hills near the river Isphavia."

7.—CARABUS CASCHMIRENSIS.

Carabus caschmirensis, Kollar & Redtenbacher, in Hügel's "Kasmir, etc." iv. 2 (1844), p. 499, t. 23, f. 4.
—— *lithariophorus*, Tatum, Ann. & Mag. Nat. Hist. xx (1847), p. 14.

Hab.—Murree. One example, ♀.

8.—CARABUS STOLICZKANUS.

Bates, Proc. Zool. Soc. 1878, p. 713.

C. cashmirensi (*Koll.*) *affinis. Maxime elongatus, angustus, niger subnitidus: thorace late sub-cordato-quadrato, angulis posticis retrorsum productis, acutis: elytris angustis, post medium perparum rotundato-dilatatis, dorso tuberculorum triplici serie, inter se carina unica separatis. Menti dente verticaliter exstanti, valde compresso; labro medio triangulariter emarginato.*

Long. 14—15 lin.

Resembles *C. caschmirensis* in the form of head, labrum, and tooth of mentum. The thorax is also similar in shape, but scarcely so broadly rounded on the anterior part. The elytra are very different both in shape and sculpture; they are narrower and more parallel in outline and much less convex, and the sculpture, instead of a triple row of narrow elongate tubercles, each row separated by a triple line of granules, consists of three distinct rows of larger, oblong tubercles, separated by a single continuous elevated line. There are, however, only two of those lines, between the 1st and 2nd and the 2nd and 3rd rows; the sutural border being an irregularly-crenated elevation, and the margin, exterior to the 3rd row, consisting of a confused coarse reticulation, with traces of a 4th row of minor tubercles.

Hab.—Murree. Two examples.

9.—CALOSOMA ORIENTALE.

Chaudoir, Ann. Soc. Ent. France 1869, p. 308.
Syn. ? *C. orientale,* Hope, Trans. Zool. Soc. l. p. 92.

Hab.—Kogyar: Sind Valley: "Dras, Karghil, and Leh." The specimens vary a little in the degree of regularity of the fine cross-striæ of the interstices; but there is no other character to indicate that they form more than one variable species.

10.—SCARITES INCONSPICUUS.

Chaudoir, Bull. Mosc. 1855, i. p. 82.

Hab.—Jhelum Valley. One example agreeing precisely with Baron Chaudoir's description above cited.

11.—SCARITES ARENARIUS.

Bonelli, Obs. Entom. 2, p. 40.
Chaudoir, Bull. Mosc. 1855, i, p. 86.

Hab.—Yangihissar. A widely-distributed species, throughout the basins of the Mediterranean and the Caspian; but not hitherto recorded from regions further east. Solsky includes the allied species, *Sc. persicus* (Chaud.), among the insects taken by Fedchenko in Turkistan. The Yangihissar examples agree better with *Sc. arenarius*, having two denticulations above the digitation of the anterior tibiæ; they are, however, rather more elongated than specimens from Algiers and Imeritia with which I have compared them. The size is 8½—9¼ lin.

12.—DYSCHIRIUS ORDINATUS.

Bates, Trans. Ent. Soc. 1873, p. 240.

Hab.—Pamir, between Sirikol and Panga. I see no definite character to separate this small species from *D. ordinatus*, hitherto known only from Japan.

13.—BROSCUS PUNCTATUS.

Dejean, Spec. Gen. Col. iii, p. 431.

Hab.—No locality, probably near Yarkand. A widely-distributed Oriental species, being recorded from Egypt, Mesopotamia, Nepaul, and China.

14.—PRISTOMACHÆRUS CHALCOCEPHALUS.

Wiedm., Zool. Mag. ii, i, p. 57.

Hab.—Jhelam Valley. One example, differing from the original Hongkong specimen only in the squarer form of both the yellow elytral spots.

Closely allied to *Pristomachærus messii* of Hongkong (Bates, Trans. Ent. Soc. 1873, p. 324). It differs a little in colour and the form of the anterior elytral spot from Wiedemann's description.

15.—CHLÆNIUS SPOLIATUS, var. INDERIENSIS.

Chlænius spoliatus, Rossi.,'var. *inderiensis*, Motschulsky, Bull. Mosc. 1864, ii, p. 340.

Hab.—Yangihissar. One example, agreeing perfectly with the above-cited description of a remarkable variety of this widely-distributed species, hitherto recorded only from the borders of lake Indiersk. The type-form occurs throughout nearly the whole Palæarctic region, from the western shores of Europe to Japan.

16.—CHLÆNIUS TENUELIMBATUS.

Ballion, Bull. Mosc. 1870, ii, p. 320.
Solsky in Fedchenko's Turkestan, Zoology, tom. ii, v. Coleop., p. 62.
Chaudoir, Monogr. Chlæniis., p. 263 (1876).

Hab.—Ladakh. Found also near Samarkand and Kodjend. I have compared the numerous examples in Stoliczka's collection with a specimen received from Russia, as taken in "Turkestan," and find no essential difference: the Turkestan specimen has a rather broader thorax, but otherwise of the same shape, so distinct from that of the following species which is subcordate with prominent and acute hinder angles.

17.—CHLÆNIUS LÆTIUSCULUS?

Chaudoir, Bull. Mosc. 1856, ii, p. 248, id., Monogr. Chlæniis, p. 264.

Hab.—Ladakh. Also in Northern Hindostan.

18.—ACINOPUS STRIOLATUS?

Zoubkoff, Bull. Mosc. 1833, 317.
P. d. l. Brulerie, Ann. Soc. Ent. Fr. 1873, p. 250.

Hab.—Sind Valley. A much damaged example, which I refer doubtfully to this species as a small variety. It is $6\frac{1}{2}$ lines long and of narrow cylindrical form, and the elytral striæ, although fine and with perfectly plane interstices, are more strongly impressed than in *striolatus*. The species occurs in the basin of the Caspian, and was taken near Tashkend by Fedchenko.

19.—DAPTUS VITTATUS.

Fischer, Ent. Russ. ii, p. 36, 46, f. 7.
Dej., Sp. Gén. iv, p. 19.

Hab.—Yangihissar. One example.

20.—DICHIROTRICHUS ALTICOLA.

Bates, Proc. Zool. Soc. 1878, p. 713.

D. amplipennis (*Bates*) *proxime affinis, differt colore pallidiori et thoracis angulis posticis rotundatis. Oblongus, supra testaceo-fulvus, capite* (*maculis rufis exceptis*) *thoracis disco macula alteraque postico-discoïdali elytrorum, nigro-œnois: palpis apice acuminatis: capite et thorace grosse subsparsim punctatis, hoc postice angustato, angulis posticis oblique rotundatis, margine postice arcuato: elytris striatis, interstitiis medio leviter culminato-convexis, biseriatim punctatis: corpore subtus nigro: antennis fuscescentibus. ♂ tarsi duo antici articulis 1—3 ovatis, 4 bilobo.*

Long. $2\frac{1}{2}$ lin.

Agrees with *D. amplipennis* (China), *D. tenuimanus* (Japan), *D. discicollis*, Dej., and others in its acuminate palpi, in which these eastern species differ from their West European congeners. The three basal dilated joints of the ♂ anterior tarsi are not triangular, but ovate, their angles being perfectly rounded. Underneath, the dilated male joints are clothed with long ragged scale-hairs, loosely arranged; but this is the case with the European species of the genus; and the statement of Schaum and others is therefore erroneous, that they are "*spongiosi*" and bring the genus within the *Anisodactylinæ* sub-family. The genus is, in fact, allied to *Ophonus*. The upper surface of *D. alticolus* is light tawny or reddish-brown, redder on the thorax and a large spot on each side of the head. The rest of the head is brassy-black. The disk of the thorax has a dusky spot, sometimes indistinct. The disk of the elytra has, posteriorly, covering interstices 3 and 4, an elongate black spot. The species is closely allied to the South Russian and Turkestan *D. discicollis*, Dej., differing chiefly in the obliteration of the hinder angles of the thorax.

Hab.—Pamir, between Sirikol and Panga.

21.—HARPALUS CÆRULEATUS.

Bates, Proc. Zool. Soc. 1878, p. 714.

Elongato-oblongus, glaber, thorace transversim quadrato, postice distincte angustato, lateribus arcuatis, angulis posticis rotundatis, basi utrinque late subcrebre punctato, margine basali bisinuato: elytris apice fortiter sinuatis, supra striatis, interstitiis planis impunctatis, ertio unipunctato.

♂. *Supra capite thoraceque nigris politis, elytris caeruleis, subviolaceis, nitidis; antennis nigris, articulo basali rufo: corpore subtus nigro, pectore medio pedibusque rufopiceis; abdomine medio nitido. Immaturo toto corpore castaneo-rufo, nitido, elytris violaceis.*

♀. *Nigro-vel rufo-castanea, raro obscuro-nigra; elytris opacis interdum violaceo-tinctis, apice fortius (ut in* II. *acneo* ♀ *) sinuatis.*

An elongate species, similar in form to *H. hospes* (Sturm), but without its punctuation. Thorax slightly narrowed behind, with hinder angles, but blunted or rather rounded at their apices. The elytra are destitute of punctuation, except the usual marginal row, and their apices are rather deeply sinuate in both sexes, but most so in the ♀. The sexual diversity in colour is constant in mature individuals, the male having the head and thorax glossy black, with violet blue elytra; the female being chestnut-red or brown, with elytra sometimes tinged with violet. Terminal spur of the anterior tibiæ lanceolate, simple.

Hab.—Yangihissar and Kogyar.

22.—HARPALUS MELANEUS.

Bates, Proc.ZZool. Soc. 1878, p. 714.

H. calceato (*Dufts.*) *forma coloreque similis, at thoracefere impunctato angulisque posticis obtusis. Oblongus, modice elongatus, niger nitidus, elytris* ♀ *opacis; antennis et palpis nigris, articulis omnibus apice piceo-rufis: thorace transversim quadrato, postice leviter angustato, angulis posticis obtusis lateribus antice modice arcuatis, basi utrinque vage via punctato, fovea lineari impresso: elytris apice paullulum sinuatis, supra convexis, simpliciter striatis, interstitiis modice convexis, tertio unipunctato.*

Long. 6 lin.

Similar in size, form, and colour to the European *H. calceatus*; convex, posterior part of elytra most so. Colour in the ♂ deep shining black above and beneath, in the ♀ the elytra opaque. The legs are more or less piceous, especially the tarsi. The antennæ and the palpi are pitchy-black, the joints in all tipped with dull rufous. The thorax is very moderately rounded anteriorly, and gradually and slightly narrowed behind to the base, the hind angles being not quite rectangular and obtuse at their apices. The base on each side is very faintly roughened and not distinctly punctured, and the fovea is rather distinct and linear. Terminal spur of the anterior tibiæ lanceolate-acute, simple.

Hab.—Sind Valley. Murree. Near Leh.

23.—HARPALUS TURCULUS.

Bates, Proc. Zool. Soc. 1878, p. 714.

Oblongus, niger, ♂ nitidus, ♀ sericinitens, antennis articulo primo rufo, palpis apice flavis : capite modice angusto, lævi ; thorace quadrato, antice prope angulos rotundato-angustato postice lateribus exacte parallelis, elytris multo angustiori, angulis posticis rectis, supra impunctato : elytris utroque sexu apice fortiter subrecte sinuatis, supra subtiliter striatis, interstitiis planis, tertio unipunctato : metasterno grosse sparsim punctato.

Long. 4 lin.

Very similar in colour in both sexes to *H. liodes;* but differing in the smaller size, narrow thorax, and strongly-sinuated apices of the elytra. The head is not notably wide, the eyes are only slightly projecting, and the forehead is remarkably even and smooth. The antennæ reach the base of the thorax; they are black and have the basal article constantly red. The thorax is parallel-sided from the base to the middle; it is then gently arcuated and nearer the head much narrowed. The elytra have the same silky gloss, plane interstices, and fine striæ as *H. liodes*. Terminal spur of anterior tibiæ long, curved, lanceolate.

Hab.—No locality. Probably near Yarkand.

24.—HARPALUS JAPONICUS.

Morawitz, Bull. Ac. St. Petersb. v. 1863, 327.

Hab.—Murree. Many examples differing in no material respect from those of China, Japan, and Formosa.

25.—HARPALUS—?

A single specimen ♀; indeterminable.

26.—HARPALUS INDICOLA.

Bates, Proc. Zool. Soc. 1878, p. 714.

Et ngato-oblongus, angustior, nigerrimus, ♂ magis, ♀ minus, nitidus, palpis et antennis fulvis : thorace quadrato, lateribus leniter arcuatis, postice longe et modice angustato, angulis posticis paullo obtusis ; basi toto subsparsim punctato et paulo rugoso, fovea utrinque obliqua : elytris convexis, apice modice sinuatis, supra striatis (♂ fortius), interstitiis vix convexis, tertio puncto conspicuo impresso : sternis et ventro lateribus grosse haud profunde punctatis.

Long. 5 lin.

Smaller and narrower than *H. melaneus*; head also much smaller or narrower. In form it approaches the European *H. tenebrosus* (Doj.), but the thorax is different in shape, the sides being more arcuated and contracted gradually behind to the base which they join at an obtuse angle. The colour is the same as in *H. melaneus*, except that the antennæ and palpi are reddish-tawny; but this is liable to variation. The elytral striæ are sharply impressed, and become deeper at the apex. The terminal spur of the anterior tibiæ is moderately long, with the basal half dilated but not dentate.

Hab.—Murree.

27.—HARPALUS MASOREOIDES.

Bates, Proc. Zool. Soc. 1878, p. 715.

Parvus, niger subsericeus, lævis, partibus oris antennisque flavo-testaceis, pedibus magis rufescentibus: thorace transverso, elytris vix angustiori, antice gradatim paullulum angustato, angulis posticis rotundatis, lævi, foveola basali utrinque oblonga, marginibus rufescentibus; elytris oblongis, apice late obtusis leniter sinuatis, supra striatis, striis minutissime punctulatis, interstitiis vix convexis, marginibus reflexis et epipleuris piceorufis: menti dente triangulari, acuto.

Long. 2½ lin.

The obtuse-angled thorax and apex of elytra, with the general form and smoothness, give this little species the appearance of a *Masoreus*. The head is small, obtuse, smooth, and polished; the eyes very slightly prominent; the frontal fovea is round and well-defined. The antennæ are rather longer than the head and thorax taken together; they are yellow, with more or less dusky on their pubescent joints. The spur of the anterior tibiæ is obtusely lanceolate, not dilated; the external angle of the apex of the tibiæ has three short and very stout, obtuse spines.

Hab.—Pamir Steppe, between Sirikol and Panga.

28.—HARPALUS LIODES.

Bates, Proc. Zool. Soc. 1878, p. 715.

Ovatus, latus, modice convexus, niger, ♂ serici-nitens, ♀ serici-opacus, antennis palpisque piceo-rufis, illis nigromaculatis, tarsis piceo-rufis: capite lato, lævigato, oculis minus prominulis; thorace valde transverso, antice angustato, postice multo latiori, angulis posticis rectis, fere impunctato: elytris ovalis, apice paullulum sinuatis, supra subtiliter striatis, interstitiis planis, tertio minute unipunctato: tibiis intermediis utroque sexu arcuatis.

Long. 5—5½ lin.

Resembles much large species of the genus *Amara*. Thorax shorter in relation to the width than in *H. brevicornis* (Germ.), or any other species of the genus known to me. The head is broad and the forehead flattened and smooth. The thorax is narrower at the apex than at the base; but the sides from the slightly dilated anterior part are slightly rounded, or nearly parallel to the hind angles, which latter are rectangular but blunt at their apex; the disc is obscurely wrinkled and there are a very few punctures in the shallow basal foveæ; otherwise the surface is impunctate. The antennæ are short and far from reaching the base of the thorax. The elytra are ovate not wider at the base than the thorax, very slightly sinuated near their apex; the striæ are very fine, faintly punctulate, and the interstices flat and impunctate throughout, except the marginal one and the customary one on the third. The colour is deep black, with a bright silky gloss in the ♂, but nearly opaque in the ♀. The abdomen is impunctate. The metasternum has a few large punctures. Terminal spur of anterior tibiæ long and lanceolate.

Hab.— No locality. Probably near Yarkand.

29.—HYPOLITHUS PERLUCENS.

Bates, Proc. Zool. Soc. 1878, p. 715.

Piceo-niger, læte iridescens, glaber, antennis, palpis, et pedibus fulvo-testaceis : capite lævissimo, post oculos angustato, mandibulis magis rectis et acutis piceo-rufis ; fovea frontali lineari versus oculum curvata : thorace quadrato, lateribus leniter fere æqualiter arcuatis, angulis posticis valde obtusis, margine postico late sinuato; supra limbo toto crebre subtiliter punctulato, disco sparsim punctulato, polito, marginibus rufescentibus : elytris fortiter striatis, interstitiis paullulum convexis, politissimis, tertio (prope striam secundam) multipunctato.

Long. 4¼ lin. ♀.

Agrees with certain species of South Africa, in the curved linear frontal fovea, and with such species as *H. glaber* (Boh.) in its naked surface. The undersurface of the insect is iridescent and glabrous, as well as the upper; the ventral segments and the deflexed margins of the elytra being more or less rufescent. The metasternum has a few shallow punctures. The legs are naked, with the exception of a few stout spines on the outer side of the tibiæ and a few setæ on their inner side and underneath the tarsi. The tooth in the emargination of the mentum is very short, but distinct.

Hab.— Jhelam Valley.

HYPSINEPHUS, nov. gen.

Bates, Proc. Zool. Soc. 1878, p. 715.

Generi Selenophoro *proxime affine. Corpus elongato-ellipticum gen.* Calatho *haud dissimile, supra glabrum. Caput antice haud obtusum, labrum et mandibulæ modice elongata. Mentum rotundato-emarginatum, edentatum. Palpi elongati ; maxillarii articulo terminali penultimo breviori, subfusiformi sed apice distincte truncato. Thorax quadratus. Elytra glabra, interstitiis tertio, quinto, et septimo (apice) pluripunctatis. Pedes elongati, validi : tarsi ♂, articulis anticis 4 dilatatis, cordatis, squamigeris, primo basi gracili apice subito dilatato, quarto breviter bilobo. Tibiæ intermediæ ♀ arcuatæ.*

A new genus is necessary for the reception of a species in Dr. Stoliczka's collection which agrees with the American *Selenophori* in its chief characters, but differs wholly in facies from that numerous group. The totally different form of the dilated tarsal joints in the male affords a good distinguishing character; the other features enumerated above having only a minor importance. The species described below has doubtless many Asiatic congeners; one I have found among the *Harpali* collected by Dr. Maack in Eastern Siberia.

30.—HYPSINEPHUS ELLIPTICUS.

Bates, Proc. Zool. Soc. 1878, p. 710.

Piceo-niger vel castaneus, ♂ nitidus, ♀ sericeo-opacus, partibus oris, antennis pedibusque testaceo-fulvis : capite mox pone oculos angustato, foveis frontalibus rotundatis; thorace quadrato, elytris angustiori, lateribus postice explanatis, arcuatis angulis posticis obtusis, supra impunctato, fovea utrinque basali vage impressa : elytris elongato-ovatis, apice modice sinuatis, striatis, interstitio tertio punctis parvis 5, quinto prope basin 2, septimo apicem versus plurimis impressis, punctis marginalibus parvis.

Long. 6 lin.

Elongate elliptical, varying from pitchy black to castaneous. The lateral margins of the thorax are gradually more and more explanated from the anterior to the posterior angles, and the base has no distinct punctuation. The punctures of the elytra are somewhat variable in number and position : there are 5 or 6 on the third interstice, mostly close to the second stria, and 2 or 3 on the fifth near the base; but in some examples the fifth interstice has a row of punctures near the apex, like the seventh. The margin has a number of minute faintly impressed punctures.

Hab.—Four examples, two without locality, one marked *a* (from the Knen-lun ?), and the fourth from the Pangong Valley : this last has the thorax distinctly more dilated behind and more rectangular hind angles than the others.

31.—HARPALUS QUADRICOLLIS.

Selenophorus quadricollis, Kollar & Redtenb. in Hügel's Kaschmir, iv, 2, p. 502.

Hab.—Between Dras and Leh. The authors above cited placed this species in the genus *Selenophorus* from the simple emargination of the mentum. M. Putzeys, in his recent monograph of the genus *Selenophorus*, has rightly restricted it to those *Harpalinæ* which have the alternate interstices of the elytra pluripunctate and other characters in addition to the simple mentum, and which belong all to America. *S. quadricollis* is very closely allied to the typical Harpali, but probably a separate genus will eventually be formed for the species with edendate mentum.

32.—STENOLOPHUS MORIO.

Ménétriés, Catal. Raisonné (1832), p. 136. Id., Insectes rec. p. Lehmann i, 25.
Solsky in Fedchenko's Turkestan, Zoology, tom. ii, v, Coleop. i, p. 88.

Hab.—Yangi Hissar; one example. The species occurs in the neighbourhood of the Caspian and in Mesopotamia near Bagdad. The Yangi Hissar specimen has a smaller and rounder thorax than is presented by Bagdad examples with which I have compared it.

33.—SPHODRUS INDUS.

Chaudoir, Bull. Mosc. 1852, i. p. 67.

Hab.—Murree; one example.

COLEOPTERA. 11

Resembles specimens from Northern India in every other respect, except that the hind trochanters are long and furcate at the apex, with one branch of the fork very short. The specimen is a female.

34.—SPHODRUS CORDICOLLIS.

Chaudoir, Bull. Mosc. 1854, i, p. 43.

Hab.—Murree; one example.
Differs from Syrian specimens by its slightly broader and more ovate elytra.

35.—CALATHUS MELANOCEPHALUS.

Lin., Fauna Suec. No. 795: Putzey's Mon. Calath., p. 58.

Hab.—Pamir, between Sirikol and Panga.
Many examples; differing from the ordinary type of Western Europe by the rather narrower and more parallel-sided thorax and elytra. According to Putzeys, alpine varieties occur which are modified in the same manner.

36.—CALATHUS ANGUSTATUS.

Koll. & Redtenb. in Hügel's Kaschmir, iv, ii, p. 500 (1844).
Syn. *C. Kollari*, Putz., Mon. Calath, p. 56.

Hab.—?

37.—CALATHUS—?

Hab.—A single example, in imperfect condition, ticketed "Sind Valley": it would probably range in the section *Pristodactyla*.

38.—ANCHOMENUS LADAKENSIS.

Bates, Proc. Zool. Soc. 1878, p. 718.

A. parumpunctato (*Lin.*) *proxime affinis, sed gracilior, thorace longiori, etc. Elongato ovatus, gracilis, supra subfusco-cupreus, capite thoraceque magis aeneis, interdum toto viridiaeneus; corpore subtus nigro nitido; femoribus nigropiceis, tibiis tarsisque rufo-piceis; antennis piceo-fuscis, articulo basali rufo: capite lævi, post oculos magis subito quam in A. parumpunctato angustato: thorace subquadrato, lateribus leniter arcuatis, angulis posticis rotundatis ibique margine explanato-reflexo, toto limbo alutaceo: elytris elongatis, margine basali utrinque fortiter sinuato, lateribus parallelis, supra acute striatis, interstitiis planis, tertio 5-punctato.*

Long 3¼ lin.

Closely allied to the common European *A. parumpunctatus*. At first sight it seems to differ only in its more slender, narrower form, and rather duller colour; but on closer examination several minor structural differences are perceived. The head is more suddenly narrowed behind the eyes. The thorax is longer, its outer borders alutaceous and rugose, and the basal line, instead of forming a regular gentle curve, is nearly straight in the middle and obliquely arcuate on each side towards the hind angle, which is more distinct than in *A. parumpunctatus*; this outline giving the appearance of a broad sinuation in the middle of the base. The striæ of the elytra are not so distinctly punctulate, and the interstices rather coarsely alutaceous or granular.

Hab.—Taken between Tangtze and Chagra in the Pangong Valley, altitude probably between 13,000 and 15,000 feet. Some specimens from the Pamir, between Sirikol and Panga. *A. parumpunctatus* is found throughout Europe and the Caucasus, and also in Western Siberia.

39.—ANCHOMENUS POLITISSIMUS.

Bates, Proc. Zool. Soc. 1878, p. 719.

A. fuliginoso (*Panzer*) *formâ subsimilis, nigro-aeneus, politissimus : capite breviter ovato, oculis vix prominulis; palpis minus elongatis, articulis ultimis acuminatis : thorace postice angustato, angulis posticis oblique rotundatis, margine prope angulum valde reflexo : elytris apicem versus valde sinuatis, supra obsolete striatis, disco utrinque haud conspicue bipunctato; pedibus aeneis, tibiis rufotestaceis.*

Long 2¼ lin.

Belongs apparently to the genus *Oxypselaphus* (Chaud.), which is not admitted by modern authors. The antennæ, however, are longer than in that group, being much longer than the head and thorax; the third joint is not pubescent and is a little longer than the first and the fourth; the basal joint is slightly rufous in front. The maxillary palpi are rather less sharply pointed at the apex than the labials, and all are pale at the tip. The thorax is quadrate-cordate; being a little rounded immediately after the anterior angles, and then gradually narrowed to the base; the lateral margin near the hind angle is remarkably and sharply elevated, and the upper edge of elevated rim has a slight notch. The striæ of the elytra are most visible at the apex, the marginal one being entire. The whole insect is highly polished, having the appearance of being varnished.

Hab.—Murree.

40.—COLPODES OVALICEPS.

Bates, Proc. Zool. Soc. 1878, p. 719.

Minus elongatus, nigro-chalybeus nitidus, elytris ampliatis, ovatis : capite parvo, ovato, oculis haud prominulis; menti dente apice sulcato-emarginato : thorace ovato capite dimidio latiori, margine laterali æqualiter explanato, subreflexo, angulis posticis subrotundatus : elytris convexis, late ovatis, apice vix sinuatis, humeris rotundatis, striatis, interstitiis planis, tertio tripunctato : metasterni episternis brevibus; antennis, palpis, pedibusque rufopiceis, femoribus nigris.

Long. 5 lin.

Differs from the great majority of the genus *Colpodes* by a combination of peculiarities,— notched tooth of mentum, short metathoracic episterna, and simply but deeply sinuated fourth joint of anterior and middle tarsi. The head appears ovate and small, owing to the unsalient eyes and the continued width and fulness far behind the eyes, the short neck close to the thorax only being contracted; the upper surface also at the neck is depressed. The palpi are not notably elongated, and the apical joints are but slightly narrowed to the apex and briefly truncated. The third antennal joint is naked and of the same length as the fourth. The thorax is widest a little before the middle and the sides are there slightly angulated; the anterior angles are prominent, the posterior very obtuse, almost rounded. The tarsi are clothed beneath with long soft hairs, longest on the fourth joint as characteristic of the genus *Colpodes*.

Hab.—Murree.

41.—ARGUTOR DIFFICILIS?

Chaudoir, Enum. Carab. Caucas. p. 136.

Hab.—A single example of an *Argutor*, from Sanju, closely allied to the common European *A. strenuus* (Panzer). From its somewhat larger size, I think it likely to be the species, or variety, above-named.

42.—MOLOPS PILIFERUS.

Bates, Proc. Zool. Soc. 1878, p. 718.

Niger, nitidus; thorace late cordato, post medium subsinuatim angustato, angulis posticis rectis; antice juxta marginem lateralem punctis decem longe piliferis lineatim dispositis: elytris elongato-ovatis, convexis, prope apicem fortiter sinuatis, supra exarato-striatis, striis 7—8 valde approximatis, 7ma uninterrupte punctatis, punctis longe piliferis, interstitiis dorsalibus planis, tertio et quinto apice pilifero-punctatis.

Long. 6—7 lin.

Distinguished from all its European congeners by the remarkable row of punctures along the seventh elytral stria, each bearing an extremely long stiff hair: a similar row of hairs accompanies the lateral margin of the thorax, at the rounded anterior part, and a group of the same is situated near the inner margin of each eye. The general shape of the insect is similar to that of *M. elatus*; but the thorax is more fully rounded anteriorly and more narrowed posteriorly, the hind angles being rectangular and not abruptly prominent. The striæ of the elytra are more sharply impressed, not distinctly punctured, and the interstices are plane.

Hab.—Murree.

43.—AMARA TRIVIALIS.

Gyllenhal, Fauna Suec. vi, 240.

Hab.—Two examples : Sind Valley ; agreeing tolerably well with West European specimens.

44.—AMARA BAMIDUNYÆ.

Bates Proc. Zool. Soc. 1878, p. 716.

A. triviali (*Dufts.*) *affinis*. *Ovata, subtus viridi-aenea, supra aenea, antennis articulis 2 basalibus rufis, pedibus nigro-vel aeneo-piceis : thorace quam in* A. triviali *et* A. spreto *breviori, basi impunctato, foveolis interiori oblonga, exteriori parva obliqua subobsoleta : elytris striis subtilibus, apice haud profundius impressis, distincte punctulatis, interstitiis planis.*

Long. 3½—4 lin.

Partakes of the characters of three species—*trivialis*, *spreta*, and *famelica*, having the shorter thorax of the last, the basal coloration of antennæ of the second, and the size and colouring of the first. But it is distinguished from all by the peculiarly fine striation of the elytra, in which the punctures are generally more conspicuous and broader than the striæ themselves. Another character is the less polished surface; owing to the minute striation, especially of the elytra, even in the male. In matured individuals the undersurface of the breast and epipleuræ of the elytra is polished brassy green, the abdomen and femora brassy black. The scutellar striole lies between the first and second striæ, arising from near the base of the latter.

Hab.—Pamir; between Sirikol and Panga. A large number of examples.

45.—AMARA AMBIGENA.

Bates, Proc. Zool. Soc. 1878, p. 716.

Breviter ovata, nigro-aenea polita, ventris apice rufo ; palpis, antennis basi, pedibus, elytrorumque epipleuris, rufis, elytris interdum castaneis : thorace brevi, antice gradatim rotundato-angustato, apud basin elytris paulo angustiori, margine basali flexuoso, angulis posticis subacutis ; foveolis basalibus utrinque duabus latis, sparsim grosse punctatis ; elytris brevibus, punctulato-striatis, interstitiis planis : menti dente elongato triangulari sed apice anguste fisso. ♂ Tibiæ posticæ intus pauciter pilosæ.

Long. 3½ lin.

Approaches the genus *Leiocnemis*, the hind tibiæ of the ♂ having only a few soft hairs on their inner edge and the thorax being narrower than the elytra and parallel-sided for a short distance from the base: the facies is also that of *Leiocnemis tartaricæ*. The frontal foveæ are narrow, deep, and flexuous. The sides of the thorax are explanated gradually after the middle. The elytral striæ are fine and equally impressed from base to tip, the scutellar striole being united to the first stria. The sterna are smooth.

Hab.—Between Tanktze and Chagra, Pangong Valley.

46.—LEIOCNEMIS HIMALAÏCA.

Bates, Proc. Zool. Soc. 1878, p. 716.

Elongato-ovata, rufo-picea vel castanea supra aeneo-tincta ; partibus oris, antennis, pedibusque flavotestaceis ; thorace brevi, transverso, lateribus fere æqualiter arcuatis, antice paulo magis quam postice angustato, angulis posticis obtusis, lateribus paululum explanatis, foveolis basalibus grosse sparsim punctatis, interiori rotundata, exteriori vage impressa carinaque obsoletissima vel nulla: elytris acute et simpliciter striatis, interstitiis planis: corpore subtus lævi, nitido.

♂. *Tibiæ intermediæ subtus medio sinuatæ, deinde paulo dilatatæ et denticulatæ.*

Long. 3½ lin.

The rudimentary bidentate undersurface of the middle tibiæ of the ♂ show a tendency towards the genus *Curtonotus*, but the facies of the species is totally unlike that group; the general appearance of the insect being that of a moderately robust *Calathus*. There is scarcely any trace of the oblique carina at the posterior angles of the thorax, and the sides of the latter form a tolerably regular curve from base to apex, without the slightest sinuation near the hind angles.

Hab.—" Dras, Kargil, and Leh."

47.—LEIOCNEMIS TARTARIÆ.

Bates, Proc. Zool. Soc. 1878, p. 716.

Oblongo-ovata, modice convexa, nigra polita, supra aenescens; partibus oris, antennis, pedibusque piceo-rufis: thorace valde transverso, lateribus fortiter arcuatis, antice et postice fere æqualiter angustato, angulis posticis distinctis sed obtusis, basi utrinque foveis duabus modice impressis fortiter punctatis: elytris punctato-striatis, interstitiis planis: prosterni apice late rotundato, marginato, meso-et metasternis punctatis; menti dente lato, magno, bifido.

Long. 3⅛ lin.

The whole surface of the body is polished, beneath black, sometimes piceous and reddish, with the elytral epipleuræ also reddish; above tinged with greenish-brassy, the elytra sometimes bright brassy-green. The frontal foveæ are sharply impressed and linear, as in many other *Leiocnemis*. The thorax is transverse, distinctly narrower than the elytra, very strongly rounded on the sides, so as to give it an almost rounded appearance; the widest part is the middle, whence it narrows almost equally towards the apex and the base, the sides joining the base without any sinuation and forming an angle which is more obtuse than rectangular; the basal foveæ are never deep and in some examples scarcely apparent except from their coarse punctuation.

Hab.—Between Yangi Hissar and Sirikol.

48.—LEIOCNEMIS FRIVOLA.

Bates, Proc. Zool. Soc. 1878, p. 717.

Parva, oblonga, subtus rufo-castanea, supra nigro-aenea, vel aenea, elytris interdum castaneis aeneo-tinctis; partibus oris, antennis, pedibusque rufotestaceis: foveis frontalibus linearibus extus acute exaratis, antice supra epistomatem continuatis: thorace transversim quadrato lateribus leniter arcuatis; postice minus quam antice angustato, mox ante basin paullulum sinuato, angulis posticis fere rectis; foveis basalibus grosse punctatis; elytris punctulato-striatis; menti dente bifido, plano.

Long. 2½ lin.

This small species has the general appearance of a *Bradycellus*. The thorax at first sight appears quadrate, but the sides are gently arcuated and just before the hind angle very slightly incurved, so as to make the hind angles rectangular; but there is some little individual variation in this respect. The frontal foveæ are linear and cross the suture to the epistome which they invade for a short distance; their outer edge (towards the eye) is deepest, and the line is there so sharply incised that their border is vertical, in some lights appearing cariniform.

Hab.—No locality. Taken in the latter part of the journey; on the Pamir or near Yarkand.

49.—AMATHITIS BADIOLA.

Bates, Proc. Zool. Soc. 1878, p. 717.

A. rufescenti (*Dej.*) *proxime affinis, at angustior, corporeque infra nigro. Oblonga, depressa, subtus nigra, supra capite thoraceque rufo-castaneis (illo obscuriori) elytris sub-fusco-badiis, interdum aeneo-nitidis, striis obscurioribus : capite minus elongato, oculis multo minus quam in* A. rufescenti *prominulis : thorace elytris angustiori, ante medium modice rotundato-dilatato, prope basin angustato, ibique lateribus obliquis, angulisque posticis vix rectangulis (sed apice acutis) ; basi toto discrete punctato, foveis modice impressis, carinaque vix elevata : elytris punctulato-striatis, interstitiis planis ; corpore subtus nigro-nitido ; partibus oris, antennis, pedibusque fulvo-testaceis. Menti dente prominulo, triangulari. ♂ tibiis posticis intus parce breviter pubescentibus.*

Long. 4 lin.

Nearly allied to *A. rufescens*, but abundantly distinct. Its smaller head, much less prominent eyes, and relatively smaller thorax, distinguish it at once, independently of the light brown colour of the elytra. The distinct equilateral triangular tooth of the mentum distinguishes it from *A. subplanata* of Putzeys.

Hab.—One of Stoliczka's latest captures. The majority of the specimens bear no locality; but one example clearer in the colour of the elytra is ticketed as from the neighbourhood of Sanju.

50.—AMATHITIS KUENLUNENSIS.

Bates, Proc. Zool. Soc. 1878, p. 717.

Valde elongatus, modice convexus, pallide ferrugineus, antennis pedibusque gracilibus, illarum articulo tertio cæteris multo longiori : thorace elytris multo angustiori, late cordato, lateribus antice fortiter arcuatis, postice sinuatim angustato, angulis posticis acutis ; supra impunctato, foveis basalibus latis, vagis, carinaque prope angulum indistincte elevata : elytris elongato-ovatis, apice paulo sinuatis, striis lævibus modice impressis.

Long. 5 lin. ♀.

A species remarkable for its very slender antennæ and long legs, apparently allied to *A. longipennis* (Chaudoir) and allies from the Altai, none of which I have seen. The mentum is scarcely toothed in the middle of its emargination, and the horny ligula is very broad and truncated at the apex. The anterior tibiæ are much dilated and compressed towards the

COLEOPTERA.

apex, which is armed with only one long spur and is fringed, as well as the outer edge, with short, strong bristles; the middle and hind tibiæ are clothed all round with long, fine bristles. The episterna of the metathorax are elongated.

Hab.—Neighbourhood of Sanju.

51.—BRADYTUS APRICARIUS.

Carabus apricarius, Payk., Monogr. Carab. p. 77.
Amara apricaria, Dej., Spec. Gen. iii, 506.
Bradytus apricarius, Stephens, Ill. Brit. Ent. i, p. 136.

Hab.—Sind Valley, Dras, Kargil, and Leh : Pamir, between Sirikol and Panga.
The Pamir and Ladakh examples agree closely with the West European form of the common Palæarctic species. One of the Sind Valley specimens is rather more elongate, and is probably the var. *parallelus* (Chaudoir) from Lenkoran on the Caspian.

52.—BRADYTUS COMPACTUS.

Bates, Proc. Zool. Soc. 1878, p. 717.

Breviter oblongo-ovatus, latus, subaeneo-niger, convexus; capite brevi et crasso, oculis parum convexis, epistomatis margine antico transversim sulcato; thorace elytris haud angustiori, postice modice angustato, angulis posticis acutis, foveolis basalibus parvis punctatis, carinaque obsoleta : elytris simpliciter striatis, interstitiis planis : palpis, antennis, pedibusque rufopiceis : episternis parumpunctatis : menti dente magno apice inconspicue emarginato.

Long. 4 lin.
Of short, broad, oblong form. Distinguished from all other species by the smooth furrow accompanying the arcuated front margin of the epistome, which itself forms a thickened rim. A further important distinctive character is the absence of the usual carina of the thorax near the hind angles, in the situation of which there is a scarcely perceptible obtuse elevation. The hind angles of the thorax are acute, the arcuated lateral margin being slightly and briefly sinuated just before the angle, and the hind margin being incurved on each side. The elytral striæ are not perceptibly punctured. Prosternum with a long smooth longitudinal furrow. Posterior tibiæ of the male on the inside with a sparse clothing of soft hairs.

Hab.—Murree.

53.—CURTONOTUS PAMIRENSIS.

Bates, Proc. Zool. Soc. 1878, p. 717.

Elongato-oblongus, angustus, rufo-castaneus, supra olivaceo-aeneus, thoracis elytrorumque marginibus reflexis, rufescentibus : capite lævi, mox pone oculos angustato ; thorace transversim quadrato ante basin subfortiter constricto, ibique lateribus parallelis et margine laterali hand interrupto, angulis posticis rectis ; base grosse subsparsim punctato, foveis utrinque linearibus ; elytris striatis, striis (versus apicem exceptis) punctatis : metasterno et ventri basi sparsim punctatis.

♂. *Tibiis intermediis post medium angustatis et acute breviter bidentatis.*

Long. 4¼—5 lin.

In form this species is narrow, with remarkably elongate elytra. The undersurface is constantly chestnut-red, together with the epipleuræ of the elytra and prosternum, the legs, antennæ, parts of the mouth, epistome, and narrow lateral rims of the elytra and thorax. The rest of the upper surface is dark, brassy-olivaceous. The thorax is of the same width anteriorly as the elytra, but is much narrowed near the base, nearly as in *C. fodinæ*, but the hind angles do not at all project. The elytral interstices are plane; the punctuation of the striæ is strongest in the striæ nearest the suture, and the edges of the interstices are there crenulated; it disappears towards the apex and becomes very faint towards the sides.

Hab.—The Pamir Steppe; between Sirikol and Panga.

54.—BEMBIDIUM (PERYPHUS) PAMIRENSE.

Bates, Proc. Zool. Soc. 1878, p. 718.

Oblongum, depressum, capite thoraceque viridi-vel aurato-aeneis, politis, elytris fulvotestaceis, vitta suturali (ante apicem abbreviata) fasciaque pone medium (interdum quoque margine et apice) aeneo-fuscis; antennis, palpis, pedibusque flavo-testaceis: thorace breviter cordato, antice fortiter rotundato, angulis posticis rectis, basi rugato, fovea utrinque profunda carinulaque obliqua: elytris striato-punctatis, interstitiis planis.

Long. 2¼ lin.

Of the flattened form of *B. andreæ*, *B. femoratum*, and allies; elytra scarcely so elongated and rather more ovate than in *B. andreæ*. Elytral striæ sometimes scarcely impressed, the exterior ones much fainter but visible, all punctate and interstices plane or slightly convex. The apical joints of the antennæ and penultimate joint of the maxillary palpi are faint ashybrown. The dusky cruciform mark on the elytra is very variable and is never very dark or clearly defined. As in the allied species, the sutural border is dilated where the transverse fascia joins it. When the lateral margins are dusky, the colour only covers the marginal interstice.

Hab.—Pamir, between Sirikol and Panga.

55.—BEMBIDIUM (PERYPHUS) PUNCTULIPENNE.

Bates, Proc. Zool. Soc. 1878, p. 718.

Subdepressum, aeneo-nigrum politum, mandibulis piceo-rufis: thorace antice leniter rotundato, postice usque ad angulos angustato, his fortiter reflexis, obtusis, margine basali utrinque prope angulum valde obliquo; supra basi et margine rugulosis, fovea oblonga: elytris punctato-striatis, striis 1—4 solum impressis, 6—7 obsoletis, interstitiis planis, minutissime sparsim punctatis.

Long. vix 2 lin.

Distinguished from all the species of the group known to me by the form of the thorax. This member is moderately elongate and subquadrate, widest near its anterior angles, where its sides are gently rounded, and after this narrowing moderately to its hind angles; but the lateral margin near these latter becomes flattened-out and reflexed, and the angle itself (which is obtuse) is a little in advance of the apparent angle, the interval between the two being oblique and curved. The basal fovea lies against the false angle, and the surface between it and the true angle is convex. The punctuation of the elytral interstices is extremely minute and in a single row; visible only in certain lights. The species has the general form of the *Peryphi* allied to *atrocœruleum*.

Hab.—No locality. Most probably the Pamir.

56.—BEMBIDIUM (PERYPHUS) TIBIALE.

Dufts., Faun. Austr. ii. 200.

Hab.—A single example from Ladak, closely allied to, if not a variety of, this European species.

57.—BEMBIDIUM sp.

Hab.—A single specimen from Kogyar, in immature condition and indeterminable. It is a *Peryphus*, with strongly rounded thorax and pale apical spot to elytra.

58.—BEMBIDIUM 4-PUSTULATUM.

Dej., Spec. Gen. Col. v. p. 186.

Hab.—Between Yangi Hissar and Sirikol. A single specimen, with much enlarged anterior elytral spot.

59.—ANTHIA ORIENTALIS.

Hope, Coleop. Manual ii, p. 163, pl. 6, f. 14.

Hab.—Jhelam Valley, one example. Agrees with Hope's description and figure and with Chaudoir's subsequent description, so far as concerns the depressed elytra and smoother thorax, but differs in the anterior spot of the elytra being transverse-oblong. It forms probably another of the numerous local forms of the *A. sexguttata*.

60.—METABLETUS TARTARUS.

Proc. Zool. Soc. 1878, p. 719.

M. truncatello *(Lin.) paullo major, magis elongatus, subaeneo-niger, nitidus, antennis et pedibus fusco-piceis : thorace quam in M. truncatello postice magis angustato, angulis posticis obtusioribus, deinde usque ad basin magis obliquis ; elytris elongatis, apice obtuse subsinuatim truncatis, obsolete striatis, impunctatis.*

Long. 1⅘ lin.

Closely allied to the European *Metabletus truncatellus*, but larger and the elytra relatively longer. The colour is a little more metallic, and the thorax differs in being more narrowed behind, with the hind angles much more obtuse and the margin thence to the base more oblique. The elytra are equally smooth, sometimes only the sutural stria is visible, and in all examples this stria is the only one sharply impressed, most so towards the apex.

Hab.—Between Yangi Hissar and Sirikol. One example, much the most feebly striated, Sind Valley.

61.—CYMINDIS GLABRELLA.

Dates, Proc. Zool. Soc., 1878, p. 719.

C. andreæ (*Ménétr.*) *affinis; at gracilior, oculis minus prominulis, elytrisque fusco-castaneis, flavomarginatis. Gracilis, glaberrima, castaneo-rufa, abdominis margine picescenti, capite obscuriori, partibus oris, antennis, pedibus, elytrorumque margine fulvo-testaceis: capite angusto, sparsim punctulato: thorace capite haud latiori, anguste cordato lateribus postice leviter sinuatis angulis posticis fere rotundatis, sparsissime punctulato: elytris basin versus angustatis, humeris rotundatis, subpunctulato-striatis, interstitiis sparsim punctulatis: palpis labialibus apice modice dilatatis, triangularibus.*

Long. 4—4¾ lin.

Allied to *C. andreæ.* Upper surface naked and shining, labial palpi moderately dilated, triangular. Eyes scarcely prominent, and punctuation of the whole upper surface very sparse and minute. General colour castaneous, but the thorax redder and the head slightly darker, the margins of the elytra (extending to the 8th striæ) are pale testaceous-fulvus, the antennæ, legs, and parts of the mouth being of a similar hue.

The species seems to be closely allied to *C. pallidula* (Chaudoir) from Lenkoran; but in that species the elytra are not wider at the base than the base of the thorax; in *C. glabrella* they are (taken together) nearly double the width.

Hab.—Ladak.

62.—CYMINDIS MANNERHEIMII.

Gebler, Bull. Acad. Petrop. 1843, 1. p. 36.

Chaudoir, Bull. Mosc. 1850; Suppl. Faune Carab. d. 1. Russie, p. 22.

Hab.—Pamir; between Sirikol and Panga: also the Pangong Valley and between Dras and Leh. By the Russian entomologists recorded as from the Tarbagatai Range. The elytral interstices are of equal breadth and punctured each in more than one row. The Pangong specimens are generally more shining in colour and with more convex and more strongly punctured elytral interstices; ?=*rufipes*, Gebler.

63.—CYMINDIS ALTAICA.

Gebler, Bull. Mosc. 1833, p. 264; id., 1847, p. 276.

Chaudoir, Bull. Mosc. 1850; Suppl. Faune Carab. d. 1. Russie, p. 21.

Hab.—Between Dras and Leh; one example agreeing with the description given of the elytra by Baron Chaudoir, l. c., *vis.*, alternate interstices narrower and with one row only of punctures.

COLEOPTERA. 21

64.—CYMINDIS sp.

Hab.—One specimen from the route between Leh and Yarkand; without legs and apparently immature.

LONGICORNIA.

1.—PRIONUS CORPULENTUS.

Proc. Zool. Soc. 1878, p. 720.

Magnus, elongatus, nigro-castaneus, supra omnino coriaceus vix nitidus: thorace parvo, utrinque acute trispinoso: elytris basin versus parallelis, compressis, deinde modice dilatato-rotundatis apice late obtusis, utrinque lineis elevatis tribus vix conspicuis: pectore toto dense fulvo hirto; abdomine politissimo: antennis 12-articulatis grosse punctatis, articulis 3—11 serratis, 5—12 apice foveo porosa,8—12 irregulariter strigosis. ♀.

Long. 2 unc.

Remarkable for the great length and bulk of the after-body (including the elytra) relatively to the head and thorax. The palpi are also longer, and their apical joint less dilated than in other species. The posterior thoracic angle is rather more produced and spiniform than in *P. asiaticus* (Falderm.), making the lateral armature 3-spinose; the middle spine is very long and acute. The thorax is narrow as well as short, and is coarsely sculptured, with the exception of a discoïdal convex area, which is more sparsely punctured; anteriorly the thorax is rather abruptly declivous. The elytra are throughout vermiculate-rugose, without mixture of punctures. The legs are long and compressed, and the tarsi, especially the claw-joint, remarkably long.

Hab.—Murree. Two examples, ♀.

2.—HESPEROPHANES CRIBRICOLLIS.

Bates, Proc. Zool. Soc. 1878, p. 720.

Cylindricus, fulvus, pilis incumbentibus cinereo-fulvis, apud elytros maculatim, vestitus: thorace rotundato, elytris multo angustiori, inæquali, lateribus medio subtuberoso, supra crebre alveolato-punctatis: scutello cinereo; elytris omnino discrete punctatis haud conspicue bicarinatis, apice gradatim angustatis, subacuminatis.

Long. 8—9 lin.

Closely allied to the European *H. griseus*, but distinguished by its smaller and more coarsely reticulate-punctate thorax, and by the elytra tapering towards the apex causing the sutural angle to be very acute. The fourth antennal joint is only a little shorter than the third and the fifth. The tawny-gray, laid pubescence is very even on the antennæ, the whole undersurface, and the legs.

Hab.—Murree.

TRINOPHYLUM, nov. gen.

Bates, Proc. Zool. Soc. 1878, p. 720.

Gen. Hesperophanes *affine, sed femoribus abrupte clavatis, oculisque minus forte granulatis. Corpus elongatum, subdepressum, breviter suberecte pubescens, crebre punctatum.*

Caput brevissimum palpis parvissimis. Thorax rotundatus, inermis. Elytra apice obtusa. Acetabula antica extus haud elongata. Prosternum angustum: mesosternum latum, subconvexum. Antennæ filiformes, corpore paullo breviores, articulo 4 to paullo abbreviato.

The facettes of the eyes are intermediate in size between those of the *Hesperophaninæ* and the *Callidiinæ*. The structure of the sterna and acetabula is very similar to that of the genus *Zamium*; but the clavate femora resemble those of *Callidium* and allied genera. The scarcely elevated antenniferous tubercles are again those of *Zamium* rather than *Hesperophanes*.

3.—TRINOPHYLUM CRIBRATUM.

Bates, Proc. Zool. Soc. 1878, p. 720.

Castaneo-fuscum, subnitidum, omnino suberecte fusco-pubescens, crebre sed discrete punctatum: thorace supra antice et postice paullo depresso, linea dorsali lævi: elytris lineis duabus indistinctis lævibus.

Long. 6½ lin.

The general colour is brownish-chestnut, and shining, notwithstanding the rather close slantingly-erect pubescence with which the whole body is clothed: the underside, antennæ, and legs are of a lighter and more reddish hue. The thorax has regularly rounded sides and is but slightly unequal on its upper surface.

Hab.—Murree. Two specimens.

4.—LEPTURA RUBRIOLA.

Bates, Proc. Zool. Soc. 1878, p. 720.

L. sanguinolentæ (*Lin.*) *affinis. Nigra, subtus sparsim fulvo-pilosa, thoracis plaga magna discoidea elytrisque rufo-opacis: capite et thorace crebre reticulato-punctatis, illo ut in L.* sanguinolenta *paullo post oculos subito et fortiter constricto: thorace medio haud conspicue dilatato modice convexo, angulis posticis modice productis: elytris sub-crebre punctulatis, apice recte truncatis, angulisque breviter dentatis ♀ .*

Long 7 lin. ♀ .

Closely allied to the European *L. sanguinolenta*, differing (♀) in the upper surface of the thorax being dark red like the elytra; a narrow anterior border and a spot in the middle of the hind border, like the whole undersurface, black: the surface of the thorax and elytra is clothed with a short erect pubescence. The elytra are wholly red, without a trace of black.

♂ . Taken in the same locality are two ♂ examples, which probably belong to this species: they are 5½ lines long. One is wholly black, and the other has the basal half (and a little more) rufo-testaceous, the rest black.

Hab.—Murree.

5.—CLYTANTHUS IGNOBILIS.

Bates, Proc. Zool. Soc. 1878, p. 721.

Cl. 4-punctato (*F.*) *proxime affinis. Nigro-fuscus, tomento cinereo-flavo vestitus, elytris utrinque maculis 5 nigris, quam in* Cl. 4-punctato *majoribus, scilicet 1 curvata post scutellum,*

1 parva humerali, 1 antico-discoidali, 1 mediana majori transversa, et 1 huic proxima longitudinali oblonga.

Long. 6 lin.

Very closely allied to the Mediterranean *Cl. 4-punctatus* (F.), the only apparent difference being the larger size of the dark elytral spots. The thorax, however, appears to be a little more cylindrical and less convex both above and on the sides, and is furnished with a number of large scattered punctures (besides the close general punctuation), most conspicuous on the sides. The spot behind the transverse median spot of the elytra is further removed from the apex than in *Cl. 4-punctatus*.

Hab.—Murree. Two examples.

EXPLANATION OF THE PLATE.

GEODEPHAGA.

FIG. 1. *Harpalus cæruleatus.*
2. ,, *liodes.*
3. ,, *indicola.*
4. ,, *melaneus.*
5. ,, *masoreoides.*
6. *Dichirotrichus alticola.*
7. *Colpodes ovaliceps.*
8. *Cicindela stoliczkana.*
9. *Hypsinephus ellipticus.*
10. *Amathitis kuenlunensis.*
11. ,, *badiola.*
12. *Molops piliferus.*
13. *Carabus stoliczkanus.*
14. *Leiocnemis tartariæ.*
15. *Curtonotus pamirensis.*
16. *Anchomenus politissimus.*
17. *Cymindis glabrella.*

LONGICORNIA.

18. *Prionus corpulentus.*
19. *Trinophylum cribratum.*
20. *Clytanthus ignobilis.*
21. *Leptura rubriola,* ♂ ♀
22. ,, ,, ♀.

PHYTOPHAGA.

By JOSEPH S. BALY, F.L.S.

The Phytophagous Coleoptera collected by Dr. Stoliczka, although few in number, and containing no striking novelties, are extremely interesting in relation to geographical distribution. The 25 species contained in the collection belong to no less than 21 genera, out of which *Nodostoma*, *Enneamera*, *Charæa*, *Macrima*, *Mimastra*, *Merista*, and *Leptorthra* (one-third of the whole) are exclusively Asiatic; *Paria* has its metropolis in America, but is sparingly represented in Japan, China, and Eastern Siberia; *Luperodes* is largely spread throughout the Asiatic continent, and is also found (according to v. Harold, whose accuracy cannot be doubted) in South America and Abyssinia; of the twelve others, five are cosmopolitan, and the rest occur abundantly in Europe. Out of the 25 species, one only, *Plagiodera versicolora*, Laich. (*armoraciæ*, Auct.), is found in Europe; seven, *Lema coromandeliana*, *Clytra palliata*, *Enneamera variabilis*, *Galleruca indica*, *Gallerucella placida*, *Merista interrupta*, and *Leptarthra collaris*, occur in various parts of British India; two, *Haltica cærulescens* and *H. viridicyanea*, have been described by myself from Japan; and one *Chrysomela angelica*, Reiche, is not uncommon in Syria; the fourteen others have not as yet been found in any other locality, and seventeen species are described for the first time in the present paper.

1.—LEMA COROMANDELIANA, Fabr., var. PRÆUSTA.

Crioceris præusta, Fabr., Ent. Syst. i, 2, p. 8; *Lema præusta*, Lac., Mon. Phyt. i, p. 340.

Hab.—Jhelam Valley. A single specimen.

2.—CLYTRA PALLIATA.

Clytra palliata, Fabr., Syst. El. ii, p. 30.

Hab.—Jhelam Valley; also various parts of India.

3.—COPTOCEPHALA DUBIA.

Baly, Cyst. Ent. ii, 1875—82, p. 370.

Subelongata, subcylindrica, nitida, subtus nigra, argenteo sericea, prothorace pedibusque fulvis; supra fulva, capitis vertice nigro; thorace lævi; scutello piceo; elytris tenuiter punctatis, fasciá communi baseos, extrorsum abbreviatá, alteráque vix pone medium nigris.

Long. 2⅓ lin.

Vertex black, impunctate, lower face fulvous, a ray of the same colour extending upwards on the vertex; front deeply excavated between the eyes, irregularly punctured; anterior margin of clypeus concave-emarginate. Thorax rather more than twice as broad as long; sides rounded, converging from behind the middle to the apex; the anterior angles obtuse, the hinder ones rounded; disc transversely convex, shining, impunctate, excavated on either side near the lateral margin. Scutellum trigonate, piceous. Elytra scarcely broader than the thorax, parallel, very finely punctured; the black markings on their surface extend from the base nearly to the middle of the disc, and again from the middle itself nearly to the apex, leaving only an irregular flavous transverse band across the middle, which sends a narrow ramus along the suture nearly to the base.

Hab.—Murree.

4.—COPTOCEPHALA DIMIDIATIPENNIS.

Baly, Cist. Ent. ii, 1875—82, p. 371.

Subelongata, subcylindrica, flava, nitida, corpore inferiori, capite, elytrorumque limbo inflexo, fulvo hirsutis, thorace lævi; elytris tenuiter punctatis, nigris, a basi ad paulo ante medium flavis.

Long. 3—3¾ lin.

Head clothed with long, erect hairs, minutely punctured; clypeus not separated from the face, its anterior margin angulate-emarginate; apex of jaws black; antennæ equal in length to the head and thorax, the basal joint thickened, pyriform, the second also thickened, short, nodose, the third small, not longer than the second, the fourth trigonate, scarcely longer than the third, the rest to the apex dilated, the fifth to the ninth transversely trigonate, the tenth and eleventh ovate; eyes large, oval, notched on the inner margin. Thorax nearly three times as broad as long; sides obtusely rounded, slightly converging in front, the hinder angles rounded, the anterior ones very obtuse; basal margin sinuate on either side the median lobe, the latter slightly reflexed, very obtusely rounded; upper surface transversely convex, remotely and very minutely punctured, a concave transverse space on and immediately in front of the basal lobe, coarsely and closely punctured. Scutellum longer than broad, subtrigonate, its apex obtuse. Elytra scarcely broader than the thorax at the base, slightly dilated posteriorly, convex, rather distantly and finely punctured. Body beneath and legs clothed with long, erect fulvous hairs.

I possess two specimens of this species, both labelled India, but without precise locality; in one of them the head is more coarsely punctured and subrugose, in all other respects it agrees with the type.

Hab.—Jhelam Valley; also India, my collection.

5.—CRYPTOCEPHALUS INTERJECTUS.

Baly, Cist. Ent. ii, 1875—82, p. 372.

Elongato-oblongus ♂, oblongus ♀, convexus, nitidus, subtus niger, pedibus nigro-piceis; supra flavus, capite hic illic parce fortiter punctato, fronte sulco longitudinali impresso; vertice, maculis duabus inter oculos, labro, antennisque nigris, his basi, sulco longitudinali, mandibulisque piceis; thorace lævi, limbo angusto et utrinque maculâ subrotundatâ nigris; scutello subcordato, nigro; elytris fortiter punctato-striatis, punctis piceis, apicem versus minus fortiter impressis; interspatiis convexis, transversim rugulosis; utrisque limbo angusto, externo ante medium excepto, maculisque quinque 2, 2, 1 dispositis nigris.

Long. 2¼ lin.

Var. A. Pygidio corporeque subtus flavis, illo maculâ cuneiformi, pectore, abdominisque disco nigris.

Var. B. Corpore nigro, antennarum basi, clypeo, faciei signaturis, thoracisque lineâ longitudinali sordide flavis.

Head rather coarsely but not closely punctured, the puncturing varying in degree in different individuals; front impressed with a distinct longitudinal groove; clypeus broader than long, trigonate; antennæ three-fourths the length of the body in the ♀, rather longer in the ♂, the three lower joints pale piceous, the rest black. Thorax rather more than twice as broad as long at the base; sides moderately rounded and obliquely converging from base to apex; basal margin concave-emarginate on either side, the outer angles produced backwards, acute; above convex, minutely but not closely punctured. Elytra slightly broader than the thorax, oblong-quadrate, convex, rather strongly punctate-striate, the punctures piceous, finer, and less strongly impressed towards the apex; interspaces faintly but distinctly convex, transversely wrinkled; each elytron with the extreme outer limb (interrupted on the lateral margin before its middle) and five large patches black; these spots are arranged as follows: two transversely below the base, the outer one oblong, covering the humeral callus and attached to the basal margin, the inner one subrotundate, placed on the inner disc; two just below the middle also placed transversely, both subrotundate, the outer one usually attached to the lateral margin; and, lastly, one apical, transversely oblong, either free or attached to the apical border; these patches are often more or less confluent, and occasionally, as in var. B, cover the entire surface of the elytron. Pygidium and body beneath clothed with griseous hairs. Apical margin of prosternum obliquely produced, deflexed, slightly emarginate, the hinder margin concave, armed on either side with a deflexed, obtuse tooth; mesosternum transverse, its apical border angulate-emarginate. Apical segment of abdomen in the ♂ impressed with a shallow fovea; the same segment in the ♀ deeply excavated, the fovea large, rotundate. Basal joint of the four anterior tarsi in the ♂ dilated, elongate-ovate, longer than the following two united.

The form of the prosternum will separate this species from any nearly allied species.

Hab.—Murree.

6.—NODOSTOMA CONCINNICOLLE.

Baly, Cist. Ent. ii., 1871—85, p. 373.

Oblongo-ovatum, convexum, pallide piceum, nitidum, pedibus antennisque fulvis; thorace transverso, lateribus ante basin acute angulatis, disco crebre foveolato-punctato; elytris nigris, fortiter punctato-striatis, interspatiis planis.

Long. 2 lin.

Var. A. Elytris piceo-fulvis, punctis piceis.

Head coarsely and deeply punctured, the punctures on the extreme vertex crowded; clypeus not distinctly separated from the face; antennæ slender, filiform, the second joint ovate, three-fourths the length of the third, the latter two-thirds the length of the fourth. Thorax more than twice as broad as long; sides abruptly diverging and acutely angled just in front of the base, thence obliquely converging to the apex, just before reaching the latter abruptly incurved, the apical angle obtuse, the hinder one armed with a lateral tooth; disc closely covered with large, round, deeply-impressed punctures; on either side are a few short, suberect griseous hairs. Scutellum longer than broad, cuneiform, its apex obtusely angulate. Elytra convex, transversely depressed below the basilar space, strongly punctate-striate; on the transverse depression, and also below the shoulder, the puncturing is confused; interspaces plane, irregularly wrinkled on the sub-basilar depression. All the thighs armed beneath with an acute tooth.

Hab.—Jhelam Valley.

7.—NODOSTOMA PLAGIOSUM.

Baly, Cist. Ent. ii., 1875—82, p. 373.

Oblongo-ovatum, piceum, nitidum, pedibus antennisque piceofulvis, his extrorsum piceis; thorace profunde et crebre punctato, lateribus pone medium obtuse angulatis; elytris fortiter punctato striatis, striis apicem versus fere deletis; sordide fulvis, limbo angusto, striarum punctis et utrinque plagâ irregulari magnâ, a basi ad paulo pone medium extensâ, ad marginem lateralem affixâ, piceis.

Long. 1⅘ lin.

Vertex and front sub-remotely punctured; clypeus coarsely and irregularly punctured, not distinctly separated from the upper face, its anterior border deeply excavate-emarginate, the emargination produced and forming two sub-acute teeth; labrum fulvous; antennæ slender, filiform, the second and third joints nearly equal in length, the fourth very slightly longer than the third, four or five lower joints obscure fulvous, the rest piceous. Thorax nearly twice as broad as long; sides diverging at the base, obtusely angled behind the middle, thence obliquely converging and very slightly rounded to the apex; disc transversely convex, very coarsely and deeply punctured. Elytra oblong, sub-acutely rounded at the apex, convex, strongly punctate-striate, the punctures near the apex much finer and nearly obsolete, interspaces plane, impunctate; the irregularly piceous patch on each elytron covers the outer disc (the humeral callus excepted) and extends from the base to just below the middle of the disc. All the thighs armed beneath with a small tooth.

Hab.—Murree. A single specimen, also India, without precise locality, my collection.

8.—PARIA CUPRESCENS.

Baly, Cist. Ent. ii., 1875—82, p. 374.

Anguste ovata, subtus cum capite picea, pedibus antennarumque basi pallidis; supra cuprea, thorace sub-conico, vage punctato; elytris regulariter punctato-striatis, interspatiis planis, impunctatis.

Long. 1¼ lin.

Vertex swollen, shining, impunctate; clypeus transverse, its anterior border emarginate; antennæ, rather more than half the length of the body, piceous, the two lower joints paler,

Thorax broader than long at the base; sides straight and obliquely converging from base to apex, the hinder angles very acute; basal margin oblique on either side, the median lobe obtusely rounded; disc subcylindrical, impressed, but not closely, with very shallow punctures. Elytra ovate, attenuated at the apex, regularly punctate-striate, the interspaces plane, each impressed with an irregular row of minute punctures; humeral callus thickened.

Hab.—Jhelam Valley.

9.—PLAGIODERA VERSICOLORA.

Chrysomela versicolora, Laicharting, Verz. Tyrol. Ins. i, p. 148 (1781).
Chrysomela armoraciæ.—Fabr.

Hab.—Jhelam Valley.

10.—CHRYSOMELA ANGELICA.

Chrysomela angelica, Reiche, Ann. Soc. Ent. France, 1858, p. 33, tab. i, fig. 8; Fairm., l. c. 1865, p. 80.

Hab.—Sind Valley; also Syria. I do not detect the slightest difference between specimens brought from Syria and those contained in the present collection.

11.—PHRATORA ABDOMINALIS.

Daly, Cist. Ent. ii., 1875—82, p. 375.

Elongata, parallela, nigro-aenea aut nigro-cyanea, nitida, pedibus abdomineque nigro-piceis, hujus segmentis ultimis duobus piceo-fulvis; thorace transverso, sat fortiter irregulariter punctato, utrinque leviter rugoso; elytris thorace latioribus, parallelis, sat fortiter punctatis, punctis subseriatim dispositis, interspatiis planis, subremote, tenuiter punctatis, infra callum humeralem transversim rugulosis.

Long. 2¾—3 lin.

Head short, transverse; vertex impressed, but not very closely, with large deep punctures, lower face more closely, but less coarsely, punctured than the vertex, sub-rugulose; in the middle, between the encarpæ, is a short longitudinal sulcation, which extends upwards from the apex of the clypeus; the latter depressed, broader than long, its upper margin obtusely angulate, its surface closely punctured, subrugose; antennæ scarcely more than half the length of the body, filiform, slightly thickened towards the apex, the basal joint thickened, the second slender, equal in length to the first and also to the fourth joints, but slightly shorter than the third; two lower joints fulvous, stained above with piceous, the third to the sixth obscure piceous, the five others slightly thickened, black. Thorax nearly one half broader than long; sides nearly straight and parallel from the base to the middle, thence obliquely converging to the apex, the hinder angles produced laterally into a large acute tooth, the anterior ones subacute; apical margin concave; upper surface irregularly punctured, the interspaces smooth and shining on the middle disc, finely rugulose on the sides. Elytra broader than the thorax, parallel, rather strongly punctured, the punctures arranged irregularly in ill-defined longitudinal rows, which, on the inner disc below the middle, approximate in pairs; interspaces plane, sparingly and very minutely punctured on the anterior disc, rugulose on the outer one below the humeral callus. Basal joint of anterior tarsus dilated, subcordate.

Hab.—Murree.

12.—HALTICA CŒRULESCENS.

Haltica cœrulescens, Baly, Trans. Ent. Soc. 1874, p. 190.

Hab.—Murree; also China and Japan.

13.—HALTICA VIRIDICYANEA.

Haltica viridicyanea, Baly, Trans. Ent. Soc. 1874, p. 191.

Hab.—Sind Valley, apparently common. I possess this species from Japan; it is probably found in the intermediate localities.

14.—ENNEAMERA VARIABILIS.

Nonarthra variabilis, Baly, Journ. of Entom. i, p. 456, tab. 21, fig. 1.

Hab.—Murree. This species is also found in Northern India.

CHARŒA, n. gen.

Baly, Cist. Ent. ii, 1875—82, p. 376.

Corpus *elongato-ovatum*. Caput *exsertum*, facie *perpendiculari;* oculis *rotundatis integris, prominentibus;* encarpis *tranversis, contiguis;* carinâ *oblongo-elongatâ, apice, acutâ;* antennis *filiformibus.* Thorax *transversus, dorso modice convexus.* Elytra *thorace latiora, confuse punctata, limbo inflexo fere ad apicem extenso.* Pedes, *femoribus posticis non incrassatis; tibiis simplicibus, apice spinâ acutâ armatis;* tarsis *posticis articulo basali sequentibus tribus longitudine fere æquanti,* unguiculis *appendiculatis.* Prosternum *angustum,* coxis *fere æquialtum;* acetabulis *anticis apertis.*

This genus at first sight bears in its facies a strong resemblance to *Aphthora*, but the slender hinder thighs at once separate it and place it amongst the Gallerucinæ.

15.—CHARŒA FLAVIVENTRIS.

Baly, Cist. Ent. ii, 1875—82, p. 376.

Elongato-ovata, convexa, subtus picea, aeneotincta, abdomine flavo; supra viridi-cyanea, antennis nigris; thorace lateribus rotundatis, disco lævi, modice convexo; elytris tenuiter confuse punctatis.

Long. 1¼ lin.

Vertex and front shining, impunctate; encarpæ transverse, contiguous; antennæ half the length of the body, second and third joints equal, the fourth nearly twice the length of the third; three lower joints nigro-piceous, stained with æneous, the rest black. Thorax broader than long; sides converging from the middle towards the base; the anterior angles slightly produced, obtuse, the hinder ones rounded, armed with a very small acute tooth; disc moderately convex, very minutely punctured, the punctures only visible under a very strong lens. Scutellum trigonate. Elytra broader than the thorax, parallel, finely but not closely punctured, the interspaces obsoletely wrinkled.

Hab.—Murree.

MACRIMA, n. gen.

Baly, Cist. Ent. ii, 1875—82, p. 377.

Corpus *anguste oblongum, convexum*. Caput *exsertum;* antennis *filiformibus, articulo primo duobus sequentibus conjunctis æquali, his brevibus, longitudine fere æqualibus;* oculis *sub-rotundatis, prominentibus;* encarpis *medio contiguis;* carinâ *obsoletâ;* palpis *maxillaribus articulis duobus ultimis conjunctim anguste ovatis, ultimo apice acuto.* Thorax *transversus, disco leviter excavatus.* Scutellum *trigonatum.* Elytra *thorace latiora, oblonga, confuse punctata, limbo inflexo fere integro, concavo.* Pedes *mediocres,* coxis *anticis elevatis, obtrigonatis, contiguis;* tibiis *apice mucronatis;* tarsis *posticis articulo primo tribus sequentibus fere æquilongo;* unguiculis *appendiculatis.* Prosternum *medio angustissimum;* acetabulis *anticis integris;* episternis *posticis a basi ad apicem angustatis.* Type, *Macrima armata*.

Macrima may be separated from *Aulacophora*, which genus it strongly resembles in outward form, by the closed anterior acetabula and by the appendiculated claws.

16.—MACRIMA ARMATA.

Baly, Cist. Ent. ii, 1875—82, p. 377.

Anguste oblonga, convexa, pallide flava, subnitida, pectore, abdominis segmentis anticis tribus basi, scutelloque nigris; thorace tenuiter punctato, utrinque leviter excavato; elytris distincte subcrebre punctatis, punctis pallide fuscis, utrisque super marginem basalem nigro maculatis.

Long. 3¼ lin.

♂. *Facie tridentatâ, dente intermedio compresso, nigro, apice deflexo; clypeo utrinque ad apicem foveolato.*

♀. *Facie tridentatâ, dente intermedio non compresso, apice acuto.*

Head exserted; vertex smooth, impunctate; face excavated between the eyes, clothed with hairs, tridentate, the middle tooth compressed and deflexed in the ♂, conical in the ♀; clypeus transverse, impressed at the apex on either side in the ♂ with a deep fovea; apex of jaws nigro-piceous; antennæ slender, clothed with coarse suberect hairs, second and third joints nearly equal in the ♂, the third one-half longer than the second in the ♀. Thorax about three times as broad as long; sides parallel and slightly sinuate behind the middle, obliquely converging from the middle to the apex, the anterior angles slightly produced obtuse, the hinder ones obtusely angulate; upper surface moderately convex, the lateral margin rather broadly reflexed, disc irregularly excavated; finely but not very closely punctured, interspaces minutely granulose-strigose. Scutellum trigonate, shining black. Elytra broader than the thorax, oblong, moderately convex, faintly excavated below the basilar space, more strongly punctured than the thorax, the punctures pale fuscous.

Hab.—Jhelam Valley.

17.—MIMASTRA GRACILIS.

Baly, Cist. Ent. ii, 1875—82, p. 378.

Elongata, attenuata, pallide flava, nitida, antennis (basi exceptis) fuscis, oculis nigris, genubus tarsisque piceis; thorace transverso, basi emarginato, disco irregulariter excavato, lateribus late marginatis, ante medium angulatis, elytris parallelis, tenuiter punctatis.

Long. 3 lin.

Head strongly exserted; encarpæ and clypeus thickened, the former bounded above by a transverse groove, trigonate, contiguous for their whole length; antennæ very slender, filiform, nearly equal to the body in length, second joint about half the length of the basal one, nearly a third shorter than the third; three basal joints pale flavous, the rest pale fuscous. Thorax transverse; sides broadly margined, nearly parallel, distinctly angled just beyond the middle, thence obliquely converging to the apex; disc broadly and irregularly excavated, impunctate. Scutellum trigonate. Elytra broader than the thorax, parallel, elongate; disc very minutely punctured, very faintly wrinkled. Outer edge of knees, together with the tarsi, pale piceous.

Hab.—Murree.

18.—AGELASTICA ORIENTALIS.

Baly, Cist. Ent. ii, 1875—82, p. 379.

Elongato-ovato, convexa, metallico-cærulea, nitida, antennis nigris; thorace elytrisque crebre punctatis, illo lateribus rotundatis.

Long. 3½—4 lin.

Encarpæ and clypeus thickened, the former pyriform, contiguous, separated from the front by a deep transverse groove; antennæ filiform, half the length of the body, the second joint short, the third one-half longer than the second, more than half the length of the fourth. Thorax nearly three times as broad as long; sides rounded, slightly converging in front; the hinder angles rounded, the anterior ones obtuse; disc closely punctured. Scutellum trigonate, shining, impunctate. Elytra rather broader than the thorax, oblong, closely punctured.

Hab.—Neighbourhood of Sanju, apparently common.

Closely allied to *A. cærulea*, it may be known from that insect by the relative lengths of the second and third joints of the antennæ.

19.—MALACOSOMA FLAVIVENTRE.

Baly, Cist. Ent. ii, 1875—82, p. 379.

Elongatum, convexum, obscure viridi-aeneum, nitidum, abdomine flavo, antennis (basi exceptis) nigris; thorace transverso, minute, subremote punctato; elytris oblongis, infra basin transversim excavatis, tenuiter punctatis.

Long. 4 lin.

Head trigonate; vertex and front smooth, impunctate, the latter separated from the encarpæ by a deep groove; encarpæ transverse, contiguous above, separated below by the narrow wedge-shaped carina, the surface of which is coarsely punctured; antennæ more than half the length of the body, moderately robust, filiform, the second joint short, the third twice the length of the second, the fourth about one-third longer than the preceding one. Thorax about one-half as broad again as long; sides moderately rounded, the anterior angles armed with an obtuse tubercle, the hinder ones acute; disc moderately convex, finely but rather distantly punctured; lateral margin reflexed. Scutellum smooth, impunctate. Elytra much broader than the thorax, oblong, convex, transversely excavated below the basilar space, the latter slightly elevated; surface finely but not very closely punctured, very sparingly clothed with short hairs: on the apical half of each elytron is a number of broad, ill-defined, longitudinal sulcations. Abdomen flavous, the apex of the terminal segment emarginate.

Hab.—Murree.

20.—LUPERODES ERYTHROCEPHALA.

Baly, Cist. Ent. 1875—82, p. 380.

Anguste oblongo-ovata, convexa, nigra, nitida, capite rufo-testaceo, ore, antennis, pedibusque piceis; thorace crebre punctato, disco utrinque leviter transversim excavato; elytris sat crebre punctatis.

Long. 2 lin.

Head exserted, vertex and front shining, impunctate; encarpæ transverse, contiguous; labrum piceous; jaws and palpi rufo-piceous; antennæ filiform, three-fourths the length of the body, second and third joints short, conjointly about equal in length to the first. Thorax twice as broad as long; sides rounded, slightly converging at the base; all the angles distinct, the anterior thickened, sub-tuberculate; disc closely punctured, distinctly excavated on either side. Elytra oblong, less closely punctured than the thorax.

Hab.—Murree.

21.—GALLERUCA VITTATIPENNIS.

Baly, Cist. Ent. ii, 1875—82, p. 380.

Elongato-oblonga, convexa, nigro-picea aut nigra, nitida, vertice rufo-piceo, abdominis segmentorum margine apicali pallide rufo-piceo; thorace excavato, rude foveolato; elytris abdomine multo brevioribus, fortiter substriatim punctatis, sordide fulvis, utrisque lineā suturali elevatâ, vittisque elevatis quatuor, utrinque abbreviatis, 1m*â et* 4t*ñ,* 2d*â et* 3t*iâ apice per paria conjunctis, nigro-piceis instructis.*

Long. 4⅛ lin.

Head sub-rotundate, vertex and front deeply and coarsely foveolate-punctate, impressed in the middle with a deep longitudinal groove, which extends downwards between the encarpæ as far as the apex of the clypeus, where it terminates in a triangular fovea; oncarpæ thickened, trigonate, smooth, impunctate; clypeus very short, thickened and forming a transverse ridge, its anterior border narrowly edged with rufous; antennæ robust, the second joint ovate, rather more than half the length of the third, the third and fourth joints equal. Thorax rather more than twice as broad as long; sides sinuate and parallel from the base to beyond the middle, thence obliquely converging to the apex, the anterior angles slightly produced, somewhat recurved, obtuse; disc excavated on either side, the middle disc impressed with a broad longitudinal sulcation which extends from base to apex; the whole surface covered with large, deep, round foveæ. Scutellum semirotundate, piceous, impunctate.

Hab.—On the road across the Pamir, from Sirikol to Panga.

22.—GALLERUCA INDICA.

Baly, Cist. Ent. ii, 1875—82, p. 361.

Ovata, postice paulo ampliata, modice convexa, nigra, subtus nitida, griseo-sericea, supra opaca; capite thoraceque rude rugoso-punctatis, hoc transverso, utrinque foveolato, medio longitudinaliter sulcato, lateribus reflexis, ante medium obsolete angulatis; elytris vage rufo-piceo limbatis, rugoso-punctatis, utrisque vittis elevatis quatuor, duabus intermediis interruptis, interdum fere omnino obsoletis, instructis.

Long. 5 lin.

Head very coarsely rugose-punctate. Thorax nearly twice as broad as long; sides parallel, slightly sinuate, obtusely angled just before the middle, thence obliquely converging to the apex, the anterior angle moderately produced, its apex rounded; disc very coarsely rugose-punctate, the middle portion with a longitudinal sulcation which extends from base to apex, either side impressed with a large fovea. Scutellum coarsely rugose-punctate. Elytra broader than the thorax, ovate, slightly dilated towards the apex, moderately convex, rugose-punctate, but less coarsely so than the head and thorax; black, sometimes tinged with piceous, the outer margin obscure rufo-piceous; each elytron with four raised vittæ, the two intermediate ones interrupted, and sometimes almost entirely obsolete; the suture also thickened.

Hab.—Murree; also Northern India, my collection.

23.—GALERUCELLA* PLACIDA.

Baly, Cist. Ent. ii, 1875—82, p. 381.

Anguste oblonga, griseo-hirsuta, subtus picea, nitida, prothorace fulvo; supra sordide fulva, subnitida, antennis, verticis plaga, thoracis maculis tribus transversim positis, scutel-

---

* *Galerucella*, Crotch, Proc. Acad. Philad. 1873, p. 55.

loque basi piceis; thorace transverso, lateribus ante basin dente subacuto armatis, ante dentem concavis, ante medium ampliatis, disco rude rugoso, bifoveolato ; *elytris profunde confuse punctatis, interspatiis granulosis.*

Long. 2 lin.

Vertex and front finely rugose-punctate, clothed with appressed griseous hairs, the middle with a large ill-defined piceous patch; encarpæ thickened, contiguous, pyriform; antennæ moderately robust, filiform, the second joint nearly equal in length to the first, about two-thirds the length of the third. Thorax more than half as broad again as long; sides diverging at the base, and armed at the apex of the diverging portion with a subacute, setiferous tooth, immediately in front of which, before the middle, they are deeply sinuate, in front they are broadly dilated, the anterior angle armed with a subacute tooth; disc coarsely rugose-punctate, broadly excavated on either side, and again more deeply, but to a less extent, on the anterior half of the middle disc; the piceous patches, placed transversely on the disc, are large but ill-defined, and cover nearly the whole of the surface. Scutellum narrowed from its base towards the apex, the latter obtusely truncate. Elytra oblong, nearly parallel, deeply and coarsely punctured, densely clothed with short suberect griseous hairs.

Hab.—Jhelam Valley, one specimen; I also possess this insect from India.

24—MERISTA INTERRUPTA.

Galleruca interrupta, L. Redtb. in Hugel's Kaschmir, iv, p. 553, tab. xxvii, fig. 4 (1844).

Hab.—Murree, a single specimen.

The transverse black patch differs greatly in extent in different individuals, in some being entirely obsolete ; in the specimen before me it is reduced to two small fuscous points placed transversely on the middle disc.

In this species, of which I possess many specimens from various parts of India, the second and third joints of the antennæ vary in relative length in the sexes : in the ♂ these joints are very short and nearly equal; in the ♀ the third joint, though short, is distinctly longer than the second.

25.—LEPTARTHA COLLARIS.

Baly, Cist. Ent. ii, 1875—82, p. 382.

Ovata, postice ampliata, nigra, nitida; thorace transverso, fulvo ; elytra fortiter sat crebre punctatis, castaneis, punctis piceis, utrisque maculá basali juxta suturam nigro-aeneá notatis.

Long. 4½—5 lin.

Vertex shining, impunctate ; encarpæ thickened, contiguous, semilunate ; antennæ nearly equal to the body in length, filiform, tapering towards the apex, second and third joints very short, equal. Thorax transverse, sides constricted behind the middle, dilated in front, the anterior angles produced, their apices obtuse; apical border concave-emarginate ; disc smooth, impunctate, thickened on either side near the anterior angle, impressed on each side the middle with a faint transverse groove. Scutellum trigonate. Elytra broader than the thorax, dilated behind the middle, moderately convex, deeply punctured, the punctures piceous, arranged without order over the general surface, placed in ill-defined longitudinal striæ near the base of the suture; on the anterior disc are several short ill-defined obsoletely raised vittæ; at the base of each elytron, close to the suture, is a small nigro-aeneous patch. Last two segments of abdomen bordered with fulvous.

In the specimen from Murree, the sides of the thorax are less dilated anteriorly, the anterior angles being less produced and at the same time more acute; the transverse depressions on the middle disc are also obsolete; in this specimen the antennæ are unfortunately broken, but the fourth and fifth joints (which remain) are slightly compressed, and are rather more robust than in the insect from Northern India; it is probably the other sex.

Hab.—Murree; in my own collection from Northern India.

HALIPLIDÆ, DYTISCIDÆ, GYRINIDÆ, HYDROPTILIDÆ, STAPHY-LINIDÆ, AND SCARABÆIDÆ (except CETONIINI).

By D. SHARP.

HALIPLIDÆ.

HALIPLUS (Munich Cat.).

1.—HALIPLUS MACULIPENNIS, Schaum.

A single individual found in the Jhelam Valley, July 1873. Differs a little from the Egyptian specimens of the species by being rather smaller, and by the punctuation of the elytra being rather less coarse and deep.

DYTISCIDÆ.

HYDROPORUS (Munich Cat.).

1.—DYTISCUS GRISEO-STRIATUS, Degeer.

A series of this species, which is in Europe alpine and boreal, was found in the Pankong Valley in September 1873. The specimens show more variation in markings than I have observed in European individuals.

AGABUS (Munich Cat.).

1.—AGABUS ABNORMICOLLIS, Ballion.

This interesting species is allied to the Corsican *A. cephalotes*; it is represented in the collection by a very mutilated female specimen, without indication of locality or date.

2.—DYTISCUS NITIDUS, Fab.

Dras, Kargil, and Leh, 15th August to 9th September 1873. Three individuals.

3.—AGABUS AMŒNUS, Solsky.

A single female of an Agabus found in the neighbourhood of Sanju I refer to this species, although it differs from Solsky's diagnosis (Fedchenko's, Turkestan, Coleoptera, p. 142) in having the ventral segments entirely black. The species has an elongate and acuminate prosternal process, which projects far back, between the middle legs.

4.—AGABUS DICHROUS, n. sp.

Oblongo-ovalis, nitidus, subtus niger, supra testaceus, vertice nigro, rufo bimaculato, antennis pedibusque testaceis, femoribus in medio late nigris; scutello fusco; elytris apicem versus vix fusco-nebulosis.

Long. 8 mm., lat. 4 mm.

This species is closely allied to the European *Dytiscus conspersus*, Marsh., but is comparatively narrower and more parallel, and the colour of the upper surface is more purely yellow and much less infuscate: the dark mark on the head is much less extended towards the front, and is deeply divided in the middle by a backward prolongation of the yellow colour. The male has the front tarsi moderately thickened, and their anterior claw is little thickened or toothed in the middle. The female I have not seen.

A single male individual was found on the road across the Pamir from Sirikol to Panga, 22nd April to 7th May 1874.

ILYBIUS (Munich Cat.).

1.—ILYBIUS CINCTUS, n. sp.

Ovalis, angustulus, parum convexus, subtus ferrugineus; supra fusco-aeneus, prothoracis elytrorumque lateribus late testaceis, subnitidus subtilissime reticulatus.

Long. 8¼ mm., lat. vix 4½ mm.

This is one of the smallest species of Ilybius, and is allied to the Japanese *I. apicalis;* it is, however, smaller and narrower than that species, and the yellow lateral stripe is continued at the extremity to the suture, and close to the suture it has one or two angular prolongations. The male has the front and middle tarsi a good deal incrassate, and their claws are nearly simple.

Two individuals (in bad preservation) from Yangihissar, April 1878.

RHANTUS (Munich Cat.).

1.—COLYMBETES PULVEROSUS, Sturm.

A female individual found at Sanju seems to be referable to this species; it has not, however, the small central mark on the thorax that exists in the European specimens of the species, and it is possible that a knowledge of the male would show it to be a distinct species.

TROGUS (Munich Cat.).

1.—DYTISCUS ROESELII, Fab.

A female specimen of this common European species was found at Yarkand, 21st to 27th May 1874.

GYRINIDÆ.

DINEUTES (Munich Cat.).

1.—DINEUTES INDICUS, Aubé.

This species is represented by two very large male individuals found in the Jhelam Valley, July 1873.

HYDROPHILIDÆ.

HYDROPHILUS (Munich Cat.).

1.—HYDROPHILUS PICEUS, Fab.

Sanju, and Yarkand.

HYDROBIUS (Munich Cat.).

1.—HYDROPHILUS BICOLOR, Payk.

Kogyar, 31st May to 2nd June 1874. Three individuals.

PHILHYDRUS (Munich Cat.).

1.—PHILHYDRUS MARITIMUS, Thoms.

Kogyar, 31st May to 2nd June 1874. Two individuals. In Europe the species is found only in brackish waters.

STAPHYLINIDÆ.

TACHYPORUS (Munich Cat.).

1.—STAPHYLINUS CHRYSOMELINUS, Lin.

On the road across the Pamir, from Sirikol to Panga. Three individuals.

TACHINUS (Munich Cat.).

1.—TACHINUS STOLICZKÆ, n. sp.

Parvulus, subdepressus, niger, elytris castaneis vel piceo-castaneis, antennis pedibusque sordide testaceis; prothorace fere impunctato, elytris parce punctatis, obsolete strigosulis- abdomine sat crebre subobsolete punctato.

Long. 6 mm., lat. 1¾ mm.

This species is closely allied to the European *T. fimetarius*, but is rather larger. The antennæ are formed as in that species, but are rather longer and paler in colour. The punctuation of the elytra is fine and very scanty, and the fine scratches are less distinct than in *T. fimetarius*: the abdominal punctuation is rather denser than it is in the European species. In the male, the dorsal plate of the last segment ends in four short stout teeth as in *T. fimetarius*; the ventral plate of the same segment is also almost similar in the two species; the ventral plate of the preceding segment has a broad notch in the middle; this is fringed, except in the middle, with very distinct pectinations, and in the middle, where the pectinations are absent, the margin has a rough or spongy appearance; the termination of the notch on each side is not acuminate: the chief differences from *T. fimetarius* are the less produced and less acuminate terminations of the notch, and the greater development of the pectinations of its margin. The dorsal and ventral plates in the female are formed as in *T. fimetarius*, except that the teeth of both plates are very much longer.

Four individuals found on the road across the Pamir, from Sirikol to Panga, between the 22nd April and 7th May 1874.

CREOPHILUS (Munich Cat.).

1.—STAPHYLINUS MAXILLOSUS, Lin.

Kogyar, 1st June 1874.

PHILONTHUS (Munich Cat.).

1.—PHILONTHUS CYANELYTRUS, Kr.

Murree. One individual.

2.—PHILONTHUS ROTUNDICOLLIS, Men.

Sanju, Pamir, Yarkand. A large series of this species exhibits considerable variation in colour.

3.—STAPHYLINUS SORDIDUS, Gray.

A single individual, without locality or date, has the elytra darker coloured than usual.

4.—PHILONTHUS STOLICZKÆ, n. sp.

P. rubido Er. *similis et affinis: angustulus, subparallelus, niger, elytris rufis, antennis fuscis, basi cum pedibus testaceo, abdominis segmentis ferrugineo-marginatis; thorace angustulo, subparallelo, serie discoidali punctorum 5, et punctis lateralibus sat numerosis; elytris rufis basi summo paulo obscuriore, crebre, fere fortiter punctatis; abdomine dense, æqualiter subtiliterque punctato, opaco.*

Long. 5 mm.

Antennæ moderately long, second and third joints rather long, sub-equal, tenth about as long as broad. Palpi yellow. Head sub-oblong, with rather numerous coarse punctures, which are wanting along the middle in front. Thorax narrower than the elytra, not narrowed in front; the punctures coarse, the lateral series at the base mixed with the dorsal series. Hindbody throughout densely and very finely punctured. Front tarsi of male a little dilated, and last ventral segment with a moderately large excision.

This seems to be a very distinct little species, and resembles in some respects the species of the genus *Actobius*, Fauvel. It is rather similar to *P. rubidus*, Er., but has the hindbody much more finely and densely punctured, and the thoracic lateral punctures, as well as those of the head, more numerous.

Yarkand, November 1873. A single specimen.

5.—PHILONTHUS PAMIRENSIS, n. sp.

Ex affinitate Staphylini tenuis, Fab. *Angustulus, haud parallelus, niger, elytris rufis, antennis pedibusque posterioribus fuscis, illarum basi pedibusque anterioribus testaceis; abdomine subtiliter punctato.*

Long. 6 mm.

Antennæ stout, distinctly thickened towards the apex; the basal joint yellow. Palpi blackish. Head oval, narrow, finely punctured at the sides behind the eyes. Thorax narrower than the elytra, a little narrowed in front, black, very shining, the dorsal series consisting of five fine punctures; the lateral punctures few and fine. Elytra about as long as the thorax, red, rather finely punctured. Hindbody narrowed towards the apex, the segments finely punctured.

The male has the front tarsi rather strongly dilated.

Though closely allied to *Staphylinus tenuis*, Fab., this species is readily distinguished from it by its black thorax.

A single individual was found on the road across the Pamir, from Sirikol to Panga, between the 22nd April and 7th May 1874.

PÆDERUS (Munich Cat.).

1.—PÆDERUS FUSCIPES, Curtis.

Jhelam Valley.

OXYTELUS (Munich Cat.).

1.—OXYTELUS NITIDULUS, Grav.

A single individual, without date or locality.

SCARABÆIDÆ (Munich Cat.).

SISYPHUS (Munich Cat.).

1.—SISYPHUS HIRTUS, Wied.

Jhelam Valley, July 1873. Three individuals.

GYMNOPLEURUS (Munich Cat.).

2.—GYMNOPLEURUS MUNDUS, Wied.

Jhelam Valley, July 1873. A single individual.

3.—COPRIS CYANEUS, Fab.

Jhelam Valley, July 1873.
Harold (Col. Hefte V, p. 56) thinks *G. indicus*, Cast., a distinct species from the Fabrician *Copris cyaneus*; but it appears to me more probable that *Gymnopleurus indicus*, Cast., and *Gymnopleurus impressus*, Cast., are merely varieties of the variable *Copris cyaneus*, Fab.

CATHARSIUS (Munich Cat.).

4.—COPRIS SABÆUS, Fab.

Jhelam Valley, July 1873.
One female specimen.

COPRIS (Munich Cat.).

5.—COPRIS SINICUS, Hope.

Murree.—The two individuals which represent this species are a very small undeveloped male, and a female; they are undoubtedly conspecific with an individual from Chosan in my own collection.

ONTHOPHAGUS (Munich Cat.).

6.—COPRIS GAZELLA, Fab.

This abundant and widely-distributed species was found in the Jhelam Valley, July 1873.

7.—ONTHOPHAGUS REFLEXICORNIS, Reiche.

A single individual of an Onthophagus found at Kogyar, 31st May—2nd June 1874, seems to be a variety of a species from Northern India, named as above in my collection.

8.—ONTHOPHAGUS ARMICEPS, Reiche.

A single individual of an Onthophagus found in the Jhelam Valley, July 1873, seems to be probably a very undeveloped male, of a species bearing the above name in my collection.

COLEOPTERA. 43

9.—ONTHOPHAGUS CONCOLOR, n. sp.

Niger, fere nudus, supra opacus, subtus sat nitidus ; prothorace peropaco, parcius subtiliter punctato, lateribus ad angulos anteriores evidenter sinuatis ; elytris subtiliter striatis, interstitiis parcius et subtiliter punctatis, punctis haud perspicue setigeris.
Long. 7-9 mm.
Mas.—Capite vertice medio breviter tuberculato, prothorace fere mutico.
Fem.—Capite medio linea curvata sat elevata, vertice medio lamina elevata (ad apicem plus minusve emarginata) brevissima.

Antennæ black, with the intermediate joints marked with red. Clypeus broadly, but very lightly, emarginate in the middle in front, its anterior part coarsely and rugosely punctured. Thorax quite sparingly punctured, the punctures most numerous near the front in the middle, quite wanting at the anterior angles. The punctures of the elytra are inconspicuous, but are rather less indistinct on the external interstice, and are there seen to be fine granulations. The pygidium is very opaque and sparingly punctured. The ventral segments are almost impunctate, the breast is sparingly punctured, and bears a few black hairs. The tarsi are pitchy.

This species has the appearance of the South African *O. giraffa*, but is readily distinguished therefrom by the diminished punctuation, and by the sinuation of the sides of the thorax near the front angles. The thorax is very slightly prominent in the middle quite near the front, and the prominent part is slightly emarginate: this thoracic development is, however, quite slight in all the specimens before me; and although it is variable, it seems to be unconnected with the sex of the individual.

Sind valley, Aug. 5—13, 1873, and Murree.

APHODIUS (Munich Cat.).

10.—SCARABÆUS SUBTERRANEUS, Lin.

A single individual of this common European species was found on the road across the Pamir, 22nd April to 7th May 1874.

11.—SCARABÆUS GRANARIUS, L.

Of this species (which is now found in most parts of the world) a single individual was found on the road across the Pamir, 22nd April to 7th May 1874.

12.—APHODIUS ÆGER, n. sp.

Scarabæo granario, Lin., *similis ; oblongus, leviter convexus, nitidus, niger, elytris piceis vel fere nigris, pedibus rufis ; clypeo medio emarginato, fronte fere mutica, prothorace subtiliter punctato, latera versus punctis majoribus crebribus, margine basali integro, angulis posterioribus sinuatis ; elytris vix subtiliter striatis, striis indistincte crenatis, 7° et 8° ante apicem conjunctis, humeris longius ciliatis.*
Long. 5-5½ mm., lat. 2½ mm.

Though rather similar to *Scarabæus granarius*, this species is readily distinguished from it by the sinuate hind angles of the thorax, and the conspicuous pale cilia of the sides of the thorax and the basal portion of the elytra. The half dozen individuals before me show no sexual differences, and are perhaps all females. If this be the case, and the male should prove to have the head trituberculate, the species may then be satisfactorily placed in Erichson's Section E., for it has all the characters assigned to that section except the sexual ones. The clypeus is much emarginate and depressed in the middle in front, the sides of the emargination are rounded, the head is coarsely and closely punctured in front, more finely on the vertex, there is an excessively indistinct transverse line on the vertex, and on the middle of it a slight gibbosity or prominence of the surface, not worth calling a tubercle. The scutellum is rather narrow and parallel-sided, and is punctured except at the apex. The striæ of the elytra are quite as distinct at the apex as they are at the base: the sides of the wing cases bear numerous white setæ, which are long and conspicuous at the shoulders, behind which they become gradually shorter and disappear altogether from the apical half.

The specimens are marked "a," indicating that the exact locality is unknown; two small specimens were, however, found at Yangihissar in April 1874.

13.—APHODIUS PARVULUS, Har.

A single individual found in the Jhelam Valley, July 1873, agrees exactly with specimens from Abyssinia of this species recently described by Baron von Harold. I have in my own collection some specimens of this species from Ajmere.

14.—APHODIUS KASHMIRENSIS, n. sp.

Niger, nitidus, sat convexus, pedibus rufo-piceis, antennis rufis clava fusca; clypeo anterius latius emarginato, et utrinque subacute prominulo; prothorace punctis magnis profundis sat numerosis, aliisque minutis, margine basali distincto; sulculo ante eum crenulato; elytris fortiter crenato-striatis, interstitiis subtilissime, sparsim punctatis.

Long. 6-6¼ mm., lat. 3½ mm.

I have seen only two specimens of this species; they seem both to be females. I believe it is an Aphodius belonging to the Section E of Erichson, although the form of the front of the clypeus suggests rather that it may prove to be an Ammœcius when the mouth can be examined. It is almost as large as *Scarabæus scybalarius*, Fab., and somewhat similar in form to that species. The head bears no distinct tubercles, but has an obsolete curved elevation some distance behind the front, and on the middle of the vertex traces of an obsolete tubercle; its punctuation is moderately coarse and close, but irregular and rather indistinct. The sculpture of the thorax consists of very large and very small punctures, the basal margin is coarse and distinct, and the groove which procedes it is coarsely punctate so as to appear crenulate. The scutellum is small and coarsely punctured; the striæ of the elytra are deep quite to the apex, the middle ones being joined together, or not extending quite to the apex.

Dras, Kargil, and Leh, 15th August to 9th September 1873.

15.—APHODIUS TENUIMANUS, n. sp.

Aphodio melanosticto Er. *persimilis; oblongus, subconvexus, nitidus, infuscato-testaceus, capite thoraceque nigris, hoc lateribus testaceis, elytris luteis, maculis dorsalibus 4 vel 5 strigaque sublaterali nigris, pedibus metasternoque medio testaceis; fronte medio vix tuberculato; tibiis anterioribus tenuibus, intus conspicue ciliatis.*

Long. 5-6 mm.

This species is so similar to *A. melanosticticus* that it will be easily recognized by comparison with that species and by noticing the points in which it differs; these are that the base of the thorax is less sinuate on each side, makes in fact scarcely any deviation from a gentle rounding of the base; the front tibiæ are long and much more slender, and have the cilia on their inner edge more numerous. I have seen only three individuals; they show no sexual difference, so that I am not quite sure whether they are males or females; but I believe the former, and in that case an important point of difference will be found in the tubercles of the head, for these specimens show only a very obsolete central tubercle and no lateral ones.

The specimens are labelled "*a*," which indicates that the exact locality where Stoliczka procured them is unknown.

RHYSSEMUS (Munich Cat.).

16.—SCARABÆUS ASPER, Fab.

Jhelam Valley, July 1873.

This common European species is apparently plentiful in Northern India. The Indian specimens are usually a little smaller and more shining, and have the imbricate granulations on the elytra rather finer than European specimens, but I can find no characters of specific value to distinguish them. There are several allied, but as yet undescribed, species found in India.

GEOTRUPES (Munich Cat.).

17.—GEOTRUPES ORIENTALIS, Hope.

Murree, two specimens.

18.—GEOTRUPES FOVEATUS, Har., var.

I refer two females of a Geotrupes from Sanju to the above European species, from which however they differ by being as large as *G. stercorarius*, Har., and by having the under-surface of a beautiful golden green colour, and clothed with a long rufous pubescence. These points would not justify me in considering the specimens as specifically distinct; moreover, I have great doubts whether *G. foveatus* is more than a small form, with diminished male distinctions, of *G. stercorarius*; and if this be so, it is probable that these Sanju individuals may really prove to be only a variety of the widely distributed and abundant *Geotrupes stercorarius*, Har.; it will not, however, be possible to consider this conclusively established until the male has been found and examined.

19. GEOTRUPES KASHMIRENSIS, n. sp.

G. stercorario (Haroldi) *persimilis, sed elytris longioribus; oblongo-ovalis, supra virides-centi-niger, nitidus, subtus purpureus, fulvo-pubescens; antennis piceo-rufis; mandibulis extus rotundatis, ad apicem leviter unisinuatis; elytris striis 14, minus distincte punctatis; abdomine etiam in medio punctato, sed illo minus pubescente; tibiarum posticarum carina tertia (ab apice) omnino carente.*

Long. 24 mm., lat. 13 mm.

This species is closely allied to *G. stercorarius*, Harold, but is rather more elongate, so as to look at first sight like *G. mutator*, Er. It is readily distinguished, however, by the characters mentioned in the above diagnosis. The front tibiæ of the male have a carina-like swelling on their lower face, which bears a broad tooth below the knee, and after that is continued in a straight line, till it terminates in a sharply-elevated tooth opposite the third marginal tooth.

Dras, Kargil, or Leh, 15th August, 9th September 1873. Two individuals.

Obs.—Jekel, in his classification of Geotrupes (Ann. Soc. Ent. Fr., 1865), gives as one of the most important characters of his sub-genus Anoplotrupes that there are only two entire carinæ on the posterior tibiæ, while in the subgenus Geotrupes (pr. d.) he states that there are three entire carinæ. In this, however, he was mistaken, for I find that in *G. spiniger*, Har., there are truly three entire carinæ; in *G. spiniger*, Har., the upper carina is less entire, for it suffers a slight double interruption, which is still more conspicuous in *G. foveatus*, Har., while in *G. mutator*, Er., only the lower half (or less) of this carina exists, and in *G. kashmirensis* it is entirely wanting. The species comprised by Jekel in his subgenus Anoplotrupes are but little concordant: thus his *Anoplotrupes G. vernalis* differs much more widely from *Anoplotrupes G. sylvaticus* than this does from Geotrupes (pr. d.); the sub-genus Anoplotrupes should therefore be entirely suppressed, for it is much less natural than the other subgenera established by M. Jekel in the able and satisfactory memoir to which I have alluded.

20.—SCARABÆUS SYLVATICUS, Panz., var.

The two individuals of this species differ considerably from any European specimens I have seen; they are proportionally narrower, on the upper side are of a black colour a little tinted with brassy, on the under side are entirely black, and the club of the antenna is black. These characters do not, however, seem to me to be of such a nature as to warrant their specific value.

Sind Valley.

TROX (Munich Cat.).

21.—TROX PROCERUS, Har.

Jhelam Valley, July 1873. A single individual.

COLEOPTERA. 47

HOPLIA (Munich Cat.).

22.—HOPLIA CONCOLOR, n. sp.

Oblonga, sat elongata, ferruginea squamulis pallide griseis, magnis, fere æqualiter cestita; tarsorum posticorum unguiculo mutico.

Long. 8 mm., lat. 4¼ mm.

The head, except in front, is rather darker than the rest of the surface; the scales it bears are pale, sub-depressed, and narrower than on the rest of the surface, and on the front part where they are scanty are setæ rather than scales. The thorax is of the same form as in *Melolontha graminicola*, Fab., but it is evenly clothed with closely set, coarse, pale ochre-coloured scales, and bears also a very few erect fine setæ. The elytra are long, and are clothed in a similar manner to the thorax, the scales being very nearly circular in form. The pygidium, propygidium, and ventral segments are also closely covered with scales but little different from those of the elytra. The legs are of a uniform reddish colour scarcely different from the elytra.

The only individual I have seen is a female; it has the antennæ 9-jointed, and the front tibiæ with two moderately prominent teeth, and an emargination above the upper one; the front claw seems quite single and without tooth; the claw of the middle foot has a very minute division some distance before the apex, while the posterior claw seems to be quite simple.

The species seems to me more allied to *Melolontha graminicola*, Fab., than to any other I know; but it is of a more elongate and parallel form, of a paler and more uniform colour, and is clothed with larger scales.

Kogyar, 31st May to 2nd June 1874.

SERICA (Munich Cat.).

23.—SERICA LÆTICULA, n. sp.

Obovata, convexa, nitidula, tantum abdomine opaco, lata, brunneo-ferruginea; prothorace fortiter punctato, elytris seriatim punctatis, seriebus leviter depressis, interstitiis planis, tantum juxta series punctatis; antennis 10-articulatis, flabelli articulo primo apice emarginato.

Long. 8½ mm., lat. 4 mm.

Similar in form to *Serica mutata*, except that it is much broader and only a little longer; the whole of the upper surface quite shining; clypeus emarginate in front, the whole of its anterior part coarsely and rugosely punctured, the hinder part sparingly but distinctly punctured. Thorax strongly transverse, evenly and coarsely punctured. Scutellum coarsely and closely punctured, with a smooth longitudinal space along the middle. Elytra with rows of punctures, which are so little depressed as scarcely to amount to striæ; the interstices hardly at all convex, and their punctures confined to the neighbourhood of the rows of punctures. Pygidium rather obsoletely punctured, shining; outer edge of hind coxa scarcely twice as long as the outer edge of metasternum, its hind margin slightly sinuate near the outer angle, which is hardly at all rounded; the punctures of the hind coxæ coarse and ocellate.

LACHNOSTERNA (Munich Cat.).

24.—LACHNOSTERNA STRIDULANS, n. sp.

Testacea, capite, thorace, scutelloque fere ferrugineis, supra opaca, opalescens, subtus abdomine inflato nitido, pectore minus dense villoso; capite brevi, fortiter punctato; prothorace sparsissime punctato, fortiter transverso, margine laterali integro, sinuato, angulis posterioribus obtusis; elytris sat crebre parum profunde punctatis.

Long, 15½ mm., lat., 8 mm.

Head very short; margin of clypeus strongly reflexed, not emarginate; the transverse suture nearly twice as distant from the vertex as from the clypeal margin; the part of the head behind the suture rather sparingly punctured with punctures of different sizes, and bearing some long erect setæ; in front of the suture the punctures are deep, and are closer together, and all of one size, and there are no setæ. The thorax is very short, the sides not greatly dilated in the middle, the hind angles obtuse, but not departing greatly from right angles; the punctuation is distant and not coarse. The scutellum is rather sparingly punctured. The elytra are finely and irregularly, and not closely, punctured, and have only indistinct traces of longitudinal impunctate spaces. The propygidium is densely and finely punctured across the middle, but coarsely punctured behind; the pygidium is shining, and coarsely punctured. The membranous border of the elytra is very small and indistinct; the epipleural line is sharply elevated in its anterior portion, and when examined with a powerful lens it is seen to be densely and finely crenulate, from the middle of the sternum to the hind margin of the first ventral segment; by strongly flexing the hind femur, and rubbing it against this line, a considerable stridulation is produced.

Murree. A single individual.

Obs.—I am acquainted with only one other species closely allied to this; it is as yet undescribed, and is labelled in my collection "*Ancylonycha pulvinosa*, Reiche, India bor." It has the same appearance as *L. stridulans*, and has, like it, the epipleural line finely crenulated, but it differs considerably in the structure of the antennæ and of the claws; in *Lachnosterna stridulans* the flabellum of the antennæ is rather long, and composed of five joints; the first leaf is, however, very short, not half the length of the second, which itself is a good deal shorter than the three following ones; the claws are divided into two rather divergent portions of equal length. In the undescribed Reichein species, the flabellum is short and composed only of three joints, and the claws of the feet are strongly dentate in the middle.

I add a short diagnosis of this insect:—

LACHNOSTERNA PULVINOSA, n. sp.

Ferruginea, elytris dilutioribus, supra opaca, opalescens, subtus abdomine inflato, medio nitido, pectore parcius villoso; capite brevi, dense rufoso-punctato; prothorace sparsim punctato, punctis in margine anteriori magnis, fortiter transverso, lateribus valde sinuatis, in medio perdilatatis, angulis posterioribus valde obtusis, margine laterali subcrenulato; elytris sat crebre subtiliter punctatis.

Long. 16 mm.

COLEOPTERA. 49

25.—LACHNOSTERNA STOLICZKÆ, n. sp.

Oblonga, picea, nitida, pectore prosternoque griseo-villosis; capite haud parvo, clypeo fortiter reflexo-marginato, anterius vix emarginato; prothorace lateribus rotundatis, anterius quam posterius magis angustato, crebrius punctato, angulis posterioribus obtusis, margine laterali serrato; elytris crebrius fortiter punctatis, areis longitudinalibus parcius punctatis, haud argute elevatis.

Long. 15—16 mm., lat. 8 mm.

Head coarsely, but not rugosely, punctured, the clypeal margin greatly reflexed, the clypeus of the same breadth as the vertex, so that its elevated side margin is continued directly backwards in a straight line along the inner margin of the eye as an elevated space which extends as far as the portion of the head which forms the summit of the vertex; this upper part of the vertex is placed on a different plane from the middle part of the head and is impunctate except at the sides; there is no trace of any transverse carina. The membranous border of the elytra is very fine, but is distinct throughout.

The male has the flabellum of the antennæ a little longer than the female; in this latter sex the front tibiæ are evidently tridentate, while in the male they are only bidentate, and even the upper of the two teeth is not very distinct: the tooth of the claws is placed quite near the base in the male, but in the female it is rather nearer the middle of the claw, and is also a little larger. The species will be readily identified by the structure of the head and the claws. I have only one allied species in my collection; it is also from Northern India and is still undescribed.

Murree. Three individuals.

BRAHMINA (Munich Cat.).

26.—BRAHMINA CALVA, Blanchard.

Murree. A single individual.

RHIZOTROGUS (Munich Cat.).

27.—RHIZOTROGUS BILOBUS, n. sp.

Antennis 10-articulatis; oblongus, colore variabilis, ferrugineus vel piceus, sub-opacus, prothorace in medio sæpius nitido, ad latera albido-pruinoso; clypeo in medio profunde emarginato; prothoracis lateribus anterius crenulatis; elytris indistincte et inequaliter punctatis, lateribus dense ciliatis; pygidio ventreque pruinosis; pectore prosternoque dense villosis.

Long. 17¼—20 mm., lat. 9-10 mm.

The head is small and very densely punctured; the clypeal suture consists of an extremely fine elevated line, the part in front of which is reddish, while behind the line the colour approaches black; the clypeus is divided into two lobes by a very deep medial incision. The thorax is without hairs on its upper surface; it is moderately closely, and hardly coarsely, punctured; it is much narrowed towards the front, and the anterior angles are not in the least prominent; the hinder angles are rounded and a little obtuse. The elytra are dull, and their surface is uneven, from some longitudinal elevations; their punctuation is irregular and indistinct; they bear no hairs, but under a good lens are seen to possess some extremely short setæ; and their sides are densely ciliate. The pygidium is of a whitish colour, that is to say, the surface is very pruinose; it is finely punctured.

The male has the flabellum of the antennæ twice as long as the female, it being quite as long as joints 2 to 7 together; the front tibiæ do not show much difference in the sexes; they are rather stout and distinctly tridentate. The ventral segments in the male are not impressed along the middle, but are a little flattened, and almost free from punctuation.

The species is remarkable on account of the divided front of the head; a trans-Caucasian species which I received some time ago from M. Deyrolle as *R. porosus* agrees with it in this respect, but has the ventral segments densely pubescent, while in *R. bilobus* these same parts are bare. The species is variable in colour and size.

Yangihissar, April 1874. About twenty individuals, mostly much injured, and probably picked up dead. Also Kogyar, 31st May to 2nd June.

RHINYPTIA (Munich Cat.).

28.—RHINYPTIA DORSALIS, Burm.

Jhelam Valley, July 1873. Two specimens.

ANOMALA (Munich Cat.).

29.—ANOMALA STRIOLATA, Blanchard.

A single individual of an Anomala found at Murree differs from the description of Blanchard's *A. striolata* only by some details of colour, the most important of which is that the under-surface and legs are purple. Specimens in my own collection, labelled India, are probably conspecific with the Murree individual, though they do not quite agree in all details. In a genus like Anomala, where many species are so extremely variable in colour, it is not advisable to make new specific names on the evidence of such slight differences.

30.—ANOMALA* STOLICZKÆ, n. sp.

Ovata, minus convexa, lætissime viridis, nitidissima, elytris subopacis, antennis nigris; capite thoracoque lævigatis; elytris seriebus duplicatis punctorum tribus, et inter eas sat crebre punctatis.

Long. 12½ mm., lat. 6½ mm.

Of a very beautiful, brilliant, golden-green colour, with the elytra pure green and but little shining. The thorax is narrowed towards the front, with the anterior angles acute and prominent, the posterior ones well marked and slightly obtuse; the raised margin is very distinct, and is wanting only from the middle both in front and behind. The scutellum is impunctate and shining, like the thorax. The punctuation of the elytra is rather fine and scanty; they have some irregular and unsymmetrical black spots, which are probably only accidental. The propygidium is rugose; the pygidium is rugose at the base, and is elsewhere

* Genus *Callistethus*, Blanchd.

sparingly punctured, but close to the sides there is also a narrow rugose band. The metasternum, except in the middle, and the hind coxæ are coarsely punctured, and sparingly pubescent; the ventral segments are smooth in the middle and very shining; at the sides each has some coarse punctuation bearing a row of coarse setæ; the side of each segment at the base is purple. The legs are brilliant green, with the tarsi approaching to purple. Of this beautiful species a single individual was found at Murree.

POPILIA (Munich Cat.).

31.—POPILIA CYANEA, Hope.

Sind Valley, August 1873. A dozen individuals.

ADORETUS (Munich Cat.).

32.—ADORETUS PLAGIATUS, Burm.

The species of Adoretus at the present moment are excessively difficult to name with certainty: the specimens I here call *A. plagiatus* agree with Burmeister's description, but are four-and-a-half or five lines long, instead of three-and-a-half. The species may readily be distinguished from *A. nudiusculus* by the deeply serrate margins of the labrum; the two species are also a little different in colour, punctuation, and pubescence.

Jhelam Valley, July 1873.

33.—ADORETUS NUDIUSCULUS, n. sp.

Testaceus, clypeo ferrugineo, fronte fusca, nitidula, parcius brevissimeque setosus; prothorace fortiter punctato, lateribus subcrenulatis, angulis posterioribus omnino rotundatis; elytris obsolete costatis, fortiter punctatis.

Long. 9¼ mm., lat. 5¼ mm.

A short and moderately broad species. Head moderately large, rather coarsely and closely, but not deeply, punctured. Labrum with narrow, but elongate, appendage, which is very finely carinate along the middle; the margins of the labrum are only indistinctly crenulate, the basal portion is punctate, and bears short hairs. The thorax has the hinder angles much rounded, the basal margins fine, but quite even throughout, being neither more nor less strongly elevated at the sides than in the middle; the side margin is indistinctly crenulate; the surface is rather coarsely, but not closely, punctured, the punctures are evenly distributed, being about as numerous and distinct on the middle as at the sides. The elytra have three indistinct longitudinal spaces free from punctures, and between these are moderately coarsely punctured.

This species is remarkable from the very slight development of the pubescence: it is perhaps more nearly allied to *A. nigrifrons* than to any other species, but it is much smaller, and the pubescence is much slighter. The only individual I have seen is no doubt a female; it has the legs quite short, the anterior tibiæ stout and tridentate.

Jhelam Valley, July 1873. A single individual.

34.—ADORETUS SIMPLEX, n. sp.

Angustulus, parallelus, sat elongatus, testaceus, densius albidosetosus, subopacus, subtus parcius setosus, nitidus; clypeo rotundato, in medio alte reflexo; prothorace basi æqualiter et tenuiter marginato, angulis posterioribus rotundatis; elytris obsoletissime costatis, crebrius punctatis.

Long. 10 mm., lat. 4¼ mm.

The labrum is shining, the basal part is rather large, and has a series of small tubercles arranged at a distance from the rather deeply serrate edge; the appendicular portion is broad, but is not distinctly carinate along the middle. The punctuation of the head seems close, but is quite obscured by the conspicuous depressed white setæ or hairs. The thorax is not very short (for the genus Adoretus); the raised margin is fine, and is not more strongly elevated at the base near the side than elsewhere; the hinder angles are rounded, but not broadly so; its punctuation is only moderately close and coarse. The elytra are rather closely punctured, and have only indistinct longitudinal costæ.

The specimen described is no doubt a male; it has the legs moderately long, and the anterior tibiæ tridentate.

I am unable to point out any near described ally for this species, although I have several closely allied undescribed Indian species in my collection.

Jhelam Valley, July 1873.

PENTODON (Munich Cat.).

35.—PENTODON TRUNCATUS, n. sp.

Nigro-piceus, nitidus, capite anterius truncato, angulis inter se distantibus tuberculo longitudinali acuto, fronte in medio tuberculis duobus minutis; prothorace fortiter punctato, basi ad angulos posteriores tenuiter marginato; elytris sat crebre haud profund punctatis, seriebus duplicatis haud distinctis.

Long. 19—20 mm., lat. 12 mm.

Head finely and densely rugose, in the middle with two minute tubercles, in front truncate and not margined; the unmargined part terminated on each side by a distinct longitudinal tubercular elevation; lateral portions of head with a thick elevated margin. Thorax with the hinder angles completely rounded, and the fine lateral margin continued along the base till the commencement of the slight sinuation on each side; the surface is shining; the punctuation is moderately fine and not close about the base and the middle; it is closer about the front and sides, and quite dense and coarse towards the anterior angles. The sutural stria of the elytra is very distinct, but only indistinctly punctured; of the three double longitudinal series of the elytra only the inner one is distinct; the punctuation of the elytra is moderately close, the surface completely shining. The pygidium is sparingly punctured, but is rugose at each side angle, and there are some obscure, fine, transverse rugæ quite at the base.

The species is similar in form and appearance to the European *Scarabæus punctatus*, but it is smaller; the punctures of the elytra are more numerous and finer, and it is rendered very distinct by the distant tubercles of the front margin of the head.

Kogyar, 31st May to 2nd June 1874. Two individuals, which are no doubt both males.

36.—PENTODON PUMILUS, n. sp.

Nigro-piceus, nitidus, capite anterius truncato, angulis inter se distantibus tuberculo longitudinali acuto, fronte in medio tuberculis duobus minutis; prothorace fortiter punctato, basi ad angulos posteriores tenuiter marginato; elytris fere dense, subrugulose punctatis, seriebus duplicatis haud distinctis.

Long. 14½—15½, lat. 9—10 mm.

This species is so extremely similar to *P. truncatus* that a repetition of the description of that species is unnecessary. *P. pumilus* is, however, scarcely half so large as *P. truncatus*, and has the thorax rather shorter and the punctuation of the upper surface closer. The female has the teeth of the front tibiæ much longer than the male, and the sculpture of the pygidium more diminished.

Kogyar, 31st May to 2nd June 1874. Two individuals.

ORYCTES (Munich Cat.).

37.—SCARABÆUS NASICORNIS, Linn.

Yangihissar, April; Kogyar, 31st May to 2nd June 1874.

38.—ORYCTES GRYPUS, Ill.

Jhelam Valley, July 1873. A single male.

CETONIIDÆ.
By OLIVER JANSON.

1.—CLINTERIA CONFINIS, Hope.

A single specimen of this common Indian species was taken in the Jhelam Valley in July 1873.

2.—CETONIA ORIENTALIS, G. & P.

Dr. Schaum has regarded this species as identical with *C. aerata*, Er., and as only a variety of *C. speculifera*, Swartz; they are, however, three quite distinct species, and have been indicated as such by Blanchard. In the Munich Catalogue *aerata* is given as a synonym under *speculifera*.

Two specimens in the collection were taken at Kogyar between 31st May and 2nd June 1874.

3.—CETONIA DALMANNI, G. & P.

Three specimens of this variable species were taken at Murree; it appears to be generally distributed over the central and northern parts of India.

COLEOPTERA.

HETEROMERA.

By FREDERICK BATES.

Family—*TENEBRIONIDÆ.*
Sub-Family—*TENTYRIINÆ.*
Group—*GNATHOSIIDES.*

SYACHIS, n. g.

Intermediate between *Ascelosodis* and *Capnisa*. At once to be separated from the former by its having the outer apical angle of the anterior tibiæ not dentiform, and from the latter by its having the antennary orbits more convex and more rounded in front; prothorax wider and more deeply emarginate in front, the sides rounded and decidedly contracted behind; elytra shining black and more or less strongly punctured, the epipleuræ being sometimes muricately punctured; prosternal process horizontal and pointed behind; mesosternum declivous and concave in front.

The lateral teeth of the *submentum* are long and pointed: the *mentum* is strongly transverse, almost flat, hexagonal, the apex triangulately notched in the middle, coarsely punctured: the last joint of the *labial palpi* is robust, more or less semi-oval (broadly truncated at the apex): the outer lobe of the *maxillæ* is furnished with a long, curved claw; the last joint of the *palpi* is elongate-oval and broadly obliquely truncated at apex (*picicornis*), or triangulate with the apex a little oblique (*himalaicus*): the *mandibles* are stout, notched at apex, and are furnished on their upper edge, before the apex, with a stout horizontal tooth, which clasps the sides of the labrum, and is sub-acute (*himalaicus*), or obtuse (*picicornis*), and is always, more strongly developed on the right mandible than the left: the *head* is short, robust, more or less wrinkled above the eyes, almost obsoletely so in *picicornis;* throat transversely impressed: *epistoma* more or less prominent, more or less distinct from the antennary orbits, rounded or truncated in front, and is, in *himalaicus*, hollowed out at the sides, leaving the mandibles almost completely exposed: the *antennæ* are short, slender, a little thickened apically, joint 3 much longer than 2 or 4: the *prothorax* is strongly transverse, convex, decidedly wider in front than the head, sides more or less rounded, apex more or less deeply emarginate, base more or less feebly sinuately rounded: *elytra* convex, more or less abruptly declivous behind, wider at base than base of prothorax: epipleuræ moderately broad, the fold expanding at the base and reaching the humeral angle, narrowly, but very distinctly, attaining the apex: *tibiæ* hispid, or spinulose, elongate-triangulate, the anterior most strongly so and finely denticulate on the outer edge, the apex simple; *tarsi* sparsely ciliate, with short spiniform hairs, the first joint of the posterior as long as the last: *intercoxal process* moderate, a little contracted anteriorly and broadly rounded at apex: *prosternal process* horizontal, a little produced and pointed behind: *mesosternum* declivous and a little concave in front: *episterna of metathorax* slightly curvedly contracted posteriorly.

I have failed to discover any really distinctive sexual characters in this and cognate genera of the old world; there are differences of *degree* in the punctuation, &c., of the abdomen, which *may be* sexual, the more strongly punctured, &c., being the male. In the North American representatives of these genera the male is distinguished by having on the first ventral segment a round patch of short, silky-golden hairs; at least it is so in the genera *Trioropkus* and *Stibia*. To this latter genus Dr. Horn denies the presence of a tooth on the upper surface of the mandible. This is evidently a *lapsus*, as so keen and accurate an observer could not have failed to detect it. With some remarkable exceptions (hereafter noticed), I have found this tooth existent in all the numerous genera I have dissected, and, I believe, it will be found all but universally present in this sub-family.

Syachis himalaicus.

Black, moderately shining; underside and legs reddish-brown, antennæ and palpi paler: labrum entire in front: head rather strongly but not closely punctured; strongly wrinkled above the eyes, the wrinkles extending nearly to the crown; epistoma prominent, strongly separated from the antennary orbits, slightly hollowed out at the sides, broad and truncated at apex; superior tooth of right mandible sub-acute: prothorax transverse, a little narrower in front than behind, subangulately rounded at the sides, front angles prominent and acute; strongly and rather closely punctured at the sides, more feebly so on the middle: elytra briefly oval, convex, abruptly declivous behind, produced at apex, shoulders rounded; the surface a little uneven, covered, but not densely, with rather large but more or less shallow punctures, the epipleuræ rather strongly muricately punctured: underside rather densely covered with large rounded punctures: flanks of prothorax very coarsely longitudinally rugose and confluently punctured: epipleural fold more or less, but never strongly, muricately punctured.

Length 3½ to 4 lines.
Dras, Kargil, and Leh.

Syachis picicornis.

Black, moderately shining; underside of body black, legs and antennæ piceous: labrum feebly emarginate in front: head moderately, not closely, and uniformly punctured, not wrinkled at the base, the punctuation a little coarser and confluent above the eyes; epistoma feebly separated from the antennary orbits, being almost continuous with them, broadly rounded in front; superior tooth of right mandible obtuse: prothorax narrower in front and more feebly emarginate than in the preceding, front angle not so prominent; sides rounded behind the middle; very finely and sparsely punctured on the disc, more strongly and closely so at the sides: elytra oblong, shoulders not rounded, gently declivous behind, the punctuation feebler than in the preceding, the epipleuræ not muricately punctured: epipleural fold smooth; punctuation of underside much feebler and less dense than in the preceding.

Length 3½ to 4 lines.
Dras, Kargil, and Leh.

Ascelosodis, Redtenb., Reis. Novar., p. 117.

Prothorax and elytra more or less ciliate at the sides.
Base of prothorax not lobed in the middle.
Elytra densely rugose punctate.

COLEOPTERA. 57

Head and prothorax with mixed punctures, *i.e.*, there are minute punctures scattered between the regular punctuation.
Antennary orbits feebly separated from sides of epistoma : punctuation on back of elytra not muricate—*assimilis*, n.s.
Antennary orbits strongly separated from sides of epistoma : punctuation on back of elytra finely muricate—*ciliatus*, n.s.
Head and prothorax simply punctured—*serripes*, Redtenb.
Elytra thinly and feebly rugose punctate.
Sides of epistoma well separated from antennary orbits—*concinnus*, n.s.
Sides of epistoma nearly continuous with antennary orbits—*Haagi*, n.s.*
Base of prothorax distinctly broadly lobed in the middle—*grandis*, n.s.
Prothorax and elytra not at all ciliate at the sides—*intermedius*, n.s.

ASCELOSODIS SERRIPES, Redtenb.

The series of examples of this species in the collection were taken by Dr. Stoliczka at Yanktze, Chagra, and Pankong Valley. Specimens have been very kindly compared with the type by Dr. Rogenhofer of the Imperial Museum of Vienna.

ASCELOSODIS ASSIMILIS.

Very close to *A. serripes*, Redtenb., from which it differs in having the head distinctly wrinkled above the eyes; the punctuation on the head and prothorax mixed, that is, there are scattered minute punctures on the spaces between the larger punctures; the hind angles of the prothorax and the humeral angles of the elytra are distinct.
Length 2¾ to 3¼ lines.
Dras, Kargil, and Leh.
These differences are rather slight, but they are constant in a large series of examples.

ASCELOSODIS CILIATUS.

Very near to the preceding, and perhaps only an extreme variety of it : it differs by its larger size, more prominent epistoma, the antennary orbits being separated from it and from the front by a deeply impressed line; the elytra entirely, though much more coarsely on the sides and epipleuræ muricate punctate, and the hairs that fringe the sides of the prothorax and elytra much larger and fuller.
Length 4 lines.
Dras, Kargil, and Leh. A single example.

ASCELOSODIS CONCINNUS.

Dark brown, shining; underside reddish-brown; legs, antennæ, palpi, labrum, and front half of the epistoma, red : head finely but not closely punctured, feebly wrinkled above the

* This species was not found by Dr. Stoliczka. A single example exists in Dr. Haag's collection.

eyes: prothorax strongly transverse, front angles prominent, sides gradually expanding from apex to behind the middle, thence strongly rounded to the base; hind angles very open and, being depressed, appearing to be broadly rounded (or obsolete) when viewed from above; sparsely punctured on the disc, more closely so at the sides; side margins reflexed: scutellum distinct: elytra somewhat oval, squarely truncated at base, humeral angle very open; not densely, and but little rugosely, punctured, the epipleuræ muricately punctured, sides fringed with hairs, longest at the shoulders.

Length 3 to 3¼ lines.

Pamir, between Sirikol and Panga.

ASCELOSODIS GRANDIS.

Broadly ovate, convex, black somewhat nitid, underside of body black, legs piceous, tarsi and antennæ paler: head strongly wrinkled above the eyes: prothorax densely punctured, confluently so at the sides, base considerably wider than apex, front angles not prominent; sides gradually curvedly expanded to near the base, whence they curve inwards to the hind angles, which are very obtuse; base rather strongly sinuate at each side, broadly lobed in the middle: elytra convex, humeral angles distinct; rather closely and regularly and slightly rugosely punctured, the epipleuræ strongly muricately punctured: margins ciliated.

Length 5½ lines.

Dras, Kargil, and Leh.

ASCELOSODIS INTERMEDIUS.

Ovate, black, a little shining, legs pitchy brown, antennæ and palpi rufescent: labrum distinctly notched in the middle of fore margin and shortly ciliate: head rugosely punctured, most strongly so above the eyes: prothorax rounded at the sides, more contracted in front than behind; apex not deeply emarginate, front angles not produced; base feebly sinuate, hind angles distinct but obtuse; finely not densely and somewhat uniformly punctured, the punctures largest and slightly rugose at the sides: elytra broadest behind the middle, uniformly but not closely or deeply punctured, and faintly rugulose: epipleuræ muricately punctured; sides not ciliate; base feebly emarginate at the middle, thence sloping to the humeral angle at each side; humeral angle distinct but open.

Length nearly 4 lines.

Dras, Kargil, and Leh.

By its habit, style of punctuation, and non-ciliated sides of prothorax and elytra, this species approaches the genus *Syachis*.

ANATOLICA MONTIVAGA.

Habit of genus *Colposcelis*. Head and prothorax finely, not closely, but uniformly, punctured: epistoma rather prominent, a little hollowed out at the sides, broadly truncated in front; mandibles without superior tooth: prothorax as long as broad, contracted behind, widest anteriorly, apex feebly emarginate, front angles depressed and rounded; hind angles very open, not prominent; basal margin gradually sloping downwards at each side from the angle to the centre, the point opposite the scutellum is consequently on a lower level than the angles : elytra more or less elongated and acuminate behind, sutural region more or less depressed; minutely, sparsely, and irregularly punctulate; smooth, or slightly irregularly wrinkled, or feebly costate; base rather strongly arcuately emarginate, but with the fold entire and continuous from the humeral angle to the scutellum: humeral angle rather strongly produced: flanks of prothorax and prosternum finely and not closely punctured, the latter very strongly thickened at each side between the coxæ; base of mesosternum very strongly and densely punctured ; base of metasternum and of first abdominal segment rather coarsely, but not closely, punctured, the rest of their surface, as well as the other abdominal segments, very finely and remotely punctured.

Length 4½ to 5¾ lines.

Yangihissar, Kogyar.

This species has completely the aspect, and many of the characters, of the genus *Colposcelis*. The third joint of the antennæ, however, is but little longer than the second; the antennary orbits are not separated from the epistoma by a broad, deep impression, this latter being less prominent, and exhibiting no tendency to become umbonate, as it does in *Colposcelis*. The form, &c., of the eye is completely that of *Colposcelis*. The prothorax is as long as it is broad in its widest part, and is slightly angulately rounded at the base. The elytra are as deeply emarginate at the base, and the humeral angle is as strongly produced, as in *Colposcelis*. The middle and hind tibiæ are distinctly hispid, the latter being also elongated and feebly flexuous in the ♂.

It is in the genus *Anatolica* that we find species with mandibles edentate on their upper margin: in some species the mandibles are thick, and provided with a distinct tooth above, which is always the most developed in the right mandible: in others (in both sexes) they are more slender, and the tooth is either but faintly indicated, or is entirely wanting: the present species is in this latter case.

MICRODERA LATICOLLIS.

Approaching *M. gracilis*, Esch., in habit, but more robust. Black, shining : head moderately, prothorax closely, elytra sparsely and minutely, punctured. Prothorax moderately convex, transverse, widest before the middle, well rounded at the sides, strongly contracted behind to the base; base broadly margined, rounded, a little sinuate at each side; apex feebly sub-sinuately emarginate; all the angles depressed; the anterior rounded, the posterior obtuse: elytra elongate, oval, and rather sharply produced at apex; epipleural fold uninterruptedly continued round the shoulders : the parapleuræ entirely, the sides of all the sterna and of the abdomen coarsely, closely, and confluently punctured, finely and remotely so on their middle.

Length 5¼ lines.

Kashgar, Yangihissar, Kogyar.

MICRODERA PARVICOLLIS.

In habit approaching *M. convexa*, Tausch, but prothorax more rounded anteriorly, broadest before the middle, thence gradually contracted to the base, which is much more strongly margined, and the elytra more broadly oval.

Head, prothorax, and elytra, minutely and sparsely punctulate: prothorax nearly as long as broad, sides well rounded anteriorly, thence rather strongly contracted to the base; base slightly rounded, its margin broad and very convex; apex very feebly emarginate; all the angles depressed and obtuse: elytra oval, the apex rather strongly produced: epipleural fold uninterruptedly continued round the shoulders: inner side of the flanks of the prothorax, and the prosternum, rugosely punctured; sides of metasternum, and of the two first abdominal segments, with a few coarse punctures; rest of the abdomen smooth.

Length 4⅔ lines.
Kogyar.

Sub-Family—*AKISINÆ*.

CYPHOGENIA PLANA.

Narrow, elongate, flattened above; black, more or less obscure. Mentum notched (but not deeply) in middle of fore margin, disc more or less plane. Head-rhomboidal, more or less sparsely punctured; epistoma transversely convex, widely and sub-triangulately emarginate in front, completely exposing the labrum and its attachment, front angles more or less acute; front transversely, sometimes triangulately, depressed; supraorbital ridge more or less distinct; cheeks (immediately behind the eyes) prominent and coarsely rugosely punctured. Prothorax quadrate, apex wider than base, front angles produced, sub-acute; sides more or less feebly sinuous (sometimes a little angulate at the middle), and with a narrowish flattened margin; base squarely truncated, or feebly—sometimes sinuately—emarginate; hind angles more or less produced (scarcely outwardly directed) and obtuse; disc irregularly foveate, more or less finely and sparsely punctured, the punctuation stronger at the sides. Elytra elongate-oval, moderately produced and rounded at apex, faintly (sometimes obsoletely) irregularly and minutely muricate punctate, the unpunctured intervals more or less faintly reticulately rugulose; keeled from behind the shoulders to the apex; this keel is not completely marginal, being placed a little within the outer edge, which is rounded, the epipleura being strongly inflexed. Last three or four joints of antennæ usually bright ferruginous, the last acutely pointed at apex.

Length 7 to 9½ lines.
Dras, Kargil, Leh, and Pankong Valley.

CYPHOGENIA HUMERALIS.

In habit approaching *C. aurita*, Pall. Black, obscure; mentum very deeply notched in front, the disc very convex. Head and prothorax very finely and dispersedly punctured, the former with a longitudinal elevated line down the middle, and depressed at each side; supraorbital carina very distinct: epistoma widely emarginate in front in the ♂, more deeply (and sub-angulately) in the ♀; front angles broadly rounded. Prothorax transverse, disc convex, median line lightly impressed, and with a short transverse depression across the middle near the base, apex broadly emarginate, front angles not produced, but somewhat acute; base feebly emarginate, hind angles acute and outwardly directed; sides well rounded anteriorly, rather broadly margined, a little reflexed, and finely transversely rugulose. Elytra depressed, gently declivous behind, the apex rather strongly produced and narrowly rounded; widest behind the middle; obsoletely punctured, and showing some faint smooth reticulations; shoulders keeled, this keel slightly obliquely extending down the elytron, but never for more than one-fourth its length. The ♂ is smaller than the ♀, and has the abdomen much more distinctly punctured. Antennæ with joints 9-10 shorter, triangulate, 11 rather small, acutely pointed at apex.

Length 10 to 12 lines.

Yangihissar.

Sub-Family—*BLAPTINÆ.*

BLAPS STOLICZKANA.

Approaching *B. mortisaga*, Linn., in habit. Elongate, depressed, acuminate behind, black, underside shining black, antennæ and palpi pitchy brown; labrum rufescent, coarsely punctured: head more or less coarsely (never densely) punctured, the base densely and finely muricate, becoming granulous: epistoma trapeziform, widely and feebly emarginate in front: prothorax slightly transverse, sides rounded anteriorly, gradually (sometimes feebly sinuately) contracted posteriorly; base closely applied to the elytra, feebly sinuate and wider than the apex, which is broadly emarginate; front angles rather broadly rounded; hind angles obtuse, slightly overlapping the shoulders, more or less coarsely punctured; the punctures more crowded and more or less reticulately confluent at the sides, and with scattered very minute punctures on the interspaces; sides feebly guttered: elytra at base a little wider than base of prothorax; sides feebly rounded, attenuate behind, the apex gradually produced forming a mucro, which, in the ♂, extends beyond the abdomen by a length equal to the fourth ventral segment; depressed, more or less gently declivous behind; more or less densely confusedly covered with smallish, somewhat shining tubercles, which, except at the base, are flattened, generally acute behind, and here and there run together, forming irregular, more or less transverse, elevated rugosities: flanks of prothorax more or less strongly undulately strigulose, and, as well as the prosternum, parapleuræ, &c., sparsely minutely tuberculate: three first abdominal segments longitudinally rugose at the sides, and transversely rugose on the middle.

Length 8½ to 10 lines.

Pamir, between Sirikol and Panga.

BLAPS INDICOLA.

Habit of ♀, *B. mortisaga*, Linn. Dull black; underside and legs shining black; antennæ, labrum, and palpi brownish black: head and prothorax very finely and not densely punctured; the latter sub-quadrate, feebly but regularly convex, widest before the middle, strongly contracted in front, more gradually behind; sides slightly sinuous before the hind angles, very narrowly channelled; front angles narrowly rounded, the hinder nearly forming right angles, and reposing on the shoulders; base feebly sinuously emarginate: elytra not wider at base than base of prothorax, elongate, acuminate behind; sides feebly expanded to behind the middle, very gently declivous behind, the apex gradually produced, forming a distinct but simple mucro, which is rather densely punctured; uniformly minutely, but not densely, granuloso-punctate, a little stronger on the epipleuræ and at the base, which is also rugulose.

Length 12 lines.

Sind Valley.

BLAPS PERLONGA.

Elongate, slender, acuminate behind, black, obscure: head and prothorax finely and not closely punctured; the latter gently convex, but little (not half a millim.) wider than long; sides gently evenly rounded and very finely margined; base but little wider than apex; base and apex truncated; front angles rounded, the hinder forming right angles: elytra elongate, widest behind the middle, attenuate behind, the apex produced, forming a distinct but not elongate mucro, very gradually declivous behind, and with distinct trace of a short costa within the apex; finely, uniformly, and not closely muricate-punctate, and faintly transversely rugolose.
Legs very long, slender.
Length 10 lines.
Yanktze to Chagra, Pankong Valley.

BLAPS LADAKENSIS.

Black, elytra a little shining; oblong-ovate: head rather closely punctured: prothorax decidedly broader at base than at apex, transverse; sides well rounded anteriorly, slightly sinuously contracted posteriorly; apex feebly emarginate, with the angles rounded; base feebly sinuously truncate, with the angles somewhat obtuse; but little convex; rather finely and not densely punctured: elytra somewhat depressed on the back, not wider at base than base of prothorax; sides gradually moderately rounded, somewhat rapidly declivous behind; apex a little produced, but not mucronate; disc irregularly, finely, and faintly muricately punctured, and intricately rugulose; apex and epipleuræ somewhat intricately covered with irregular flattened tubercles, which are pointed behind: legs and antennæ rather short and robust.
Length 7½ to 8¼ lines.
Yanktze to Chagra, Pankong Valley.

BLAPS KASHGARENSIS.

Elongate, black, elytra more or less nitid: head and prothorax finely remotely punctured, sometimes becoming obsoletely so on the latter: prothorax regularly convex; sides well rounded anteriorly, very gradually (and but little) contracted posteriorly, finely margined; base decidedly wider than apex, very feebly sinuately truncated; hind angles somewhat obtuse; apex feebly emarginate, the angles well rounded; median line faintly impressed on the disc: elytra more or less elongate-ovate, somewhat rapidly declivous behind; apex produced and terminating in a very distinct, pointed mucro, which, in the ♂, extends beyond the abdomen by a length nearly equal to the third ventral segment; convex, a little depressed down the suture; very finely (sometimes almost obsoletely) seriate-punctate, the punctures simple, the intervals also finely punctured, and more or less feebly irregularly convex. Legs rather elongate. Hind margin of first ventral segment in the ♂ a little emarginate at each side, leaving in the centre a more or less triangulate plate, and with a villose tuft of rufous hairs; it is also more or less (sometimes very strongly) coarsely transversely rugose (with traces of a callosity) in the middle, and longitudinally rugose at the sides; the second and third segments being likewise rugose, but never so strongly.

The ♀ is relatively broader than the ♂, the legs not quite so long, the punctuation a little more distinct, the elytral mucro shorter, the abdomen finely rugose-punctate, hind margin of the first segment entire.

Length 9 to 13 lines.

Kashgar, Yangibissar.

Prosodes trisulcata.

♂. Elongate, parallel, pitchy brown; approaching castaneous on the prothorax, legs, and antennæ. Head and prothorax finely, irregularly, and sparsely punctured, the punctuation a little stronger on the sides of the latter: prothorax gently convex, uneven at the sides, a well marked rounded fovea near the hind angle, and several faint irregular foveate depressions on the disc; lateral margins faintly ruguloso; sides slightly rounded anteriorly, gradually and but little contracted posteriorly, widest before the middle; base truncate, hind angles forming right angles; apex very feebly emarginate, the angles depressed and narrowly rounded: elytra not wider at base than base of prothorax, elongate, sub-parallel, depressed on the back, gradually attenuated, and gently declivous, behind, margins reflexed at the apex; on each elytron two dorsal, broad, nearly smooth, costæ; the suture is also thickened; the lateral margin is likewise somewhat costiform; the intervals, which are broader than the costæ, form three shallow furrows, which are somewhat densely irregularly granulous; both furrows and costæ become effaced at the base: the epipleuræ are very broad, vertical, smooth and shining; the fold is also smooth, broad at base (where it attains the humeral angle), it gradually, obliquely, and sinuously narrows to the apex, and has a slightly flexuous elevated line running down its centre: underside pitchy nitid: abdomen feebly longitudinally rugose at the sides: prosternal process terminating behind in a small reflexed mucro: legs rather slender; femora finely muricately punctured; tibiæ more strongly and closely so, and shortly hispid; hind tibiæ feebly flexuous; first joint of hind tarsi as long as the last.

♀. Larger, more robust, less parallel, the punctuation, &c., stronger; the sides of prothorax slightly sinuate before the hind angle: elytra more abruptly declivous behind: hind tarsi shorter.

Length ♂ 8¼, ♀ 10 lines; width of elytra across the middle ♂ 2¾, ♀ 4 lines.

Dras, Kargil, and Leh.

Prosodes vicina.

Very close to the preceding, from which it differs by its broader form; the prothorax underside, legs, &c., shining black; the apex of the prothorax distinctly emarginate; the sides more narrowed anteriorly, the angles more broadly rounded: the elytra a little more gradually declivous behind; the dorsal costæ more elevated, narrower, and, especially in the ♂, punctured and rugose; the epipleuræ are also distinctly ruguloso: the antennæ and legs are stouter; the middle joints of the former sub-moniliform; and the prosternal process is more feebly mucronate behind.

Length ♂ 10 lines, ♀ 11 lines; width of elytra across the middle ♂ 3¼ lines, ♀ 3⅜ lines.
Sind Valley.

CŒLOCNEMODES, n. g.

Habit of *Cælocnemis*, Mann. *Submentum* rather strongly ¦pedunculate; the sinus very open, exposing the base of the maxilla, its outer angle feebly dentiform: *mentum* broader than long, nearly plane, contracted posteriorly, sides and front angles broadly rounded from near the base, coarsely rugosely punctured, nearly covering the *ligula*, which is strongly transverse, and, apparently, broadly emarginate in front: last joint of *labial palpi* ovoid and truncate at apex; that of the *maxillary* feebly securiform: *mandibles* very thick, notched at apex: *antennæ* having the first seven joints coarsely punctured and hispid; joint 3 elongate, equal to 4-5 united; 4-7 equal, obconic; 8-11 perfoliate, setose, clothed, except at the base, with a fine silky yellowish pubescence; 8-10 moniliform, scarcely wider than long; 11 longer and tapering to an acute point: *head* transverse, sub-quadrate, but little contracted behind the cheeks; not prolonged behind the eyes: *antennary orbits* sub-angulate, prominent: *epistoma* rather short, gradually narrowed to the front, which is broadly emarginate with the angles distinct: *labrum* strongly transverse, pilose, nearly entirely visible, very feebly emarginate in front, the angles rounded: *eyes* very narrow, flat, obsoletely faceted, anterior margin entire: *prothorax* moderately convex, a little wider than long; sides well rounded, somewhat abruptly contracted near the base, narrowly but distinctly channelled and transversely rugulose, finely margined; base and apex not margined, the latter arcuately emarginate, the angles broadly rounded, the former truncated, the angles distinct but not prominent, and reposing on the shoulders of the elytra: *scutellum* broadly triangular, penetrating between the elytra: *elytra* regularly convex, scarcely wider at base than base of prothorax; oblong-ovate, rapidly declivous behind, the apex produced but scarcely mucronate; shoulders depressed; sides a little sinuous near the base, gradually expanding to behind the middle: *epipleuræ* broad; the fold attaining the humeral angle, moderately broad, very gradually narrowed to the apex, which it attains: *prosternal process* closely curved round the coxæ, and broadly truncated behind: *mesosternum* declivous, faintly concave: *metasternum* very short between the coxæ; the episterna narrow and but little contracted posteriorly; epimera distinct: *intercoxal process* broad, truncated in front, angles rounded: *legs* moderate; *femora* thickened outwardly, the 4 posterior a little compressed, the anterior with a strong tooth on the upper edge near the apex; all the *tibiæ* rounded, the anterior not denticulate at outer edge, and having a curved excision near the base on its inner side; the intermediate the same but more feebly: *tibial spurs* short, but stout: *tarsi* channelled and briefly spinose beneath, the first joint of the posterior much shorter than the last, upper surface granulous.

Notwithstanding its peculiar *habit*, this genus unquestionably belongs to the *Blaptides*. As in this genus, so in many species of true *Blaps*, the ligula is almost entirely concealed by the mentum. The structure of the antennæ is entirely that of the genus *Blaps*, and most closely approaches the form as seen in *B. mortisaga*, Linn. The short epistoma, leaving the labium almost entirely uncovered, the unmargined base and apex of the prothorax, the hind angles reposing on the shoulders of the elytra, also manifestly approach this genus to *Blaps*.

COLEOPTERA. 65

The form, &c., of the elytral epipleuræ and its fold is nearly identical with what obtains in most of the species of *Prosodes*. The dentate anterior femora have their counterpart in the genus *Dila*, which, however, is of so widely different a habit that one cannot institute even a comparison between the two genera. The peculiar and exceptionally narrow, flattened, and obsoletely facetted eyes are also found in the genus *Dila*. The emargination at the base of the anterior tibiæ, and the abbreviated first joint of the hind tarsi, are the most exceptional characters, and show certain affinities in the direction of the *Scaurides*.

CŒLOCNEMODES STOLICZKANUS.

Obscure brownish black: head uneven, coarsely, but not deeply, punctured, with smaller punctures on the interspaces; the punctuation finer and somewhat granulous behind and confluent on the front: epistoma separated from the front by an impressed arched line: labrum lightly punctate and, together with the palpi, rufescent: prothorax having a broad transverse sinuate impression near the hind margin, and with two or three foveated depressions on the disc; covered with small granules, the sides being confluently granulose punctate: scutellum densely and minutely punctulate: elytra rather closely sub-seriately punctured, faintly transversely rugulose, and granulose; the suture a little thickened and smoother: epipleural fold sparsely muricately punctured: flancs of prothorax coarsely undulately rugose: abdomen punctured: prosternal process finely sulcate down the centre: legs moderately muricately punctured.

Length 10½ lines.
Murree.

Sub-Family—*PIMELIINÆ*.

TRIGONOSCELIS SETOSA.

Black, sometimes a little nitid; more or less broadly oblong-ovate: head strongly transverse, remotely punctured, briefly setose: epistoma and labrum more strongly and closely punctured; the former short, gradually obliquely contracted to the front, which is broadly emarginate and with the angles distinct; the latter feebly sinuately rounded in front, and densely ciliate with longish bright rufous hairs: prothorax transverse, quadrate, more or less gently convex, sometimes depressed on the disc (immature individuals?); median line distinct, or not; apex truncated; front angles small, but prominent, acute, directed forwards; base, which is scarcely wider than the apex, sinuate, strongly so at the middle; hind angles a little outwardly directed; sides more or less feebly rounded before the middle, very feebly sinuate before the front angles, more strongly so towards the base; near the base is a more or less distinct, broad, transverse, somewhat sinuate, impression; granulous, the granules not densely placed, especially on the disc, where they are also smaller; each granule furnished with a short black seta; everywhere finely margined: scutellum small, almost petiolate, generally pubescent, or covered by the hairs which fringe the lower edge of the prothorax: elytra more or less depressed above, more or less broadly oval, widest before the middle, or not, more or less gradually declivous behind; the apex rather strongly produced, and (conjointly) somewhat narrowly rounded; base wider than base of prothorax, appearing more or less strongly emarginate at each side as the shoulder is more or less advanced, which, in some examples, is as strongly so as in *Diesia karelini*, Fisch.; usually with an angulate depression behind the scutellum; variously granulose, each granule bearing a setiform hair, which is longer or shorter, black or fuscous brown; these granules are minute on the disc, larger (almost tuberculiform) at the sides, more or less distinctly seriately arranged down the elytra, moderately intervalled both longitudinally and transversely; at two-thirds the width of the elytron, from the suture, is placed a more or less distinct row of rather larger and more closely-placed granules, with sometimes indications of two others, one between it and the suture, the other between it and the margin; the margin is closely, sub-serrately granulose, or tuberculose: starting from the humeral angle a more or less flexuous elevated line obliquely traverses the epipleura; this line is always granulose at the base, and sometimes more or less minutely interruptedly so along its entire length; above this line the epipleura is sparsely granulose, the granules somewhat large at the base and apex, minute and more remote between; the part of the epipleura below the line is more or less smooth, except at the apex, where are a few small granules. Underside and legs moderately, closely, and uniformly (a little largest on the femora) granulose, and clothed with a fine cinereous pubescence: the four hind tibiæ are hispid, and outwardly fringed with long fuscous hairs; the front tibiæ are strongly triangular, the outer apical angle dentiform, the outer edge finely numerously spinose or dentate[1]; tibial spurs long and powerful, the inner one considerably longer than the outer: the four hind tarsi are fringed with long fuscous (sometimes becoming a little rufescent) hairs at the sides, and with a tuft of bright fulvous hairs at their tips beneath: antennæ black, clothed with short hairs, the last joints ferruginous and naked, except for a few long setæ arising from near their base; last joint short, free, acuminate at apex: prosternal process horizontal, feebly convex, granulose, more or less prominent and rounded behind.

Length 7 to 9¼ lines; width of elytra across the middle, 3½ to 4¼ lines.

Kashgar to Kogyar.

A variable species, and showing affinities with the genus *Diesia*.

TRIGONOSCELIS LACERTA.

Ovoid, black: head large, strongly transverse, a little tomentose,[2] feebly remotely punctured, granulose behind: epistoma broadly emarginate in front, angles distinct: labrum black, middle of front emarginate, angles rounded, distinctly punctured: prothorax rather strongly transverse, quadrate, depressed on the disc, and rather broadly so down the median line, apex very faintly sinuate, angles somewhat prominent and acute; sides moderately rounded anteriorly, sinuately contracted behind, base not wider than apex, sinuate, moderately so at the middle; hind angles slightly outwardly directed; irregularly covered (sparsely on the disc) with rather large, round or oval, flattened tubercles, these largest on the disc, smaller and closer at the sides, the interspaces having a few scattered minute pointed tubercles: scutellum triangular, sub-petiolate, placed on the mesonotum: elytra oval, regularly convex, gently declivous behind, apex a little produced and (conjointly) rounded: on each elytron five irregular

[1] It is *really* tubercled with spines arising from their summit; and when these spines get rubbed or broken off the edge *appears* to be dentate.
[2] Most probably, in fresh examples the *entire* surface is covered with a dull yellowish tomentum.

rows of from seven to nine generally rather large rounded and flattened tubercles, with scattered minute granules on the interspaces, the apex being more closely tubercled; the margins rather closely set with smallish, oblique, pointed setiferous, tubercles; the carina which traverses the epipleura is tubercled at base and apex and minutely denticulate between; the upper portion of the epipleura is studded (especially on its upper edge) with setiferous tubercles, which are largest at the apex; the lower portion is sparsely minutely granulose: underside and legs densely tomentose; four hind tibiæ hispid, not fringed outwardly with long hairs; front tibiæ strongly triangulate, the outer edge irregularly shortly spinose; hind tarsi not compressed, and with a few longish hairs outwardly; the first joint as long as the last: inner spur of the four hind tibiæ nearly twice the length of the outer.

Length 7¼ lines.
Yangihissar.

The following four species of *Pterocoma* form a distinct group in the genus, and are distinguished by the third joint of the antennæ much elongated; the prosternum protuberant in the front, leaving a more or less triangulate open space between it and the head[1]); its process enormously produced, sometimes nearly entirely overlapping the mesosternum; the elytra have each three (with the exception of *Pt. semicarinata*) crenated, or tubercled, costæ, besides the marginal, the third uniting with the marginal just behind the shoulder.[2] They all have the true *Pterocoma habit*; and the four hind tibiæ are rounded.

PTEROCOMA TIBIALIS.

Black, somewhat nitid; the four hind tibiæ distinctly rufescent: antennæ slender, pilose, and setose: legs also slender, spinose, and pilose: labrum notched in the middle of fore margin: head feebly punctured and pilose: antennary orbits reflexed, rounded anteriorly: prothorax convex, slightly sinuately emarginate in front; the angles acute and prominent; base strongly sinuate, the angles small and somewhat outwardly directed, sides feebly rounded; finely, especially on the disc, and sparsely tuberculate, the tubercles erect, pointed, and setiferous: scutellum minute: elytra with a more or less strong depression behind the scutellum; the first costa distinctly continued along the base to the scutellum; the intervals between the costæ nearly smooth, minutely remotely granulous, feebly concave, with a few long decumbent hairs: epipleuræ a little rougher, finely rugose granulate, the hairs shorter and denser: marginal costa closely set with long pointed teeth: prosternal process coarsely corrugated.

Length 4½ to 6⅓ lines; width of elytra across the middle 2⅜ to 3⅝ lines.
Neighbourhood of Sanju.

PTEROCOMA SERRIMARGO.

Smaller than the preceding, dull brownish black: four hind tibiæ not distinctly rufescent. Antennæ shorter and stouter: labrum not notched in middle of fore margin: apex of prothorax not at all sinuate, the angles scarcely so prominent; the base not so strongly sinuate as in

---

[1] Lacerdaire has given the name of "*mentonnière*" to this form of prosternum.
[2] The same is found in *Lasiostola pubescens*, Pall.

the preceding: scutellum larger: elytra not depressed behind the scutellum; the first costa not distinctly continued along the base to the scutellum; the intervals from near the base clothed with a cinereous pubescence: prosternal process relatively broader, sparsely granulose.
Length 5¼ lines; width of elytra across the middle 3½ lines.
Kogyar.

PTEROCOMA CONVEXA.

More narrowly ovate, and more convex, than the preceding; black, a little nitid; thinly clothed with fine cinereous pubescence and setose; legs tomentose, finely setose, and pilose: labrum entire in front: head and prothorax *at bottom* very finely, densely, and rugulosely punctulate, and with scattered small setiferous tubercles, which are largest on the sides of the latter; apex of prothorax truncated; the base sinuate: elytra not depressed behind the scutellum; base a little emarginate at each side, rapidly declivous behind; the intervals with scattered minute punctures mixed with very small setiferous tubercles; the first costa strongly continued along the base to the scutellum: underside more densely tomentose and studded with small setiferous tubercles: prosternal process not quite so strongly produced, and more pointed behind than in the preceding, and rugosely tuberculate.
Length 5 lines; width of elytra across the middle, 3 lines.
No locality given.

PTEROCOMA SEMICARINATA.

Very broadly ovate, the elytra being almost rotundate; black, slightly shining: labrum emarginate in front, the angles very broadly rounded: head with a few rather large, shallow, scattered punctures, most perceptible on the epistoma, and with some small setiferous tubercles clustered above the eyes; and others, still smaller, flattened, and not setiferous, on the crown: prothorax slightly sinuate in front, the angles acute and prominent; the sides studded with setiferous tubercles, the disc having a few flattened tubercles which are distinctly umbilicate: each elytron with a single costa placed half-way between the suture and the shoulder, and extending but little beyond the half the length of the elytron; this costa is furnished with tubercles which are a little flattened at top, clustered two or three together at the base they gradually thin out into single ones, and become smaller, behind; it is also strongly continued along the base to the scutellum, the tubercles here being largest of all; between this costa and the side are indications of two other costæ, the outer one being decidedly the most distinct, these are composed of distant, very small, setiferous tubercles, there are also a few exceedingly minute tubercles scattered on the intervals near the base, each carrying a short seta; the marginal carina is composed of a double row of closely-set bluntish tubercles, which gives to the margin a finely-crenulated appearance: there is no trace of pubescence on the upper surface; the underside is thinly pubescent, the abdomen neither granulose nor tuberculate: prosternum very coarsely, deeply, and confluently punctured; its process very strongly produced, smooth and polished.
Length 6 lines; width of elytra across the middle 4¼ lines.
Yangihissar.

OCNERA SUBLÆVIGATA.

Habit of *O. imbricata*, Fisch. Black, more or less obscure, legs, etc., fuscous, antennæ and tarsi sometimes rufescent and clothed with ferruginous hairs, palpi and labium usually reddish. Head large, minutely and sparsely muricately punctured, and with minute simple punctures on the interspaces; epistoma with a few coarse punctures at the sides and front: labrum coarsely punctured anteriorly: prothorax quadrate, a little transverse, front angles slightly prominent, sides very feebly rounded anteriorly, slightly, and sinuously, contracted posteriorly; sparsely furnished with small, round, somewhat flattened umbilicate granules, and with some minute punctures scattered on the interspaces : elytra not wider at the base (which is sinuous) than the base of the prothorax, regularly oval and convex, gradually declivous behind, the apex a little produced; on each elytra are seven more or less conspicuous rows of varying, but never large, granules, these being generally somewhat oblique, and pointed behind; the 2nd, 4th, and 6th rows are the most apparent (the 6th being the most distinct of all), the others being more or less (especially at the base) confused with the granules scattered in the intervals; the 6th row is also the most continuous (and the tubercles are placed more closely together), extending from the shoulder to near the apex; the others are abbreviated behind, where they are represented by minute distant granules; the 4th and 6th converge towards the base and are united at the shoulder; the extreme outer margin is finely serrate; the intervals are plain, sparsely and very minutely granulose: the epipleuræ are loosely granulose: the flanks of the prothorax are sparsely granulose ; the pro- and meso-sterna are rather strongly granulose; the metasternum with its flanks, and the flanks of the mesosternum are very minutely dispersedly granulose, and clothed with a fine cinereous tomentum : the abdomen, except in the middle, is more or less coarsely punctured, and with scattered granules : the prosternal process is broad, horizontal, and triangulate behind: the femora are rugose and granulous and hispid ; the tibiæ closely hispid, the posterior feebly sinuous; the joints of the tarsi beneath are furnished at the apex with a tuft of bright fulvous hairs.

Length 9 to 10 lines.

Kashgar, Yangihissar.

Sub-family—*PEDININÆ.*

Group—*PLATYSCELIDES.*

BIORAMIX, n. g.

♂ . Head strongly transverse, front declivous to the epistomial suture; epistoma very short, broadly rounded, or truncated, in front; the angles distinct, or not; marked off from the front by a well-impressed arched line : third joint of antennæ as long as 4-5 united, or a little shorter : prothorax variable, always transverse, not closely applied to base of elytra; sometimes very feebly convex and slightly narrowly depressed at the margins, or regularly convex direct from the margins; apex strongly emarginate with the angles prominent and loosely embracing the head (*asidioides*) ; or very feebly emarginate, the angles depressed and more closely embracing the head ; front angle sometimes acute, or more or less rounded; base either truncate, or feebly emarginate, or sinuate, the angles prominent, or not, and either rectangular, obtuse, or rounded; sides sometimes more or less regularly rounded, or rounded in front and sub-parallel, or sinuate (*asidioides*), behind: scutellum rather large; transverse; angulate, or rounded, behind; more or less exposed: elytra variable; more, or less (*asidioides*), elongate-oval, regularly convex, or depressed on the back (*asidioides*), more or less strongly declivous behind; shoulders prominent, or not, the angles distinct, or rounded; diffusely, or seriately, punctured; in the latter case (*asidioides*) the punctures are rather large and somewhat oblong, the intervals being more finely punctured and alternately feebly costiform; these punctures are more (*asidioides*), or less (sometimes only to be detected at the apex), visibly setiferous, the setæ being very short, and erect only at the apex: epipleuræ continuous with the sides, or (*asidioides*) vertical and marked off from the sides by a well-defined costa, which extends from the humeral angle to just within the apex; epipleural fold more or less broad, always attaining the humeral angle, which it sometimes reflects, and gradually somewhat curvedly narrowed from the base to near the apex, which it does not quite attain: prosternal process more or less horizontal, generally contracted and more or less vertical behind: intercoxal process truncate, or rounded, in front. The legs are less robust, and much less scabrous, and more finely pilose, than in *Platyscelis*: the outer apical angle of the anterior tibiæ is not dentiform; the hind tibiæ are straight, or slightly curved (*asidioides*). The oral organs, and the tarsi (except that the middle joints of the intermediate pair have the basal angles well rounded) do not materially differ from the same parts in the genus *Platyscelis*.

♀. All the tarsi simple. Form generally a little more robust and convex, the elytra more rounded at the sides, less nitid, the punctuation, &c., fainter, and the antennæ a little shorter and stouter.

BIORAMIX PAMIRENSIS.

♂. Elliptic oval, black, a little nitid, underside and legs brunneous, antennæ and palpi rufescent: head finely irregularly punctured in front, coarsely punctured behind the eyes, finely closely muricate punctate and pubescent behind; broadly rounded in front; epistoma very short, the suture arched and well impressed; labrum strongly transverse, very feebly emarginate in middle of front margin, the angles broadly rounded, finely and densely punctured: third joint of antennæ nearly as long as 4-5 united: prothorax finely, sharply, and somewhat uniformly punctured, gently convex direct from the lateral margins, not closely applied to base of elytra, truncated at base and apex; sides contracted anteriorly, sub-parallel from before the middle to the base and finely bordered, base and apex still more finely bordered, but only at each side; hind angles rectangular; front angles depressed, slightly obtuse: scutellum strongly transversely triangular, densely punctured: elytra scarcely wider at base than base of prothorax, shoulders broadly rounded, leaving a distinct open angle between them and the prothorax; sides very feebly rounded, attenuate and gently declivous behind; moderately but very distinctly punctured, and showing here and there slight indications of a longitudinal seriate arrangement, faintly irregularly rugulose and alutaceous; thinly hispid at the apex and sides; epipleuræ very narrow, rounded; the fold broad, gradually curvedly contracted from humeral angle to near the apex; the punctuation, &c., as on the elytra above, but more closely and less cleanly: underside somewhat closely and very finely corrugated, and appearing granulous on the flanks: abdomen finely imbricately rugulose, pilose, first segment with a depression at the middle of its hind margin: intercoxal process rounded in front: front and middle tibiæ stout, moderately expanded outwardly, the front being also trigonal, the outer edge sharp and a little sinuous; hind tibiæ larger than the others, and more feebly

expanded outwardly: three middle joints of intermediate tarsi broader than long; hind angles rounded.
Length 5 lines.
Pamir from Sirikol to Panja.

BIORAMIX OVALIS.

Oval, less elongate, and relatively broader than the preceding: head more closely, and slightly rugosely, punctured, not densely sub-muricately punctured, nor pubescent behind: prothorax more transverse, a little less convex; sides more rounded, distinctly, and slightly sinuately, contracted behind, a little depressed at the margins, foveolate at each side the middle, the punctuation not so clean, the angles distinctly more obtuse: elytra more rounded at the sides, more convex, more abruptly narrowed, and more strongly declivous behind; punctuation much finer and little less regular, with sometimes faint indications of costæ: epipleural fold less distinctly (sometimes obsoletely) punctured; front tibiæ a little more compressed, not distinctly sinuate at outer edge: last ventral segment with a faint depression in the middle of its upper margin.
Length $4\frac{1}{4}$ to $4\frac{1}{2}$ lines.
Dras, Kargil, and Leh.

BIORAMIX PUNCTICEPS.

Very near the preceding: differs in having the epistoma distinctly squarely truncated in front, more densely punctate and rugose: prothorax still more strongly transverse, front angles more obtuse, sides strongly rounded behind, effacing the hind angles: the elytra very faintly punctured, the shoulders still more strongly rounded, consequently the middle of the base of prothorax only impinges on the elytra: epipleural fold very finely rugulose, but not visibly punctate: anterior tibiæ distinctly more compressed, more triangulate, the outer apical angle a little produced: front and middle tarsi (especially the latter) distinctly narrower and more pilose.
Length 4 to $4\frac{1}{2}$ lines.
Dras, Kargil, and Leh.

BIORAMIX ASIDIOIDES.

Very distinct from the three preceding species by its larger size, broader and more depressed form, distinctly seriately punctate elytra, etc.
Oblong, oval, black, slightly nitid; head sub-angulate in front, somewhat coarsely, but not closely, punctured and a little rugose, more finely and closely so, and pubescent, behind: epistoma very short, but the sides are well distinguished from the antennary orbits, the angles being very distinct and nearly rectangular, the apex squarely truncated: third joint of antennæ as long as 4-5 united: prothorax transverse, somewhat depressed, its base rather closely applied to the base of elytra, moderately punctured, the punctures each bearing a short decumbent hair; apex arcuately emarginate, the angles sub-acute; base considerably wider than apex, sinuate, the angle sub-acute and somewhat outwardly directed, although

reposing on the shoulders of elytra; sides gradually expanded from apex to the middle, thence slightly and sinuately narrowed to the base; margins irregularly depressed, and transversely rugose; a faint depression at each side the disc, and another within each hind angle: scutellum small, triangular: elytra a little depressed above, more convex posteriorly and somewhat rapidly declivous behind; base a little wider than base of prothorax, slightly sinuate, shoulders slightly rounded, sides gradually, but feebly, rounded to the middle, gradually narrowed behind, on each elytron eight rows of punctures, more or less effaced at base, sides, and apex; intervals finely punctured, faintly transversely rugulose, the alternate ones a little convex, especially towards the apex; very finely and shortly hispid, most distinctly so at sides and apex; lateral margins costiform: epipleuræ distinct; the fold broad, gradually curvedly narrowed from humeral angle to near the apex, faintly rugulose punctate: last ventral segment with a broad depression in middle of front margin: front tibiæ trigonal, gradually, but not strongly, expanded outwardly; outer edge a little sinuate: hind tibiæ slightly curved: middle tarsi moderately expanded: intercoxal process truncated in front.

Length 6 lines.
Sind Valley.

CHIANALUS, n. g.

Closely related to *Bioramix*; differs in having the head longer and narrower, the epistoma distinctly larger, and more produced anteriorly; the elytra costate and clothed with short erect hairs; the epipleural fold continued to the apex; the anterior tibiæ finely denticulate down the outer edge, the outer apical angle very strongly dentiform; the intermediate tarsi very feebly dilated, the joints being distinctly longer than wide. In both sexes the last abdominal segment has a semi-circular depression in the middle of the basal margin.

CHIANALUS COSTIPENNIS.

♂. More or less oblong-ovate, dark brown, head and prothorax still darker, and nitid: head but little wider than long, rather strongly and somewhat closely (except on the crown) punctured, finely densely granulose punctate behind; the front is declivous to the epistomal suture, which is well marked and sub-angulate; epistoma distinctly produced beyond the base of antennæ, broadly rounded anteriorly; labrum strongly transverse, closely punctured, nearly entire in front, the angles rounded: prothorax transverse, but little convex, sides well rounded, a little uneven; the margins narrowly irregularly depressed; finely and not closely punctured, lightly (almost obsoletely) impressed down the median line, usually with a slight foveate depression at each side, and a distinct depression at each side at the base, half-way between the middle and the angles; apex narrower than base, lightly emarginate, front angles strongly depressed, lightly rounded; base faintly sinuately truncate, hind angles obtuse: scutellum small, transversely triangular: elytra more or less oval, sides more or less rounded, generally widest at the middle, suture costiform, and on each elytron four stout prominent, rounded costæ, the second and third united before the apex and continued thence as but one; running down each interval is also another costa, narrower and much less prominent; both costæ and intervals are finely granulose-punctate, and transversely rugulose, and the entire surface is moderately densely covered with short erect hairs: epipleural fold broad, very gradually narrowing

COLEOPTERA. 73

in a curve from the shoulders to the apex, which it narrowly attains, regularly but not closely
covered with very minute granules, but without trace of hairs: flanks of prothorax rather
closely undulately rugose and sparsely granulose: abdomen punctured and finely imbricately
corrugated: third joint of antennæ as long as 4-5 united: tibiæ densely hispid, compressed,
triangulate (the four anterior most strongly); the posterior straight.

♀. More convex and robust; the elytral costæ all sub-equal.

Length ♂ 5¼ lines— ♀ 5½ to 6 lines.

Width of elytra across the middle ♂ 2¼ to 2½ lines— ♀ 3 to 3¼ lines.

Dras, Kargil, and Leh.

MYATIS, n. g.

In this genus the head is again very short and transverse; the epistoma is excessively
short, very broadly and squarely truncated in front, almost on a level with the insertion of the
antennæ; the impressed line, or suture, arcuate: prothorax gently convex, somewhat variable
as to its form, &c.; generally it is curvedly contracted in front, sub-parallel, or faintly sinuate,
behind; the apex very feebly emarginate; the front angles obtuse; base slightly sinuately
truncate, the angles acute: elytra oblong, gently convex, sides very feebly rounded, narrowed,
and moderately declivous, behind; shoulders more or less oblique, the angle more or less
acutely prominent, sometimes dentiform: epipleural fold moderately broad, gradually narrowed behind, not reaching the apex of elytra: the legs are slender; the outer apical angle of
the anterior tibiæ acutely dentiform; the first joint of the anterior tarsi is relatively longer
than in the preceding genera, and the three first joints of the intermediate tarsi are (although
provided with a small brush beneath) scarcely at all dilated: the intermediate tibiæ in the ♂
are thickened outwardly, and, as well as the posterior tibiæ, are densely fringed within
with silky golden-yellow hairs: the pro- and meso-sterna are not nearly so convex, or
protuberant, as in the other genera of the group: the prosternum between the coxæ is
thickened at each side, and terminates behind in a short reflexed mucro: the elytra are
finely minutely hispid, most distinctly so at sides and at apex.

MYATIS HUMERALIS.

Oblong, pitchy brown, head and prothorax nitid: the former rather finely punctured,
the punctuation a little closer, somewhat confluent, and pubescent at the sides and base;
slightly irregularly foveolated between the eyes: epistoma broadly and squarely truncated,
and densely ciliate, in front; the suture well marked: labrum rather closely punctured,
strongly pilose: third joint of antennæ nearly as long as 4-5 united: prothorax gently convex,
very nearly as long as broad, finely evenly punctured, sides delicately margined, curvedly
expanded in front to before the middle, thence very slightly incurved to the hind angles,
which are prominent, sub-acute, and somewhat outwardly directed; apex truncated, front
angles a little depressed and rounded; base a little sinuate at each side, broadly and very
gently rounded in the middle: scutellum strongly transverse, generally concealed by the short
dense hairs which fringe the base of the prothorax: elytra broader at base than base of prothorax, oblong, sides feebly expanded to behind the middle, somewhat rapidly declivous be-

hind, minutely and not closely granulose-punctate, with distinct indications of striæ, intervals very faintly rugulose, and run over with very delicate sub-reticulate lines; very faintly hispid; humeral angle very prominent, dentiform : epipleural fold finely sparsely granulous; underside and legs of a lighter brown: abdomen closely, finely, sub-muricately corrugated, and thinly clothed with a long yellowish pubescence; the last joint in the ♂ with a depression at the middle of its upper margin.

Length 4¼ to 4½ lines.
No locality given.

MYATIS QUADRATICOLLIS.

Brown, of a much lighter shade than the preceding: head somewhat strongly punctured, more distinctly foveolated between the eyes: prothorax distinctly transverse, dull-reddish castaneous, clouded with dark brown, less evenly convex, irregularly foveolately depressed at each side near the border, sides more contracted posteriorly, the hind angles not produced nor outwardly directed; base not distinctly sinuate at each side; the punctuation distinctly coarser at the sides: punctuation of elytra a little less clean and less distinct; distinctly hispid at sides and apex, this very fine, short, and of a golden-yellow colour; humeral angle prominent, but not dentiform: underside and legs of a paler reddish brown.

Length 4¼ lines.
Between Leh and Yarkand.

MYATIS VARIABILIS.

Varying from light reddish to very deep dark brown: head less distinctly foveolated between the eyes than in the preceding: prothorax distinctly less transverse, and more uniformly brown, the punctuation stronger, the median line nearly always distinct and quite smooth, more regularly and evenly convex, more rounded at the sides; the hind angles are rectangular, or are a little outwardly produced: the elytra are more parallel; the humeral angle is more or less distinct, but never prominent, the punctuation, &c., is a little closer and stronger; they are also more distinctly and uniformly hispid: the underside and legs vary from very dark to pale-reddish brown.

Length 3¼ to 4½ lines.
Between Yangihissar and Sirikol, and Sirikol and Sanju.

These three species are very close to each other, and I strongly suspect they really constitute but one intensely variable species.

Sub-Family—*OPATRINÆ.*

Group—*OPATRIDES.*

OPATRUM KASHGARENSE.

This species has been submitted to M. Miedel, who returns it as a new species belonging to the *rusticum* (*Oliv.*) group.

Oblong, brown, little nitid: head broadly and sinuately rounded in front: epistoma short, a little convex on the middle, notched, but not sharply angularly, in the middle of the front margin, the angles being well rounded; antennary orbits outwardly angulately produced beyond the eyes; finely granulose (the granules black), and thinly clothed with short scale-like hairs of a golden-yellow color: prothorax gently convex, rather deeply curvedly emarginate in front, front angles not produced, sub-acute; sides a little reflexed, gently regularly rounded; base a little wider than apex, sinuate; hind angles produced, acute, directed backwards; the surface more distinctly and regularly granulose, &c., than the head: scutellum semi-circular, finely granulose and pubescent: elytra a little wider at base than base of prothorax; oblong, slightly widest behind the middle; shoulders distinct, very finely transversely rugulose; punctate-striate, the punctures being rather large; intervals a little convex, very finely and not at all densely granulose, each granule furnished with a short scale-like hair, as in the prothorax, &c.: underside thinly clothed with a fine greyish-yellow pubescence: flanks of prothorax rather strongly granulose, meso- and meta-sterna and their flanks more finely so: abdomen finely granulose-punctate, and transversely rugulose: metasternum as long as the first ventral segment: prosternum closely curved round the coxæ: anterior tibiæ expanding outwardly, finely muricately punctured and shortly setose: last joint of all the tarsi elongate: antennæ reddish, thickening outwardly, joint 3 nearly as long as 4-5 united, 8-10 transverse and gradually broader, 11 large, ovoid.

Length 4¼ lines.
Kashgar.

OPATRUM OCHTHEBIOIDES, Fauvel.

Dras, Kargil, and Leh.

PENTHICUS (LOBODERUS) GRACILIS.

I have submitted this species to M. J. Miedel, of Liege, who for the past five years has been engaged on a critical examination of the *Opatrides*: he returns it to me as a species distinct from the *rufescens* of Mulsant, and has furnished me with the following differential characters:—

Than *rufescens*—larger: prothorax more contracted posteriorly, the sides consequently are sub-angulated in the middle, front angles more pointed; the punctuation, &c., different; in *gracilis* the prothorax is somewhat closely, uniformly, and finely punctured on a very minutely granulose ground, and at each side the disc are some irregular foveate depressions: in *rufescens* the middle of the prothorax is somewhat sparsely covered with fine, but well-marked, punctures on a smooth ground, the punctuation on the sides being stronger and closer: the elytra in *gracilis* are more gradually (*longuement*) attenuated behind, but not more pointed at the apex; very faintly sulcated, the intervals somewhat sparingly covered with very minute granules and showing a line of small shallow punctures; there is also a line of very minute punctures down by the suture: in *rufescens* the elytra are visibly although very finely, punctate-striate, the intervals being finely, transversely, unequally rugulose: the abdominal segments in *gracilis* are somewhat thinly covered with very small granules, arranged in almost transverse lines; whilst in *rufescens* they are well punctured: the legs and antennæ are similar in both species, except that joints 3 to 7 of the latter are more elongate in *gracilis*.

Length of *gracilis*, 4½ lines.
Length of *rufescens* 2¾ to 3¼ lines.
Kogyar.

Sub-Family—*HETEROTARSINÆ.*

Group—*PHOBELIIDES.*

LYPROPS INDICUS, Wiedm.

Jhelam Valley.

Sub-Family—*HELOPINÆ.*

Group—*ADELIIDES.*

LÆNA LACORDAIREI, Marseul.

Sind Valley.

Family—*CISTELIDÆ.*

Sub-Family—*CISTELINÆ.*

Group—*CISTELIDES.*

ALLECULA (DIETOPSIS) COSTIPENNIS.

Elongate, narrow, chocolate-brown, head and prothorax of a little deeper tint; underside with a reddish tinge, and shining: head closely and finely punctured, pubescent; a foveate depression between the eyes: epistoma long, convex, expanding anteriorly, apex squarely truncated; labrum strongly transverse, entire and ciliate in front, with the angles rounded: last joint of maxillary palpi very broadly cultriform: antennæ sub-filiform, joints 3-11 of nearly equal length, obconic, apex of 11 narrowly rounded: prothorax convex, transverse, narrowed in front, sides parallel, front angles broadly rounded, base lightly sinuate, the angles obtuse; finely and uniformly punctured, lightly impressed down the median line, a foveate impression at each side the median line, broadly impressed at each side at the base: scutellum large, rounded behind, closely punctured, and lightly keeled down the centre: elytra at base broader than the base of prothorax; shoulders well rounded; strongly crenate-striate, the intervals convex, sparsely and very minutely punctulate, each puncture bearing a very fine, minute, pale decumbent hair: the sterna are all very finely and densely punctured and transversely rugulose, their flanks rather closely punctured, the punctures rounded and well marked: abdomen and legs very finely uniformly punctured and pubescent: tarsi with the third and fourth joints of the two front pairs lamellated, the penultimate joint only in the hind pair.

Length 5 lines; width of elytra across the middle 1½ lines.

Murree.

Group—*CTENIOPIDES.*

HYPOCISTELA, n. g.

Near *Cteniopus*, from which it differs in having the third joint of the antennæ but little more than half the length of the fourth, and, as well as joints 3-6, obliquely truncated at apex: the palpi slender, the last joint, both of labial and maxillary, elongate, oval, and not

COLEOPTERA. 77

truncated at tip: the eyes larger, more approximate beneath, and very coarsely facetted: the prothorax not curvedly narrowed to the front, and decidedly narrower at base than the base of the elytra.

HYPOCISTELA TENUIPES.

Pale testaceous, legs yellow, antennæ palish brown, eyes and tips of mandibles black, head fuscous behind. The entire upper surface is uniformly and very minutely punctulate and rugulose, and finely pubescent: elytra delicately striated: flanks of prothorax, breast, and abdomen, clouded with fuscous.

Length $3\frac{1}{4}$ lines.
Kogyar.

Family—*LAGRIIDÆ*.
Sub-Family—*LAGRIINÆ*.

LAGRIA INDICOLA.

Form, size, and colour of *L. glabrata*, Oliv. The eyes are silvery grey with an oblique fuscous spot above: antennæ moderately stout, filiform, last joint elongate, straight, cylindrical, and pointed at apex: prothorax broader at base than at apex, very feebly rounded at the sides, somewhat shining piceous, the front and hind margins reddish; a broad transverse impression before the base; feebly punctate, and, together with the head, clothed with a longish fuscous pilosity: elytra delicately striated, distinctly uniformly punctured, and irregularly transversely wrinkled: underside, femora, and antennæ, pitchy brown: tibiæ and tarsi paler.

Length $4\frac{1}{4}$ lines.
Murree.

Family—*MELOIDÆ*.
Sub-Family—*MELGINÆ*.

MELOË SERVULUS.

Small, black, with a faint bluish tinge on the elytra: antennæ shining black, compact, a little thickened towards the apex, joints obconic, 5-6-7 shorter than 3-4 or than 8-10, 10 somewhat cylindric, 11 elongate and tapering to the apex: head large, convex, distinctly and rather uniformly, but not closely, punctured: prothorax rather small, transverse, quadrate, all the angles rounded, base arcuately emarginate, punctured like the head, and with a distinct foveate depression at each side the disc: elytra faintly reticulately rugulose, somewhat scrobitulate on the epipleuræ; dehiscent from one-third their length, and somewhat gradually curvedly contracted to the apex, which is narrowly rounded; base emarginate at each side, shoulders obliquely rounded: abdomen above faintly transversely rugulose, and very sparingly minutely punctulate.

Length $3\frac{3}{4}$ lines.
No locality given.

K

Sub-Family—*CANTHARINÆ.*

Group—*MYLABRIDES.*

MYLABRIS SIDÆ, Fab., Marseul.

Sind Valley. Murree.

MYLABRIS MACILENTA, Marseul.

Murree.

Group—*CANTHARIDES.*

CANTHARIS ANTENNALIS.

Sind Valley.

EPICAUTA HAAGI.

♂. Head dull red, with a large, smooth, blood-red callosity at the base of each antenna; strongly and closely punctured, a short, fine, elevated line running down the middle of the crown; scantily clothed with fine black hairs at the sides and behind; epistoma more or less clouded with black, broadly and feebly sinuately truncated in front, less densely punctured than the head; labrum entirely black, sinuous in front, punctured like the epistoma: antennæ, excepting the two basal joints which are red above, black, strongly depressed, the joints longitudinally excavated on their inner side; joint 3 elongate, triangulate, 4-6 much shorter, and becoming gradually narrower, 3-7 more or less strongly obliquely emarginated at apex, with the inner angle produced, 8-10 of nearly equal length, but becoming gradually narrower, truncated at apex, 11 longer and narrower than 10, cylindric and rounded at apex, the outer joints densely clothed with cinereous pubescence: prothorax black, a little nitid, slightly longer than wide, convex, a strong depression at the middle of the base; sides a little rounded before the middle, strongly narrowed anteriorly from before the middle, very gradually, and but little, contracted posteriorly; closely and deeply punctured, and pilose; sides, apex, and median line usually clothed with dull yellowish-white decumbent hairs: elytra dull black; base considerably wider than base of prothorax, divaricate nearly from the base, the apex obliquely rounded; very finely and densely granulose and transversely rugulose, clothed with short decumbent black hairs; the margins entirely bordered with a line of dull yellowish-white interwoven hairs, and there is also, in fresh examples, a dorsal stripe of the same: underside and legs shining black, and, except the last ventral segment, more or less thickly clothed with hairs of the same character as those that border the elytra; last ventral segment triangulately excised at apex: anterior femora with the usual sericeous hairy spot at the emargination near the apex; the front tibiæ are also emarginated at the middle within and excavated down the inner side, this is filled in with the like silky-golden pubescence: tarsi simple, the first joint of the anterior swollen on the inner side.

♀. Smaller; the callosities on the head feebler: antennæ shorter, slightly attenuated outwardly; joint 3 elongate, 4-6 much shorter, 3-7 more or less slightly obliquely truncated at apex, 8-10 equal, cylindric, 11 longer, rounded at tip: front legs as in the ♂ but weaker: last ventral segment more feebly excised at apex.

Length ♂ 9 to 10 lines, ♀ 7¼ lines.

Murree.*

* Dr. Haag, who is at the present time engaged on a monograph of this group, has examined and pronounced this species to be undescribed. Mr. C. O. Waterhouse of the British Museum also states that it is quite distinct from any species described by himself (in Trans. Ent. Soc. London 1871, pp. 405-8), or by Hope.

COLEOPTERA. 79

Group—*SITARIDES.*

SITARIS (CRIOLIS) PECTORALIS.

Shining testaceous, tips of mandibles, eyes, scutellum, meso- and meta-sterna, and their flanks, black; middle and hind coxæ shining black : antennæ filiform, last 7 joints fuscous black, last joint elongate and tapering to a point: head broadly triangulate, convex, smooth, faintly punctate : the epistoma is separated from the front by a deeply-impressed arched line, and is in a lower plane than the front: labrum impressed on the disc, notched in front: eyes strongly transverse, rather narrow, not prominent: prothorax convex, transverse, sides rounded, and broadest, in the middle; somewhat abruptly and strongly contracted anteriorly, less so posteriorly; faintly punctured : scutellum closely punctured, narrowly rounded behind ; the part placed on the mesonotum is broad, faintly costate down the middle, and with a thickened border at each side of a cinnamon-brown colour : elytra somewhat of a pale-cinnamon colour; tapering gradually behind, and dehiscent from about a third of their length ; thinly clothed with a fine, silky, greyish pubescence, and with two or three slightly flexuous costæ, the outer one less distinct : upper division of the tarsal claws closely finely pectinated.

Length 4¾ lines.
Kogyar.

COLEOPTERA HETEROMERA.

EXPLANATION OF PLATE II.

Fig. 1. *Spachis himalaicus.*
 „ 2. *Ascelosodis ciliatus.*
 „ 3.   „   *grandis.*
 „ 4.   „   *intermedius.*
 „ 5. *Anatolica montivaga.*
 „ 6. *Microdera parvicollis.*
 „ 7. *Cyphogenia plana.*
 „ 8.   „   *humeralis.*
 „ 9. *Blops perlonga.*
 „ 10.   „   *indicola.*
 „ 11. *Prosodes triauleata.*
 „ 12. *Cœlocnemodes stoliczkanus.*

Fig. 13. *Trigonoscelis lacerta.*
 „ 14. *Plerocoma serrimargo.*
 „ 15.   „   *semicarinata.*
 „ 16. *Bioramix asidioides* ♂.
 „ 16. *Idem, anterior tarsus.*
 „ 17. *Chianalus costipennis.*
 „ 18. *Myatis humeralis.*
 „ 19. *Penthicus (Loboderus) gracilis.*
 „ 20. *Hypocistela tenuipes.*
 „ 21. *Meloë sertulus.*
 „ 22. *Epicauta haagi* ♂.

COLEOPTERA

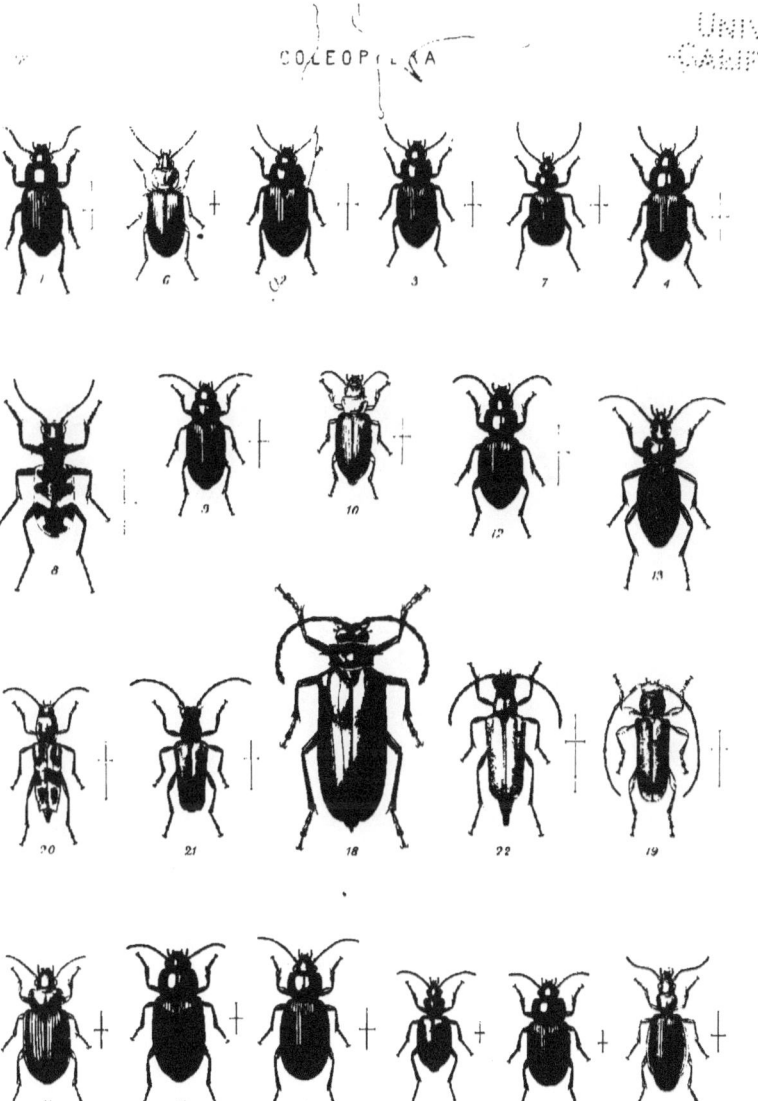

COLEOPTERA, HETEROMERA.

UNIV. OF CALIFORNIA
Plate II.

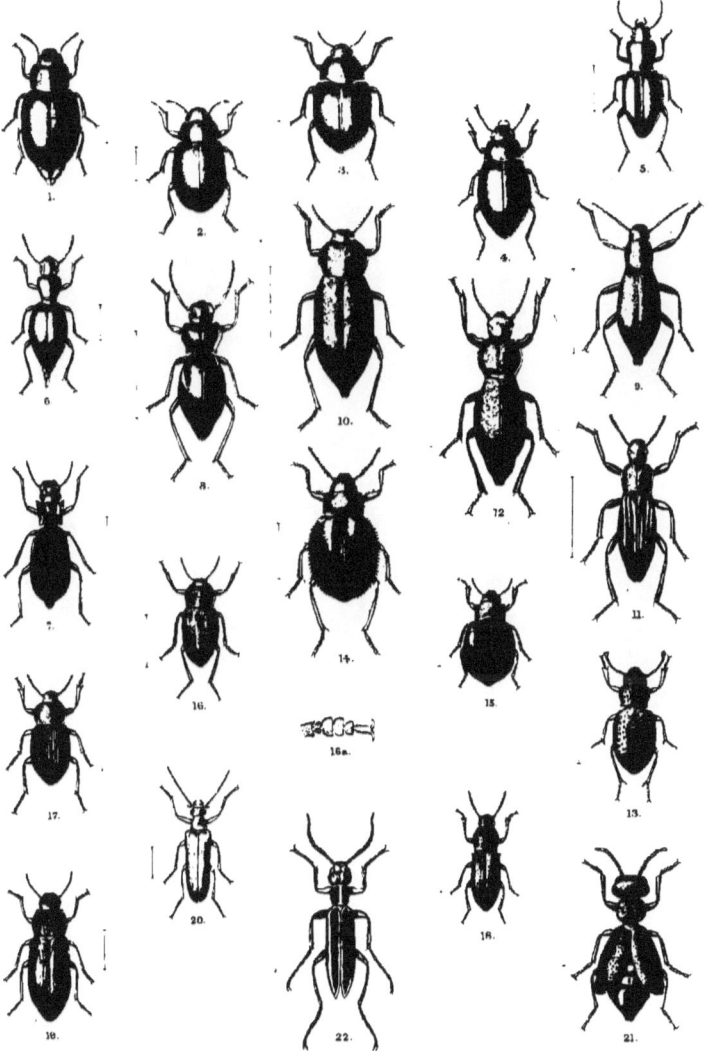

SCIENTIFIC RESULTS

OF

THE SECOND YARKAND MISSION;

BASED UPON THE COLLECTIONS AND NOTES

OF THE LATE

FERDINAND STOLICZKA, Ph.D.

HYMENOPTERA.

BY

FREDERICK SMITH,
ZOOLOGICAL DEPARTMENT, BRITISH MUSEUM.

Published by order of the Government of India.

CALCUTTA:
OFFICE OF THE SUPERINTENDENT OF GOVERNMENT PRINTING.
1878.

SCIENTIFIC RESULTS

OF

THE SECOND YARKAND MISSION;

BASED UPON THE COLLECTIONS AND NOTES

OF THE LATE

FERDINAND STOLICZKA, Ph.D.

HYMENOPTERA.

BY

FREDERICK SMITH,
ZOOLOGICAL DEPARTMENT, BRITISH MUSEUM.

Published by order of the Government of India.

CALCUTTA:
OFFICE OF THE SUPERINTENDENT OF GOVERNMENT PRINTING.
1878.

CALCUTTA:
PRINTED BY THE SUPERINTENDENT OF GOVERNMENT PRINTING,
8, HASTINGS STREET.

SCIENTIFIC RESULTS

OF

THE SECOND YARKAND MISSION.

HYMENOPTERA.

BY FREDERICK SMITH, *Zoological Department, British Museum.*

THE collection made by Dr. Stoliczka while attached to the Second Yárkand Expedition contains sixty-three species, only nine of which appear to have been previously described; among them are species belonging to the families, *Andrenidæ, Apidæ, Formicidæ, Myrmicidæ, Scoliadæ, Pompilidæ, Sphegidæ, Larridæ, Eumenidæ, Vespidæ, Tenthredinidæ,* and *Ichneumonidæ.*

Tribe—ANTHOPHILA (Latr.)

Division 1—SOLITARIÆ.

Family—*ANDRENIDÆ.*

1. PROSOPIS FERVIDUS.

Femina.—*P. atra, fronte maculata, tibiis omnibus flavo-annulatis; abdominis segmentorum marginibus rufo-testaceis.*

Black; the head closely and finely punctured; a yellow line on each side of the face along the margin of the eyes; the flagellum of the antennæ fulvous, slightly fuscous above. Thorax punctured above and shining; the metathorax rugose in the middle of its base; the collar, tubercles, tegulæ, the anterior tibiæ and tarsi, the intermediate and posterior tibiæ at their base, and their tarsi, yellow; the wings hyaline and iridescent; their nervures towards the base pale testaceous, beyond, fuscous.

Abdomen shining, very finely and closely punctured; the apical margins of the segments rufo-testaceous. Female, length 2¾ lines.

Hab.—Sind valley, Káshmir. Taken in August.

The genus *Prosopis* has a wide geographical distribution. Species occur both in the Old and New World; in Europe not less than forty species are found. The genus also occurs in Egypt, at Natal, and in the Cape of Good Hope; in Australia it is plentiful, and it has been found in New Zealand. In the New World it appears to be most plentiful in the United States, Mexico and California; but in tropical localities only two or three species have, to my knowledge, been discovered.

The habits of these bees, as far as those of the European species have been observed, are to form their burrows in dead sticks, in the pith of which they excavate their tunnels.

2. LAMPROCOLLETES PEREGRINUS, Fig. 5.

Femina.—*L. capite thoraceque nigris, abdomine chalybeo, alis fusco-hyalinis.*

Head and thorax black; the face with silvery white pubescence, as well as the thorax anteriorly and posteriorly above; the head beneath, a band between the wings, and the legs covered with black pubescence; wings fusco-hyaline, the anterior wings darkest, and having a violet iridescence; at their base they are subhyaline. Abdomen ovate and of a dark shining steel-blue; the apical segments with black pubescence. Female, length 5 to 5½ lines.
Hab.—Yangihissár, Eastern Turkestan. Taken in April.

All the species of this genus, previous to that here described, have been from Australia; about twenty have been described.

3. ANDRENA FAMILIARIS, Fig. 3.

Mas.—*A. atra, pallide villosa, abdominis segmentis tribus basalibus pallide rufo-marginatis.*

Black; the face and cheeks with a pale fulvous pubescence; the flagellum of the antennæ fusco-ferruginous beneath; the thorax has a similar pubescence; the coxæ and femora rufo-piceous; the tibiæ and tarsi pale testaceous yellow; the wings hyaline and iridescent; the nervures and the stigma pale rufo-testaceous. Abdomen oblong-ovate, with a thin short pale pubescence; the apical margins of all the segments testaceous, those of the three basal segments pale ferruginous; beneath entirely pale rufo-testaceous. Male, length 4½ lines.
Hab.—Neighbourhood of Yárkand. Taken in May.

4. ANDRENA FLORIDULA,[1] Fig. 4.

Mas.—*A. nigra, cinerascenti-pilosa, tibiis tarsisque posticis fulvis; abdomine nitido, segmentorum marginibus testaceis.*

Black; the head, thorax and legs with long thin cinereous pubescence; the flagellum of the antennæ fulvous beneath; the mandibles ferruginous at their apex. Thorax; the wings hyaline and iridescent; the nervures and stigma rufo-testaceous; the legs dark rufo-piceous the posterior tibiæ and tarsi fulvous, and clothed with silvery pubescence. Abdomen oblong; ovate; the apical margins of the segments testaceous, and with a thin fringe of whitish pubescence; the apex fulvo-testaceous. Male, length 4 lines.
Hab.—Drás, Kárgil, and Leh, all in Ladák; August to September.

This is a genus the geographical distribution of which is very extensive; the species are found in all parts of Europe, ranging north into Lapland. Numerous species have been found in

[1] This name is misprinted *floricula* on the plate.

the Azores, in Madeira, Cape de Verd Islands, and in Egypt; a few have occurred in Northern India, China, and Japan. In Australia and New Zealand they appear to be very rare; no species has, to my knowledge, been found in South America, but they are found in Mexico, and are plentiful in North America.

All the species appear to construct their nests in tunnels excavated in the ground.

Family—*APIDÆ*.

5. OSMIA LABORIOSA, Fig. 6.

Femina.—*O. nigra, pube fulva vestita ; abdominis segmentis rufo-marginatis ; pedibus ferrugineis.*

Black; head shining and finely punctured; the scape of the antennæ, the anterior margin of the clypeus, and the mandibles, ferruginous; the latter tridentate, the teeth black; the face and the vertex with fulvous pubescence. The thorax has a fulvous pubescence, which is usually more or less abraded above; the mesothorax closely punctured, more strongly so than the head; the legs bright ferruginous, with the coxæ black; wings fulvo-hyaline at their base, and fuscous beyond the base of the first submarginal cell; the tegulæ and nervures at the base of the wings ferruginous, becoming nigro-fuscous beyond. Abdomen closely punctured; the basal segment above, and the apical margins of the second and third segments broadly ferruginous; all the segments fringed with fulvous pubescence; beneath, densely clothed with fulvous pubescence. Female, length 4½ lines.

Hab.—Taken in May, in the neighbourhood of Yárkand.

This genus is numerous in species, but it appears only to be found, in any abundance, in temperate climates; nearly one hundred species are known, about half of these are European; several are found in North Africa, and they occur plentifully in North America.

6. MEGACHILE RESCINDUS.

Femina.—*M. pallide pubescens, abdomine subcordato, segmentorum marginibus pallide fulvis, subtus fulvo-villosis.*

Black; the face covered with fulvous pubescence, that on the cheeks paler; the mandibles with three ferruginous blunt teeth. Thorax; the pubescence on the disk short and thin, at the sides it is more dense, long and pale fulvous, that on the legs is very pale and glittering above; on the basal joint of the tarsi beneath it is bright fulvous; the claws of the tarsi ferruginous; wings sub-hyaline, the nervures fusco-ferruginous. Abdomen; a little pale fulvous pubescence on the apical margin of the basal segment; on the following segments it is fulvous, and very dense and bright on the segments beneath. Female, length 6 lines.

Male.—Black, with the anterior legs ferruginous; their coxæ armed with an acute black tooth; a dark stain on the femora and tibiæ behind, the tarsi dilated and fringed behind with very pale curled pubescence. The face covered with dense yellowish white pubescence; on the thorax above it is thinner, shorter and brighter; wings hyaline, with a faint cloud at their apical margin; the metathorax has a cinereous pubescence. Abdomen; the segments fringed with pale fulvous pubescence; the margin of the apical segment emarginate; the emargination denticulate. Length 6½ lines. In this species the head is a little wider than the thorax, and narrowed behind the eyes.

Hab.—Taken in May, in the neighbourhood of Yárkand.

7. MEGACHILE FULVA.

Femina.—*M. atra, thorace abdomineque hirsutis fulvo-aureis, pedibus ferrugineis.*

Black; the mandibles, scape of antennæ, and the legs ferruginous; the coxæ, trochanters, and tips of the mandibles, black; densely covered with fulvous pubescence, sparingly so on the vertex and basal margins of the intermediate abdominal segments; the pubescence on the legs is short and thin; the anterior wings flavo-hyaline towards their base, beyond which they are fuscous; the nervures ferruginous at the base of the wings, beyond the stigma they are rufo-fuscous; the tegulæ ferruginous. Female, length 7½ lines.

Taken on the Yárkand Expedition; the precise locality not known.[1]

8. MEGACHILE DENTIVENTRIS.

Mas.—*M. pallide pubescens; abdominis apice denticulato; tarsis rufis.*

Black; the face densely covered with white pubescence, that on the cheeks is also white, but shorter and less dense; on the vertex it is pale fulvous; the mandibles stout and bidentate the teeth rufo-piceous, the apical one black at the tip. The thorax and legs with white pubescence; the anterior femora and tibiæ in front, and all the tarsi bright ferruginous; the anterior coxæ armed with a stout spine; wings hyaline, the anterior pair faintly clouded at their apical margin; the nervures ferruginous; the tegulæ black. Abdomen; the apical margins of the segments fringed with very pale fulvous-white pubescence; that at the sides, and beneath, is white; the apical segment with four teeth on its margin. Male, length 5 lines.

Hab.—Neighbourhood of Yárkand. Taken in May.

9. MEGACHILE SERRATA.

Mas.—*M. pallide pubescens, abdomine oblongo, ano inflexo, spinuloso.*

Black; the face with pale fulvous pubescence, that on the cheeks cinereous; the antennæ fulvous beneath. Thorax clothed above with pale fulvous pubescence; at the sides, beneath, and on the legs, it is cinereous; the tarsi ferruginous, with the basal joint of the intermediate and posterior pairs, black above; the anterior coxæ dentate; wings hyaline, the anterior pair slightly clouded at the apex, the nervures ferruginous, the tegulæ black. Abdomen oblong, obtuse at the apex; the two basal segments with pale fulvous pubescence; the apical margins of the segments fringed with pale pubescence; the apical segment clothed with short pale pubescence, its margin serrated; beneath, the apical segment is produced into a large triangular process, acute at its apex. Male. Length 5 lines.

Hab.—Neighbourhood of Yárkand. Taken in May.

[1] Probably, like some other specimens without labels, this may have been from the hills south of Yárkand.

10. MEGACHILE VIGILANS.

Femina.—*M. pallide pubescens; abdomine subtus argenteo-villosulo, segmentorum marginibus dorsalibus pallido-fasciatis.*

Black; the face clothed with dense white pubescence; the mandibles with four blunt teeth. The pubescence on the thorax and legs whitish, that on the tarsi beneath fulvous; wings hyaline, the nervures black. Abdomen subcordate, the basal segment deeply concave, the metathorax rounded and fitting into the cavity, the abdomen curving upwards, the apical margins of the segments with fasciæ of white pubescence; beneath, densely clothed with silvery-white pubescence. Female, length 4½ lines.

Hab.—Drás, Kárgil, and Leh, all in Ladák. (August and September.)

This genus is perhaps the most numerous in species of all the genera of bees; it is also the most cosmopolitan; about three hundred species are known; they occur both in temperate and tropical climates; about fifty are known to inhabit India, China, and the islands of the Eastern Archipelago; they are abundant in Australia, also in both North and South America.

A large number are, from their habit of lining their nests with pieces of leaf, popularly called leaf-cutting bees, but their habits vary; nests of Indian species prove that some species belong to the section of mason-bees, their nests being constructed of agglutinated particles of sand or mud; of the habits of the Australian species, we are at present ignorant.

11. ANTHIDIUM VIGILANS, Fig. 7.

Femina.—*A. atrum, capite thoraceque flavo-variegatis; abdominis maculis lateralibus flavis.* Mas.—*A. abdominis inflexi lateribus fasciculato-pilosis, ano septemdentato.*

Black; the head and thorax very closely punctured and subopaque, the abdomen shining and more finely punctured. The clypeus, base of the mandibles, and a line on the posterior margin of the vertex, interrupted in the middle, yellow. A stripe on each side of the thorax in front, and an interrupted line on the posterior margin of the scutellum yellow; the femora at their apex beneath, and the tibiæ and tarsi outside, yellow; wings sub-hyaline, the marginal cell with a fuscous stripe at its anterior margin. Abdomen; each segment with a transverse yellow lateral macula; beneath, clothed with bright pale fulvous pubescence. Female, length 5¼ lines.

The male is considerably larger than the female and is much more pubescent, but is marked with yellow in the same manner, the yellow stripes on the abdomen being broader and forming interrupted bands; the segments have at their lateral margins a floccus of whitish glittering pubescence; the apical segment is tridentate, the lateral teeth yellow, the central one smaller and black; there is also a tooth at the lateral margins of the fifth and sixth segments. Male, length 7 lines.

Hab.—The locality of the male is the neighbourhood of Yárkand, and although the precise locality of the female is not ascertained, there is a sufficient general resemblance between the sexes to justify uniting them as one species.

The genus *Anthidium* has a wide geographical distribution; species are found in Europe, Arabia, Syria, Algeria, Cape of Good Hope, Sierra Leone. About six species are known from

India, but I have not seen any from China, the islands of the Eastern Archipelago, nor from Australia; the known number of species is about one hundred.

12. CROCISA HISTRIO, Fabr.

Nomada histrio, Fabr., Ent. Syst., ii, 345.
Melecta histrio, Latr., Hist. Nat. Crust. et Ins., iii.
—— Fabr., Syst. Piez., 385.
—— Spin., Ins. Ligur., i, 153.
Crocisa histrio, Latr., Gen. Crust. et Ins. iv, 172.
—— St. Farg., Hym., ii, 454.
—— Eversm., Bull. Mosc., xxv, 104.

Hab.—Taken in the neighbourhood of Yárkand, also at Yangihissár. Found also in Southern France, Russia and Algeria.

13. CROCISA INTRUDENS, Fig. 8.

Femina.—*C. nigra, capite, thoraceque, pedibusque albo variegatis, scutello emarginato.*

Black; the face and cheeks covered with dense snow-white pubescence, on the vertex it is shorter and thinner; the clypeus porrect, the mandibles ferruginous. The anterior margin and sides of the thorax covered with white pubescence; the mesothorax with three spots anteriorly, the central one oblong, the other two ovate; also two quadrate spots posteriorly, of white pubescence; the scutellum deeply emarginate and having a little white pubescence in the emargination; the tibiae white outside; wings hyaline, their apical margins slightly clouded. Abdomen, a broad band of white pubescence at the base and a narrower one on the apical margin of the segments, all slightly interrupted in the middle. Female, length 3½ lines.

Hab.—Neighbourhood of Yárkand. Taken in May.

This genus of bees is not numerous in species; only about twelve are at present known, but their distribution is extensive; they have occurred in Europe, North Africa, Natal, South Africa, Ceylon, India, in various islands of the Eastern Archipelago, China, and Australia.

14. ANTHOPHORA VIGILANS.

Mas.—*A. nigra, pallide villosa, thorace flavescente, abdominis segmentis pallido-marginatis.*

Black; the mandibles, labrum, anterior margin of the clypeus, and a central line uniting with it, white; the tips of the mandibles rufo-piceous, and two minute black spots at the base of the labrum; the pubescence on the face yellowish white, the thorax with similar pubescence above; the posterior tibiae and tarsi with white pubescence. Abdomen; the apical margins of the segments pale testaceous and having fasciae of white pubescence. The wings clear hyaline. Male, length 5 lines.

Hab.—Drás, Kúrgil, and Leh, all in Ladák.

HYMENOPTERA. 7

15. ANTHOPHORA SENEX.

Mas.—*A. atra, pallide villosa, facie antice labroque flavis; pedibus intermediis elongatis.*

Black; the face as high as the insertion of the antennæ, the labrum, and scape of the antennæ in front, yellow; the pubescence on the head pale fulvous, whitish on the clypeus. Thorax pubescent; the pubescence faintly yellowish, that on the legs long and ragged; the tarsi testaceous, except the basal joint; the intermediate legs elongate, the fifth joint densely fringed with black pubescence, forming a thick brush; the apical joint of the tarsi rufo-piceous. Abdomen thinly covered with pale pubescence; the margins of the segments pale testaceous. Male, length 5¼ lines.

Hab.—Neighbourhood of Yárkand and Yangihissár. Taken in April.

This genus is cosmopolitan, not less than one hundred and fifty species are known; of these twelve are from India.

16. XYLOCOPA NITIDIVENTRIS,[1] Fig. 10.

Femina.—*X. nigra, thorace supra pube flava decorata, tibiarum posticarum apicibus tarsisque omnibus pube ferruginea vestitis, alis nigro-fuscis iridescentibus.*

Black; the pubescence on the head black; very closely punctured and opaque. Thorax, clothed above with bright pale yellowish pubescence; on the sides, beneath, and on the femora and tibiæ, it is black; that at the apex of the posterior tibiæ and on all the tarsi, bright fulvo-ferruginous; that on the anterior tarsi mixed with a little black on the first joint; wings fuscous, palest towards their base, and having a violet iridescence in certain lights. Abdomen very smooth and shining, and also very convex, being subglobose; beneath, the apical margins of the segments are narrowly ferruginous, the two sub-apical ones being fringed with bright ferruginous hairs. Female, length 7¼ lines.

Hab.—Taken in May, in the neighbourhood of Yárkand.

17. XYLOCOPA DUBIOSA, Fig. 9.

Mas.—*X. nigra, fulvo-pubescente; facie antice labioque flavis; abdomine ovato, convexiusculo, segmentorum marginibus fulvo fasciatis.*

Black; the head, thorax, and the base of the abdomen densely clothed with fulvous pubescence; the face below the insertion of the antennæ and the labrum, yellow; the anterior margin of the latter black; the tarsi ferruginous; wings fulvo-hyaline, slightly fuscous towards their apical margins; the nervures ferruginous, the costal nervure blackish. Abdomen ovate, truncate at the base, convex, shining and finely punctured; the apical margins of the segments with a narrow fringe of fulvous pubescence, more or less interrupted in the middle; the sixth and seventh segments covered with fulvous pubescence. Male, length 7¼ lines. Probably the male of *X. nitidiventris*.

Taken in April, at Yangihissár, Eastern Turkestan.

[1] Represented on the plate as *X. dubiosa*. ♀

18. XYLOCOPA CONVEXA.

Femina.—*X. nitida, nigra ; alis nigro-fuscis violaceo splendide micantibus ; abdomine convexo.*

Black and shining; the head not closely but rather finely punctured; the front with short dense black pubescence; the margins of the clypeus raised and shining, and with a central shining carina; the flagellum, except the basal joint, obscurely fulvous beneath. The disk of the mesothorax and the base of the scutellum smooth, shining, and impunctate; the sides, beneath, and the legs, with black pubescence; wings with a beautiful blue, violet and green iridescence. Abdomen very convex, with fine distant punctures; the sides and apex with black pubescence. Female, length 9 to 10 lines.

Hab.—Kugiar, 90 miles south of Yárkand. Taken in May and June.

Xylocopa has an universal distribution; in my monograph of the genus, published in 1874, one hundred and twenty-three species are registered; the number has been slightly increased since that time. In India twenty species have been found.

Division 2—SOCIALES.

19. BOMBUS VALLESTRIS.

Operaria.—*B. hirsutus, ater, thorace supra abdominisque fascia basali flavis, segmento secundo et apice ferrugineo-fulvis.*

Black; the head sub-rotundate, with black pubescence; the thorax above, the sides, and beneath the wings, densely clothed with pale yellowish-white pubescence; beneath and on the legs it is black; the four apical joints of the tarsi ferruginous; wings fuscous, the nervures black. Abdomen; at the extreme base a fringe of pale yellowish white pubescence, on the second segment and also on the three apical ones it is ferruginous. Worker, length 6¼ lines.

Hab.—Sind valley, Káshmir. Taken in August.

The male exactly resembles the worker, having also black pubescence on the face.

20. BOMBUS LONGICEPS.

Operaria.—*B. hirsutus, ater, thorace dorso pallide fulvo ; abdominis segmentis analibus tribus rufo-fulvis.*

Black; the head elongate, the clypeus smooth and shining; the tips of the mandibles rufo-piceous; the flagellum of the antennæ obscurely fulvous beneath. Thorax above, and at the sides beneath the wings, densely clothed with bright fulvous pubescence; beneath, and on the legs, it is black; the four apical joints of the tarsi ferruginous; the wings hyaline. Abdomen; the three basal segments with black pubescence, usually more or less abraded towards the base, which is smooth and shining; the apical segments fulvo-ferruginous. Worker, length 9 lines.

Hab.—Drás, Kárgil, and Leh, Ladák. (August and September.)

The male of this species is clothed exactly the same as the female, the face having also black pubescence.

HYMENOPTERA.

21. BOMBUS ALTAICUS.

Bombus altaicus, Eversm. Bull, Mosc., xix, 436, tab. 4, fig. 1.

Hab.—Tankse, Pangkong valley, Ladák; also in Asiatic Russia.

22. BOMBUS BIZONATUS.

Femina.—*B. hirsutus, ater; thorace antice, scutello, abdomineque basi flavis, medio nigro-fasciato, apice pallido.*

The head clothed with black pubescence, the labrum fringed with fulvous; the clypeus naked, smooth and shining. Thorax clothed with pale fulvous pubescence and having a black pubescent band between the wings; the pubescence on the legs and on the body beneath, black; the apical joints of the tarsi with short pale pubescence, that on the basal joint beneath rich fulvous; wings sub-hyaline, the nervures black. The first and second segments of the abdomen with pale fulvous pubescence, the third with black, and the three apical ones with very pale fulvous. Female, length 7¼ lines.

Hab.—No locality indicated.

23. BOMBUS OPPOSITUS, Fig. 11.

Femina.—*B. hirsutus, ater; thorace abdomineque supra fulvis.*

Black; the head clothed with black pubescence; the clypeus naked, smooth and shining. Thorax, clothed above with rather short, rich fulvous pubescence, that on the sides, beneath, and on the legs is black; the pubescence on the apical joints of the tarsi is pale fulvous, on their basal joint within it is bright fulvous, outside it is black; wings subhyaline, their nervures black. Abdomen, clothed above with rich fulvous, beneath with black pubescence. Female, length 9 lines.

Hab.—No precise locality indicated.

The genus *Bombus* is widely distributed; its number of species amounting to little short of one hundred. *Bombi* are found both in the Old and New World, a few species occurring in the Tropics; the genus has not been observed to penetrate Africa beyond Algeria, and it has not been found either in Madagascar, Australia or New Zealand. In the Old World it has been found in Lapland, Siberia, Kamtschatka, China, Japan, India and Java. In great Britain twenty species occur. In the New World, it has been found in Greenland, Boothia Felix, and at the Great Bear Lake, within the Arctic Circle. Mexico has produced some of the most beautiful species of the genus; North America is rich in species; in South America several fine ones occur.

Tribe—HETEROGYNA.

Family—*FORMICIDÆ.*

24. CAMPONOTUS BASALIS, Fig. 1.

Femina.—*C. niger, thorace subtus, pedibus abdominisque basi castaneo-rufis.*

Shining black; the anterior margin of the clypeus and the mandibles rufo-piceous; the flagellum of the antennæ ferruginous. Thorax ovate; the mesothorax and scutellum dark rufo-fuscous; beneath, the pro-and metathorax and also the legs castaneo-rufous. Abdomen, subglobose; the scale and petiole, and the two basal segments castaneo-rufous, their apical margins black, the second most broadly so; beneath, these segments are entirely castaneous. Female, length 5 lines.

Hab.—Sind valley, Káshmir. Taken in August.

25. CAMPONOTUS BACCHUS.

Femina.—*C. capite abdomineque nigris, metathorace pedibusque pallide ferrugineis.*

Formica Bacchus, Smith, Cat. Hym. Ins., Pt. VI, Formicidæ, p. 21. ♀
Componotus Bacchus, Mayr, Novaræ Voy., Form., p. 27.
———— *fervens,* Mayr, lib. cit., nec Smith, Cat. Hym. Ins., p. 241.

In the same bottle in spirit a male ant accompanied the female; its size and general appearance justify me in considering it to be the male of *C. Bacchus*; it is 4½ lines long, black, with the antennæ and legs pale furruginous, the scape being darker than the flagellum; the mandibles, palpi and post-scutellum are also pale ferruginous; the wings flavo-hyaline; the nervures pale rufo-testaceous; the stigma fuscous.

Hab.—Jhilam Valley, Punjab Hills. Ceylon, Calcutta, and Islands of the Eastern Archipelago.

This genus is cosmopolitan; its species are numerous, new kinds occurring in almost every collection made in little frequented places; any attempt to calculate the number of species would be an impossibility; until each species has been collected from its nest, and all the different kinds of sexes carefully ascertained, the number of specific forms cannot be ascertained; workers of several sizes and forms occur in nests of many species, and if captured at large, are doubtless described as distinct; the number of species doubtless amounts to hundreds.

26. FORMICA SIMULATA.

Operaria.—*F. rufo-ferruginea, lævissime cinereo-micans; fronte vertice et abdomine nigrofuscis; squama subtriangulariter rotundata, margine supero rotundato; area frontali opaca.*

The head red, with the vertex and the front, as far as the insertion of the antennæ, more or less rufo-fuscous; the clypeus with a longitudinal sharp carina in the middle; its anterior margin rounded and entire; mandibles ferruginous, with their teeth black; the antennæ ferruginous, with the flagellum, beyond the third joint, fuscous. Thorax and legs bright blood-red. Abdomen covered with fine cinereous pile; in some examples fusco-ferruginous at the base; with a few scattered pale setæ; the scale of the peduncle red, its superior margin rounded. Worker, length 3 lines.

Hab.—On the road across the Pámir, from Sirikol to Panja; also at Yárkand; April and May.

This species closely resembles the *Formica sanguinea* of Europe, particularly small workers of that species; the specimens were collected in spirit; therefore it is probable the entire insect would, when living, be covered with a fine pile.

27. FORMICA FRATERNA.

Operaria.—*F. rufo-fusca, sparse pilosula; mandibulis, antennarum scapis, flagellorum basi, pedibusque pallide rufescentibus; area frontali opaca.*

The insect covered with a fine grey pile; the anterior part of the head rufo-testaceous, the mandibles ferruginous; the scape of the antennæ and a few of the basal joints of the flagellum pale ferruginous; the legs and scale of the peduncle entirely of that colour; the scale rounded above. The base of the abdomen in some examples more or less tinged with ferruginous; the abdomen with a few scattered pale setæ. Worker, length $2\frac{3}{4}$ to 3 lines.

Hab.—No precise locality indicated.

This species very closely resembles the *Formica cunicularia* of Europe, but its pale legs give it a different aspect.

28. FORMICA DEFENSOR.

Operaria.—*F. rufo-ferruginea, lævissime cinereo-micans; fronte abdomineque fusco-nigris; squama subtriangulariter, margine supero rotundato.*

Head, thorax, legs, antennæ, and squama rufo-ferruginous; the apical half of the flagellum and the front above the insertion of the antennæ, fuscous; an impressed line from the anterior ocellus to the clypeus, the latter with a sharp central carina; the frontal area opaque; the teeth of the mandibles black; the head and the flagellum with fine cinereous pile.

The abdomen fusco-ferruginous at the base, and the extreme apex pale ferruginous; the abdomen covered with a fine cinereous pile; beneath fusco-ferruginous. Worker, length $2\frac{3}{4}$ lines.

Hab.—On the road across the Pámir, from Sarikol to Panja. (April and May.)

29. FORMICA CANDIDA.

Femina.—*F. nigra nitidissima; mandibulis, antennis, pedibusque rufescentibus; squama lata, subtriangulariter, margine supra rotundato.*

Shining black; the mandibles, antennæ, and legs ferruginous; the latter slightly fuscous above, as are also a few of the apical joints of the flagellum above. The head, the width of the thorax; the clypeus with a sharp central carina; the frontal area semiopaque; the mandibles stout, and with longitudinal punctures. The metathorax semiopaque, and with a fine cinereous pile, which also covers the squama, legs, and thorax on the sides and beneath. Abdomen oblong-ovate, very smooth and shining, and with a sprinkling of pale hairs at the apex; the extreme apex pale testaceous. Female, length $3\frac{1}{2}$ lines.

Hab.—On the road across the Pámir, from Sarikol to Panja. (April and May.)

Family—MYRMICIDÆ, (Sm.)

30. MYRMICA CURSOR.

Femina.—*M. sordide rubra; capite thoraceque longitudinaliter striatis; abdomine fusco-nigro, nitido; mandibulis, antennis, pedibusque pallide ferrugineis.*

Obscure ferruginous, with the head sometimes nearly black, or with the sides more or less ferruginous; the antennæ with the scape and a few of the basal joints of the flagellum pale ferruginous, the rest fusco-ferruginous; the head longitudinally but irregularly striated, the striæ at its sides formed of confluent punctures. Thorax longitudinally striated, oblong-ovate, and having a longitudinal ferruginous space above, enclosed by a black margin; the sides and beneath ferruginous; the metathorax with two short, stout, acute, compressed spines; the legs pale ferruginous. Abdomen globose, smooth and shining; the first node of the petiole wedge-shaped when viewed sideways, and coarsely rugose, the second node globose and sub-rugose. The scape of the antennæ in this species is bent and slender at the base as in the *M. ruginodis* of Europe, which it closely resembles; it is, however, a rather smaller insect. Female, length 2¾ lines.

Hab.—No precise locality indicated.

31. MYRMICA LUCTUOSA.

Mas.—*M. niger, nitidiusculus; mandibulis, antennis, pedibusque, necnon capite thorace-que, sordide pallescentibus; alis hyalinis, nervis rufo-pallidis.*

Black; head and thorax longitudinally striated; the metathorax with transverse curved striæ, and with two stout compressed spines, its apex obscurely rufo-piceous; the club of the antennæ paler than the rest of the antennæ. The wings pale fulvo-hyaline and iridescent; the stigma and nervures pale ferruginous; the legs long and slender, with the apical joints of the tarsi pale testaceous. Abdomen smooth and shining, and with a few scattered pale hairs; the nodes of the abdomen rugose, the first oblong, the second globose. Male, length 2¾ lines.

Hab.—Murree (Mari), Punjab hills.

32. MYRMICA BREVICEPS.

Femina.—*M. sordide rubida; mandibulis, antennis, pedibusque pallide ferrugineis; capite thoraceque longitudinaliter profunde striatis; abdomine rufo-nigro, nitido.*

Rufo-ferruginous; the head strongly longitudinally striated; a small smooth shining space at the base of the clypeus, which is deeply longitudinally grooved; the mandibles striated. Thorax above with a black patch on each side of the mesothorax, and another at the anterior portion; the metathoracic spines short, stout and acute. The first node of the abdomen longitudinally rugose, the second transversely so; the abdomen smooth and shining, ferruginous at the base; the rest dark rufo-fuscous, nearly black, and with scattered erect pubescence. Female, length 2½ lines.

Hab.—No locality or date.

This insect very closely resembles two or three of the British species, particularly *Myrmica sulcinodis*; its head is, however, shorter than that of the British insect.

33. CREMATOGASTER APICALIS.

. Femina.— *C. pallide castaneo-rufus, lævis, nitidus; abdomine apicem versus nigrescente; alis hyalinis.*

Rufo-castaneous, smooth, shining and impunctate; the eyes, margins of the mandibles, and the anterior margin of the clypeus narrowly black; the clypeus finely longitudinally striated. The mesothorax with a few longitudinal dark lines; the scutellum convex, and rounded behind; the metathorax truncate, not spined, and paler than the mesothorax; wings clear hyaline, the nervures and stigma pale rufo-testaceous. Abdomen oblong-ovate, the base castaneo-rufous, from thence becoming gradually darker to the apex. Female, length 3½ lines.

Hab.—Jhilam valley, Punjab hills. Taken in July.

34. DORYLUS (TYPHLOPONE) LÆVICEPS, Fig. 2.

Operaria.—*T. rufo-testaceus, capite castaneo, mandibulis nigris.*

Head oblong, subquadrate, rather wider anteriorly, the flagellum slightly fuscous; very smooth and shining, and with a very faintly impressed central longitudinal line, and also a few delicate scattered punctures. Thorax oblong, divided in the middle by a transverse suture, flattened above and having a few fine punctures. Abdomen, the node of the peduncle incrassate, subquadrate above; oblong-ovate. Worker, length 3¼ lines.

Hab.—Jhilam valley. Taken in July.

The insects described under the generic name *Typhlopone* of Westwood are now discovered to be the workers of *Dorylus*; the late Dr. Jerdon observed them issuing from the nests, in company with males of *Dorylus*; Dr. Gerstaecker has described the female of *Dorylus* in the Stett. Ent. Zeits. for 1863, under the generic name *Dichthadia*, pointing out its affinities of structure with those of the male, upon which the genus was founded; and proving demonstratively the affinities of the genera.

35. MUTILLA SUSPICIOSA.

Mutilla suspiciosa, Smith, Journ. Proc. Linn. Soc., II, 84 ♂ (1857).

Hab.—Jhilam valley. Taken in July.

This species resembles the male of *M. sexmaculata*, but it differs from it by having the wings entirely dark brownish-black; the abdomen is red, with the base and apex black; it has been found in Borneo, Batchian, Celebes, Amboyna, and Bouru.

36. MUTILLA SEXMACULATA, Swederus.

Femina.—*M. nigra, thorace supra rufo, abdomine utrinque serie trium macularum albidarum.*

Mas.—*M. nigra, thorace antice cinereo, abdomine rufo, basi nigro, alis fuscis basi hyalinis.*

The male of this species was discovered by Sir John Hearsay, who captured the sexes *in coitu*; it, as well as the female, varies greatly in size. I suspect it is the *Mutilla rufogastra* of St. Fargeau.

M. Radoszkovsky, in his Monograph of the *Mutillidæ* of the Old World, says that the insect I have named *M. sexmaculata* of Swederus is not that species; I have carefully compared the insect with the descriptions given both by Swederus and by M. Radoszkovsky, and am quite satisfied that my quotation is correct, although it is said by the above author to be synonymous with a species described by him under the name *M. tretraops*, with the description of which it in no way agrees.

Tribe—FOSSORES, (Latr.)

Family—*SCOLIADÆ*, (Leach).

37. SCOLIA HAEMORRHOIDALIS.

Scolia haemorrhoidalis, Fabr., Ent. Syst. ii, 230, Syst. Piez., 240.
„ „ Klug, Weber & Mohr, Beitr. i, 24.
„ „ Spin., Ins. Ligur., i, 74.
„ „ Latr., Gen. Crust. et Ins., iv, 105.
„ „ Vand., Lind. Hym., Eur., 18.
„ „ Brullé, Exped. Sc. de Morée, iii, 370.
„ „ St. Farg., Hym., iii, 522.
„ „ Burm., Mon. Scolia, 18.
„ „ Smith, Cat. Hym. Ins., Pt. iii, 110.
„ „ Sauss. et Sich., Cat. des Espèc. Scol., 50.

Hab.—France, Spain, Albania, Hungary, Russia, Siberia, Asia Minor, India. The locality of Dr. Stoliczka's specimens has not been recorded.

This insect belongs to the division of the genus in which the anterior wings have three submarginal cells and one recurrent nervure, and is a " Triscolia" of Saussure.

Family—*POMPILIDÆ*.

38. POMPILUS ARROGANS.

Femina.—*P. ater, abdominis basi rufo-ferruginea, alis anticis fuscis.*

Black; the head subopaque and finely longitudinally rugulose; the anterior margin of the clypeus slightly rounded, the mandibles rufo-piceous at their apex. Thorax slightly shining above, and having a thin grey sericeous pile; the metathorax concave in the middle posteriorly; the anterior wings fuscous, palest at their base; the posterior pair hyaline, with their apex slightly clouded; the second submarginal cell subquadrate, the third longer than the second and slightly narrowed towards the marginal cell; the tarsi thickly spinose, the tibiæ with a few scattered spines; the tips of the claws of the tarsi ferruginous. Abdomen; the three basal segments ferruginous, the apical margin of the third and the following segments black. Female, length 6 lines.

Hab.—Drás, Kárgil, and Leh, in Ladák. Taken in August and September.

39. POMPILUS ATRIPES.

Femina.—*P. niger ; abdomine ferrugineo, apice nigro ; alis fuscis ; antennis crassis.*

Black; the head smooth and shining, very finely punctured; the antennæ much thicker in the middle than is usual in the genus. Thorax finely pilose, shining and finely punctured

above; the posterior margin of the prothorax angulated; the metathorax smooth and shining, with a central impressed line not quite extending to the apex; wings fuscous, clearer towards their base; legs entirely black, as are also the spines and calcaria that arm the tibiæ at their apex; tibiæ only very slightly spinose. Abdomen smooth and shining; the three basal segments ferruginous. Female, length 3¼ lines.
Hab.—Murree, Punjab hills.

40. POMPILUS DIVISUS.

Mas.—*P. niger, abdominis segmento secundo tertioque ferrugineis; alis subhyalinis.*

Black; the head and thorax slightly pubescent, shining and punctured; the antennæ obscurely fulvous beneath beyond the third or fourth joints. The posterior margin of the prothorax rounded; the metathorax, with four longitudinal carinæ, two lateral and two approximating in the middle, strongly punctured; wings fulvo-hyaline, the nervures ferruginous; the tibiæ and tarsi ferruginous, the former slightly fuscous above, as are also the apical joints of the posterior tarsi. Abdomen shining, the second and third segments ferruginous, and more distinctly punctured than the following ones, which are very smooth and shining; the apex rufo-fuscous. Male, length 4 lines.
Hab.—Sind valley, Káshmir. Taken in August.

41. PRIOCNEMIS RUFO-FEMORATUS.

Femina.—*P. niger, abdominis basi rufo; femoribus posticis rubris; alis apicibus fuscis, puncto albo ornatis.*

Black; the head slightly shining, and, as well as the mesothorax, very finely and very closely punctured; the metathorax with a central longitudinal impressed line not quite extending to the apex, and having a fine transverse striation; the wings fusco-hyaline, darkest in the middle of the anterior wings, which have beyond the third submarginal cell a large hyaline spot; the posterior margin of the prothorax angular; the posterior femora bright ferruginous, the tibiæ serrated exteriorly. Abdomen bright ferruginous to the apex of the third segment; the rest black. Female, length 4 lines.
Hab.—Drás, Kárgil, and Leh, in Ladák.
This species in general aspect exactly resembles *Priocnemis agilis* of Europe, but its transversely striated metathorax distinguishes it from that species.

42. MYGNIMIA ALECTO.

Femina.—*M. nigra, alis nigro-fuscis violaceoque splendide micantibus.*

Black; the abdomen shining, covered with a changeable violet and purple pile observable in certain lights; the clypeus emarginate; the mandibles shining, their apex rufo-piceous. Thorax; the posterior margin of the prothorax arched; the sides of the metathorax rounded, its apex obliquely truncate; the coxæ greatly enlarged; wings dark blackish-brown, with a purple gloss. Abdomen smooth and shining. Female, length 10 lines.

Hab.—Yangihissár, Eastern Turkestan. Taken in April.

The *Pompilidæ* are found in all parts of the globe; little short of five hundred species belonging to the various genera of which the family is composed have been enumerated.

Family—*SPHEGIDÆ*.

43. AMMOPHILA SPINIPES.

Femina.—*A. nigra, alis fulvo-hyalinis, metathorace rugoso, abdomine antice rufo.*

Black; the head wider than the thorax, shining and strongly punctured; the mandibles with a ferruginous spot in the middle, the palpi rufo-piceous. Thorax; the pro and mesothorax shining and strongly punctured, as well as the scutellum; the mesothorax with a deeply impressed line in the middle anteriorly, extending to the middle of the disk; the metathorax opaque and rugulose; wings fulvo-hyaline, the apical margins with a slight fuscous cloud; the nervures and stigma ferruginous; the tegulæ rufo-piceous; the tibiæ and tarsi thickly spinose, the claws of the tarsi ferruginous. Abdomen; the first, second, third and base of the fourth segment of the abdomen ferruginous; the petiole not quite as long as the first segment. Female, length 8 lines.

Hab.—Drás, Kárgil, and Leh, in Ladák. Taken in August and September.

This genus is one of universal distribution; about eighty species are described; some twenty species are found in Europe, and about the same number are at present known from Africa; twelve are described from India; North and South America both possess numerous species, but only three or four have been brought from Australia.

Family—*LARRIDÆ*.

44. LARRADA AURULENTA.

Sphex aurulenta, Fabr., Ent. Syst., ii, 213, ♀.
Pompilus auratus, Fabr., Ent. Syst. Supp., 250.
Liris aurata, Fabr., Syst. Piez., 228.
„   „   Dahlb., Hym. Europ., i, 135.
Tachytes opulenta, St. Farg., Hym., iii, 246.
Lyrops auratus, Guer., Icon. Reg. Anim., iii, 440.
Larrada aurulenta, Smith, Cat. Hym. Ins., pt. iv; Sphegidæ, Larridæ and Crabronidæ, p. 276.

Hab.—Yangihissár, in Eastern Turkestan; also Madras, Bombay (India); China; Philippines; Sumatra; Borneo; Java; Celebes; Bachian; Bouru.

This genus is widely distributed; Europe has five species at present known; others are found in India, China, Borneo, in most of the islands of the Eastern Archipelago, New Guinea and Australia; species also occur in Africa; others are found both in North and South America.

HYMENOPTERA. 17

Tribe—**DIPLOPTERA.**

Family—*EUMENIDÆ.*

45. PTEROCHILUS ALBO-FASCIATUS, Fig. 12.

Femina.—P. niger, capite thoraceque albo-variegatis; abdominis segmentis albo-marginatis.

Black; head as wide as the thorax, strongly and closely punctured; the clypeus porrect and rugosely punctate, its anterior margin truncate; the mandibles tridentate, the teeth obscurely rufo-piceous; a small subovate white spot above the eyes on the vertex. The prothorax white above; the mesothorax and scutellum coarsely punctured; a white line crosses the post-scutellum and curves up towards the insertion of the posterior wings, which are fusco-hyaline and darkest along the foreborder of the anterior pair; legs black. Abdomen subovate; the basal segment campanulate and having a broad white fascia on its apical margin, which is slightly notched in the middle; the second segment has also a broad fascia, which is curved up laterally towards the base of the segment; the three following segments have each a similar broad white fascia. Female, length 5¼ lines.

Hab.—Yangihissár, Eastern Turkestan. Taken in April.

This is a genus of small extent; about twenty species are described; it occurs in Europe, a single species being found in Russia, another in Germany, and one in Switzerland; species have been found in Algeria, the Cape of Good Hope, Chili, and the United States.

Family—*VESPIDÆ.*

46. POLISTES CHINENSIS, Fabr.

Polistes chinensis, Fabr., Syst. Piez., 270.
    ,,   ,,   Sauss., Mon. Guépes Soc., 56, tab. 7, fig. 2.

Hab.—Neighbourhood of Yárkand. Taken in May. Hong-Kong; Shanghai.

Specimens from Yárkand are frequently more marked with yellow spots than any I have seen from China, whilst others exactly resemble Chinese ones, which do not appear ever to have any yellow spots on the mesothorax, which the Indian examples usually have. The genus *Polistes* is cosmopolitan, but no species has been found in the United Kingdom; species occur in South Europe, in India, China, Borneo, in the Islands of the Eastern Archipelago, in Africa, Australia and Tasmania, Brazil, Central America, Mexico, and North America; from seventy to eighty species are known

47. VESPA GERMANICA.

Vespa germanica, Fabr., Syst. Piez., 256.
  ,, ,, Pauz., Faun. Germ., 49, 20, ♀ .
  ,, ,, St. Farg., Hym., i, 515.
  ,, ,, Ratz., Forst. Ins., Bd. iii, 51.
  ,, ,, Smith, Zool., ix., Append. clxxvii.
  ,, ,, Sauss., Mon. Guépes Soc., 116, pt. xiv, fig. 4.

Hab.—Europe. Found at Sanju, and in its neighbourhood; also at Yangihissár, Eastern Turkestan.

No specific difference can be found that will separate this insect from the European species; the sexual organs of the male are precisely the same as those of *V. germanica*. About fifty species of this genus are known; they are widely distributed, and are insects that are almost universally known. Twelve species inhabit Europe; twenty are Asiatic, and ten are found in Mexico and North America.

Family—TENTHREDINIDÆ.

48. HYLOTOMA FUMIPENNIS.

Femina.—*A. corpore flavo; capite, antennis, tarsisque nigris; alis fuscis.*

Head shining black, antennæ pubescent; abdomen paler than the thorax; wings fuscous and iridescent, nervures and stigma blackish-brown; the legs pubescent, the coxæ and femora reddish-yellow, the tibiæ fuscous, the tarsi black. Female, length $3\frac{1}{4}$ lines.

Hab.—Jhilam valley, Punjab hills. Taken in August.

49. ALLANTUS PROVIDUS.

Femina.—*A. niger, capite thoraceque opacis, abdomine nitido; tibiis, femoribus posticis ferrugineis.*

Head and thorax opaque, and covered with strong confluent punctures; wings subhyaline, and having a fuscous cloud on the anterior pair that occupies the two marginal and four submarginal cells; the stigma and nervures black; the posterior femora and tibiæ pale ferruginous; the anterior femora towards their apex, and the tibiæ in front, pale; abdomen glossy black. Female, length 6 lines.

Hab.—Murree, Punjab hills.

50. ALLANTUS MULTICOLOR.

Femina.—*A. capite thoraceque cupreo-nigris, abdomine purpureo, corpore flavo-maculato.*

Head and thorax obscure brassy-black; abdomen dark purple above; the entire body yellow beneath, the legs black with yellow markings. Head semiopaque, the cheeks, clypeus, and mandibles testaceous-yellow; the teeth of the latter acute and black. Thorax semiopaque; the scutellum, tegulæ, and costal nervure, and the cell, yellow; the wings hyaline and iridescent; the stigma and nervures blackish-brown; the anterior and intermediate coxæ and femora yellow beneath; the tibiæ and tarsi yellow, the former with a dark stain at or near their apex; the posterior coxæ beneath and the basal half of the tibiæ yellow. Abdomen; the lateral margins of the second segment, the apical, as well as the lateral margins of the third segment, yellow; the sixth, seventh and eighth segments narrowly yellow. Female, length 6 lines.

Hab.—Murree, Punjab hills.

This species resembles *Allantus flavomaculatus* of Cameron, described in the " Transactions of the Entomological Society of London" for 1876, but it is rather smaller; the femora are black, and the abdomen has a purple tinge.

HYMENOPTERA. 19

51. ALLANTUS SIMILLIMUS.

Femina.—*A. niger, abdominis cingulo basali flavo, tibiis tarsisque pallide ferrugineis, alis flavo-hyalinis, antice fuscis.*

The scape of the antennæ, the following joint, the labrum, mandibles, and palpi pale yellow. Thorax; the posterior margin of the prothorax slightly interrupted in the middle, pale yellow; the wings pale fulvo-hyaline, a slight fuscous cloud occupying the marginal cell and extending a little beyond into the submarginal cells; the costa and stigma fulvous, the nervures ferruginous; the coxæ black, the legs pale ferruginous, with the femora black above. Abdomen with a whitish fascia at the base; the fourth and fifth segments with their apical margins narrowly whitish; the two apical segments with their margins more broadly yellowish white. Female, length 4¾ lines.

Hab.—Murree, Punjab hills.

The species resembles *Allantus trochanterinus* of Cameron, described in the "Transactions of the Entomological Society" for 1876, but it differs from it in having the scape pale, independently of other differences.

52. ALLANTUS TERMINALIS.

Mas.—*A. capite thoraceque nigris, abdomine purpureo nitido.*

Head emarginate behind, covered with confluent punctures and subopaque; the mandibles, labrum, and clypeus, also a triangular space above it, and a narrow line round the lower margins of the eyes, white; the antennæ fulvous beneath. Thorax black above and at the sides; beneath, the anterior and intermediate legs beneath and also the posterior coxæ beneath, white; wings hyaline, their nervures black. Abdomen purple above; the margins of the segments beneath white laterally. Male, length 4 lines.

Hab.—Sind valley, Káshmir. Taken in August.

53. MACROPHYA OPPOSITA.

Femina.—*M. nigra, capite thoraceque opacis, abdomine nitido, pedibus anticis pallidis.*

Black; the head wider than the thorax, covered with confluent punctures and opaque; the clypeus, mandibles, and labrum white; the palpi testaceous; the antennæ fulvous beneath. Thorax opaque and punctured, the same as the head; wings fuscous, with a violet iridescence, the nervures and stigma black; the anterior and intermediate tibiæ rufo-testaceous beneath. Abdomen shining black. Female, length 5 lines.

Hab.—Sind valley, Káshmir. Taken in August.

54. TENTHREDO SIMULATA.

Femina.—*T. nigra, abdominis medio, et tibiis tarsisque pallide ferrugineis ; alis hyalinis, nervuris stigmatibusque nigris.*

Black; the clypeus, labrum, and mandibles pale reddish-yellow; the teeth of the latter black; the antennæ obscurely fulvous beneath. Thorax; the wings hyaline and iridescent, the nervures and stigma black; the tibiæ and tarsi reddish-yellow, the coxæ spotted with yellow. Abdomen; the three basal segments black above; the four following segments reddish-yellow, the rest black. Female, length 5 lines.

Hab.—Sind valley, Káshmir. Taken in August.

55. TENTHREDO FALLAX.

Femina.—*T. luteo-testacea; vertice, antennarumque basi maculis nigris; alis hyalinis.*

Pale ochraceous-yellow; the abdomen with a slight silky gloss; a minute black spot on the scape, another on the basal joint of the flagellum, and an oblong black macula on each side of the ocelli on the vertex ; the wings hyaline and iridescent; the nervures fuscous, the stigma and costal nervure testaceous. Female, length 5 lines.

Hab.—Sind valley, Káshmir, and Murree, Punjab hills. Taken in August.

56. TENTHREDO NIGRO-MACULATUS.

Femina.—*T. lutea ; capitis vertice thoracisque maculis dorsalibus, necnon abdominis linea interrupta nigris ; alis hyalinis.*

The insect pale luteous; the antennæ black above, with the apex of the joints, the scape beneath, and the flagellum beneath beyond the third joint, luteous; the inner margin of the eyes near their summit, and a large spot on the vertex, enclosing the eyes, black. Thorax; a triangular black spot anteriorly, an oblong one on each side opposite the tegulæ, and an oblique one on each side behind them ; the wings hyaline with the nervures fuscous ; the costal nervures and the stigma pale testaceous ; the legs with a narrow black line above. The longitudinal black line on the abdomen above is slightly interrupted by a very narrow pale margin on each segment ; the entire insect pale luteous and immaculate beneath. Female, length 4 lines.

Hab.—Sind valley, Káshmir. Taken in August.

This insect closely resembles the *Tenthredo scalaris* of Europe, and probably, when living, is green marked with black ; the European insect changes, more or less, from green to pale yellow, particularly when collected in spirit.

Family.—*ICHNEUMONIDÆ.*

57. ICHNEUMON BIMACULATUS.

Femina.—*I. niger ; pedibus, abdominis segmentis tribus basalibus ferrugineis ; metathorace bimaculato ferrugineo.*

Black ; the basal half of the antennæ pale ferruginous ; a line on the face close to the inner margin of the eyes, which is narrow above the insertion of the antennæ, and which expands into a large macula below it, a transverse line on the clypeus, and the mandibles, ferruginous. Thorax ; two oblong spots on the mesothorax, a minute one on the scutellum, the tibiæ and tarsi, ferruginous ; the posterior tibiæ fuscous at their apex ;

wings subhyaline and iridescent; the nervures and stigma black. The three basal segments of the abdomen ferruginous, the rest shining black. Female, length 5 lines.
Hab.—Murree, Punjab hills.

58. CRYPTUS INSIDIATOR.

Femina.—*C. niger, abdominis basi femoribusque rufis, alis fusco-hyalinis.*

Head, antennæ, thorax, coxæ, trochanters, tibiæ and tarsi black; the femora and three basal segments of the abdomen ferruginous; the apical segments black, with a purple gloss; the wings fusco-hyaline, the anterior pair darkest; the metathorax truncate posteriorly, the margin of the truncation somewhat arched inwardly; the lateral angles acute, or subdentate. Female, length 7¼ lines; of the ovipositor, 4 lines.
Hab.—Sind valley, Káshmir. Taken in August.

59. CAMPOPLEX LONGIPES.

Femina.—*C. niger, abdomine pedibusque ferrugineis, scutello albo.*

Black; a line at the inner orbits of the eyes below the insertion of the antennæ; the clypeus and mandibles yellowish-white, the latter ferruginous at their apex; the antennæ ferruginous. The thorax opaque, with the scutellum white; wings hyaline and iridescent; the nervures and tegulæ ferruginous; the costal nervure fuscous; the legs ferruginous with the coxæ, and trochanters black; the apex of the posterior tibiæ fuscous. Abdomen ferruginous, curved downwards, and petiolated; a black spot on the petiolated segment near its apex; the ovipositor ferruginous. Female, length 5 lines.
Hab.—Sind valley, Káshmir. Taken in August.

60. PANISCUS UNICOLOR.

Femina.—*P. ferrugineus, antennarum apicibus fuscis.*

Ferruginous; the eyes, ocelli, and apical portion of the antennæ fuscous; the claws of the tarsi black; wings hyaline and iridescent; the costal nervure and stigma pale ferruginous, the other nervures ferruginous. Abdomen falcate, smooth, and shining. The areolet of anterior wings oblique and triangular. Female, length 5 lines.
Hab.—Murree, Punjab hills.

61. PANISCUS QUADRILINEATUS.

Femina.—*P. rufus, capite thorace dorso quadrilineato.*

Antennæ, thorax, and legs rufo-fulvous; the abdomen rufo-ferruginous. The head yellow, fulvous behind; eyes and ocelli black. Thorax smooth and shining, two longitudinal lines on the disk of the mesothorax and a line at the lateral margins, yellowish; the scutellum triangular, with its margins raised; wings hyaline and iridescent, the nervures ferruginous, the stigma yellow. Abdomen darker than the thorax, becoming fusco-ferruginous towards the apex. Female, length 7 lines.

Hab.—Neighbourhood of Yárkand. Taken in May.

The yellow markings are much brighter in some examples than in others, and the abdomen also varies in brightness.

62. OPHION DENTATUS.

Femina.—*O. rufus, metathorace lineis elevatis transversis, lateribusque unidentatis.*

Pale ferruginous; eyes, ocelli, and tips of the mandibles black; wings hyaline and iridescent; the mesothorax smooth and shining, the metathorax with two arcuate transverse carinæ, the second being at the margin of the posterior truncation and terminating laterally in an acute angle or tooth. Abdomen falcate, smooth, shining, and with a fine short sericeous pubescence, or pile. Female, length 8 lines.

Hab.—Sind valley, Káshmir. Taken in August.

63. OPHION ALBOPICTUS.

Femina.—*O. rufus, capite, scutello, lineis 4 dorsalibus maculisque lateralibus albis.*

Head white; eyes and ocelli black, the antennæ and mandibles, rufous. Thorax rufous; two central longitudinal lines, the lateral margins, the scutellum and numerous spots on the sides beneath the wings, white; the wings hyaline and iridescent, the nervures ferruginous, the stigma pale rufous. Abdomen rufous, with some pale whitish spots at the sides towards the apex; the two basal segments beneath white. Female, length 6 lines.

Hab.—Neighbourhood of Yárkand. Taken in May.

HYMENOPTERA.

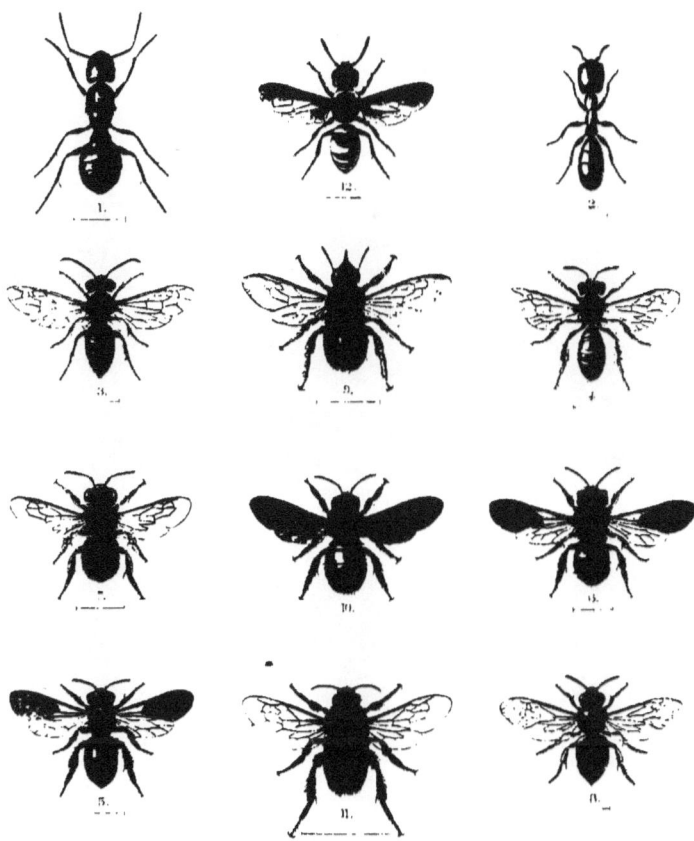

SCIENTIFIC RESULTS

OF

THE SECOND YARKAND MISSION;

BASED UPON THE COLLECTIONS AND NOTES

OF THE LATE

FERDINAND STOLICZKA, Ph.D.

NEUROPTERA.

BY

ROBERT McLACHLAN, F.R.S., F.L.S.

Published by order of the Government of India.

CALCUTTA:
OFFICE OF THE SUPERINTENDENT OF GOVERNMENT PRINTING.
1878.

CALCUTTA:
PRINTED BY THE SUPERINTENDENT OF GOVERNMENT PRINTING,
8, HASTINGS STREET.

SCIENTIFIC RESULTS

OF

THE SECOND YARKAND MISSION.

NEUROPTERA.
By ROBERT McLACHLAN, F.R.S., F.L.S.

PSEUDO-NEUROPTERA.
Family—*ODONATA*.
Sub-Fam.—LIBELLULINA.

LIBELLULA QUADRIMACULATA, L.

Two males of rather small size (expanse of wings 65—69 *mm.*), from Yárkand, 22nd May; indicated as "very common on the jheel (marsh)." Both pertain to the var. *prænubila*, Newman, in which the apex of the wings has a fuscous spot or band; also 1 male and 1 female of the typical form from Yangihissar, in April.

The insect is spread over all the temperate and cold regions of the Northern Hemisphere, and is occasionally of migratory habits.

Sub-Fam.—ÆSCHNINA.

ANAX PARTHENOPE, De Selys.

One ♀ from Srinagar, 28th July.

A widely-distributed, but probably not very abundant, species. In Europe it extends northward to Paris, and is found also in Central Germany, Austria, Hungary, Italy, Turkey, &c.; also in Asia Minor and Western Turkestan, and in Algeria.

This female has the wings tinged with smoky in the middle, as in the form from Algeria.

Sub-Fam.—AGRIONINA.

SYMPYCNA FUSCA, Van der Linden.

One ♀ from Yangihissar, 18th April.

Occurs also over the greater part of Europe (but not in the British Isles), and in Siberia, Asia Minor, Western Turkestan, Algeria and Morocco.

AGRION PULCHELLUM, Van der Linden.

5 ♂ and 4 ♀ from Yárkand, 22nd May.

Distributed over the greater part of Europe; occurs also in Asia Minor, Mingrelia, and Western Turkestan.

There is also (in spirits) a larva of some species of *Libellulidæ*, together with larvæ and 'nymphs' of a species of *Agrionidæ*, all from Yárkand, taken in November.

Family—*EPHEMERIDÆ*.

EPHEMERA, sp.

There is a fragment of a male imago of a species of this genus in spirits, from the Jhelum valley, not determinable.

Family—*PERLIDÆ*.

Of this family there are 3 males and 1 female of a large species of *Perla*, and four or five of a small species (with two ocelli) in spirits, from the Jhelum valley—from Kohala to Baramula; a small pinned *Perla* (nearly destroyed) from Tankse, Pankong valley, to Chagra, and a *Nemoura*, in spirits, from Murree, in the Punjab.

It is useless to attempt to describe these with any chance of success. The ♀ of the large *Perla* (in very bad condition) has a deep triangular notch on the margin of the egg-valve; the head and thorax without markings.

PLANIPENNIA.

Family—*MYRMELEONIDÆ*.

MYRMECÆLURUS PUNCTULATUS, Steven.

One ♀ from Leh, 6th September.

Occurs also in Hungary and South Russia. The example from Leh does not materially differ from others in my collection from Sarepta. Possibly the black markings on the head and thorax (always variable) are rather less pronounced.

Family—*CHRYSOPIDÆ*.

CHRYSOPA VULGARIS, Schneider.

One ♀ from Ighiz Yar, Eastern Turkestan, 18th May, appears to belong here.

It differs from ordinary examples in the dividing veinlet of the third cubital cellule in the anterior wings being interstitial; but this is a not infrequent aberration in European examples.

The species is of very wide distribution in Europe, and is also known from Asia Minor and Western Turkestan. I have seen individuals from the islands of Madeira and St. Helena that did not appear to differ.

CHRYSOPA BIPUNCTATA, Burmeister.

One example, either from Yárkand or Kugiar, appears to pertain to this Japanese species, which is probably nothing more than a local form of the common European *C. septempunctata*, Wesmaël.

CHRYSOPA, sp.

One ♂ from Karghalik near Yárkand, 29th May.

Allied to *C. vulgaris*, but distinct. It would be injudicious to describe it as new from this single example, especially as it belongs to a section of the genus in which the characters are so little obvious.

TRICHOPTERA.

Family—*LIMNOPHILIDÆ*.

STENOPHYLAX MICRAULAX, n. sp.

3 ♂, 4 ♀, Leh, in August.

Brownish-testaceous above, yellowish-testaceous beneath. Head small; eyes very prominent; ocelli very large, those of the disk encircled with fuscous, with which colour the anterior margins of the disk are bordered; hairs blackish; the posterior warts not prominent : on the face the raised lateral margins have two large, oval, prominent warts, furnished with blackish hairs, and there are four smaller warts forming the corners of a quadrangle on the median portion ; labrum very long (for the genus) ; maxillary palpi ordinary, the basal joint very short, the two others (♂) gradually clavate; a large and prominent triangular horny lobe at the base of the maxillæ; labial palpi small, the second joint broadly triangular. Antennæ rather shorter than the wings, moderately slender, testaceous, with rather broad, brownish annulations on the upper side. Pronotum well developed, its anterior edge semi-circular with a median excision ; the disk is concave, but the edges are thickened and raised and clothed with long fuscous hairs. Meso- and metanota broadly fuscous or blackish at the sides. Legs testaceous, moderately stout; spines deep black ; anterior and intermediate tibiæ with a conspicuous fuscous semi-annulation in the middle and at the apex externally; tarsal joints slightly fuscescent at the tips externally. Abdomen fuscescent above, testaceous beneath. Anterior wings broad, the apex elongately parabolic or elliptical : pale brownish-grey with numerous very indistinct paler spots, the membrane very finely granulose, with minute fuscous hairs ; the hairs on the neuration short and fine, fuscous; in the narrow area below the inferior branch of the upper cubitus, and in the post-costal basal cellules and area, are longer black hairs arranged somewhat in tufts, and at the extreme base are a few longer brown erect hairs ; the costal margin (in the ♂ only) near the base is turned under for a space of about 4 *mm.*, forming a deep narrow groove on the under side, filled with black hairs and conspicuously dark: neuration rather strong, testaceous ; radius sharply bent before its termination; discoidal cell extending to near the base of the wing, its upper edge nearly straight, the lower slightly curved; all the apical cellules broad at the base, the 2nd very broad, truncate, 4th also truncate, 3rd bi-angulate. Posterior wings pale greyish sub-hyaline, with sparse, minute blackish hairs on the membrane; neuration pale; discoidal cell rather shorter than in the anterior; 1st apical cellule much narrower than the 2nd to 4th, which are very broad; upper branch of cubitus furcating about on a level with the middle of the discoidal cell.

In the male the anal parts are arranged as follows :—The 8th dorsal segment is very large, testaceous, rather thickly clothed with long and strong testaceous hairs springing from small tubercles; viewed from above its margin appears to be straight, with several strong testaceous spines in the middle placed closely together; but viewed in front (or from beneath) the median portion is seen to be strongly turned under, forming a triangle, closely set with black tubercles. What appear to be superior appendages are band-shaped, flattened, slightly curved, truncate processes, little prominent, and inserted so far inferiorly as to cause a doubt as to their true value. (It may be that they represent the intermediate appendages, and that the superior are only indicated by a tubercle projecting slightly beyond the margin above them). The 8th ventral segment is rounded on the margin, and from it proceed two short, broad, excessively hairy lobes, divided by a suture, and each excised on its margin, apparently belonging to the 9th ventral segment; internally each of these lobes is very concave, and lying in them are what appear to be the very short obtuse inferior appendages. What appear to be the penis-sheaths are sub-cylindrical processes, curved strongly inward in a forcipate manner and nearly touching at the tips, which are somewhat thickened, blackish, and furnished with short spines. The penis lies between them, and is strong and rather short.

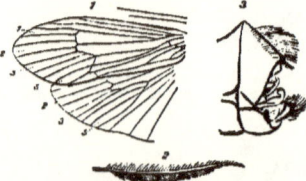

Stenophylax micraulax, McLachlan, male.

Fig. 1. Neuration of wings.
2. Groove in costa of anterior wings, more enlarged.
3. Apex of abdomen, from side.

In the female the apex of the abdomen is very obtuse. The 8th dorsal segment broad (concealing the 9th in the dry insect), its margin slightly rolled inwards, and fringed with yellow hairs; the 7th ventral segment forms a kind of pouch, the 8th with a concave space, 9th in the form of a short open tube.

Length of body ♂ 11—12 $mm.$, ♀ 12—13 $mm.$ Expanse ♂ 36 $mm.$, ♀ 44 $mm.$; greatest breadth of anterior wings ♂ 6½ $mm.$, ♀ 7 $mm.$

Pending the discovery of some method for satisfactorily dividing *Stenophylax*, this insect must be placed therein. It differs from any species known to me in the curious groove near the base of the costal margin in the anterior wings of the male, and also in the anal parts, which almost defy intelligible description.

The external aspect is somewhat intermediate between the groups of which the European *S. stellatus* and *S. concentricus* are representatives.

PLATYPHYLAX, n. sp.

Two females from the same locality as the last; it is useless to describe them without more examples in better condition, and of the other sex.

NEUROPTERA.

Family—*SERICOSTOMATIDÆ*.

DINARTHRUM INERME, n. sp.

10 ♂, 4 ♀, the latter in spirits, Loh, 7th September.

Male brownish, clothed with greyish-brown pubescence. Basal joint of antennæ rather longer than the head and entire thorax united, its basal portion black, but the apical portion brown; somewhat compressed laterally, nearly straight, but with a slight bend in the basal portion to about the middle, unarmed, but the basal half beneath has a very dense fringe of thickened black hairs; this portion above, and all the apical half, are furnished with long outstanding grey hairs: thread of the antennæ longer than the wings, pale-yellow, very distinctly annulated with brown up to the tips. Maxillary palpi long and slender, with a small terminal joint; the basal portion clothed with long and dense thickened[1] and ordinary grey hairs, intermingled (but with no short 'scales'), the terminal joint with ordinary hairs only. Labial palpi small and slender, pale-yellowish. Legs pale yellow. Anterior wings greyish; the costal margin for more than half its length from the base has a very dense inturned fringe of thickened blackish-grey hairs; the membrane lightly clothed with short greyish pubescence, and with numerous small, deep, black 'scales'; but there is a broad median longitudinal space free from 'scales,' limited inferiorly by a narrow groove extending from base to apex; apical fringes greyish, very long on the apical portion of the inner margin: neuration pale; discoidal cell short; nerves very irregular below the groove, forming large cellules. Posterior wings slightly paler than the anterior, with a few scattered black 'scales' on the costal portion, but otherwise with only slight and very short greyish pubescence; fringes very long and greyish; discoidal cell very short.

Dinarthrum inerme, McLachlan, male.

Fig. 1. Head, &c., from side.
2. Neuration of wings.

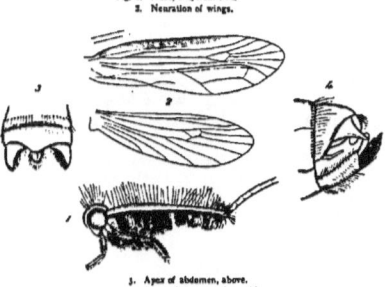

3. Apex of abdomen, above.
4. Apex of abdomen, from side.

The 9th dorsal segment of the abdomen rather broad, brown, its margin produced in a triangular form, fringed with yellowish hairs; from each side of it proceeds a large yellow triangular plate with the apex considerably produced and sub-acute, apparently connected with two yellowish median parts, little prominent, and separated one from the other. In-

[1] Under the microscope, with a high power, each of these thickened hairs has a peculiar rugose appearance.

ferior appendages long and stout, slightly curved, directed upward, yellow with concolorous hairs; at the apex is a dense brush of spiniform yellow hairs, perhaps concealing a smaller apical joint; from without this brush projects a flattened obtuse process, perhaps connected with the appendage, or perhaps distinct from it, and for its greater length lying in its concave inner side. Penis placed far internally, slender, slightly geniculate, yellow.

In the ♀ the neuration and palpi are regular, and in details almost precisely as in *D. pugnax* (*vide* my Revision and Synopsis of the *Trichoptera* of the European Fauna).

Length of body 5½—6 *mm*. Length of basal joint of antennæ 3 *mm*. Expanse 18—21 *mm*.

The genus *Dinarthrum* was established by me in the *Journal of the Linnæan Society, Zoology*, vol. xi, p. 116 (1871), for an insect from North India described as *D. ferox*, in which the extraordinary basal joint of the antennæ of the ♂ has a very strong basal tooth. Later on, in 1875, I described another species in the *Neuroptera* of Fedtschenko's Travels in Turkestan, page 30 (and more recently in Part V of my Monographic Revision and Synopsis of the *Trichoptera* of the European Fauna, page 279, pl. xxx, 1877), as *D. pugnax*, in which the said joint has two such teeth. In *D. inerme* there is no tooth. All the species bear considerable external resemblance one to the other, and are only separable by structural characters. The form is very curious, and as is usual in this section of *Sericostomatidæ*, the sexes differ greatly in appearance and structure: the nearest ally amongst true European insects is the genus *Lasiocephala*.

SUMMARY.

Only about 15 species of *Neuroptera* (in the broad sense) have been seen by me, *viz.*, four species of *Odonata* (dragon-flies), one of *Ephemeridæ*, three of *Perlidæ*, one of *Myrmeleonidæ*, three of *Chrysopidæ*, and three of *Trichoptera*.

The general aspect is European. All the *Odonata* are European, and two of them occur in Britain. The ant-lion (*Myrmecælurus*) is a species of Eastern Europe. The *Chrysopidæ* have nothing peculiar about them. The genus *Dinarthrum* in the *Trichoptera* was orginally founded on an Indian species, but I have since seen another species from Turkestan, so that the genus should probably be regarded as more Central Asian than Indian.

SCIENTIFIC RESULTS

OF

THE SECOND YARKAND MISSION;

BASED UPON THE COLLECTIONS AND NOTES

OF THE LATE

FERDINAND STOLICZKA, Ph.D.

RHYNCHOTA,

BY

W. L. DISTANT.

Published by order of the Government of India.

CALCUTTA:
OFFICE OF THE SUPERINTENDENT OF GOVERNMENT PRINTING.
1879.

SCIENTIFIC RESULTS

OF

THE SECOND YARKAND MISSION;

BASED UPON THE COLLECTIONS AND NOTES

OF THE LATE

FERDINAND STOLICZKA, Ph.D.

RHYNCHOTA,

BY

W. L. DISTANT.

Published by order of the Government of India.

CALCUTTA:
OFFICE OF THE SUPERINTENDENT OF GOVERNMENT PRINTING.
1879.

CALCUTTA:
PRINTED BY THE SUPERINTENDENT OF GOVERNMENT PRINTING,
8, HASTINGS STREET.

SCIENTIFIC RESULTS

OF

THE SECOND YARKAND MISSION.

RHYNCHOTA:

By W. L. DISTANT.

THE Hemiptera collected by Dr. Stoliczka, though not numerous, are interesting in the details of their geographical distribution, and conform, I believe, in that respect, to the other portions of the fauna of the districts traversed by the expedition. Two faunas are represented in the collection, one the Indian or Oriental, rather strongly by the insects collected at Murree, the remaining Hemiptera from the other localities being almost wholly Palæarctic.

The following is the analysis of the portion of the collection obtained at Murree :—

Dalpada confusa, n. sp.	A genus belonging principally to the Indian region, extending to Java and the Philippines, but represented also in Madagascar.
———— *tecta*, Walk.	Eastern Bengal Province (Blanford).
Palomena viridissima, Poda	Palæarctic.
———— *reuteri*, n. sp.	The genus extends to the Japanese sub-region of the Palæarctic region.
Bagrada picta, Fab.	Indian Province (Blanford). Palæarctic, Persian sub-region.
Menida distincta, n. sp.[1]	Genus represented chiefly in Indian and Ethiopian regions, and by one species in the Japanese sub-region of the Palæarctic.
Prionaca exempta, Walk.	Eastern Bengal Province (Blanford).
Acanthosoma proxima, Dall.	———————————————— ? type collected by General Hardwicke (no locality).
A. forfex, Dall.	———————————————— ? Northern India ? (Brit. Mus.).
A. aspera, Walk.	———————————————— ?
Urostylis fumigata, Walk. var.	
Cletus puncliger, Dall. var.	Indian region.
Lygæus (Spilostethus) militaris, Fab.	Palæarctic and Indian regions.
Arocatus piloculus, n. sp.	Genus represented in Palæarctic, Indian, and Australian regions.
Phytocoris stoliczkanus, n. sp.[2]	Genus Palæarctic.
Calocoris stoliczkanus, n. sp.	————————.
———— *forsythi*, n. sp.	————————.
Euacanthus extrema, Walk.	Eastern Bengal Province (Blanford),? N. India (Brit. Mus).

[1] This species was also collected in the Sind Valley.
[2] This species was also collected in the Jhelam and Sind Valleys.

The strong Indian affinities of these 18 species collected at Murree may be seen as under:—

Common to	Indian Region	9
,,	Indian and Palæarctic Regions	2
,,	Indian, Ethiopian, and Palæarctic	1
,,	Indian, Australian, and Palæarctic	1
,,	Palæarctic	5

18 species.

In discussing the Indian relationship of the Hemipterous fauna, I have followed the sub-regions or provinces of Mr. Blanford (Ann. and Mag. Nat. Hist., 4th Ser., Vol. 18, pp. 280—2, 1876). As regards the sub-regions of the Palæarctic area, I shall prefer to follow Mr. Sclater (Address Biol. Soc. Brit. Assn., Bristol, 1875).

The following are the 13 species which are also found in the Cis-Atlantean or Mediterranean sub-region:—

Zicrona cærulea, Lin.	Kugiár, Eastern Turkestan.
Carpocoris nigricornis, Fab.	
Dolycoris baccarum, Lin.	Sind Valley, Kashmir.
Eurydema festiva, Lin.	Yangihissar, E. Turkestan.
Comptopus lateralis, Germ.	Sind Valley, Kashmir.
Therapha hyoscyami, Lin.	Kugiár.
Lyg. (Spilostethus) militaris, Fab.	Murree, Punjab hills.
——— *saxatilis*, Scop.	Sind Valley, Kashmir.
——— *(Graptolomus) equestris*, Lin.	Kugiár and neighbourhood of Sánju, E. Turkestan.
Gonianotus marginepunctatus, Wolff.	Pámir road, Sarikol to Panja.
Coriscus ferus, Lin.	Yárkand.
Notonecta glauca, Lin.	
Corixa geoffroyi, Leach	

The following 4 species have been recorded from the sub-region of the Atlantic Islands, principally from Madeira and Teneriffe:—

Dolycoris baccarum, Lin.	Sind Valley, Kashmir.
Comptopus lateralis, Germ.	
Lyg. (Spilostethus) militaris, Fab.	Murree, Punjab hills.
Gonianotus marginepunctatus, Wolff.	Pámir road, Sarikol to Panja.

This list is, however, very poor and inadequate, owing to the little knowledge we yet possess as to the Hemiptera of the Atlantic Islands. It would be futile to carry the analysis of this region further, owing to the paucity of record.

HEMIPTERA-HETEROPTERA.

Family—*PENTATOMIDÆ*, Stål.

Sub-family—*CYDNINÆ (CYDNINA)*, Stål.

1. ÆTHUS MAURUS, Dall.

Æ. maurus, Dall., List, pt. 1, p. 118, 18 (1851).—Walk., Cat. Het. 1, p. 158 (1857).—Stål, Enumerat. Hemip., pt. 5, p. 26, 1876.

Hab.—Jhelam Valley, July 1873.
Distributed generally throughout Hindustan.

2. ÆTHUS, sp.

Allied to *Æ. pygmæus*, Dall., of which it is probably only a slight variety. The collection contains only one specimen.
Hab.—On the road across the Pámir, from Sarikol to Panja, April to May 1874.

Sub-fam.—*ASOPINÆ (ASOPIDA)*, Stål.

3. ZICRONA CŒRULEA, Lin.

Cimex cæruleus, Lin., Syst. Nat., ed. 10, i, p. 445, 38 (1758).—Stål, Enumerat. Hemip., pt. 1, p. 36, (1870).

Hab.—Kugiár, South Yárkand, May to June 1874.
This wide-ranging species extends throughout the whole of the Palæarctic and Oriental regions.
Europa tota, India orientalis, Java, Borneo, Malacca (Stål.); Astracan (Jacovlev); N. W. Siberia (Sahlberg); Bagdad (Coll. Brit. Mus.); Morocco, Japan (Coll. Distant).

Sub-fam.—*PENTATOMINÆ (PENTATOMINA)*, Stål.

4. DALPADA CONFUSA, Dist. Fig. 1.

Dalpada confusa, Dist., Trans. Ent. Soc. Lond., 1879, p. 121.

Luteous, thickly covered with green punctures. Head emarginate in front with the sides reflexed, and some small indistinct ochreous markings at base. Antennæ pitchy, each joint luteous at base, basal and apical joints smallest, 2nd shorter than 3rd, 3rd and 4th subequal. Rostrum just passing posterior coxæ, with the tip pitchy. Pronotum somewhat transversely gibbous at base in a line with lateral angles, after which it is abruptly deflexed towards head, lateral angles prominent, subacute, lateral margins denticulated for about half their length from apex; the punctuation is very dense along the lateral margins and at pronotal angles. Scutellum somewhat gibbous at base, deflexed towards apex, where it is more sparingly punctured. Corium with a faint impunctate longitudinal line on disc, extending from base to about two-thirds its length, rather widened at apex. Membrane extending beyond apex of abdomen, pale fuscous with the nervures dark brown for half the length from base, followed by a row of four brown spots and a marginal row of six spots of the same colour, the two outer ones being long and linear. Under side of body luteous, with the pectoral and abdominal margins broadly punctured with green, sparingly on abdomen, and more densely on prosternum; legs luteous, thickly spotted with brown; tarsi luteous, apical joint pitchy.

♂. Long. 14 mill.; lat. pronot. ang. 6½ mill.
♀. Long. 15 to 16 mill.; lat. pronot. ang. 7½ mill.
Hab.—Murree.

I have compared the above with all the congeneric types of Dallas in the British Museum and of Hope in the collection at Oxford, from which it is quite distinct in general structure. Dr. Stoliczka collected a fine series of both sexes.

5. DALPADA TECTA, Walk.

Dalpada tecta, Walk., Cat. Het. 1, p. 224, 1867.

Hab.—Murree. The type was from Sylhet.

6. PALOMENA VIRIDISSIMA, Poda.

Cimex poda, Mus. Gr. 56, 10. *Pal. viridissima,* Stål, Hem. Fab. 1, p. 25. Muls. et Rey., Pun. Fr. 277, 1, 1866. *Pent. dissimilis,* Dall., List. 1, p. 241, 20, 1851.

Hab.—Murree.
This Palæarctic form is generally distributed throughout Europe. N. W. Siberia (Sahlberg).

7. PALOMENA REUTERI, Dist. Fig. 2.

Palomena reuteri, Dist., Trans. Ent. Soc., Lond., 1879, p. 122.

♂. Green; head, anterior border of pronotum, basal half of scutellum, and membrane bronzy. Head obscurely rugulose, very thickly and strongly punctured with black, median lobe slightly shorter than the lateral ones. Rostrum luteous, with the tip black. Antennæ luteous, apical joint somewhat fuscous, 3rd joint distinctly longer than the 2nd, rather shorter than the 4th, 5th longest. Pronotum obscurely rugulose, very thickly and strongly punctured with black, with two slightly waved lateral linear impunctate foveæ situated a little behind the anterior margin, lateral angles somewhat prominent and rounded. Scutellum thickly covered with deep black punctures, slightly rugulose at base. Corium thickly and deeply punctured with black. Abdomen above black, thickly and finely punctured with the connexivum luteous, punctured with black. Body beneath pale luteous, slightly clouded with greenish. Legs greenish, tarsi luteous.

♀ Second joint of antennæ distinctly longer than the 3rd, 2nd and 4th subequal. Abdomen, beneath, with some irregular obscure black markings.

Long. 11 to 12 mill.; exp. pronot. ang. 6 to 7 mill.

Hab.—Murree.

Allied to *P. viridissima*, Poda, but differs in its smaller size and shorter antennæ; it is also more straightened and narrowed than in that species, and the structure of the pronotum is different.

8. CARPOCORIS NIGRICORNIS, Fab.

Cimex nigricornis, Fab., Ent. Syst., IV, 94, 59.

Hab.—Kugiár, May to June 1874.

This is a common Palæarctic form. N. W. Siberia (Sahlberg); Astracan (Jacovlev); Algeria (Lucas); Morocco (Coll. Distant).

9. DOLYCORIS BACCARUM, Lin.

Cimex baccarum, Lin., Faun. Sv., 249, 928.—*Mormidea baccarum*, Fieb., Eur. Hem., 335, 1.—*Pentatoma baccarum*, Hahn, Wanz. Ins., fig. 152.—*Cimex verbasci*, De Geer, Mem. iii, 257, 4 (1773).

Hab.—Sind Valley, August 1873.

Common to Palæarctic region. Madeira (Wollaston); Tunis (Coll. Brit. Mus.); Morocco (Coll. Distant). Algeria (Lucas); N. W. Siberia (Sahlberg); Astracan (Jacovlev).

10. EURYDEMA WILKINSI, Osch. in litt., Fig. 4.

Eurydema wilkinsi, Dist., Trans. Ent. Soc. Lond., 1879, p. 123.

Pale luteous, somewhat thickly and coarsely punctured. Head with the anterior portion of the submarginal lateral borders and a large triangular marking at base; pronotum with two large discal subquadrate linear markings elongated exteriorly; scutellum with the base and two central forked lines extending therefrom to about middle, and two spots on lateral margins a little before apex; corium with two claval streaks; a linear spot on middle of outer margin; a transverse-waved fascia, extending from base of membrane for two-thirds across corium, and a rounded sub-apical spot, shining green. Abdomen above luteous, apical segment black, connexivum with a row of large green spots. Underside of body pale luteous. Abdomen with a marginal row of spots situated on the outer edge of each segmental suture, and a submarginal row of transverse slightly-waved linear markings, situated on middle of each segment, greenish-black. Sternum with some irregular markings of the same colour. Legs pale luteous, streaked with greenish-black, and femora obscurely annulated with the same colour near apex. Antennæ black, 2nd joint about as long as 1st and 3rd together, 4th somewhat dilated, about equal in length to 5th. Rostrum luteous pitchy at base and apex.

Long. 7 mill.

Hab.—Yangihissar, April 1874.

I have retained the unpublished name under which, Dr. Reuter informs me, this species has been sent from Turkestan. In most specimens the markings on the pronotum are not perfectly subquadrate, but disjointed. I have, however, thought it best to describe the specimen submitted to Dr. Reuter and returned as above.

11. EURYDEMA FESTIVA, Lin.

Cimex festiva, Lin., Syst. Nat., 723, 57.—*Strachia festiva*, Hahn, Wanz. Ins., fig. 93.—*Eurydema ornata*, Sahlb., Mon. Geoc. Fenn., 24, 1.

Hab.—Yangihissár, April 1874. Kugiár, May to June 1874. Sind Valley, August 1873. A common Palæarctic form. Madeira (Wollaston); N. W. Siberia (Sahlberg); Astracan (Jacovlev); Algeria (Lucas).

11a. E. FESTIVA, var. HERBACEA, H. Sch.

Eurydema herbaceum, H. Sch., Cont. Panz. F. G., 115, 12, and Nom. Ent., 1, 55, and 91 (1835). Hahn, Wanz. 3, F. 239 (1835).

Hab.—Sind Valley, August 1873.

12. BAGRADA PICTA, Fab.

Cimex pictus, Fab., S. Ent., p. 715, 93 (1775), Spec. 2, p. 359, 127 (1781). Wolff, Ic., 1, p. 17, F. 17 (1800).—*Strachia picta*, Dall., List. 1, p. 259, 5 (1851). Stål, Enumerat. Hemip., pt. 5, p. 88.

Hab.—Murree.
This species, with few exceptions, has hitherto been only received from Bengal. Bombay (Coll. Dist.); Bagdad (Coll. Brit. Mus).

13. MENIDA DISTINCTA. Fig. 3.

Menida distincta, Dist., Trans. Ent. Soc. Lond., 1879, p. 122.

Luteous, covered with strong greenish-black punctures. Head luteous, with the lateral margins and four longitudinal furrowed punctured lines greenish-black. These lines are much more distinct on the ante-ocular portion of the head. Eyes dull ochreous. Antennæ pilose with the 2nd joint shorter than the 3rd, 4th and 5th subequal, rather longer than 3rd; first 3 joints luteous, apex of the 1st and apical half of the 3rd, black, 3rd and 4th joints black, narrowly luteous at base. Rostrum luteous, apical joints pitchy. Pronotum with an anterior submarginal line of greenish-black punctures and two irregular transverse ocellated punctured marks of the same colour on anterior portion of the disc. Scutellum with a large central subbasal greenish-black spot, and two small and somewhat indistinct ones of the same colour situated on the lateral margins a little before apex. Membrane transparent, whitish. Abdomen above black, connexivum luteous, spotted with black. Underside of body and legs luteous, sparingly and distinctly punctured with black. Tarsi pitchy.
Long. 6 mill.
Hab.—Murree and Sind Valley, Kashmir.
Menida histrio, Fab., is the nearest allied species.

14. PRIONACA EXEMPTA, Walk.

Prionaca exempta, Walk., Cat. Het., 3, p. 569.

Hab.—Murree.
N. Hindostan (Coll. Brit. Mus.); Assam (Coll. Distant).

Sub-Family—*ACANTHOSOMINÆ (ACANTHSOMINA)*, Stål.

15. ACANTHOSOMA PROXIMA, Dall.

Acanthosoma proximum, Dall., List., 1, p. 303, 2 (1851).

Hab.—Murree.
The type in the British Museum without a locality was presented by General Hardwicke, and is probably from N. India.

16. ACANTHOSOMA FORFEX, Dall.

Acanthosoma forfex, Dall., List., 1, p. 308, 16 (1851).

Hab.—Murree.
N. India (Coll. Brit. Mus.).

17. ACANTHOSOMA RECURVA, Dall.

Acanthosoma recurvum, Dall., List., 1, p. 310, 19 (1851).—*Clinocoris recurvus*, Stål, Enumerat. Hemip, p. 5, p. 114 (1876).

Hab.—Sind Valley, August 1873.
N. India (Coll. Brit. Mus.).

18. ACANTHOSOMA ASPERA, Walk.

Acanthosoma aspera, Walk., Cat. Het., p. 2, p. 395, 17 (1867).

Hab.—Murree.
N. India (Coll. Brit. Mus.).

Sub-Family—*UROSTYLINÆ (UROSTYLINA)*, Stål.

19. UROSTYLIS FUMIGATA, Walk. var.

Urostylis fumigata, Walk., Cat. Het., 3, p. 413 (1867).

Hab.—Murree. The type was from Sylhet.

Family—*COREIDÆ*, Stål.

Sub-Family—*COREINÆ (COREINA)*, Stål.

Division **Gonoceraria**, Stål.

20. CLETUS PUNCTIGER, Dall. var.

Gonocerus punctiger, Dall., List. 2, p. 494, 3 (1852).

Hab.—Murree.
The type was from China. Malacca (Walker).

Sub-Family—*ALYDINÆ (ALYDINA)*, Stål.

21. CAMPTOPUS LATERALIS, Germ.

Coreus lateralis, Germ., Reise. Dalm., 491 and F. Ins. Eur., 8, 21. H. Sch., Wanz., v, fig. 549, 1839.

Hab.—Sind Valley, August 1873.
This species has a wide Palæarctic range and is a somewhat common European form. Teneriffe (Coll. Brit. Mus.); Madeira (Wollaston); Morocco (Coll. Distant); Astracan (Jacovlev).

Sub-Family—*CORIZINÆ (CORIZINA)*, Stål.

22. THERAPHA HYOSCYAMI, Lin.

Fieb., Eur. Hem., 232 (1861). *Corizus hyosciami*, Hahn, Wanz., 1, fig. 10.—*Cimex hyoscyami*, Lin., Faun. Sv. 252, 945.

Hab.—Kugiár, south of Yárkand, May to June 1874.
A well-known European species. N. W. Siberia (Sahlberg); Astracan (Jacovlev); Morocco (Coll. Distant).

Family—*LYGÆIDÆ*, Stål.

Sub-Family—*LYGÆINÆ (LYGÆINA)*, Stål.

23. LYGÆUS (SPILOSTETHUS) MILITARIS, Fab.

Lygaeus militaris, Stål, Hem. Afr., 2, 133, 13 (1865).—*Cimex militaris*, Fab., S. Ent., p. 717, 103 (1775).

Hab.—Murree.
Europa meridionalis, Africa borealis et media, India orientalis, Insulæ Philippinæ (Stål.); Madeira (Wollaston); Teneriffe, Canaries, Mauritius, Bagdad, N. India (Coll. Brit. Mus.); Morocco (Coll. Distant); Algeria (Lucas).

24. LYGÆUS (SPILOSTETHUS) SAXATILIS, Scop.

Cimex saxatilis, Scop., Ent. Carn., 128, 371 (1763).—*Lygæus saxatilis*, Dall., List., 2, p. 544 (1852).— *L. (Spilostethus) saxatilis*, Stål, Hem. Fab., 1, 75, 1868.

Hab.—Sind Valley, August 1873.
Generally distributed throughout Europe. Algeria (Lucas).

25. LYGÆUS (GRAPTOLOMUS) EQUESTRIS, Lin.

Cimex equestris, Lin., F. Sv., 253, 946 (1761).—*Lygæus equestris*, Fieb., Eur. Hem., 166, 5.

Hab.—Kugiár, south of Yárkand, May to June 1874. Neighbourhood of Sánju, south-east of Yárkand.
Europa tota, Africa borealis, Sibiria orientalis (*Mus. Holm.*, Sahlberg); Bagdad, Japan (Coll. Brit. Mus); Algeria (Lucas).

26. AROCATUS PILOSULUS. Fig. 5.

Arocatus pilosulus, Dist., Trans. Ent. Soc., Lond., p. 123, 1879.

Testaceous, pilose. Head with the central portion black. Antennæ black, pilose, 2nd joint rather the longest, 3rd and 4th subequal. Rostrum pitchy. Pronotum obscurely punctured, distinctly rugulose on posterior portion; anterior portion crossed by a transverse black submarginal band, and an obscure pitchy band on posterior border. Scutellum pitchy, with the tip red, and with two large round foveæ at base. Corium suffused with dull-pitchy shadings. Membrane pitchy opaque, outer border pale transparent. Under side of body testaceous, strongly suffused with pitchy shadings. Sternum with a submarginal row of three black spots, placed one on prosternum, one on mesosternum, and one on metasternum. Legs pitchy, pilose. The corium is more densely pilose than other parts of the upper surface.
Long. 6 mill.
Hab.—Murree.

Sub-Family—*MYODOCHINÆ (MYODOCHINA)*, Stål.

27. LAMPRODEMA BREVICOLLIS, Fieb.

Lamprodema brevicollis, Fieb., Eur. Hem., 185 (1861).

Hab.—Tanktse to Chagra, Pankong Valley, Ladák, September 1873.
The type was from Dalmatia, and Mr. Edward Saunders kindly compared these specimens for me with insects in his own collection received from the Continent.

28. GONIANOTUS MARGINEPUNCTATUS.

Lygæus marginepunctatus, Wolff, Ic. Cim., 150, t. 15, fig. 144 (1804).

Hab.—On the road across the Pámir, from Sarikol to Panja, April to May 1874.
A not uncommon European species. Madeira (Wollaston); Algeria (Lucas).

Family—*CAPSIDÆ*.

29. PHYTOCORIS STOLICZKANUS. Fig. 6.

Phytocoris stoliczkanus, Dist., Trans. Ent. Soc., Lond., p. 124, 1879.

Uniform pale ochraceous. Head with a V-shaped mark, consisting of small transverse striæ, commencing from near base of antennæ. First joint of antennæ almost as long as head and pronotum together. Pronotum with two slightly raised transverse callosities extending across and occupying the anterior border. Scutellum with the base somewhat raised and gibbous, a waved transverse cordate line near base, and a faint pale longitudinal median line near apex. Hemielytra sparingly clothed with a few minute blackish hairs. Membrane with bright prismatic reflexions.

Long. 6 mill.

Hab.—Murree, Jhelum Valley, and Sind Valley.

30. CALOCORIS STOLICZKANUS. Fig. 7.

Calocoris stoliczkanus, Dist., Trans. Ent. Soc., Lond., 6, p. 124, 1879.

Ochreous clouded with brown, and sparingly clothed with pale yellowish pile. Antennæ brownish, 2nd, 3rd, and 4th joints with the apices pitchy. First joint robust, 2nd somewhat suddenly thickened towards apex, 3rd and 4th very slender, 4th not much more than half the length of the 3rd. Cuneus somewhat paler in colour than corium, brownish and pilose at base, and with a small pitchy spot at apex. Membrane pale fuscous clouded with brown. Underside pale obscure ochreous, clothed with fine pale yellow pile, and a somewhat obscure stigmatal row of small brown spots. Legs mutilated. The pronotum is faintly angulose, and the scutellum somewhat more plainly strigose.

Long. 8 mill.

Hab.—Murree.

I have placed this species in the genus *Calocoris*, though Dr. Reuter writes to me, "*Calocoris*, vel n. gen."

31. CALOCORIS CHENOPODII, Fall.

Phytocoris chenopodii, Fall., H., p. 77, 1.—*Calocoris chenopodii*, Fieb., Eur. Hem., 255 (1861).

Hab.—Sind Valley, August 1873.
Europa tota, Dauria (Sahlberg).

32. CALOCORIS FORSYTHI. Fig. 8.

Calocoris forsythi, Dist., Trans. Ent. Soc. Lond., p. 125, 1879.

Brownish testaceous. Antennæ with the 1st joint not quite so long as head and pronotum, 2nd slightly and gradually thickened towards apex, 3rd pale luteous at base, 4th wanting.

Head with a deep central longitudinal incision between the eyes. Pronotum rugulose, faintly anteriorly and more distinctly towards posterior border. Hemiclytra slightly pilose, somewhat paler towards costal margin, and with extreme outer margin somewhat obscure pitchy. Membrane pale fuscous, somewhat clouded. Scutellum obscurely and transversely strigose. Underside of body castaneous. Fore-legs ochraceous, tibiæ with a longitudinal row of small brownish spots. The rest of the legs wanting.

Long. 7 mill.
Hab.—Murree.

Family—*NABIDÆ*, Fieb.

33. CORISCUS FERUS, Linn.

Cimex ferus, Lin., Faun. Suec., 256, 962 (1761).—*Nabis ferus*, Fieb., Eur. Hem., p. 161, 9 (1861). Reut., O. V. A. F. 29, 6, p. 72, 5 (1872).

Hab.—Yarkand and neighbourhood.

Palæarctic species. America borealis, New Jersey, Wisconsin, California (*Mus. Holm.*, Stål); Europa tota, North-Western Siberia (Sahlberg); Algeria (Lucas).

Family—*REDUVIIDÆ*, Stål.

Sub-Family—*REDUVIINÆ (REDUVIINA)*, Stål.

34. REDUVIUS (HARPISCUS) REUTERI, DIST. Fig. 9.

Reduvius (Harpiscus) reuteri, Dist., Trans. Ent. Soc. Lond., p. 125, 1879.

Black, shining, trochanters and bases of femora sanguineous.

Allied to *R. morio*, Kol. Dr. Reuter, who has kindly compared the two species for me, writes :—" R. (Harpisco) morioni, *colore similis, sed major et in omnibus latior, magis nitidus, pedibus pilis exsertis longis destitutis, capite pronoti lobo postico tantum paullo longiore, trochanteribus basique ipsa femorum rufis divergens.* Obs.—*Gula nigra, ut in* R. morione."

Long. 20 mill.
Hab.—Sind Valley.

35. REDUVIUS (RHINOCORIS) IRACUNDUS, Scop.

Cimex iracundus, Scop., Ent. Carn., p. 130, 378 (1763).—*Harpactor iracundus*, Fieb., Eur. Hem. 153 (1861).

Hab.—Sind Valley, August 1873. European form.

Sub-family—*PIRATINÆ (PIRATINA)* Stål.

36. PIRATES (LESTOMERUS) AFFINIS, Serv.

Peirates affinis, Serv., Ann. Sc. nat., 23, p. 216, 2 (1831). *Lestomerus affinis*, A. and S., Hist., p. 323, 2 (1843).

Hab.—Jhelam Valley, July 1873.
India orientalis (Mus. Holm); Borneo (Mus. Leiden); Cochin-China (*Coll. Signoret,* Stål); Assam (Coll. Distant).

Family—*HYDROBATIDÆ*, Stål.

37. GERRIS (LIMNOTRECHUS) SAHLBERGI. Fig. 10.

Gerris (Limnotrechus) sahlbergi, Dist., Trans. Ent. Soc. Lond., p. 125, 1879.

Head thickly covered with olivaceous pubescence, with a small black spot on vertex. Antennæ ochraceous, 1st joint longest, 2nd and 3rd shortest and subequal, 4th rather longer than 3rd, thickly covered with greyish pile. Pronotum ochraceous, pubescent, anterior 3rd, lateral borders, and a central longitudinal line, olivaceous; the last is testaceous on anterior portion of pronotum. Hemielytra brownish testaceous, with the nervures olivaceous. Under side of body covered with greyish pile, except lateral borders, apex, and central portion of abdomen, which parts are ochraceous. Legs ochraceous, fore femora with an outer longitudinal black fascia.
Long. 10 mill.
Hab.—Neighbourhood of Leh.
Dr. Reuter, who has done me the favour of examining the species, reports—"L. thoracico *affinis et segmentorum genitalium maris structura similis, differt autem pronoto breviore, postice brevius et obtusius producto, angulis dentiformibus segm. abdominalis sexti brevissimis, vix productis, tibiis, præsertim posticis, brevibus, tarsis posticis his tantum ⅓ brevioribus.*"

38. GEN. (?) ORIENTALIS. Figs. 11 and 12.

Halobates (?) *orientalis*, Dist., Trans. Ent. Soc. Lond., p. 126, 1879.

Brownish ochraceous, finely pilose. Antennæ with the 1st joint curved, robust, and about the length of head and pronotum together; remaining joints more slender, 2nd and 3rd subequal, 4th a little shorter than 3rd. Pronotum with a median pale longitudinal line and a large rounded fovea on posterior portion of disc. The rostrum is 5-jointed; the first two joints are very robust and somewhat fused together, the 2nd minute and much shorter than the 1st, the 3rd much the longest and rather less robust than 1st and 2nd, 4th small, slender, and black, 5th ochreous, very slender and hair-like, and rather shorter than 4th. Sternum clothed with greyish pile.

The eyes are large, semi-globular, and castaneous, situated at base of lateral margins of head. The pronotum is about the length of the head, but broader, truncate in front and rounded behind; mesonotum and metanotum hardly distinguishable, much longer than pro-

notum, and gradually and regularly widened posteriorly. Legs ochreous, fore femora much thickened.
Long. 7 to 8 mill.
Hab.—Jhelam Valley.
I have refrained for the present from making a new genus for the reception of this species. It is in many respects allied to *Halobates* and cannot be included in the genus *Gerris*. The figures will show its anatomical peculiarities.

Family—*NEPIDÆ*, Burm.

39. RANATRA, sp. ?

Too mutilated for determination.
Hab.—Yárkand.

Family—*NOTONECTIDÆ*, Stål.

40. NOTONECTA GLAUCA, Lin.

Notonecta glauca, Lin., Faun. Sv. 244, 903, Sahl., Not. Faun. et. Fl. Fenn., Forh., XIV, 273, 1. *N. fabricii*, Fieb., Eur. Hem. 101, 2. *N. marmorea*, Fab., Syst. Rhyn., p. 103, 3 (1803).

Hab.—Yárkand.
Europa tota, Asia et America borealis, N. W. Siberia (Sahlberg); Algeria (Lucas).

41. ENITHARES, sp. ?

Allied to *E. indica*, Fab., if not a variety of that species.
Hab.—Jhelam Valley.

Family—*CORISIDÆ*, Fieb.

42. CORISA HIEROGLYPHICA, L. Duf.

Corisa hieroglyphica, L. Duf., Hem., 86, 2, fig. 85, 87. *Corisa hieroglyphica*, Fieb., Eur. Hem., 93, 15 (1861).

Hab.—Yárkand.
Palæarctic form. Astracan (Jacovlev).

43. CORISA GEOFFROYI, Leach.

Corisa geoffroyi, Leach, Class. Lin. Tr., 12, 7. *Corisa geoffroyi*, Fieb., Eur. Hem., 91, 6 (1861)= *dentipes*, Thom. (Sahlberg).

Hab.—Yárkand.
Astracan (Jacovlev); Algeria (Lucas).

HEMIPTERA-HOMOPTERA.

Family—*CERCOPIDÆ.*

Sub-family—*CERCOPINÆ (CERCOPINA)* Stål.

44. COSMOSCARTA DORSIMACULA, Walk.

Cercopis dorsimacula, Walk., List. Homop. Ins. III, p. 658, 31 (1851).

Hab.—Jhelam Valley.
N. Bengal, N. India, Cachar (Types, Brit. Mus.)

Sub-family—*APHROPHORINÆ (APHROPHORINA)* Stål.

45. PTYELUS COSTALIS, Walk.

Ptyelus costalis, Walk., List. Homop. Ins. III, p. 707, 13 (1851). *Ptyelus concolor*, Walk., *ib.*, p. 715, 26. Stål, Ofv. vet. Ak. Forh., 1862, p. 493.

Hab.—Dras, Kargil, and Leh, August to September 1873.
N. India (Types, Brit. Mus).

46. CLOVIA NEBULOSA, Fab.

Cercopis nebulosa, Fab., Ent. Syst., IV, 50, 14; Syst. Rhyn., 94, 3. *Ptyelus quadridens*, Walk., List. Homop. Ins. III, p. 711, 19 (1851). *Ptyelus guttifer*, Walk., *ibid.*, p. 712, 21. *Clovia nebulosa*, Stål, Hem. Fab., 2, p. 16, 1869; Sign., Rev. and Mag. Zool., 1853, tom. 5, p. 183. Stål, Ofv. vet. Ak. Forh., 1862, p. 493.

Hab.—Sind Valley, August 1873.

Family—*JASSIDÆ.*

Sub-family—*CENTROTINÆ.*

In his Hem. Af. 4, pp. 82-83 (1866), the late Dr. Stål gave a "*Conspectus subfamiliarum*" of his family "*Jassida.*" In that work he placed the genus *Oxyrhachis*, Germ., in his sub-fam. "*Membracida,*" owing no doubt to the dilated fore tibiæ of the insects comprised in that genus. Subsequently, however, Ofv. vet. Ak. Forh., 1869, p. 280, he placed it in his sub-fam. "*Centrotida,*" adding "*Conspectus generum, centrotidum mundi antiqui—vide* Hem. Af., IV, pp. 86-89." Although there is an error in this statement, the genus *Oxyrhachis* clearly belongs to the *Centrotidæ*, in which I have placed it.

47. OXYRHACHIS, sp.

Hab.—Jhelam Valley. One spirit-specimen too much damaged to be described.

RHYNCHOTA.   15

Sub-family—*PROCONIINÆ (PROCONIINA)* Stål.

48. EUACANTHUS EXTREMUS, Walk.

Tettigonia extrema, Walk., List. Homop. Ins., III, p. 761 (1851). Sign., Ann. Ent. Fr. Ser. 3, 1, p. 663, pl. 21, fig. 4 (1853). Stål, Ofv. vet. Ak. Forh., 1862, p. 495.

Hab.—Murree.
The type was from N. India.

Sub-family—*JASSINÆ (JASSINA)* Stål.

50. BYTHOSCOPUS STRAMINEUS, Walk.

Acocephalus stramineus, Walk., List. Homop. Ins., III, p. 847 (1851).
Bythoscopus indicatus, Walk., List. Homop. Ins. Suppl., p. 266, 1858. Stål, Ofv. vet. Åk. Forh., 1862, p. 494.

Hab.—Sind Valley, August 1873.
The types were from Java, N. China, and Celebes ; it is, however, a commonly received Indian species.

The remaining specimens of *Homoptera* contained in the collection, mostly somewhat minute species, are so damaged by immersion in spirit as to be undeterminable and of little value as museum-specimens. I should certainly pause before describing insects in this condition, as colour is obliterated and good figures could not be made. There are two small species of *Ricania* and one of *Nephesa* ; the rest call for little comment.

Explanation of the Plate.

Fig. 1. Dalpada confusa, Dist., p. 3.
 „   2. Palomena reuteri, Dist., p. 4.
 „   3. Menida distincta, Dist., p. 6.
 „   4. Eurydema wilkinsi, Dist., p. 5.
 „   5. Arocatus pilosulus, Dist., p. 9.
 „   6. Phytocoris stoliczkanus, Dist., p. 9.
 „   7. Calocoris stoliczkanus, Dist., p. 10.
 „   8. ———— forsythi, Dist., p. 10.
 „   9. Reduvius (Harpiscus) reuteri, Dist., p. 11.
 „ 10. Gerris (Limnotrechus) sahlbergi, Dist., p. 12.
 „ 11. Halobates? orientalis, ♂, viewed from above, enlarged, p. 12.
 „ 11a. The same, from below, more highly magnified.
 „ 11b. The anal appendages of the same, from above.
 „ 11c. The same, from below.
 „ 11d. The same seen vertically.
 „ 11e. The rostrum.
 „ 12a. & 12b. The anal appendages of the female.

HEMIPTERA.

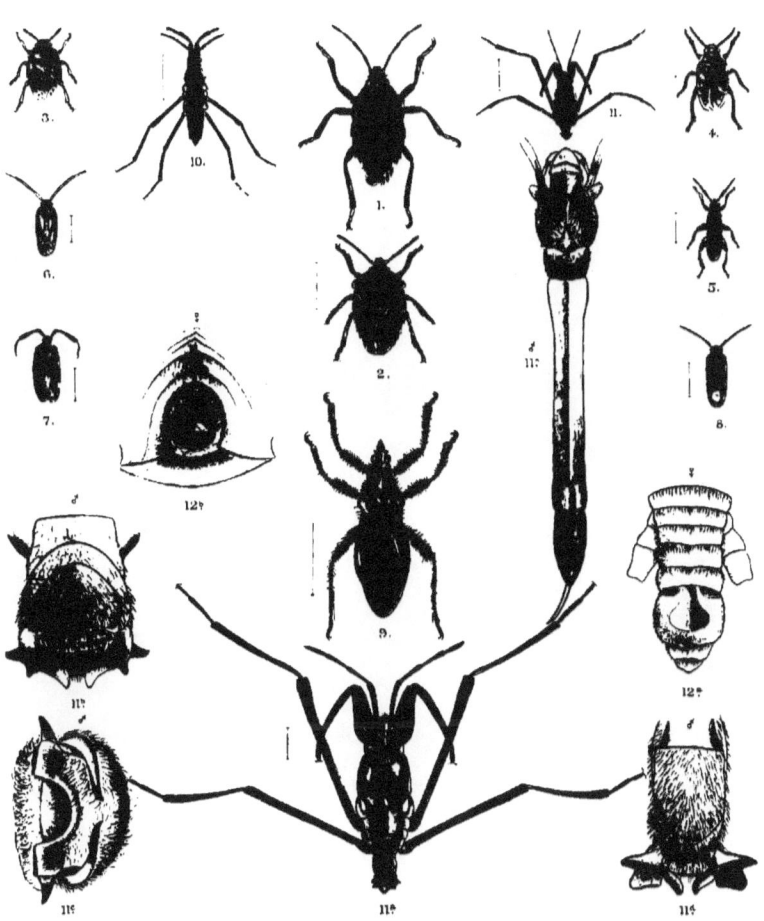

SCIENTIFIC RESULTS

OF

THE SECOND YARKAND MISSION;

BASED UPON THE COLLECTIONS AND NOTES

OF THE LATE

FERDINAND STOLICZKA, Ph. D.

ARANEIDEA,

BY THE

REVD. O. P. CAMBRIDGE, M.A.,C.M.Z.S.,

Honorary Member of the New Zealand Institute, &c.

Published by order of the Government of India.

CALCUTTA:
PRINTED BY THE SUPERINTENDENT OF GOVERNMENT PRINTING, INDIA.
1885.

SCIENTIFIC RESULTS

OF

THE SECOND YARKAND MISSION;

BASED UPON THE COLLECTIONS AND NOTES

OF THE LATE

FERDINAND STOLICZKA, Ph. D.

ARANEIDEA,

BY THE

REVD. O. P. CAMBRIDGE, M.A.,C.M.Z.S.,

Honorary Member of the New Zealand Institute, &c.

Published by order of the Government of India.

CALCUTTA:
PRINTED BY THE SUPERINTENDENT OF GOVERNMENT PRINTING, INDIA.
1885.

SCIENTIFIC RESULTS

OF

THE SECOND YARKAND MISSION.

ARANEIDEA.

BY THE REV. O. P. CAMBRIDGE, M.A., C.M.Z.S.,
Honorary Member of the New Zealand Institute, &c.

INTRODUCTORY REMARKS.

THE spiders collected by the late Dr. Stoliczka in the Yárkand expedition can by no means be considered a full, nor perhaps even a fair, representation of the *Araneidea* of the extensive area over which the expedition passed.

Mr. A. O. Hume informs me that this area may be subdivided into five well-marked regions, and suggests that the spiders found in each should be distinctly (*i.e.*, I conclude, separately) treated. Now, judging from the contents of the collection, I should have thought that the regions might have been considered as two only,—that is, (1) from Murree to Cashmere, including the latter as well as the former; and (2) the whole of the rest of the area travelled over by the expedition, and comprising the neighbourhood of Leh, the route from Tantze to Chagra and Pankong valley, and from Yárkand to Bursi, as well as Yárkand and neighbourhood, Káshghar, the hills west of Yárkand, and the Pamir.

In the first of these regions—Murree and Cashmere—more than half of the whole number of spiders were collected; the total number of species is 131; the number from this region is 69. The leading character of the spiders of this region is European, with a few more distinctly tropical and subtropical species, such as *Idiops designatus*, *Episinus algiricus*, *Phycus sagittatus*, *Meta mixta*, *Chorizoopes stoliczkæ* and *O. congener*, *Cyrtarachne pallida*, *Diæa subargentata*, *Monastes dejectus*, *Sarotes regius* and *S. promptus*, *Sparassus fugax*, *Ocyale rectifasciata*, *Philodromus medius*, and *Oxyopes jubilans* and *O. prædicta*. The leading character of the second region is also European, but with a decided subalpine feature, and no trace scarcely of anything tropical or even subtropical, excepting perhaps *Prosthesima cingara*, *Sparassus flavidus*, and *Hæbe benevola*. Of the 69 species found in the Cashmere regions, three only were found in the other regions mentioned; and one species only, *Drassus dispulsus*, occurred generally throughout the whole area travelled over,—*i.e.*, in all the five regions specified by Mr. Hume.

In the systematic list of species following the present descriptions, figures are added notifying in which of Mr. Hume's regions each species occurred. Supplementary lists are also appended of the spiders found in each separate region, with figures showing in what other regions, if any, each occurred. From these lists, it will be seen that one species only (that mentioned above) was common to all the five regions; three others were common to four of the regions; four others common to three regions; and fourteen others common to two of them;—sixty-six of the species being, as above observed, found only in region 1 of Mr. Hume, corresponding exactly to the first region indicated, as it seems to me, by the character of the spiders themselves.

The regions named by Mr. Hume are:—

(*1*) "*Cashmere including Murree and the road thence to Cashmere.*"—This comprises the spiders noted in my descriptions as *Murree, Murree to Sind valley*, and *Sind valley*.

(*2*) "*Ladakh, from the Zojeela Pass to the head of the Pankong Lake.*"—This comprises the spiders noted in my descriptions as *Neighbourhood of Leh*, and *Tontze to Chagra*, and *Pankong valley*.

(*3*) "*The mountain masses between the head of the Pankong Lake and the plains of Yárkand*," comprising only the spiders noted as *Yárkand to Bursi*, there being no spiders in the collection labelled as having been obtained during the forward journey from the Pankong Lake to the plains of Yárkand.

(*4*) "*The plains of Yárkand*," comprising the spiders noted as *Yárkand and neighbourhood*, and *Yárkand.*" Excepting the three species mentioned as subtropical in my second region, there were no spiders, in this region 4 of Mr. Hume, differing in character from the general run of those in his regions 2, 3, and 5.

(*5*) "*The high country west of Yárkand, the hills leading up to the Pamir, the Pamir and Wokhan.*"—This comprises the spiders noted as *Káshghar, between Yangihissár and Sirikol, Yangihissár, road across the Pamir from Sirikol to Panjah and back*, and *hills between Sirikol and Aktalla*.

It will be seen from the above that my first region corresponds exactly with region 1 of Mr. Hume, and that my second region includes Mr. Hume's regions 2, 3, 4, and 5.

The localities noted for each species in my descriptions are those written (I suppose by Dr. Stoliczka himself) upon the several bottles in which the spiders were contained. No attempt had been made to separate the species in each bottle, nor, with one exception, is there anything in Dr. Stoliczka's diary referring intelligibly to the separate contents of the bottles. Dr. Stoliczka's notes on the spiders are very few, and of the most general description. In the one exceptional instance (Diary, p. 3, dated 19th July 1873, *Tinali*), the note refers to the capture of a "great number of spiders, chiefly *Thourcus*" (probably a misprint for *Thomisus*, though there were very few *Thomiside* in this bottle) "and *Sphassus* [*Sphasus*]; among the latter I recognised *Sphasus viridanus*." Now, there was not a single example of *Sphasus* in any one of the bottles, excepting in one, which contained no label nor any other clue either to the locality or its contents; the mention therefore of *Sphasus* is thus important, and fixes the locality in which the contents of this unlabelled bottle were collected. The *Sphasus viridanus* alluded to is a Calcutta species, described by Dr. Stoliczka himself (Journ. Asiat. Soc., Bengal, vol. xxxviii, p. 220, pl. xx, fig. 1), but is quite distinct from either of the three species found in this bottle (*vide* remarks on these species, *infra*).

Out of the 132 species in the collection, I can only determine 23 as certainly identical with European species already described, leaving 109, which I believe to be new to science.

This appears to be a large proportion of undescribed species, but no more than might be expected from a district hitherto wholly (so far as I am aware) unknown to arachnologists. The researches of Alexis Fedtschenko, Reise in Turkestan, lately (in respect at least to the *Araneidea*) published by Kronenberg, give 146 species, of which 101 are identified with known European species. Excepting the Latin descriptions of new species, this work is written in the Russian language, with which I have, unfortunately, no acquaintance whatever. Eight only of the spiders described or recorded by Kronenberg appear to me identical with those contained in Dr. Stoliczka's collection. These are *Epeira tartarica*, Kron.; *Epeira cornuta*, Clk.; *Epeira cucurbitina*, Clk.; *Tetragnatha extensa*, Linn.; *Pachygnatha clerckii*, Sund.; *Erigone dentipalpis*, Wid.; *Theridion tuberculatum*, Kron.; and *Xysticus cristatus*, Clk. At first sight it might seem remarkable that so large a proportion of the collection made by Fedtschenko in Western Turkestan should be identical with European species, while so small a proportion out of those collected by Dr. Stoliczka are similarly identical; but when it is borne in mind that more than half of Dr. Stoliczka's collection was made in Murree and Cashmere, we need not be surprised at these results, for, indeed, a comparatively small collection only can be said to have been made in Eastern Turkestan, and that chiefly on the high mountain ranges and during the winter and early spring months; these months being probably there, as in other analogous districts, the least favourable for the fullest development of the *Araneidea*.

I have observed that the number of species contained in Dr. Stoliczka's collection cannot be by any means considered a full representation of the spiders inhabiting the country travelled over. The season of the year had probably much to do with this paucity of species, but more than anything else it may be accounted for when we remember the number of irons Dr. Stoliczka had in the fire, embracing the whole field of the zoology, as well as of the geology, of the districts visited; instead, therefore, of being surprised at the smallness of the arachnological results of the expedition, we must, under the circumstances, wonder at their extent. We may look forward now with great interest to future collections made in the north of India, on the southern slopes of the mountain ranges of Cashmere, and in the plains adjoining, where we should expect the tropical character of the spiders to become far more marked, though probably still with a great diversity in the species from those found in the more central regions of India. No materials, however, exist, so far as I am aware, for any comparison upon these points; indeed, the materials for comparison with any Indian spiders are, as yet, comparatively small, and but little has been hitherto published upon them.

Order—ARANEIDEA.

Family—THERAPHOSIDES.

Genus—IDIOPS, Perty.

1.—IDIOPS DESIGNATUS, sp. n., Pl. I, Fig. 1, ♂.

Adult male: length 5⅔ lines; to the end of falces 6½ lines; length of cephalothorax 3 lines (nearly); breadth rather over 2¼.

Cephalothorax round-oval, truncated at each end and rather flattened; it is of a bright red-brown colour, and the normal indentations are strongly marked. The caput is a little elevated above the general level, being rather the highest near the occiput, across which is a well-defined, transverse curved depression (or indentation); the convexity of the curve is directed forwards, and its ends merge in those of the thoracic indentation, which is also curved (the convexity of the curve directed backwards) and deeply impressed; these two curved indentations enclose a well-defined, somewhat roundish, smooth, and shining area; a portion of the surface of the caput on each side, in front of this, is transversely rugulose, and, together with the rest of the thoracic surface, more or less, though not very thickly, covered with minute tubercular granulosities.

The *eyes* are of moderate size and disposed in three transverse rows, 2, 2, 4, forming two widely separated groups, each group placed on a tubercular elevation. Those of the first, or foremost, row constitute one group close to the fore margin of the caput; these appear to be rather the largest of the eight, and are separated from each other by about an eye's diameter; those of the hinder group (consisting of the second and third rows) form a narrow transverse elongate oval; the eyes of the second row are also separated by an eye's diameter, and the length of the row is little, if anything, different from that of the first; the hinder row is curved (the convexity of the curve being directed backwards); the eyes of this row appear to be smaller than the rest, the middle pair being of an irregular form and yellowish-white colour (the colour of the rest is dark), and considerably further from each other than each is from the lateral eye of the same row on its side, and the length of the line formed by those two, hind-central, eyes is a little greater than that of the second (or middle row).

The *legs* are tolerably strong and of moderate length; their relative length is 4, 1, 2, 3; they are of a bright yellow-brown colour, deepening into red-brown on the tibiæ (and on some other parts) of those of the first and second pairs; their armature consists of hairs bristles, and spines, but neither of these in any great abundance. The spines are chiefly on the tibiæ and metatarsi; those underneath the first and second pairs (particularly the first) are the most conspicuous: the tibiæ of the first pair are considerably but gradually enlarged at their fore-extremity on the inner side, the enlargement terminating with a long, strong-curved, blunt-pointed spur or spine. A little on the inner side behind the base of this spur, is a short and strongish denticulation; the metatarsi of the first pair are rather abruptly bent towards their fore-extremity, and slightly enlarged in a bluntish angular form on the inner side. The tarsi are devoid of any scopula, each ending with three claws; those of the upper pair are strong and pectinated; the inferior one is small and inconspicuous.

The *palpi* are long, rather strong, similar to the legs in colour, and furnished with hairs and bristles, those beneath the radial joint being the longest and most numerous. This joint is long, more than double the length of the cubital joint, and nearly equals that of the tibiæ of the first pair of legs; it is of a rather tumid form, and is bent downwards near its anterior extremity, on the outer side of which there is a strong oblique indentation extending underneath, and margined above with a somewhat tuberculous ridge armed with short, strong, tooth-like spines; the digital joint is short, expanded laterally at its fore-extremity, which is also somewhat indented, and armed with a few spines; and the palpal organs are, as usual, simple, though characteristic in detail, consisting of a roundish corneous bulb prolonged into a long, tapering, slightly sinuously-curved, bifid spine, whose extremity, when in its position of rest, is directed outwards and backwards.

The *falces* are moderately strong and bristly, and have near their extremity, on the inner side, a prominence, armed with strong tooth-like spines: their colour is similar to that of the cephalothorax.

ARANEIDEA. 5

The *maxillæ* are moderately long, cylindrical in form, and their fore-extremity, on the inside, terminates in a moderate-sized angular point.

The *labium* is of a somewhat quadrate form, though well rounded at its apex and convex on its outer surface. The colour of the *maxillæ* and *labium* is like that of the legs.

The *abdomen* is short, rather broader behind than before, considerably convex above, particulaly towards the fore part; it is furnished with hairs and a few scattered prominent spines on the upper side, which is of a dark-brownish colour, the under side being of a paler yellowish-brown. The spinners are four in number in the usual position, and pale yellow in colour; those of the superior pair are three-jointed, and tolerably strong but short; those of the inferior pair, short, small, and one-jointed.

Hab.—Murree, between June the 11th and July the 14th, 1873.

This spider appears to belong to the genus *Idiops* as restricted by Professor A. Ausserer in his work upon this family.

Though allied to *I. syriacus*, Cambr., it is certainly distinct from that species, of which, however, the male has not yet been discovered.

Family—*FILISTATIDES*.

Genus—*FILISTATA*, Latr.

2.—FILISTATA SECLUSA, sp. n., Pl. I, Fig. 2, ♀.

Immature female: length 2¼ lines.

In its general form, structure, and appearance this spider is similar to *F. testacea*, Latr., and some other nearly allied species of the genus.

The *cephalothorax*, *legs*, *palpi*, and other fore parts are yellow; the *cephalothorax* has a narrow, blackish marginal line, and occasionally there is a blackish longitudinal marking on the caput behind the *eyes*; these are in the ordinary position and differ little, if at all, in their relative size from those of the species before mentioned.

The *legs* are furnished with hairs, bristles, and some spines, the latter not being very sharp pointed.

The markings of the *abdomen* furnish a very distinctive character in the present species: it is of a dull yellowish colour, with a strong, well-defined, dark, rusty-reddish, longitudinal, median band; this band tapers towards its hinder part, where it is broken into somewhat angular patches; these are continued laterally by some more or less conspicuous oblique lines of the same colour, forming, in fact, the series of chevrons (or angular markings) more or less observable on the hinder half of the abdomen in the greater part of the *Araneidea*; the under side is slightly suffused along the middle, with dull rusty red.

Hab.—Leh, August or September 1873; Pankong-valley, 15th to 21st September 1873.

Family—*DYSDERIDES*.

Genus—*DYSDERA*, Latr.

3.—DYSDERA CYLINDRICA, sp. n., Pl. I, Fig. 3, ♂.

Adult female: length 6½ lines; length of cephalothorax 2 lines.

The *cephalothorax* and *falces* are of a bright, reddish liver-coloured brown; the *legs* and *palpi* are reddish-orange coloured; the *maxillæ, labium,* and *sternum* bright orange-brown, and the *abdomen* dull clay-coloured. In these respects there is, therefore, little or no difference between the present and many other species of *Dysdera*, nor is there any remarkable difference either in the form of the cephalothorax or in the position of the eyes; the surface of the former, although not marked with any distinct punctures, is not glossy; the normal grooves and indentations, though visible, are very slightly defined and it is uniformly but not greatly convex; the fore part is broadly truncated, and the caput is a little constricted at the lateral margins. The cephalothorax is remarkably small, and short, compared to the length of the abdomen.

The *eyes* (six in number) are placed round a slight tubercular elevation close to the fore part of the caput, the height of the clypeus being not more than equal to the diameter of one of the foremost eyes; those of the posterior row (four) are equal in size, contiguous to each other, and form, as nearly as possible, a straight transverse line; immediately in front of each lateral eye of this row, is another larger one contiguous to it, and forming an oblique line in relation to the hinder row, so that the row consisting of the two anterior eyes is rather shorter than the hinder row.

The *legs* are moderate in length and strength, and their relative length is 1, 4, 2, 3. They are furnished very sparingly with hairs, and these are chiefly on the under side; those of the third and fourth pairs have also a few short, fine spines; the tarsi are very short and terminate with two curved, pectinated claws, beneath which is a small compact claw-tuft, behind this the tarsi and the anterior portion of the metatarsi are thickly fringed underneath with hairs.

The *palpi* are similar to the legs in colour, and are furnished with hairs and bristles, the fore part of the digital joint being rather thickly clothed with them, and its extremity is furnished either with *two* small claws, or else with a short curved denticulation springing from the base of the ordinary claw; the hairs and bristles surrounding this part make it difficult to ascertain this exactly.

The *falces* are moderately long, porrected, and rather hollowed on the inner side of their fore half; their length does not exceed half that of the cephalothorax, and the front surface near their base is furnished with a few minute tubercular granulations.

The *maxillæ* are rather long, strong, excavated on the side towards the labium, obliquely truncated at the extremity, and convexly rounded on the outer side above the point where the palpi are articulated; at this point, which is nearly about the middle, the maxillæ are very strong.

The *labium* is rather more than two-thirds of the length of the maxillæ, and is very broad at its base, a little way above which there is a transverse suture or indentation; its sides above this are hollowed; the apex is also hollowed, or strongly, and roundly, indented.

The *sternum* is oval, obtusely pointed behind, truncated before, and strongly impressed at the points between the insertions of the legs.

The *abdomen* is large and of an oblongo-cylindrical form; this character alone distinguishes it at once from all other described species known to me. It is thinly clothed with very short hairs, and the spinners are short and inconspicuous.

The male differs in no essential respect from the female, except in being rather smaller, and, of course, in the smaller size of the abdomen, which, however, preserves the same cylindrical oblong form; the palpi are very like those of *Dysdera cambridgii*, Thor. (*D. erythrina*, Bl.), but the palpal organs are of a much more elongated form.

Hab.—Murree, between June 11th and July 14th, 1873.

ARANEIDEA.

Family—*DRASSIDES*.

Genus—*DRASSUS*, Walck.

4.—DRASSUS TROGLODYTES, C. L. KOCH.

Drassus troglodytes, C. L. Koch, Die Arachn. VI, p. 35, Taf. 189, figs. 455, 456.

Hab.—Examples of this widely-dispersed species were contained in the collection from the following localities: Yárkand to Bursi, May 28th to June 17th, 1874; between Sirikol and Aktallah, 8th to 13th May 1874; Tantze to Chagna and Pankong valley, 15th to 21st September 1873; Yárkand and neighbourhood, November 1873.

5.—DRASSUS INFLETUS, sp. n., Pl. I, Fig. 4, ♀.

Adult female: length 3½ lines.

The *cephalothorax* is of a rather elongate-oval form, narrowing gradually to the fore-extremity, which is truncate; the lateral impressions of the caput are very slight; looked at in profile, the fore-part of the caput slopes very little forwards, and the hinder (or thoracic slope) is short, abrupt, and rather rounded. The normal indentations are ill-defined, and the central thoracic groove is indicated by a short red-brown line; the colour of the cephalothorax is yellow-brown, and it is covered with a grey pubescence, among which are some dark hairs.

The *eyes* are in two transverse, slightly curved, and very nearly concentric, curved rows, close to the fore margin of the caput; they are of moderate size, not greatly different in this respect, and pretty compactly grouped together; those of the front row are very near together, but the interval between the two central eyes of this row is rather greater than that between each and the lateral eye nearest to it; the interval between the laterals of the two rows is nearly, if not quite, equal to the diameter of the largest of them, which appears to be that of the front row; those of the hind-central pair are contiguous to each other, oblique, of an oval form and pearly lustre, and each is separated from the hind-lateral eye nearest to it by an interval equal to its own diameter; the fore-central eyes are the largest of the eight, and the spot on which they are seated appears to be a little prominent; the height of the clypeus is no more than equal to the diameter of one of the last-mentioned eyes.

The *legs* are tolerably strong but rather short; their relative length is 4, 1, 2, 3; they are of a brownish-yellow colour, deepening to reddish-brown on the metatarsi and tarsi, and clothed with greyish pubescence mixed with darker hairs, bristles, and spines; the last chiefly on those of the third and fourth pairs; beneath the two claws with which the tarsi terminate is a small claw-tuft.

The *palpi* are short, pretty stout, and similar to the legs in colour and clothing; the radial and digital joints deepening to red-brown.

The *falces* are tolerably long and strong, directed a little forwards; their colour is like that of the cephalothorax, and they are furnished in front with hairs, bristles, and greyish pubescence.

The *maxillæ* are strong, slightly curved, and inclined towards the labium, as well as broadly impressed across the middle; the basal portion is broad and rather convex, and its colour is darker than that of the cephalothorax.

The *labium* is oblong, rounded at the apex, and similar to the maxillæ in colour.

The *sternum* is like the cephalothorax in colour, and of a regular oval form, pointed behind; it, like the maxillæ and labium, is clothed with fine brownish hairs.

The *abdomen* is of a short oval form, blunted at each end, and tolerably convex above it is of a pale clay-colour, covered thinly with yellowish-brown hairs; the four exterior spinners are moderately long, and of nearly equal length; the genital aperture is rather large and of characteristic form.

Hab.—Between Yangihissár and Sirikol, March 1874.

6.—DRASSUS INTEREMPTOR, sp. n., Pl. I, Fig. 5, ♂.

Adult male: length 3⅔ lines.

The *cephalothorax* is very similar in form to that of *D. infletus*, though rather narrower in front; its colour is yellow-brown, and it is clothed pretty thickly with grey pubescence.

The *eyes* are rather small, but placed in the usual two transverse curved rows; the foremost row, which is the shortest, is nearly straight, the hinder one considerably curved and the curves of both have their convexities directed backwards. The eyes of the hinder row are equidistant from each other, those of the central pair of this row being rather the smallest of the eight; those of the fore-central pair are the largest, and form a line longer than the hind-centrals, the interval separating them being about equal to an eye's diameter, and each is very nearly contiguous to the lateral of the same row next to it. The eyes of each lateral pair are placed a little obliquely, and are rather nearer together than those of the hinder row are to each other; the longitudinal diameter of the trapezoid formed by the four central eyes is considerably greater than the transverse one; the height of the clypeus is about equal to the diameter of one of the fore-central eyes.

The *legs* are strong and of tolerable length, and rather lighter in colour than the cephalothorax; their relative length is 4, 1, 2, 3, and they are pretty thickly clothed with sandy-grey hairs (among which are some of a browner hue), bristles, and spines; some of the latter are beneath the metatarsi and tibiæ of those of the 1st and 2nd pairs, but the greater number are on the third and fourth pairs. Each tarsus terminates with two claws, beneath which is a small claw-tuft; and beneath the tarsi is a scopula extending a little way underneath the anterior portion of the metatarsi.

The *palpi* are short, tolerably strong, and similar in their colour and armature to the legs. The humeral joint has several black spines on its upper side; the cubital joint is stronger and a little longer than the radial; the latter is furnished with longish bristly hairs, and expands at its fore-extremity, which is prolonged on the outer side into a tolerably strong, rather tapering, reddish-brown apophysis, terminating in an obtuse, flattened, corneous point; the digital joint is large, oval, and of a browner hue than the rest; the palpal organs are prominent and well developed; they are of a yellowish colour, traversed near the middle by a distinct yellow-brown spine-like fillet or band, close in front of which is a strong, curved, tapering, reddish-yellow-brown corneous process, with another very similar, but smaller, in front of it; a third, smaller still and apparently obtuse, being in front again, just below the fore-extremity of the joint.

The *falces* are neither very long nor strong; their direction is nearly vertical, and they are similar in colour to the cephalothorax; their front surface is clothed with greyish pubescence and some brown hairs and bristles.

The *maxillæ* are strong, considerably bent towards the labium, over which their extremities almost meet, and broadly impressed across the middle; their colour is rather darker than that of the cephalothorax.

The *labium*, owing to some foreign matters adhering to it, could not be very distinctly seen, but its form appeared to be oblong, rounded at the apex, and its colour like that of the maxillæ.

The *sternum* is oval, pointed behind, like the maxillæ in colour, and clothed with grey pubescent hairs.

The *abdomen* is about equal in length to the cephalothorax, of an oblong-oval form, not very convex above; it is of a somewhat mottled clay-colour, with an oblong, brownish, dorsal marking on the anterior half of the upper side, produced behind into a narrow brown-pointed stripe: the fore-extremity of this dorsal marking is strongly suffused with rusty red brown. The middle of the upper side of the abdomen has four small red-brown impressed spots in the form of a square, whose fore-side is rather the shortest; it is clothed, but not very densely, with coarsish dark brown hair. The spinners are strong, those of the inferior pair being double the length of those of the superior.

Hab.—Neighbourhood of Leh, August or September 1873.

7.—DRASSUS INVISUS, sp. n., Pl. I, Fig. 6, ♀.

Adult female: length rather more than 5 lines.

This spider is nearly allied to *D. interlisus*, which it resembles in form, general colouring, structure, and appearance: it is however smaller, and the colour of the caput is much less rich, being but little darker than the thorax, which is a dull yellow-brown; the whole of the cephalothorax is covered with a sandy-grey pubescence; and there is a dark line running down the middle of the caput from the hind-central pair of eyes to the thoracic indentation.

The *eyes* are also different in their position from those of *D. interlisus*, those of the hind-central pair being placed obliquely to each other, and those of the fore-central pair nearer together and further from the laterals.

The *falces* are less strong, and the apex of the *labium* does not reach so nearly to the extremity of the maxillæ.

The *abdomen* is of a rather short, oblong-oval form, tolerably convex above: it is of a dull-yellowish hue, thinly clothed with fine hairs: along the middle of the fore-half on the upper side, is a slightly darker, but clearly defined, oblong marking, which has its hinder part tapered off to a sharp point, and an angular point on each side where the tapering portion begins. There are also four small dark blackish-brown oblique spots on the fore-half, forming a rectangle whose length is about double its breadth: two fine parallel brownish lines run on the under side from the genital aperture to a little distance from the spinners, and from each of the inferior pair of spinners a similar line runs a little obliquely to a point in a line (in a transverse direction) with the termination of the two other lines just mentioned: the spinners are short and strong, those of the inferior pair being the strongest and a little the longest: the genital aperture is small and of a very simple form.

Hab.—Between Sirikol and Aktalla, between the 8th and 31st of May 1874.

8.—DRASSUS INTERPOLATOR, sp. n., Pl. I, Fig. 7, ♂.

Adult male: length 4½ lines.

The *cephalothorax* is oval, truncated and narrowest before, and tolerably constricted on the lateral margins of the caput; the profile line slopes gradually forwards from the beginning of the hinder slope of the thorax; its colour is yellowish-brown radiated with darker stripes, which follow the directions and lines of the thoracic and other normal indentations; the whole surface is pretty thickly clothed with yellowish-grey pubescence.

The *eyes* are in the usual two transverse curved rows, the hinder one of which is the longest and the most curved; those of the hind-central pair are separated by more than a diameter's distance from each other, and are thus rather nearer to each other than each is to the lateral of the same row on its side; those of the fore-central pair are slightly the largest of the eight, and rather further from each other than each is from the fore-lateral eye on its side; those of each lateral pair form an oblique line, and are divided by an interval of nearly about an eye's diameter. The height of the clypeus is equal to the diameter of one of the fore-central eyes.

The *legs* are rather long and not very strong; their relative length is 4, 1, 2, 3; they are of a dull yellow-brown colour, clothed with sandy-greyish pubescence, and other hairs and spines, the latter are for the most part long and rather strong, and, besides a small claw-tuft under the two terminal tarsal claws, each tarsus has a scopula (though not a very dense one) underneath it.

The *palpi* are rather short and not very strong; the lengths of the cubital and radial joints are about equal; the latter increases in strength gradually to the fore-extremity, at the outer side of which there is a small tapering apophysis, whose point ends with a small, slightly curved, corneous-looking claw or nail; the direction of this apophysis is rather away from the digital joint. The radial joint is furnished with strong bristles, and a long spine on the outer side towards the hinder extremity; the digital joint is of an elongate-oval form, and equals in length the radial and cubital joints taken together; the palpal organs are not complex; the surface of the main lobe is traversed and surrounded by two red-brown, corneous-looking fillets, resembling closely applied spines, and there is dark red-brown, corneous prominence near the fore-extremity of these organs; the digital joint is dark yellowish-brown, and hairy, and has a strongish spine on its outer margin; the colour of the other joints of the palpi is similar to that of the legs.

The *falces* are moderately long and strong, and their direction is rather forwards; they are of a dark red-brown colour and furnished with hairs and bristles.

The *maxillæ* are tolerably long and strong, slightly curved and inclined towards the labium, and strongly impressed in an oblique direction across the middle; their extremities are rather rounded, and their colour is yellowish red-brown, pale whitish at the extremities.

The *labium* is of an oblong form, truncated at the apex, and similar to the maxillæ in colour, its length being nearly about two-thirds that of the maxillæ.

The *sternum* is of a dull brownish-yellow colour, and of an oval form, pointed at its hinder extremities, and depressed between the insertions of the legs.

The *abdomen* is of a rather narrow-oval form, and moderately convex above; it is of a dull brownish clay-colour, thinly clothed with hairs, and has an oblong, dull-brown, median longitudinal marking, whose hinder extremity is gradually produced into a sharp point on the fore-half of the upper side, where also four small brown spots form a square, whose fore-side is rather the shortest; the two hinder ones of these spots are in a line with the point of the oblong marking; the spinners-are rather long and strong, those of the inferior pair being much the strongest and nearly double the length of the superior pair; their colour is brownish-yellow.

This species is nearly allied to *D. lapidicolens*, Walck.

Hab.—Hills between Sirikol and Aktalla, between the 8th and 13th of May 1874, and on the road across the Pamir from Sirikol to Panja and back between April the 22nd and May the 7th, 1874.

9.—DRASSUS DISPULSUS, sp. n., Pl. I, Fig. 8, ♂.

Adult male : length 4½ lines; adult female, 5 lines.

This spider, which is allied to *D. lapidicolens*, Walck., is very similar in its general form structure, and appearance to *D. interpolator* ; it is, however, of a generally brighter hue.

The *cephalothorax* is of a brownish-yellow colour, the normal indentations of a darker hue; the thoracic indentation being deep red-brown. The *falces*, *maxillæ*, and *labium* are reddish yellow-brown, the *legs* and *sternum* yellow, and the *abdomen* pale straw-yellow. The *cephalothorax* is covered with greyish-yellow pubescence.

The *eyes* are of moderate size, and not very unequal; they are in the usual position, but the hinder row is not so much curved as in *D. interpolator* ; those of its central pair are much nearer together than each is to the lateral of the same row on its side; they are of an oval form, placed very slightly obliquely and less than their longest diameter's distance from each other; those of the fore-central pair are further from each other than each is from the lateral eye on its side, with which it is nearly, but not quite, in contact. The interval between the fore-centrals is nearly about a diameter, and these eyes form a line rather longer than that formed by those of the hind-central pair : those of each lateral pair are obliquely placed, and are separated by an interval equal to the diameter of the foremost of them.

The *legs* are rather long and slender, armed with longish spines, especially on the tibiæ and metatarsi of those of the two hinder pairs ; their relative length is 4, 1, 2, 3. Beneath the two terminal claws of each tarsus is a small claw-tuft, with a scopula of blackish hairs along the under sides of the tarsi, and of the first and second pairs of the metatarsi also.

The *palpi* (♂) are rather short, the humeral and cubital joints are yellow, the radial and digital joints suffused with yellow-brown, the latter being the darkest : the cubital and radial joints are of equal length; the latter expands a little at its anterior extremity, which is produced (on the outer side) into a rather long, not very strong, slightly tapering apophysis : this apophysis is nearly straight, but a little divergent from the digital joint, and its extreme point is bifid ; there is also another shorter, angular prominence, or projection, at the extremity of this joint, on the inner side. The digital joint is elongate-oval, equal in length to the cubital and radial joints together. The *palpal* organs are simple but rather prominent, their fore-extremity has a somewhat truncated appearance, and is broken up into several corneous spines and processes.

The *falces* are neither very long nor strong ; they are straight, and their direction is but a little forwards.

The *maxillæ* and *labium* are similar in form to those of *D. interpolator*.

The *abdomen* is of an elongate-oval form, rather truncated before; it is very thinly furnished with hairs, and in some examples an oblong dull marking, pointed at its hinder extremity, is faintly traceable on the fore-half of the upper side, where there are usually also six small dull spots, in three successive, transverse pairs, forming an oblong parallelogram; those of the middle pair are the nearest together.

The *spinners* are long, but not very stout nor very unequal in length; those of the inferior pair are the largest and strongest: their colour is like that of the legs.

The female resembles the male in colours and general structure, but is rather larger; there is, however, some little variation in size in different individuals of both sexes; the form of the genital aperture, which is rather small, is simple, but, as usual, quite characteristic.

Hab.—Káshghar, December 1873; Tanktze to Chagna and Pankong valley, between the 15th and 21st of September 1873. Between Yangihissár and Sirikol, March 1874; near Leh, August and September 1873. Yangihissár, April 1874. Yárkand and neighbourhood, November 1873. Road from Yarkand to Bursi, May 28th to June 17th, 1874; and road across the Pamir from Sirikol to Panja and back, April 22nd to May 7th, 1874. Hills between Sirikol and Aktalla, May 8th to 13th, 1874; and the Sind Valley, August 5th to 13th, 1873.

It is thus the most widely spread and numerously represented species of this family contained in the collection, occurring in all the five districts traversed.

10.—Drassus interlisus, sp. n., Pl. I, Fig. 9, *♂*.

Adult female: length 6¾ lines.

The *cephalothorax* of this fine species is of an oblong-oval shape, tolerably convex above, broadly truncated at the fore-extremity, and but very slightly constricted on the lateral margin of the caput; the profile line is very nearly level from the hinder slope to the occiput, whence it slopes forwards in a more rounding form; its colour is a bright reddish yellow-brown, deepening gradually to the caput, the fore part and sides of which are black red-brown: the whole of the *cephalothorax* is pretty densely clothed with short yellowish-grey pubescent hairs; the normal indentations are not very strongly defined, and the height of the clypeus is about equal to the diameter of one of the fore-central eyes.

The *eyes* are rather small, not very different in size, and placed in the two usual transverse curved rows, the hinder row being the longest and most curved: they are not very closely grouped together, and those of the fore-central pair are seated on a slight but perceptible prominence. These two eyes are nearly two diameters distant from each other, and are much more widely separated from each other than each is from the lateral eye of the same row on its side, with which it is nearly, but not quite, contiguous; those of the hind-central pair are oval, not obliquely placed, but with their longer diameter in a directly transverse direction; they are very near together, but not quite contiguous to each other, and each is separated by a distance nearly equal to twice its longer diameter from the lateral eye of the same row on its side; the eyes of each lateral pair are placed in an oblique line, and are rather widely separated. All the eyes, excepting those of the fore-central pair, are rather depressed or sunken into the surface of the caput.

The *legs* are strong and moderate in length, their relative length being 4, 1, 2, 3; they are yellow, deepening to red-brown on the tarsi, and are furnished with hairs, bristles, and spines, the latter almost entirely on the tibiæ and metatarsi of the third and fourth pairs; the hairs are mostly of a grey pubescent kind. Each tarsus terminates with two curved, pectinated claws, beneath which is a claw-tuft, and the undersides of the tarsi, as well as a portion of the metatarsi, are furnished with a scopula.

The *palpi* are short; their colour is yellow, deepening to dark red-brown on the digital joint, which is double the length of the radial; it is furnished thickly with dark hairs and some black spines, and terminates with a short slightly curved black claw. The cubital is similar to the radial joint in length.

The *falces* are strong and rather long; their direction is forward, and their profile arched; they are of a deep, black red-brown colour, and are clothed pretty thickly with a greyish pubescence, besides other hairs and bristles.

The *maxillæ* are long and strong (especially at the insertion of the palpi), curved and inclined considerably towards the labium, obliquely impressed across the middle, rounded at their outer extremity, and obliquely truncated on their inner extremity: their colour is slightly less dark than that of the *cephalothorax*, and they are tipped with whitish yellow.

The *labium* is long, reaching almost to the inner extremity of the *maxillæ*; it is of an oblong form, rounded at the apex, depressed along the sides, and its colour is like that of the *maxillæ*.

The *sternum* is heart-shaped, similar in colour to the hinder part of the *cephalothorax*, and clothed with hairs.

The *abdomen* is of an oblong-oval form, of a yellowish-clay colour, and thinly clothed with brownish hairs; the spinners are short but strong, and of a yellow-brown colour, those of the superior pair being a little shorter than those of the inferior; the genital aperture and the process connected with it are of peculiar and characteristic form, and of a deep blackish red-brown hue.

The male differs in no respect of colour, general form, and character from the female. The *palpi* of the male are rather short, but strong; the humeral joint is much bent and flattened on its inner side, where it curves round the *falces*, enlarging also to the anterior extremity; the cubital joint is short, strong, tumid on the upper side, and has, at its outer extremity, a long, curved, pointed, red-brown apophysis, the point being recurved or sinuous; this apophysis reaches as far as the fore-extremity of the radial joint, which is shorter, darker coloured, and less strong than the cubital, and is furnished with two nearly black apophyses; one on the outer side is strong, curved, and has its obtuse point directed upwards; the other is shorter, straighter, more pointed, and placed near the middle of the fore-extremity, to which it is nearly perpendicular; the digital joint is large, of an oval form, dark red-brown in colour, hairy, and in length exceeds the radial and cubital joints taken together; the *palpal* organs are simple but well developed, consisting of a strong, somewhat cylindric, corneous lobe, the fore-extremity of which is broken into several not very prominent processes.

Hab.—Káshghar, December 1873; Yárkand, 21st to 27th May 1874; between Yangihissár and Sirikol, March 1874; neighbourhood of Leh, August or September 1873; Yárkand to Bursi, between May 28th and June 17th, 1874.

I have had some hesitation in describing this spider as a *Drassus*. It appears to be nearly allied to *Hypsinotus*, L. Koch, but the length of the labium distinguishes it readily from that genus; for the present, therefore, I include it in the genus *Drassus*, with which, at any rate, it is very nearly allied.

11.—DRASSUS INVOLUTUS, sp. n., Pl. I, Fig. 10, ♀.

Adult female: length 4⅜ lines.

This spider has an exceedingly *Clubiona*-like appearance, but the inclination of the maxillæ to the labium and the transverse impression of the former, as well as some other characters, distinguish it at once from the spiders of that genus.

The *cephalothorax* is oval, truncated at each end, but narrowest before; its colour is rather a bright yellow-brown, deepening towards the fore part of the caput; the normal indentations, especially those which divide the caput and thorax, are suffused with brown; and the thoracic indentation is shown by a short, deep red-brown line. The marginal constrictions on each side of the caput are very slight, and the profile line, including the hinder or thoracic slope, forms a pretty even, arched line; the upper side is thus tolerably convex, and its surface is thinly clothed with greyish sandy pubescence.

The *eyes* are rather small, and in the ordinary two, transverse, curved rows, of which the hinder one is the longest, and the most curved. The four central eyes form a rectangle, whose longitudinal is rather greater than its transverse diameter; those of the hind-central pair are oval, very little, if at all, oblique, and separated from each other by an interval equal to their longest diameter, and no more than half the length of that which separates each from the lateral eye of the same row on its side. Those of each lateral pair form an oblique line, and are wide apart, though rather nearer together than the hinder one is to the hind-central eye next to it; those of the fore-central pair (which are the largest of the eight) are separated by an eye's interval, and are farther apart than each is from the fore-lateral eye on its side; the height of the clypeus is slightly greater than the diameter of one of the fore-central eyes.

The *legs* are moderately strong, but not long; their relative length is 4, 1, 2, 3, and they are of a yellow colour, furnished with hairs and spines; these latter are, nearly all, on those of the third and fourth pairs; the two terminal tarsal claws have a small claw-tuft beneath them, and there is a scopula underneath the metatarsi and tarsi of the first and second, and under the tarsi of the third and fourth pairs.

The *palpi* are moderate in length and strength, similar to the legs in colour, and furnished with hairs and a few spines.

The *falces* are moderately long and strong, straight, and a little projecting in their direction; they are of a red-brown colour, furnished with bristles in front, and armed with two small teeth, close together at the inner corner of the fore-extremity.

The *maxillæ* are strong, inclined towards the labium, broader than usual near their extremities, and strongly impressed across the middle. They are of a red-brown colour, pale yellowish-white at the extremities.

The *labium* is oblong, its length being nearly about half that of the maxillæ, which it resembles in colour, with a pale margin at the apex.

The *sternum* is oval, pointed behind, and with depressions between the insertions of the legs; it is of a light brownish-yellow colour, suffused with a rather darker hue towards the margins.

The *abdomen* is oval, pointed at its hinder extremity; it is of a dull clay-yellow colour, thinly clothed with hairs; along the middle of the fore-half of the upper side is an oblong, dull-brownish marking, which tapers to a point at its hinder extremity, near which, on either side, is a short, oblique, faint line directed backwards. Six small, dull-brown spots in three transverse pairs, also form a long rectangle on the fore-half, the foremost side of the rectangle being rather shorter than the hinder one. The intermediate pair of these spots is (as is usually the case) nearer together than the foremost pair. The spinners are tolerably long, but not very strong; those of the inferior pair are rather the longest and strongest. The genital aperture is small, and of a transverse, oblong-oval form, margined with deep red-brown.

Hab.—Sind Valley, August 5th to 13th, 1873.

12.—DRASSUS LAPSUS, sp. n., Pl. II, Fig. 11, ♀.

Female (not quite adult) : length rather over 3½ lines.

Although not adult, this spider has sufficiently characteristic specific marks to entitle it to description.

It resembles *Drassus involutus* very nearly in colours, but the relative position of the eyes is quite different.

The *cephalothorax* is oval, shorter than that of *D. involutus*, and constricted laterally at the caput; it is of a pale yellow-brown colour, pretty thickly clothed with short greyish pubescence.

The *eyes* are in two transverse rows; the hinder row slightly curved, the front row shortest and less curved than the hinder one; the convexity of the curves is directed backwards; the height of the clypeus does not exceed, even if it quite equals, the diameter of one of the fore-central eyes; those of the hind-central pair are oval, oblique, and very near together, though separated by a distinct interval, and each is separated from the hind-lateral on its side by an interval equal to the diameter of the latter; those of the fore-central pair are nearly a diameter's distance from each other, each being very nearly, if not quite, contiguous to the fore-lateral on its side; those of each lateral pair are placed obliquely, and are separated by rather less than the diameter of the hind-lateral eye; the four central eyes form a regular quadrangular figure, whose longitudinal diameter is considerably greater than its transverse one.

The *legs* are strong, and moderately long; their relative length is 4, 1, 2, 3. They are slightly lighter-coloured than the cephalothorax, and are furnished with hairs (some of these are of a greyish hue), slender bristles, and spines; these last are tolerably strong, not very long, and almost entirely confined to the tibiæ and metatarsi of the third and fourth pairs, whereon they issue from small red-brown tubercles, on the upper, as well as the under, side of the joints; the only spines on the legs of the first and second pairs are one or two longish ones of a bristle-like nature on the upper side of the femora, and a single short strong one on the under side, close to the hinder extremity of the tibiæ of the second pair; there is a small black claw-tuft beneath the two terminal tarsal claws, and a thin scopula beneath the tarsi and metatarsi of the first and second pairs.

The *palpi* are similar in colour to the legs; the digital joint is longer than the radial, and has, besides hairs and bristles, a few stoutish spines.

The *falces* are short, and not particularly strong; they are directed a little forwards, and are of a reddish yellow-brown colour, with some prominent black bristles in front.

The *maxillæ* are of moderate length and strength, curved over the labium; impressed along the middle, and, with the *labium*, which is of an oblong-oval form, similar to the falces in colour.

The *sternum* is oval, pointed behind, and similar in colour to the cephalothorax.

The *abdomen* is of an oblong-oval form, rounded behind and truncated before; it is of a straw-yellow colour, thinly clothed with hairs, some of which are blackish-brown, and most numerous at, and below, the fore-extremity of the upper side; on the fore-half of the upper side, four impressed spots form a quadrangular figure whose interior side is rather less than its posterior one, and whose longitudinal is greater than its transverse diameter. The spinners are tolerably strong, but not very long; those of the inferior pair are the longest and strongest. Such traces of it as were visible indicated that the genital aperture would be of small size.

Hab.—Yangihissár, April 1874.

Genus—GNAPHOSA, Latr.

13.—GNAPHOSA STOLICZKÆ, sp. n., Pl. II, Fig. 12, ♂.

Adult male: length 4¼ to 4¾ lines.

Cephalothorax oval, rather broad and truncated before, but only slightly constricted on the margins at the fore part of the caput; the hinder slope is rather abrupt, and the profile line has a slight slope all the way to the eyes. The colour is a dull orange yellow; the normal grooves and indentations (which are not very strongly marked) are of a more dusky hue, the thoracic indentation forming a red-brown line. The surface is clothed with sandy-grey pubescence.

The *eyes* are of tolerable size, and placed, as usual, in two transverse, slightly curved rows. The convexity of the curve of the hinder row, which is the longest, is directed forwards, so that the interval between the eyes of each lateral pair is as great as that between the eyes of the fore and hind-central pairs. Those of the hind-central pair are narrow-oval, placed obliquely, and separated by a rather less interval than their longest diameter, and each is, as nearly as possible, the same distance from the lateral eye of the same row, on its side, as the latter is from the fore-lateral eye opposite to it. Those of the fore-central pair are placed on a slight prominence, and are the largest of the eight. They are separated from each other by an interval of rather less than an eye's diameter, forming a line perceptibly longer than that formed by those of the hind-central pair. Each fore-lateral eye is very near to the fore-central on its side, but not contiguous to it. The clypeus, in height, exceeds the diameter of one of the fore-central eyes, and is furnished with a few strong prominent black bristles.

The *legs* are strong and moderately long, their relative length being 4, 1, 2, 3. They are a little paler than the cephalothorax, and are clothed thinly with a greyish sandy-coloured pubescence, besides other hairs, bristles, and spines. Excepting a very few on the upper sides of the femora of all the legs, the spines are confined to the tibiæ and metatarsi of those of the third and fourth pairs. The two terminal tarsal claws appear to vary in the number of their pectinations, which do not exceed three or four at the most, and which in the third and fourth pairs seem to be fewer than in the first and second. Beneath these claws is a small claw-tuft; and the tarsi of the first and second pairs have a scopula underneath them.

ARANEIDEA. 17

The *palpi* are short and moderately strong, similar to the legs in colour, and furnished with hairs and some long bristles. The radial and cubital joints are short, but, as nearly as possible, of equal length and strength. The former terminates at its fore-extremity, on the outer side, with a small, tapering, sharp-pointed, curved, reddish-brown, corneous-looking apophysis. The digital joint is elongate-oval, rather stouter than the radial, but not quite so long as this and the cubital together. The palpal organs are simple, and not very prominent, with a curved, red-brown, tapering, sharp-pointed spine directed forwards at their fore-extremity near the inner side; and about the middle of their fore-extremity is another spine, much smaller, and of a somewhat crooked form.

The *falces* are strong and of moderate length; their direction is a little forwards; and they are of a deep rich red-brown colour, clothed in front with long, strong, spinous bristles.

The *maxillæ* are curved, and inclined towards the labium, and their width, across the middle, is much increased by a development of that part, resembling a large semi-circular lobe which gives them a somewhat sub-triangular form. They are also strongly bent, or impressed transversely, across the middle, and their colour (excepting at the extremities, which are pale yellowish) is like that of the falces.

The *labium* is oblong-oval, rounded at the apex, which is of a pale-yellowish hue; the colour of the rest being like that of the maxillæ.

The *sternum* is of a slightly heart-shaped, oval form, of a reddish yellow-brown colour, impressed between the insertions of the legs, and clothed with hairs.

The *abdomen* is of an oblong-oval form, rounded behind, rather truncated before, and moderately convex above. It is of a straw-yellow colour; the normal oblong, longitudinal marking on the fore-half of the upper side is generally obsolete; now and then it is slightly traceable, and a small patch at its fore-extremity is of a yellow-brown hue. The whole abdomen, above and below, is clothed with greyish sandy pubescence, mixed thinly on the upper side, chiefly, with long, nearly erect, tapering, strongish, black-brown bristles. The spinners are very unequal in size, those of the inferior pair being much the longest and strongest.

The *female* resembles the male in colour and in all other general characters, but differs in size (being 5 to 6 lines in length), and in having rather longer legs. The genital aperture is small, of characteristic form, and edged with red-brown.

This fine and very distinct species, which I have dedicated to its discoverer, the late Dr. Stoliczka, was found in the following localities.

Hab.—Between Yangihissár and Sirikol, March 1874; from Yárkand to Bursi, May 28th to June 17th, 1874; also at Yangihissár, April 1874; and Káshghar, December 1873.

14.—GNAPHOSA PLUMALIS.

Gnaphosa plumalis, Cambr., P. Z. S. 1872, p. 225, pl. xv, fig. 3.

Hab.—An immature female, which I have no doubt is of this species, was found on the route from Yárkand to Bursi, May 28th to June 17th, 1874.

15.—GNAPHOSA MŒRENS, sp. n., Pl. II, Fig. 13, ♂.

Adult female: length 3 to 3¼ lines.

The whole of the fore part of this spider is of a dull yellow-brown colour; the falces,

c

maxillæ, and labium being, however, darker than the rest; the falces, indeed, are dark red-brown.

The *cephalothorax* is oval, truncated both before and behind, and slightly constricted on the margins at the fore part of the caput. The normal grooves and indentations are distinct, but not very strongly marked; the hinder slope is rather abrupt, but convexly rounded, and the profile line of the upper part is as nearly as possible level, the fore part of the caput (including the ocular area) rounding, and sloping a little forwards. The surface is clothed with greyish pubescence, mixed with more erect and darkish hairs and bristles. The lateral margins are bounded by a black-brown line.

The *eyes* are in the ordinary position, forming two transverse curved lines; the convexity of the curve of the hinder row is directed forwards. This row is the longest and much the most strongly curved, the foremost row being, in fact, almost straight, its convexity being rather directed backwards; thus the interval between the eyes of each lateral pair is considerably greater than that between the fore and hind-central pairs. They are seated on blackish tubercles; those of the hind-central pair are oval, oblique, divided by an interval equal to the length of their longest diameter, and, with the fore-central pair, from a square whose fore side is very slightly the shortest; each of the hind-central eyes is separated from the hind-lateral next to it by more than twice its longest diameter; the fore-laterals are the largest of the eight, and each is separated from the fore-central next to it by less than the diameter of the latter; the interval between the fore-centrals being rather greater than this diameter; the height of the clypeus is about equal to the space, taken in a longitudinal line, between the fore and hind-central pairs.

The *legs* are strong and moderately long; their relative length being 4, 1, 2, 3. They are pretty thickly furnished with hairs, bristles, and spines; the last are the longest, and are most numerous on those of the third and fourth pairs; there are, however, several spines on the under sides of the tibiæ and metatarsi of the first and second pairs also; beneath the two terminal claws is a small claw-tuft, and there is a thin scopula beneath the tarsi and a portion of the metatarsi of the first and second pairs.

Palpi rather short and slender; the radial joint is rather longer than the cubital, and the digital joint is longer than the radial, and slightly suffused with reddish brown. They are furnished with hairs, bristles, and a few slender spines, and terminate with a single curved claw.

The *falces* are strong, moderate in length, rather prominent near the base in front, and furnished with long prominent bristles and hairs.

The *maxillæ* are curved, and considerably inclined to the labium; and are enlarged in a rather semi-circular form at the outer side, so as to be very broad across the middle, where they are also strongly impressed.

The *labium* is of an oval form, truncated at its base, and rounded at the apex.

The *sternum* is oval, pointed behind, and depressed between the insertions of the legs.

The *abdomen* is oblong-oval, truncated before, rounded behind, and not very convex above, but projecting a little over the base of the cephalothorax; it is of a deep mouse-brown colour with three or four transverse bars of different lengths, and tending to run one into the other, formed by silky-grey pubescence on the fore-half of the upper side; these are succeeded by several transverse, blackish, but inconspicuous, angular lines or chevrons towards the spinners. A silky-grey pubescence appears to be also more or less dispersed on the hinder part; and the whole upper side is more or less speckled with black points, some of which, on the fore-half, may be traced in two longitudinal central lines converging backwards, as is indicating the position of the normal, but here obsolete, dorsal marking; the fore margin if furnished beneath with a tuft of long, strong, upturned bristles; the under side is of a uniform yellowish mouse-brown colour: the spinners of the inferior pair are yellow-brown, and much the largest and strongest.

The male resembles the female in all general characters, colours, and markings, and differs but very little in size. The *palpi* are short but moderately strong. The radial joint is a little shorter than the cubital, and has its fore extremity, near the outer side, produced into a tolerably strong, rather long, tapering-pointed apophysis, the point spinous-looking, red-brown, and slightly bent or curved; this apophysis is about equal to the joint itself in length. The digital joint is large, of an elongate-oval form, hairy, and rather longer than the radial and cubital joints together; the palpal organs are well developed, but simple, with a tolerably strong, sharp pointed, slightly curved process of a brightish red-brown colour situated on their inner side, and directed to their fore extremity.

Hab.—Hills between Sirikol and Aktalla, May 8th to 13th, 1874; between Yangihissár and Sirikol, March 1874; and from Yárkand to Bursi, May 28th to June 17th, 1874.

Genus—*PROSTHESIMA*, L. Koch.

10.—PROSTHESIMA CINGARA, Camb.

Prosthesima cingara, Cambr., P. Z. S., 1874, p. 382, pl. li, fig. 10, ♀.

The female only of this spider has been described. The following is a description of the male.

Adult male: length 2 lines.

Cephalothorax oval, broadest towards the hinder part, whence it narrows gradually to the fore extremity; its upper side is flattened-convex above, and a little highest at its posterior extremity; it is smooth, of a deep, rich brown colour, and thinly clothed with hairs.

The *eyes* are in two very slightly curved rows, the curves directed backwards, and the front row the shortest; those of the hind-central pair are small, oval, but not placed obliquely, near to each other, but separated by a distinct interval, less than that which divides each from the hind-lateral on its side; the fore-lateral eyes are the largest of the eight, and the hind-centrals the smallest; the fore-centrals are divided by an interval rather greater than a diameter, and each is almost contiguous to the fore-lateral on its side; the interval between each hind-lateral eye and the hind-central next to it is nearly about the same as that which divides the eyes of each lateral pair. The height of the clypeus is less than half that of the facial space.

The *legs* are moderate in length and strength, the femora strongly incrassated on their upper sides; they are furnished with hairs, bristles, and spines, the last chiefly on the tibiæ and metatarsi of the third and fourth pairs. Their colour is deep blackish-brown, that of the metatarsi and tarsi being of a dull yellowish-brown hue.

The *palpi* are moderately long and strong and of a brownish-yellow colour; the radial is rather less than the cubital joint in length, and has its fore extremity on the outer side prolonged into a strong, tapering, pointed, dark red-brown, straight, and rather prominent apophysis, as long as, if not longer than, the joint itself; the digital joint is large, oval, hairy, and of a yellowish-brown colour; its length exceeds that of the radial and digital joints together; the palpal organs are well developed, and consist of several characteristic corneous processes and spines.

The *falces* are moderate in length and strength, and are rather paler coloured than the cephalothorax; they are straight, and project a little forwards, being also rather roundly prominent near their base in front, and furnished with bristles and hairs.

The *maxillæ* and *labium* are similar to the falces in colour, and their form is normal.

The *sternum* is oval, blunt-pointed behind, and like the maxillæ in colour.

The *abdomen* is of an oblong-oval form, rounded behind, truncated before, and flattened convex above; it is hairy and of deep sooty-brown colour, approaching to black, with a large shining, deep-brown coriaceous patch on the fore part of the upper side, of which it covers the whole width, but is narrower and rounded at its hinder part. The spinners are rather short, but tolerably strong; those of the superior pair are the longest and strongest.

The female resembles the male in colours and general structure; the genital aperture is characteristic, consisting of an oblong opening slightly constricted across the middle, and edged strongly on the lower side with red-brown, below which are two round, shining, boss-like corneous-looking markings.

Hab.—Yárkand, May 21st to 27th, 1874; hills between Sirikol and Aktalla, May 8th to 13th, 1874; and route across the Pamir from Sirikol to Panja and back, April 22nd to May 7th, 1874.

Genus—*MICARIA*, C. L. Koch.

17.—MICARIA CONNEXA, sp. n.

Adult male: length not quite 2 lines.

This spider is very closely allied to *Micaria pulicaria*, Sund., which it resembles in size and general characters. It may, however, be distinguished by the absence of the converging lines of white hairs on the cephalothorax, which is also deeper-coloured, and by the shorter and rather narrower form of the digital joint of the palpus; the radial joint is shorter than the cubital, and has a very sharp-pointed, tapering, rather prominent apophysis at its extremity on the outer side; the corresponding apophysis in *M. pulicaria* being much shorter and less sharp pointed.

The *abdomen* is black, covered with iridescent scales, reflecting green, purple, and reddish golden hues, but there were no white transverse lines or spots visible. The cephalothorax is of a deep rich red-brown hue, thinly clothed with grey hairs and iridescent scales.

A female example had a largish semi-circular white spot of white hairs about the middle of the upper side of the abdomen, and another on each side, the three forming a straight line traversing the upper side of the abdomen. In other respects it resembled the male. Probably, different examples of this species would present the same varieties in respect to the white spots and markings on the abdomen as are characteristic of *M. pulicaria*.

Hab.—Hills between Sirikol and Aktalla, May 8th to 13th, 1874.

18.—MICARIA PALLIDA, sp. n.

Immature male: length 2¼ lines.

This spider is allied to the preceding, but its colours and markings will at once distinguish it.

The *cephalothorax* is of ordinary form and of a brightish yellow-brown colour, clothed with greyish and iridescent hairs and scales.

The *eyes* are in two nearly concentric curved rows, like those of M. *connexa* and others, the front row being the shortest.

The *legs* are moderately long and tolerably strong; their relative length is apparently 4, 1, 2, 3. They are similar in colour to the cephalothorax, the sides of the tibiæ being suffused a little with brown, and the tarsi have the appearance of being annulated with the same colour. They are clothed with grey and iridescent hairs; the former disposed somewhat in longitudinal lines.

The *palpi* are not very long; their colour is pale yellow; the cubital joint is shorter than the radial, and the digital is of a long, narrow-pointed, oval form. Being immature, these remarks on the palpi would, perhaps, not be strictly applicable to those of the adult spider, though the proportions of the several joints would probably be the same as in the immature state.

The *falces* are tolerably long, strong, perpendicular, similar in colour to the cephalothorax, and furnished with greyish hairs and dark bristles.

The *abdomen* is of a dull yellow-brown hue, clothed thinly with iridescent, scaly hairs. It has two parallel, transverse, slight constrictions near the middle of the upper side; an indistinct longitudinal median brown marking, pointed at its posterior extremity, occupies the fore-half of the upper side, followed towards the spinners by a longitudinal series of several less distinct, transverse, curved, brown lines, the convexity of the curves directed forwards. These markings would be probably invisible, except when in spirits of wine.

Hab.—Found on the route across the Pamir from Sirikol to Panja and back, April 22nd to May 7th, 1874.

Genus—*CLUBIONA*, Latr.

19.—CLUBIONA DELETRIX, sp. n., Pl. II, Fig. 14, ♂.

Adult male 2⅔ lines: adult female, 3⅔ lines.

In colours and pattern this spider is very like *Clubiona compta*, C. L. Koch, and is still more nearly allied to *C. robusta*, L. Koch (an Australian species). It is, however, smaller than the latter, and in the palpi differs from both.

The *cephalothorax* is of ordinary form, and its colour is brownish-yellow, tinged slightly with reddish-brown towards the fore part of the caput. The normal indentations are dusky; the junction of the caput and the thoracic segments is marked with a short, fine, longitudinal red-brown line, and the surface is thinly clothed with greyish-sandy pubescence.

The *eyes* are of tolerable size, though not very unequal. They occupy the whole of the width of the upper fore part of the caput, and, when seen from above and behind, are placed in the usual two curved lines, the convexities of which are in opposite directions, and enclose a somewhat oval area. The hinder row is much the longest, and the fore-central pair are rather the largest of the eight. Those of the hind-central pair are considerably further from each other than each is from the hind-lateral on its side, the interval somewhat exceeding two diameters. Those of the fore-central pair are separated by about half a diameter, and are rather farther from each other than each is from the fore-lateral on its side; each fore-lateral eye is separated from the hind-lateral next to it by an interval similar to that which separates the two fore-central eyes; and each of the hind-central eyes is separated by a similar interval from the hind-lateral and fore-central eyes on its side; the front row, taken with the hind-lateral eyes, form a long, regularly curved line, the eyes of which are near together, and the intervals as above explained, not very different. The clypeus is very low, scarcely equalling half the diameter of one of the fore-central eyes.

The *legs* are tolerably long, but not very strong; their relative length is apparently 4, 2, 1, 3. Their colour is yellow; the tibiæ, tarsi, and metatarsi of the first and second pairs being slightly tinged with reddish-brown. They are furnished with hairs, bristles, and longish dark spines. Each tarsus ends with two curved, pectinated claws, beneath which is a small compact claw-tuft.

The *palpi* are short and similar in colour to the legs. The cubital and radial joints are of the same length, and the radial and digital joints are together greater in length than the humeral joint. The radial has, at its fore extremity on the outer side, a small, tapering, pointed, red-brown apophysis, whose direction is rather upwards. The digital joint is of tolerable size, of an elongate oval form, rounded at its base, and pointed at its fore extremity, which is densely clothed with a large patch of short, pale mouse-coloured, pubescent hairs. The palpal organs consist of a large, very prominent, oval lobe, at the fore extremity of which is a slender, coiled, filiform black spine springing from a strongish red-brown corneous process. Besides a minute filiform, slightly curved spine behind this coiled one, the large lobe has a broadish, yellow-brown, glossy, rather bent fillet running along its outer side, probably indicating the course of an internal duct.

The *falces* are moderately long, strong, somewhat subangularly prominent near their base in front, furnished with some strongish prominent bristles in front, and of a dark red-brown colour.

The *maxillæ* and *labium* are of the normal form, and of a reddish yellow-brown colour.

The *sternum* is oval, pointed behind, and its colour is yellow.

The *abdomen* is of somewhat narrow-oval form, and moderately convex above. Its colour is a dull luteous yellow, marked, more or less distinctly, with dark red-brown markings on the upper side. An elongated longitudinal marking pointed at its hinder extremity, occupies the middle of the fore half; and on the hinder half are several more or less imperfect angular bars or chevrons of the same colour; the vertices of these chevrons are usually obsolete, and their extremities are dilated and run together, so as to form two more or less diffused, lateral longitudinal, broken bands, or rows of spots and patches, which converge to the spinners: the sides have, at times, also some irregular, oblique lines of red-brown markings. The spinners are of moderate length, those of the superior pair being much more slender than those of the inferior.

The *female* is rather larger than the male, but does not differ in colours and markings. The form of the genital aperture is well defined and characteristic, but its peculiarities can only be shown satisfactorily by a figure.

Hab.—Murree to Sind valley, July 14th to August 5th, 1873.

20.—CLUBIONA LATICEPS, sp. n., Pl. II, Fig. 15, ♀.

Adult female: length 4¼ lines; length of cephalothorax 2 lines, breadth 1¼.
This spider is allied to *Clubiona deletrix*, but may be at once distinguished by the

absence of any markings on the abdomen, as well as by its larger size and broader cephalothorax. It is also nearly allied to *Clubiona cambridgii*, L. Koch, found in New Zealand, but may be distinguished from that species too by the same characters. From *C. holosericea* Degeer=*C. deinognatha*, Camb., it differs by its generally robuster form, less prominent falces, and less broad cephalothorax.

The foregoing remarks will give a general idea of this spider; the following is a more detailed description.

The *cephalothorax* is of a bluff-oval form, round behind, and truncated in front. The caput is constricted on its lateral margins, but is evenly and well rounded above. The ocular area is broad, and is a little prominent over the clypeus, which is almost obsolete. It is of a reddish yellow-brown behind, deepening into dark red-brown on the fore part of the caput, and is clothed with a short sandy pubescence: the normal indentations are of a deeper hue than the rest.

The *eyes* are rather small, but in the usual position. Those of the hinder row are equal in size. Those of the hind-central pair are farther from each other than each is from the hind-lateral on its side, and each is separated by nearly about the same interval, from the fore-central eye nearest to it. Those of the fore-central pair are the largest of the eight, and are divided by a diameter's distance; and from each of them the fore-lateral on its side is separated by rather less than a diameter. Those of each lateral pair are placed very obliquely, and are separated by an interval only a little less than that which divides the fore and hind-central pairs.

The *legs* are strong, moderately long, and of a dull orange-yellow colour; those of the first and second pairs being tinged with red. Their relative length appears to be 4, 2, 1, 3; and they are furnished with hairs, slender bristles, and strongish spines. Each tarsus ends with two curved pectinated claws, beneath which is a compact claw-tuft, followed, over the under surface of the joint as well as over some portion of that of the metatarsus, by a scopula of short compact hairs.

The *palpi* are short and slender, furnished with hairs and strong bristles. The radial joint is shorter than the digital, but longer than the cubital: the colour of the palpi is like that of the legs. The digital joint is suffused with reddish, rather enlarged at its anterior extremity, which is pretty thickly furnished with hairs, particularly on the upper side.

The *falces* are long and powerful, perpendicular, rather prominent near their base in front, where they are furnished with strong, prominent bristles: their colour is deep red-brown.

The *maxillæ* and *labium* are of the usual form, and a little lighter in colour than the falces; the inner extremities of the former and the apex of the latter being whitish yellow.

The *sternum* is oval, clothed with hairs, and of a yellow-brown colour.

The *abdomen* is oval, of a brownish clay-colour, thickly clothed with sandy and greyish pubescence, together with a few longer, erect, scattered, black and yellowish hairs. The spinners are moderate in length and rather strong; those of the inferior pair are the largest and strongest. The form of the genital aperture is characteristic.

Hab.—Murree, June 11th to July 14th, 1874.

21.—CLUBIONA LAUDATA, sp. n., Pl. II, Fig. 16, ♂.

Adult male: length rather under 2 lines.

The *cephalothorax* of this small species is broad-oval, truncate before, and the lateral constriction of the caput is very slight; it is glossy, and of a brownish-yellow colour, rather deepening towards the fore margin. The clypeus is almost obsolete.

The *eyes* are small, not greatly differing in size; they are in the usual position, and occupy the whole width of the forepart of the caput; the two rows are rather nearer together than in the two former species, but the relative position of the various eyes is very similar. The interval between those of the hind-central pair is double that between each and the hind-lateral eye on its side.

The *legs* are moderate in length and strength; their colour is yellow, and they are furnished with hairs and a few spines, but the latter had been mostly broken off.

The *palpi* are short, and their colour is yellow, excepting the digital joint, which is brown; the radial is rather shorter than the cubital joint and has a moderate-sized, red-brown, pointed, tapering, slightly curved apophysis at its extremity on the outer side. The digital joint is oval, of moderate size, and slightly exceeds in length the radial and cubital joints together; the palpal organs consist (so far as I could ascertain) of a simple, large, oval, convex lobe, apparently surrounded on its outer margins by a long, slender, filiform spine.

The *falces* are strong and massive, a little projecting, roundly prominent near their base in front, and similar in colour to the cephalothorax.

The *maxillæ* and *labium* are of normal form and similar in colour to the cephalothorax.

The *sternum* is oval, pointed at its hinder extremity, and its colour is yellow.

The *abdomen* is rather small, and its form oval; its colour is a dull yellowish, thickly mottled and streaked above, and on the sides, with rusty red. The fore half of the upper side has an elongate longitudinal marking of a deeper rusty red-brown than the rest, bisected along its length by a fine, but not very clear, yellowish line. The spinners are pale yellowish, of moderate size and length, those of the inferior pair being a little the stoutest.

The *female* is rather larger, but resembles the male in colours and markings, except that the abdomen is less marked and streaked with rusty red; the form of the genital aperture, which is quite small, is characteristic.

Hab.—Road from Yárkand to Bursi, May 28th to June 17th, 1874.

Genus—*CHIRACANTHIUM*, C. L. Koch.

22.— CHIRACANTHIUM ADJACENS, sp. n., Pl. II, Fig. 17, ♂.

Adult male: length rather over 3 lines.

The form of the cephalothorax is of the ordinary type; in fact, this spider in its general form, structure, and appearance, bears a close resemblance to *Chiracanthium nutrix, C. carnifex*, and other allied species. It must, however, be premised that, the spider now described having been preserved in spirit of wine for a long time, its present colours are in all probability very unlike those of the living spider, in which perhaps the abdomen may have been of a more or less green hue.

The *cephalothorax* is of a dull brownish-yellow colour clothed with sandy-grey pubescence; the normal indentations are distinct, and a largish depression surrounds the thoracic junction.

The *eyes* are rather small, in two transverse rows occupying the whole width of the fore part of the caput; the hinder row is curved, the convexity of the curve directed backwards; the front row is shortest and nearly straight, those of the hind-central pair are rather nearer together than each is to the hind-lateral eye on its side, the distance between these being equal to perhaps a little over two diameters. Those of the fore-central pair are also nearer to each other than each is to the fore-lateral on its side, being less than two diameters' distance from each other; those of each lateral pair are seated rather obliquely on a tubercle and are contiguous to each other; the interval between each fore-central eye and the hind-central opposite to it is rather greater than the diameter of the former, the height of the clypeus being less.

The *legs* are long, slender, and tapering; they are of a rather dull straw-yellow colour, all furnished with hairs and spines, and their relative length is 4, 1, 2, 3. Each tarsus ends with two claws hidden by a dense black claw-tuft, from which a thin scopula extends backwards beneath the joint, and some little way also along the under side of the metatarsi, where it merges among the ordinary hairs with which those parts are furnished.

The *palpi* are short and moderately strong, their colour is similar to that of the legs, except that the digital joint is dark brown; the humeral joint is rather longer than the cubital and radial joints together, the latter being double the length of the cubital, furnished with long bristly hairs, and terminating at its fore extremity with two apophyses; one of these on the outer side is tolerably long, of a deep red-brown colour, and corneous nature, sinuously bent, tapering, single-pointed, the point sharp, but not attenuated; the other apophysis is on the inner side, short, obtuse, rounded at its extremity, and margined with red-brown; the digital joint is large and hairy, the basal half roundish, the fore half somewhat cylindrically attenuate, the division between the two portions being (as usual) well marked by a sudden notch-like, or angular, depression on the outer side; the normal spur, directed backwards from the hinder part of the digital joint, is of a deep black red-brown colour; it tapers to a sharp point and is directed outwards, crosses the outer radial apophysis, its length being nearly about equal to that portion of the outer margin of the digital joint where the abrupt indentation divides it from the semi-cylindrical extremity. The palpal organs consist of a large roundish bulb, from the outer side of the fore part of which a tapering spine coils round to the base, where it ends in a filiform point; and along the middle is a rather long, pale, corneous process, broad, roundish, blunt, and reddish-brown at its fore extremity, which extends well beyond the bulb and has a semi-diaphanous membranous point in connection with it.

The *falces* are straight, moderately long, strong, and not very much porrected; they are roundly prominent near their base in front, when looked at in profile, and roundly cut away for a very little space on the inner side of the fore extremity; their colour is a deep, rich, shining red-brown.

The *maxillæ* are of the ordinary form, inclined a little towards the labium, which is oblong and truncated at the apex, the corners being a little rounded off; the colour of the labium is dark red-brown, the apex narrowly margined with pale whitish-yellow; the fore half of the maxillæ is of a less deep red-brown colour, the basal portion being yellow.

The *sternum* is heart-shaped, yellow, and depressed between the points of the insertion of the legs.

The *abdomen* is oval, broadest in the middle, and projects fairly over the base of the cephalothorax; it is clothed thinly with hairs, and is of a dull, luteous-yellow colour, thickly spotted with small, yellowish-white, cretaceous-looking spots, leaving the normal elongate macula distinct on the forehalf of the upper side; this macula is rather pointed at its hinder extremity. The spinners are rather small, of moderate length, and of a brownish-yellow colour.

The *female* is rather smaller, but resembles the male in general colours and form; the spiracular plates are of a deep red-brown colour and the genital aperture is small, of a transverse-oval shape margined with red-brown, and on each side of it is a longitudinal row of several short transverse red-brown lines, the rows converging forwards.

Hab.—Murree, June 11th to July 14th, 1873.

23.—CHIRACANTHIUM APPROXIMATUM, sp. n., Pl. II, Fig. 18, ♀.

Adult females: length a little over 4 lines.

In colours, form, and general structure, this spider is exceedingly like *Chiracanthium adjacens*, Cambr. The falces, however, project rather more forward, and the second or terminal joints of the spinners of the superior pair are longer. The cephalothorax, legs, palpi, and sternum are of a uniform straw-yellow colour; the falces, maxillæ, and labium are dark brown, the base of the maxillæ yellowish; and the abdomen is of a dull clay-colour, obscurely marked with whitish cretaceous-looking spots. The spiracular plates being of the same colour as the rest of the abdomen, furnish also a good specific character, those of *C. adjacens* being dark reddish-brown; the form and size of the genital aperture are also quite different, being very small, of a transverse, somewhat, oblong form, edged narrowly with reddish-brown, and divided across the middle by a broadish pale septum.

Hab—.Murree to Sind Valley, July 14th to August 5th, 1873.

Genus—*AGRŒCA*, Sund.

24.—AGRŒCA DEBILIS, sp. n., Pl. II, Fig. 19, ♀.

Adult female: length nearly 2¼ lines.

This spider scarcely differs in form and structure from *Agrœca brunnea*, Bl.

The *cephalothorax* is yellow, thinly clothed with brownish hairs. The normal converging indentations are dusky, and the junction of the caput with the thoracic segments is marked by a short, fine, longitudinal, red-brown line.

The *eyes* are of moderate size, and placed in two tranverse, curved rows, the convexity of both being directed backwards, but the hinder row is the longest and the most strongly curved of the two; they differ but little in size, and are all seated on black spots; those of the hind-central pair are rather further from each other than each is from the hind-lateral on its side, the latter interval being nearly about equal to an eye's diameter; the eyes of the fore-central pair are contiguous to each other, and each is separated from the hind-central eye opposite to it, by an eye's diameter, and from the fore-lateral on its side by a distinct, but very small, space. The height of the clypeus, in the middle, is equal to the diameter of one of the fore-central eyes.

The *legs* are tolerably long and strong, of an immaculate yellow colour, and are furnished with hairs and spines; the spines on those of the first and second pairs are long, strong, and consist of two (parallel) rows beneath the metatarsal and tibial joints; each tarsus ends with two rather weak and apparently non-denticulate claws, beneath which is a small, blunt, yellow-brown, corneous-looking projection, furnished with several bristly hairs turned upwards in opposition to the tarsal claws. The relative length of the legs appears to be 4, 1, 2, 3.

ARANEIDEA. 27

The *palpi* are moderate in length, and similar in colour and armature to the legs; the digital is double the length of the radial joint, and terminates with a weak, curved, black claw.

The *falces* are of moderate length and strength, straight, perpendicular, and obliquely cut away on the inner sides towards the extremity; their colour is yellow, and they are furnished with hairs and longish bristles.

The *maxillæ* are rather strong, moderately long, straight, somewhat rounded at their extremities, and similar to the falces in colour.

The *labium* is short, rounded at the apex, and of a yellowish-brown colour.

The *abdomen* is oval, truncated before, rounded and broadest behind; it is thinly clothed with hairs of a clay-yellow colour, marked above and on the sides with blackish brown; these markings were very much broken and fragmentary in the example described, but they appear to define faintly the ordinary oblong, median, longitudinal marking on the fore half, and some angular bars on the hinder half of the upper side, with some oblique lines on the sides. The spinners are very short; the second joints of those of the superior pair are barely perceptible. The genital aperture appears as a densely black, glossy patch in the centre of a largish yellow-brown, convex area, and is rather densely clothed with bristly hairs, whose points converge over the black aperture.

Hab.—Káshghar, December 1873.

25.—AGRŒCA FLAVENS, sp. n., Pl. II, Fig. 20, ♀.

Adult female: length 3¼ lines.

The whole of the fore part of this spider is of a dull, slightly brownish-yellow colour, the *labium*, however, being dark yellow-brown with a yellowish apex, and the *abdomen* of a straw-yellow thinly clothed with yellowish-grey hairs. In form and structure this species resembles *Agrœca debilis*; the normal indentations on the *cephalothorax* are well marked, and of a deeper hue than the rest, and it is clothed with hairs similar to those on the abdomen. The *eyes* are small, placed in two, nearly straight, transverse rows, and seated on black spots; the front row is considerably the shortest; those of the hind-central pair are very slightly farther from each other than each is from the hind-lateral eye on its side, and each is separated from the fore-central opposite to it by no more than, if quite so much as, the diameter of the former; those of the foremost row are very near to each other, the interval between those of the fore-central pair being slightly greater than that which divides each from the fore-lateral on its side; those of each lateral pair are placed very obliquely, and have an interval between them of nearly the diameter of the hinder one; the fore-laterals are the largest of the eight; the height of the clypeus is equal to the diameter of one of the fore-central eyes.

The *legs* are tolerably long and strong, furnished with hairs and rather long, strongish spines; each tarsus ends with two curved claws, apparently without any denticulations, below them being a rather less conspicuous, corneous, yellow-brown prominence than in *Agrœca debilis*, but furnished with similar upturned bristly hairs; the relative length of the legs appears to be 4, 1, 2, 3; the spines on the two first pairs are in two parallel longitudinal rows beneath the metatarsal and tibial joints.

The *palpi* are moderately long and strong; the radial and digital joints are yellow-brown; the latter are the darkest, and are distinctly less than double the length of the former; they are furnished with hairs, bristles, and spines, and terminate with a small black claw.

The *maxillæ* are slightly inclined to the *labium*, and the latter is slightly hollowed or notched at the apex.

The *abdomen* is of a pale straw-yellow colour; on the fore half is a very faintly visible, narrow, elongated marking prolonged posteriorly into a line-like point, and of a slightly darker hue than the rest of the abdomen; a little in front of the middle are six reddish-brown impressed spots, three in each of two curved lines, whose convexities are opposed to each other so as to form a figure like a dice-box; the intermediate spot of each line is much nearer the anterior than the posterior one, and the interval between the two anterior spots is equal to that between the two posterior ones. The form of the abdomen is somewhat oblong-oval, truncated before and rounded behind. The genital aperture consists of two somewhat roundish, large, black, contiguous spots in a transverse line, but much obscured by numerous bristly hairs of a sandy-grey hue; the spinners of the inferior pair are double the length of the superior one, and all are of a yellow-brown colour.

Hab.—Yárkand, May 21st to 27th, 1874.

Genus—*TRACHELAS*, L. Koch.

26.—TRACHELAS COSTATA, sp. n., Pl. II, Fig. 21, ♀.

Adult female: length $2\frac{1}{4}$ to $2\frac{3}{4}$ lines.

The *cephalothorax* is short-oval, rather longer than broad, truncated before, moderately, and pretty uniformly, convex above, and constricted on the lateral margins of the caput; the normal indentations are distinct and rather darker coloured than the rest, which is of a yellow-brown colour; the clypeus is less in height than the diameter of the fore-central eyes. The eyes are of moderate size, and do not differ greatly in that respect; those of the fore-lateral pair are perhaps slightly the largest of the eight; they are placed as in *Clubiona*, but the area is shorter and broader, the eyes being more closely grouped together. The hinder row is straight, the front row much shorter and curved, the convexity of the curve directed forwards. Those of the hind-central pair are rather less than double as far from each other as each is from the hind-lateral eyes on its side; the interval between each and the hind-lateral being scarcely the diameter of the former; the interval between those of the fore-central pair is less than half a diameter, and each is almost, but not quite, contiguous to the fore-lateral eye on its side. The eyes of each lateral pair are placed obliquely and are separated by about one-third of the diameter of the hind-lateral eyes. The interval between each of the hind-central eyes and the fore-central opposite to it, is equal to the diameter of the latter.

The *legs* are rather short, strong, furnished with hairs and slender bristles only; their relative length appears to be 4, 1, 2, 3, though the difference in length is not great. They are of a brownish-yellow colour, lighter than the *cephalothorax*, and faintly annulated with dusky brown. Each tarsus ends with two curved pectinated claws, beneath which is a small, not very dense, claw-tuft; and beneath the tarsi and metatarsi are some short, stout hairs of uniform length, but scarcely amounting to a scopula.

The *palpi* are short, similar to the legs in colour and armature; the digital joint is about half as long again as the radial, and terminates with a very fine, curved, and almost imperceptible claw.

The *falces* are short but strong, straight, and nearly perpendicular; they are roundly prominent at their base in front; their fore surface is granulose and bristly, and their colour like that of the cephalothorax.

The *maxillæ* are short, convex, and broad; their extremities, where they are obliquely and rather roundly truncated, being the broadest.

The *labium* is short, broad, and of a somewhat oblong-oval form, the apex being very slightly indented or hollowed; the colour of the *labium*, as well as of the *maxillæ*, is like that of the falces.

The *sternum* is heart-shaped, uniformly convex, slightly punctuose, furnished with short bristly hairs, and similar to the legs in colour.

The *abdomen* is oval, more convex above than in spiders of the genus *Clubiona* in general, and projects over the base of the cephalothorax; it is of a dull clayey-brown colour; the fore half of the upper side has a deep brown, longitudinal, central marking, enlarged in the middle, sharp pointed at its posterior extremity, and followed to the spinners by about six angular deep-brown bars, or chevrons, which decrease in length, from the first to the last, just above the anus; the angles of these chevrons are directed forwards; that of the first touching the pointed extremity of the central longitudinal markings on the fore half. The sides of the abdomen are more or less covered with brown striated markings; the spinners are rather short, moderately strong, and those of the superior and inferior pairs are of about equal length. The genital aperture is of peculiar and characteristic form.

Hab.—Murree, June 11th to July 14th; and near Leh, August and September, 1873.

Family—*DICTYNIDES*.

Genus—*DICTYNA*, Sund.

27.—DICTYNA ALBIDA, sp. n.

Adult female: length less than 1¼ lines.

This spider belongs to the *Dictyna variabilis* (Koch) group.

The *cephalothorax* is depressed on the sides and hinder part, and the caput is rounded on the upper side, but not raised above the usual level; it is of a dull yellowish colour, with a rather irregular, but distinct, marginal stripe, immediately above which, on each side, is a broad yellowish-brown, longitudinal band; the whole is covered, but not densely, with coarse white hairs; the height of the clypeus is less than half that of the facial space, being not much more than equal to the diameter of one of the fore-central eyes.

The *eyes* are small and placed in two transverse curved rows near together; the hinder row is considerably the longer; those of the hinder row are equidistant from each other, the centrals being slightly the largest of the eight; those of each lateral pair are placed obliquely, and are very near to each other, but not quite contiguous; the interval between the fore-centrals is considerably greater than that between each and the lateral eye next to it; the latter interval being scarcely equal to the diameter of one of the fore-central eyes, which are the smallest of the eight; the interval between the fore- and hind-central pairs is equal to the diameter of one of the hind central eyes. The fore-central eyes form very nearly a square, the posterior side being rather the longest.

The *legs* are short and slender, their relative length appears to be 4, 1, 2, 3; they are of a pale, dull yellow colour, furnished with hairs; and the metatarsi of the fourth pair have calamistra on their upper sides.

The *palpi* are rather short, slender, and similar to the legs in colour.

The *falces* are moderate in length and strength; they project a little forwards, and (looked at from in front) are curved, the curves directed outwards, leaving a slightly oval space between them; their colour is dull brownish-yellow.

The *maxillæ*, *labium*, and *sternum* are of normal form, and rather paler in colour than the *falces*.

The *abdomen* is oval, not very convex above, but projects considerably over the hinder part of the *cephalothorax*; it is of a dull brownish-yellow colour, covered with cretaceous white spots and small patches; four small red-brown spots form nearly a square on the middle of the upper side. In front of the ordinary spinners is a short, transverse, supernumerary mammillary organ, divided across the middle by a perceptible suture.

The genital aperture is small and inconspicuous.

Hab.—Between Yangihissár and Sirikol.

Family—AGELENIDES.
Genus—ARGYRONETA, Latr.
28.—Argyroneta aquatica.

Argyroneta aquatica, Clerck, Sv. Spindl., p. 143, pl. 6, tab. 8.

I can find no difference between examples contained in Dr. Stoliczka's collection and those found in Europe.

Hab.—Yárkand and neighbourhood, November 1873.

Genus—TEGENARIA, Latr.
29.—Tegenaria, sp.

An immature female, too much damaged to be recognisable beyond its generic affinity.

Hab.—Yárkand to Bursi, May 28th to June 17th, 1874.

Genus—CŒLOTES, Bl.
30.—Cœlotes tegenarioides, sp. n.

Immature male (almost adult) : length 6½ lines.

This spider is exceedingly like a *Tegenaria* in its general form and appearance.

The *cephalothorax* is round behind, but constricted laterally at the caput, and its margins are depressed; it is of a yellow-brown colour, and hairy, and has the normal indendations well marked.

The *eyes* are of moderate size, and not greatly different in that respect : they are placed in two transverse curved rows; the front row is the shorter and less curved, the convexity of the curves being directed backwards. The eyes of the hind-central pair are a little nearer to each other than each is to the hind-lateral eye on its side, being separated by rather less than two diameters; those of the fore-central pair are distinctly larger than those of the hind-central; they form a line slightly less, though of very nearly equal length, to that formed by the hind-centrals, but the interval between them is smaller, being scarcely equal to a diameter; and each is separated from the fore-central eye on its side by an equal interval; the eyes of each lateral pair are placed obliquely on a tubercle, and are separated by a distinct though small interval. The four central eyes form as near as possible a square, and the height of the clypeus equals half that of the facial space.

The *legs* are moderate in length, tolerably strong, and of a yellow-brown colour; the femoral joints faintly annulated with a lighter hue.

The *palpi* are short, hairy, and like the legs in colour; the radial is rather longer and stronger than the cubital joint; the digital is of great size and tumidity, its fore extremity rather pinched in to a point. The palpal organs are not developed.

The *falces* are straight, strong, and massive, very prominent at their base in front, and project a little forwards; they are of a deep reddish-brown colour, and furnished with strong bristles.

The *maxillæ* are strong, distinctly curved over the labium, rounded at their extremity on the outer side, and obliquely truncated on the inner side; they are of a yellowish colour, strongly tinged with yellow-brown along the inner side.

The *labium* is of an oblong-oval form, slightly truncated at the apex, and of a yellow-brown colour; the apex is tipped with yellowish, and, with the maxillæ, is covered with bristly hairs.

The *sternum* is oval, pinched in to a short, sharpish point behind, and broadly truncated before; it is hairy, like the maxillæ and labium, and of a yellow-brown colour, paler along the middle.

The *abdomen* is large, hairy, considerably convex above, mostly so at the fore extremity, where it projects well over the base of the cephalothorax; it is thickly spotted, mottled, and marked with dull yellowish-white and black-brown markings and spots, presenting a faint indication of an irregular, elongate, posteriorly pointed, median dark marking, tinged with yellow-brown along the middle of the fore half, followed towards the spinners by some indistinct, broken, angular bars or chevrons. The spinners are not very long: those of the superior pair are three-jointed, the terminal joint being no longer than the one next to it; those of the inferior pair are shorter but stronger.

Hab.—Murree, June 11th to July 14th, 1874.

31.—CŒLOTES SIMPLEX, sp. n.

Adult female: length slightly over 4 lines.

This species is very nearly allied to *Cælotes tegenarioides*, but may be distinguished not only by its much smaller size, but by the absence of any trace of annulation on the femora of the legs and by the small size of the fore-central eyes, which are the smallest of the eight, and form a line distinctly shorter than that formed by those of the hind-central pair: the interval also between the eyes of each lateral pair is rather greater than in *C. tegenarioides*, and the height of the clypeus is less than half that of the facial space. On the abdomen, also, the pattern is more distinct, shewing the transverse angular bars on the hinder-half of the upper side more clearly; the spinners are rather longer, and the genital aperture is a small oval opening at the hinder part of a largish, convex, yellow-brown, somewhat roundish, corneous-looking patch.

Hab.—Murree, June 11th to July 14th, 1874.

Family—*THERIDIDES.*

Genus—*EPISINUS*, Walck.

32.—EPISINUS ALGIRICUS.

Episinus algiricus, Luc., Explor. en Algérie, Arachn. p. 269, pl. 17, fig. 11.

This spider is exceedingly closely allied to *Episinus truncatus*, Walck., and I have but little doubt that the examples contained in Dr. Stoliczka's collection are of the same species as that described by Lucas in his great work on the spiders of Algeria. Be this as it may, however, these examples are decidedly distinct from, though very closely allied to, European examples of *Episinus truncatus* found in England. Among other distinctions, may be specially noted, the generally more yellow-brown hue of the present spider, and the far less distinct, though generally similar, pattern on the cephalothorax and abdomen; also the absence of a short, but distinct, yellow, longitudinal line running backwards from between the hind-central pair of eyes to the occiput; this line is distinctly visible in all the numerous British examples before me, but no trace of it exists in the present spider. The legs are pale yellowish, only faintly annulated with brown; and some distinct white spots forming a horse-shoe (the open side in front) round the lower extremity of the abdomen close to the base of the spinners, are larger and more conspicuous, especially the foremost of them; the corners also of the upper side of the hinder extremity of the abdomen are more conically gibbous, and an unfailing specific difference is presented in the different form of the genital aperture; this is a small, simple, nearly round, dark-coloured opening in the present spider; while in *E. truncatus* it is by no means so simple, and covers a much larger oblong area. In the latter species, the under side of the abdomen is dark, with, almost always, three longitudinal parallel white lines; while in the present spider it is pale and almost uniformly covered with white cretaceous spots.

M. Eugene Simon, indeed (*Aran. nouv. ou peu connus du midi de l' Europe*, Mém. Liége, 1875), concludes *Episinus algiricus*, Luc., to be identical with *E. truncatus*, Walck.; but he gives no proof of this, beyond the fact that he has taken numbers of *E. truncatus* in Morocco, Spain, and Corsica. M. Lucas, however, appears to have had no doubt of the distinctness of the spider he describes as *E. algiricus*.

Hab.— Murree, June 11th to July 14th, 1877.

Genus—*THERIDION*, Walck.

33.—THERIDION RIPARIUM.

Theridion riparium, Blackw., Spid. Great Brit. and Irel., p. 162, pl. xii, fig. 115.

An adult female of this spider, which, although in very bad condition, appeared to be indistinguishable from European examples, was found in Dr. Stoliczka's collection.

Hab.—Murree, June 11th to July 14th, 1873.

34.—THERIDION LEPIDUM, sp. n.

Adult female: length very nearly 2 lines.

The whole of the fore part of this pretty spider is brownish-yellow. The *cephalothorax* is of ordinary form; it has a brownish-black marginal line, or border, and a longitudinal, median, blackish-yellow-brown band extending from the eyes to the hinder margin (where it is narrower than at its commencement), and divided longitudinally by a fine paler line.

The *eyes* are small, and in the usual four pairs, occupying the whole width of the upper side of the fore part of the caput. Those of the hind-central pair are a little nearer to each other than each is to the hind-lateral eye on its side; those of the front row are equidistant from each other; and those of each lateral pair are contiguous to each other, and placed obliquely on a slight tubercle. The four central eyes form a square. The clypeus is impressed immediately below the fore-central eyes, but prominent towards its lower margin, and its height exceeds half that of the facial space.

The *legs* are tolerably long, and rather slender; they are furnished with hairs and numerous bristles; many of the latter, especially of those beneath the metatarsi and tarsi being of a spine-like character. Their relative length is 1, 4, 2, 3, those of the first pair considerably the longest.

The *palpi* are short and slender; the cubital joint is half the length of the radial, and the digital is nearly double the length of the radial. Their armature is like that of the legs.

The *falces* are rather weak, moderate in length, and straight, but a little projecting.

The *maxillæ* are long, of normal form, and their extremities are even with the extremities of the falces.

The *labium* is short, but of ordinary form.

The *sternum* is heart-shaped.

The *abdomen* is almost globular above, and projects over the base of the cephalothorax; it is of a dull luteous colour with a broad median dentated white band along the middle of the upper side, prolonged to the spinners in a narrow white stripe; the upper part of the sides have also an irregular longitudinal white band connected with the median band by three oblique narrow white stripes or lines; and close to the base of each of these oblique lines, *i.e.*, where they join the lateral white bands, is a distinct black spot. There is also another black spot at the hinder termination of the lateral band, with another on each side immediately and close above the base of the spinners, and one underneath immediately in front of the spinners. All these black spots, which are very distinct and characteristic, form two longitudinal lines converging to the spot under the spinners, which are also surrounded by some white spots; the lower part of the sides, and a broad longitudinal band on the under side are more or less mottled with small white spots.

The *male* resembles the female in colours and markings, but its abdomen is far less convex above, and the first pair of legs are much longer; the fore-extremities also of the femora and tibiæ of those of the first and second pairs are of a reddish-yellow-brown.

The *palpi* are short; the humeral joint is enlarged and tumid towards its hinder extremity, and the radial is much, and broadly, produced at its outer extremity, where it is fringed with a single row of strong bristles; the digital joints are of moderate size, oval, and bristly with their convex sides turned towards each other. The palpal organs (which are thus directed outwards) are rather complex, but possess no very markedly prominent processes.

Hab.—Sind Valley, August 5th to 13th, 1874.

This spider is nearly allied to *T. nervosum* (Walck.), which it resembles in the general character of its markings, though its colours are quite different.

35.—THERIDION SUBITUM, sp. n.

Adult female: length 2 lines.
The whole of the fore part of this spider is of a dull orange-yellow colour.
The *cephalothorax* is of ordinary form, though rather shorter in proportion to its length than in some other species of the genus; its lateral margins are slightly suffused with whitish.
The *eyes* are of moderate size, in the usual four pairs, and tolerably closely grouped together; those of the hinder row are equidistant from each other, and those of the fore-central pair are a little further from each other than each is from the fore-lateral eye on its side; the four central eyes form nearly a square whose fore side is rather shorter than the hinder one; the height of the clypeus exceeds half that of the facial space.
The *legs* are slender and not very long; they are a good deal denuded of armature, but are apparently furnished with hairs and bristles, many of the latter being of a slender, spine-like character. A small portion at the extremity of the tibiæ of the first and fourth pairs is red-brown.
The *palpi* are short and slender.
The *falces* are weak, slender, straight, and slightly projecting.
The *labium* and *maxillæ* are of normal form, and their colour is yellow-brown.
The *sternum* is nearly triangular.
The *abdomen* is of large size, very convex above, and projects considerably over the base of the cephalothorax; the hinder part also projects over the spinners, and the upper surface is broad, the fore part presenting, on either side, the appearance of a kind of prominent shoulder; it is of a yellow-brown colour, completely covered above and on the sides with irregular, but closely-fitted, yellowish-white spots, the interstices of which have the appearance of fine yellowish-brown reticulations. Along the middle of the forepart is a dull brown narrow band with a blunt prominent point continued into a fine irregular line: there are also several dull-brown lines running backwards from its hinder extremity, which is rather enlarged; the spinners are short, compact, of a yellow-brown colour, and surrounded by a dark-brown band. The colour of the genital aperture, and of the orifice of the breathing organs, is red-brown.
Hab.—Murree, June 11th to July 14th, 1874.

36.—THERIDION CONFUSUM, sp. n.

Adult female: length 1¾ lines.
This spider is very nearly allied to *Theridion denticulatum* (Walck.), which it resembles in size and form, but is, I think, quite distinct. It is altogether of a browner hue, and, although the markings on the abdomen are very similar, there is an important difference in the median, longitudinal, dentated band, the hinder part of which is, in the present spider, merely a narrow, undenticulate, pale stripe.
Hab.—Murree, June 11th to July 14th, 1874.

37.—THERIDION EXPALLIDATUM, sp. n.

Adult female: length 1¾ lines.
The *cephalothorax* of this spider is of ordinary form; its colour is a pale brownish-

yellow margined with brown, and with a longitudinal median band of a rusty reddish-brown hue, as broad as the hinder row of eyes, where it begins, but thence tapers gradually to the hinder extremity of the cephalothorax.

The *eyes* are in the ordinary position; those of the hind-central pair are nearer together than each is to the hind-lateral eye on its side, while those of the fore-central pair are further from each other than each is from the fore-lateral on its side. The four central eyes form a square, and those of each lateral pair are seated contiguously and obliquely on a small tubercle; the clypeus is strongly and sharply impressed immediately below the eyes, but prominent at its lower margin, and its height exceeds half that of the facial space.

The *legs* are rather short, slender, of a pale, dull yellowish colour, with a slight black-brown marking beneath the extremities of each joint, and are furnished with hairs and somewhat spine-like bristles.

The *palpi* are slender, short, and similar to the legs in colour and armature.

The *falces* are not very long nor strong, but a little projecting; they are of a dull yellowish colour suffused with brown.

The *maxillæ* and *labium* are of normal form, and similar in colour to the legs; the labium, however, is suffused with brown.

The *sternum* is triangular, and its colour is like that of the legs, with a marginal blackish line.

The *abdomen* is large and globular, and projects considerably over the base of the cephalothorax; it is of an almost uniform chalky-white colour, with the faintest possible traces of a longitudinal, median, denticulate band on the upper side, having some oblique lateral lines issuing from it; this pattern is rendered just visible by being of a rather clearer white colour than the rest; the median longitudinal line of the upper side has also a dull brownish, broken line, from which finer, oblique, lateral lines issue here and there; the spinners are surrounded by a dull brownish circular band on which are several rather conspicuous white spots. The genital aperture is of a transverse oval form edged with dark brown, and placed at the hinder part of a roundish prominence.

This spider is evidently nearly allied to *Theridion simile*, C. L. Koch.

Hab.—Murree to Sind Valley, July 14th to August 5th, 1873.

38.—THERIDION TUBERCULATUM.

Theridion tuberculatum, Kronenberg, Reise in Turkestan von Alexis Fedtschenko, Moscow, 1875, p. 9, pl. v., fig. 40.

This little white *Theridion* may readily be distinguished from *T. expallidatum* by a small sub-conical, somewhat tubercular eminence on the hinder part of the upper side of the abdomen. The cephalothorax has a narrow longitudinal median brown stripe of which the anterior portion is bifid; and the abdomen, which is of a dull yellowish-brown colour thickly covered with cretaceous-white confluent spots, has an irregular, somewhat sub-dentate, longitudinal, median, dull brownish band, emitting backwards a few fine oblique lines of the same colour. The whole of the fore part of this spider is of a dull pale yellow hue; the legs are long, slender, and furnished with hairs, bristles, and slender bristle-like spines.

Hab.—Murree, June 11th to July 14th, 1873.

39.—THERIDION INCERTUM, sp. n.

Adult male: length 1¼ lines.

The *cephalothorax* is short-oval, slightly constricted laterally at the caput, which is broad and of a somewhat truncated appearance; in the middle longitudinal line, the caput has a broadish ridge-like character, which runs far back to a deep transverse indentation at the thoracic junction. Its colour is a rather rich yellow-brown, except the hinder part of the caput, which is paler, and its surface is granulose and covered with bristly hairs.

The *eyes* are of tolerable size in two nearly equal transverse rows occupying the whole width of the fore part of the upper side of the caput. The hinder row is, as nearly as possible, straight, the front row curved. The eyes of the hind-central pair are considerably nearer together than each is to the hind-lateral eye on its side; the interval between the former being equal to a diameter, and that between the latter, to nearly two diameters. The eyes of the fore-central pair are seated on tubercles, and rather prominent, forming a line rather less than that formed by the hind-centrals: the intervals between the eyes of the front row appear to be as nearly as possible equal. The eyes of each lateral pair are seated, slightly obliquely, on a strongish tubercle, and are contiguous to each other. The fore-central eyes form, as nearly as can be, a square, and the height of the clypeus (which is impressed below the eyes and prominent at its lower side) is more than half that of the facial space.

The *legs* are moderately long, excepting those of the first pair, which are much the longest: their relative length is 1, 2, 4, 3. The first pair are strong, especially the femora, and, with those of the second pair, are of an orange-yellow colour, the fore part of the femora and tibiæ being of a deeper and richer orange than the rest; all are furnished with hairs and bristles, those of the first pair having numerous spine-like bristles, and a single longitudinal row of short, strongish, somewhat denticular spines along the under side; those of the third and fourth pairs are of a paler hue than the others.

The *palpi* are short, similar to the legs in colour, and (except the digital joint, which is large, reddish yellow-brown, and with its fore extremity considerably produced) slender; the cubital joint is very short; the radial also is short, but spreads out at its outer extremity into a very large and apparently bifid production: the *palpal* organs are well developed and prominent but tolerably simple in structure; with a slender curved filiform spine, and a small, straight, corneous process at their fore extremity; they are directed outwards, the convex sides of the digital joints being directed inwards.

The *falces* are strong, of moderate length, and similar to the cephalothorax in colour.

The *maxillæ* are tolerably long and strong, much curved, and almost meeting over the *labium*, which is short and with a somewhat pointed apex. The colour of these parts is like that of the falces.

The *sternum* is heart-shaped, broad, and truncate at its fore extremity; it is of a rather orange-yellow colour, and its surface is granulose.

The *abdomen* is short, considerably convex above, and projects well over the base of the cephalothorax; looked at from above, it is of a somewhat heart-shape. It is of a yellow-brown colour, a good deal marked and spotted with white on the upper side; these markings do not appear to follow any very distinct pattern, but a more or less broken marginal band, most complete on each side of the fore part, and least traceable behind, may be seen. Four round red-brown impressed spots form almost a square a little in front of the middle, and across this part most of the white spots occur.

This is in some respects rather an aberrant form of *Theridion*, but it is most nearly allied to *T. pulchellum*, Walck.

Hab.—Murree, 11th to July 14th, 1873.

ARANEIDEA. 37

Genus—*STEATODA*, Thor.

40.—STEATODA NIGROCINCTA, sp. n.

Adult female: length 2¾ lines.

The *cephalothorax* is of a short-oval form: the caput is slightly constricted on the sides and the normal indentations are well marked, that at the junction of the caput and thoracic segments being very strong, giving rather a crushed appearance to this part, and with a transverse direction. The colour is yellow-brown, darker in the direction of the indentations, and in a wedge-shaped form behind the eyes, but rather lighter towards the lateral margins. The surface is glossy and (apparently) devoid of hairs, but covered with minute red-brown granulosities.

The *eyes* are of moderate size, and do not differ much in this respect: they are in the ordinary position; those of the hind-central pair are rather nearer together than each is to the hind-lateral eye on its side; those of the front row, which is the shortest, appear to be divided by as nearly as possible equal intervals; those of each lateral pair are seated a little obliquely, and contiguously, on a slight tubercle. The four central eyes form a square.

The *legs* are moderate in length and strength; their relative length is 1, 4, 2, 3, but the difference between those of the first and fourth pairs is not much. They are of a rather orange yellow-brown colour, deepening in hue towards the extremities, and are furnished with hairs and bristles which spring from minute tubercular red-brown granulosities.

The *palpi* are slender, and similar to the legs in colour.

The *falces* are neither very long nor strong; they are straight, perpendicular, similar to the cephalothorax in colour, and granulose.

The *maxillæ* and *labium* are of normal form, and rather lighter in colour than the falces.

The *sternum* is somewhat heart-shaped, broadly truncated, in a rather hollowed line, at its fore-extremity, and of a pale orange-yellow colour.

The *abdomen* is large, of a short-oval form, very convex above, and projects considerably over the base of the cephalothorax; it is of a dull luteous-brown colour, sprinkled with white cretaceous spots, which are principally gathered into a longitudinal median-line and a somewhat dentated line on each side of the upper part; the median white line has some black spots and markings on each side of it, those on the hinder half forming a series of broken transverse angular bars. There are also black markings on each side of the lateral white borders; that below each is broad, and continued quite round in front, where it meets the other from the opposite side; the upper side of the abdomen has also some other black spots and points scattered over it; the middle of the under side has a largish square area of white cretaceous spots, bearing two strong parallel patches of deep red-brown on its fore part. The spinners are short, compact, and of a pale-yellowish colour.

The *male* resembles the female in general characters and colours; it is, however, smaller and paler, and the legs of the first pair are longer; the palpi are like the legs in colour; they are long and slender, the radial joint is double the length of the cubital, which is of a bent form, and the former is somewhat clavate, and has its extremity on the outer side broadly produced, but in close contact with the digital joint; the digital joint is of moderate size, of oval form, with its fore extremity pinched in to a point; the palpal organs are well developed and rather complex, with some whitish prominent membrane just above their fore extremity.

Hab.—Murree, June 11th to July 14th, 1873; and route from Yárkand to Bursi, May 28th to June 17th, 1874.

41.—STEATODA SORDIDATA, sp. n.

Adult female: length 2½ lines.
In form and structure this spider resembles *Steatoda nigrocincta*. The *eyes*, however, are smaller and more closely grouped, and the line formed by two fore-centrals is longer than that formed by the hind-central pair. The whole of the fore part is of a dull dark-brown colour, tinged with yellow. The abdomen is similar, but without the yellowish tinge; a broad longitudinal median band, as well as a narrower, lateral, dentated one on each side, meeting round the fore extremity, are formed by white cretaceous spots; and a similar line longitudinally bisects the under side; four small impressed black spots form a square (whose fore side is shortest) on the fore half of the upper side.

Hab.—Hills between Sirikol and Aktalla, May 8th to 13th, 1874.

Genus—*DREPANODUS*, Menge.

42.—DREPANODUS MANDIBULARIS.

Theridion mandibulare, Lucas, Explor. en Algérie, p. 260, pl. 17, fig. 1.
Pachygnatha mandibularis, Cambr., Spid. Pal. and Syr., P. Z. S., 1872, p. 294.
Steatoda mandibularis, Sim., Bull. Soc. Ent. Fr., 1873, p. 222.
Epeira diversa, Bl., Ann. & Mag. Nat. Hist., October 1859.

Hab.—Examples of the female of this puzzling spider were found in Dr. Stoliczka's collection, taken *en route* from Yárkand to Bursi, May 28th to June 17th, 1874.

The difficulty of assigning to it its correct systematic place is very evident from the synonyms above quoted. Mons. Eugène Simon has, however, lately suggested to me that it is nearly allied to *Drepanodus obscurus*, Menge, and, entirely agreeing with this, I have placed it here in that genus.

Genus—*PHYCUS*, Cambr.

The short broad form and very laterigrade appearance of the typical example of this genus led to the conjecture that it belonged to the family *Thomisides* (*vide* O. P. Cambridge, 'On some new Genera and Species of Araniedea,' in Proc. Zool. Soc., 1870, p. 742, pl. xliv, fig. 9. Subsequent examination leads me to conclude that its proper place is in the family *Theridiides* (not far from the genus *Euryopis* (Thor.), but certainly not among the *Orbitelariæ*, as conjectured by Dr. T. Thorell (Syn. Eur. Spid., p. 600).

43.—PHYCUS SAGITTATUS, sp. n.

Adult female: length 2 lines.
The *cephalothorax* is short, broad behind, and narrowing gradually forwards, but without much lateral constriction at the caput: this portion is large, bluff in front, considerably elevated, rising gradually but quickly from the thorax. The clypeus rather exceeds in height half that of the facial space: it is round on the lower margin, and full and rather prominent, projecting gradually from just beneath the front row of eyes. The colour of the cephalothorax is a deep yellowish-brown, and behind each hind-lateral eye is a strongish, curved,

spine-like bristle directed forwards; other bristles evidently belong to this part, but they had been rubbed off.

The *eyes* are of tolerable but nearly uniform size, and form a large, transverse, crescent-shaped area on the front and sides of the upper part of the caput; the two rows have the convexity of their curve directed forwards, the front row being much the more strongly curved, and its eyes rather larger than those of the hinder row. The eyes of the hinder row (which is the longer) are very nearly, if not quite, equally separated, the interval rather exceeding two diameters of one of the central pair; the interval between the eyes of the fore-central pair, which is of a black colour, is double that between each and the fore-lateral eye on its side, and the line formed by them is very little longer than that formed by those of the hind-central pair. The four central eyes form a rectangle, whose transverse is greater than its longitudinal diameter; the eyes of each lateral pair are seated obliquely, but not quite contiguously, on a large, black, and slightly tubercular spot.

The *legs* are short, tolerably strong and tapering; those of the fourth pair are the longest; the rest differ very little in length, perhaps that of the first pair a little exceeds that of the second, the third pair being slightly the shortest. They are furnished with hairs, and a double, divergent row of longish fine spines beneath the tibiæ, tarsi, and metatarsi; each tarsus ends with three curved claws, which spring from a small prolongation (apparently a distinct articulation) of the tarsus. The colour of the legs is a deep brown, but paler along the upper sides, the hinder extremities of the femora being of a pale-yellowish hue.

The *palpi* are short but tolerably strong; they are of a dull-yellowish hue, suffused with brown beneath and on the sides; the digital joint ends with a curved, and apparently pectinated, claw.

The *falces* are short, straight, perpendicular, moderately strong, and the fang is slender; their length does not exceed the height of the clypeus, and their colour is yellowish-brown.

The *maxillæ* are small, short, and greatly inclined to the labium, over which their extremities almost meet.

The *labium* is very short, and somewhat pointed at its apex; its colour, with that of the maxillæ, is a dull pale-yellowish, suffused, over all except their extremities, with brown.

The *sternum* is heart-shaped, and similar to the cephalothorax in colour.

The *abdomen* is of great size, heart-shaped, very convex above, and projects greatly over the cephalothorax, which it (when looked at from behind) almost entirely conceals. Its colour is a dull dark-brown, the upper side being densely covered with small, silvery, slightly yellowish-white metallic spots; leaving a large, transverse, somewhat oblong, brown area at the fore extremity, and a large arrow-headed brown marking in the middle: the point of this marking is directed backwards, going off into a fine yellow-brown line nearly to the spinners; and from the middle of its fore extremity a short brown stripe runs forward into the oblong patch of the same colour, and is crossed, close to it at right angles, by another brown line, which connects the foremost pair of four small, deep-brown, impressed spots; the hinder pair of these spots are placed just on the fore margin of the arrow-headed marking, which has, on its lateral margin, some other deep-brown spots and blotches, with a few small, silvery spots on its fore part. The under side has two very distinct transverse lines of silvery spots; and on either side of them are some pale, parallel, and slightly oblique streaks and lines of similarly coloured spots; the spinners are short, those of the inferior pair being considerably the stouter and rather the longer.

Hab.—Murree to Sind Valley, July 14th to August 5th, 1873.

SECOND YÁRKAND MISSION.

Genus—*ERIGONE*, Sav.

44.—ERIGONE ATRA.

Neriene atra, Blackw., Lond. and Edinbr. Phil. Mag. 3 ser. iii, p. 195.

——— *longepalpis*, Blackw., Spid. Great Brit. and Irel. p. 274, pl. xix, fig. 188.

Hab.—An example of the male, identical with British and other European specimens, was contained among the spiders found on the route from Yárkand to Bursi, May 28th to June 17th, 1874.

45.—ERIGONE DENTIPALPIS.

Erigone dentipalpis, Wid.-Westr., Aran. Suec. p. 199.

Although only a portion of this spider was found in the collection, I feel no doubt of its specific identity with the *E. dentipalpis* (Westr.) of Europe.

Hab.—Sind Valley, August 5th to 13th, 1873.

Genus—*PACHYGNATHA*, Sund.

46.—PACHYGNATHA CLERCKII.

Pachygnatha clerckii, Sund.-Westr., Aran. Suec. p. 144.

An adult male of this species, differing in no respect from European examples, was found in the collection.

Hab.—Káshghar, December 1873.

Genus—*LINYPHIA*.

47.—LINYPHIA CONSANGUINEA, sp. n.

Adult male: length 2¼ lines. Adult female: 2¾ lines.

This spider is, in size, colour, and markings, almost exactly like the well-known European form *Linyphia peltata* (Sund.); one description would, in fact, suit both these species. The present, however, may be distinguished by several good specific characters.

In the male, the *falces* are shorter and less divergent, but stronger and much more prominent in front when looked at in profile; in which position also the extremities are directed distinctly backwards; they are likewise granulose, furnished with short spine-like bristles, and armed with a short, strong, pointed, deep red-brown tooth on the inner side towards the fore extremity.

The *palpi* have the cubital joint very short, with a strong, tapering, spine-like bristle directed forwards from its fore extremity on the upper side; the radial joint is double the length of the cubital, and enlarges gradually to the fore extremity, where, on the upper side, is a spine-like bristle like that on the cubital joint; the digital joint is not very large, and the palpal organs are complex, with corneous processes and spines, somewhat like those of *L. peltata*, but bolder; and from their base, on the outer side, issues a slender, longish, pale-coloured, rather prominent spiny process, whose extremity is sharp-pointed and rather shortly curved; this process by its size and shape distinguishes the males of the two species at a glance.

The *female* may be distinguished by the abdomen in the present species (when looked at in profile) having a higher elevation about the middle, the general curve of that of *L. peltata* being here of a somewhat humped nature; the form of the genital aperture also differs a little : in *L. peltata* it is of a simple, transverse, oval form; in the present its form is somewhat bluntly sub-triangular.

Hab.—Murree, June 11th to July 14th, and Murree to Sind Valley, July 14th to August 5th, 1873.

48.—LINYPHIA ALBIPUNCTATA, sp. n.

Adult female: length rather over 2 lines.

This spider is nearly allied to *Linyphia leprosa*, Ohl., and *L. minuta*, Bl., to which it bears considerable resemblance both in size, form, and colours. The whole of the forepart is yellow-brown, the sternum and falces being the darkest : the upper part of the caput also, with the normal indentations, is rather darker than the rest of the cephalothorax; the legs are distinctly annulated with dark brown, and furnished with hairs, bristles, and spines; the palpi have a similar armature, but are not so distinctly annulated.

The *eyes* are in the ordinary position and seated on tubercular black spots, but the ocular area is more prominent than usual, and has the appearance of a longish, oblong, tubercular platform; those of the posterior row appear to be equally divided from each other, the interval being less than a diameter; the fore-centrals are the smallest of the eight, and are separated by about half a diameter, each being divided from the fore-lateral eye on its side by rather less than the diameter of the latter; those of each lateral pair are placed rather obliquely, but not quite contiguously. The height of the clypeus, which is very projecting, equals half that of the facial space; the four central eyes form a rectangle whose longitudinal is greater than its widest transverse diameter, and its posterior side longer than its anterior one.

The *falces* are moderately long and strong, very slightly divergent, nearly perpendicular; their profile is curved, and each has three tolerably strong, sharp teeth at the extremity on the inner side.

The *sternum* has its surface slightly tuberculose.

The *abdomen* is very convex above, and projects well over the base of the cephalothorax; it is of a rather dark yellow-brown colour, marked with deeper brown, and thickly and minutely specked with white on the upper part and sides; the only traceable pattern is a longitudinal dark-brown line along the middle of the fore part of the upper side, followed to the spinners by a series of transverse angular lines, formed by the more regular disposition of some of the white spots; probably a series of specimens would show transverse angular brown lines, with perhaps an enlargement like a spot or blotch at each of their extremities; on the sides also there is a trace of a similar white horizontal curved line meeting the one on the opposite side a little above the spinners; and above it is a horizontal brown patch. These markings are all very similar to those of *Linyphia leprosa*, though less distinct. The genital aperture in the two species, and the process connected with it, are, however, totally dissimilar; in the present spider, instead of being exceedingly prominent, and rather complex, it is only slightly prominent and very simple in its structure.

Hab —Murree, June 11th to July 14th, 1873.

49.—LINYPHIA STRAMINEA, sp. n.

Adult female : length a little over 1 line.

The whole of the fore part of this small spider is of a pale straw-yellow colour, and in its form and general structure it is very like *Linyphia ericæa*, Bl.

The *eyes* are on strong, black, tubercular spots, and rather closely grouped together in two slightly curved rows, of which the hinder one is a very little longer than the front; the two hind-central eyes are slightly further from each other than each is from the hind-lateral eye on its side; and the fore-central eyes are the smallest of the eight, near together but not contiguous to each other; those of each lateral pair are placed slightly obliquely, and touching each other; the four centrals form a square whose fore side is considerably shorter than the hinder one.

The *legs* are very slender and rather long, furnished with hairs and a few fine spines.

The *palpi* are short, slender, and similar in colour and armature to the legs.

The *falces* are tolerably long, not very strong, straight and perpendicular.

The *maxillæ, labium,* and *sternum* are normal.

The *abdomen* is very convex above, and projects considerably over the base of the cephalothorax; it is of a dull straw-colour, speckled thinly with small, white, cretaceous-looking spots. The form of the genital aperture is very simple indeed, and has a very slight, and scarcely at all prominent, process connected with it.

Hab.—Murree, June 11th to July 14th, 1873.

50.—LINYPHIA PERAMPLA, sp. n.

Female, scarcely adult : length rather over 3 lines.

This fine species is very nearly allied to *Linyphia collina*, L. Koch, found in the French Jura mountains; but it may be, I think, distinguished by a total absence of the small white spots dispersed over the abdomen of that species, as well as by a stronger and bolder pattern.

The *cephalothorax* is of ordinary form ; and the normal indentations are strongly marked ; it is of a light brownish-yellow colour, the caput dark yellow-brown, and the thoracic portion has a broadish brown marginal border.

The *eyes* are rather small, but in the usual position ; those of the hind-central pair are rather nearer to each other than each is to the hind-lateral on its side; the four central eyes form very nearly a square whose anterior is shorter than its posterior side ; the eyes of each lateral pair are placed obliquely, and contiguously, on a slight tubercle. The height of the clypeus is equal to half that of the facial space.

The *legs* are rather long, slender, and their relative length is 1, 2, 4, 3 ; they are of a yellow-brown colour, the tibiæ and femora annulated with dark brown, the former indistinctly, the latter distinctly ; they are furnished with hairs, bristles, and a few not very long, slender spines.

The *palpi* are short, slender, of a pale brownish-yellow colour, furnished with an armature similar to that of the legs.

The *falces* are moderately long and strong, straight, perpendicular : the basal portion in front, dark brown ; the rest, yellow-brown, with a reddish tinge near the fang ; and there are three teeth on the inner side near the fore extremity.

The *maxillæ* and *labium* are of normal form, and of a brownish-yellow colour, the latter being the darker.

The *sternum* is heart-shaped, broadly truncated at its fore extremity; its colour is yellow-brown, suffused with deeper brown, furnished with long bristly hairs, and with a kind of oval gibbosity at its hinder extremity.

The *abdomen* is large and very convex above, projecting considerably over the cephalothorax, and clothed with short grey hairs; it is of a pale, dull brownish-yellow hue, marked along the middle of the upper side with a series of strong, well-defined, blackish-brown angular bars of a W form, the extremities of these lines uniting with oblique lateral lines of black-brown spots and markings.

Hab.—Sind Valley, August 5th to 13th, 1873.

51.—LINYPHIA PUSILLA.

Linyphia pusilla, Sund., Sv. Spindl. Deskr. Vet.-Aksd. Haudl. f. 1829, p. 214.

Hab.—Two females of this spider contained in the portion of the collection made at Yárkand, May 21st to 27th, 1874, and on the route thence to Bursi, May 28th to June 17th, 1874, differ in no respect from European examples of the same species.

Family—EPEIRIDES.

Genus—*META*, C. L. Koch.

52.—META MIXTA, sp. n.

Adult female : length 3¼ lines.

This pretty spider is nearly allied to *Meta (Tetragnatha) decorata*, Bl., but it is smaller and the abdomen, although its hinder extremity projects a little over the spinners, is not drawn out behind into anything of a caudal form; the fore extremity also is smooth and rounded at the shoulders, instead of being somewhat humped or gibbous on those parts; the genital aperture also differs in its form.

The *cephalothorax, legs,* and *palpi* are pale yellow; the *falces, maxillæ*, and *labium* rather suffused with brown, or reddish-brown; and the *sternum* dark reddish-brown; the extremities of the tibiæ of the legs are dull reddish-brown. The *abdomen* is of a cylindric oval form, rather narrower behind than before; the upper part, and a portion of the sides, are metallic and silvery in colour, and marked, longitudinally, with five dull brownish lines or stripes. A median stripe, and another on each side of it, not reaching so far forwards as the median one, meet at the hinder extremity; these three longitudinal stripes are connected on either side by three oblique lines of the same colour, issuing from the central stripe and running obliquely backwards into the lateral one ; two of them are rather near together, not far behind the middle of the abdomen; the third is much nearer the fore part; and, quite at the fore extremity, is a transverse, dull-brown, slightly curved line; the remainder of the sides and the under side are dull yellowish-brown; and on each side of the under part is a longitudinal, slightly bent, silvery stripe from the spiracular openings to the spinners, followed by a spot of the same kind close to the outer side of the inferior pair of spinners;

these two sub-abdominal stripes are, in *Meta decorata*, Bl., replaced by the whole of that part being silvery, whereas the intermediate space between the silvery stripes, in the present spider, has only a few silvery specks on its fore half. The space between the spinners and the upper side has also some silvery spots upon it.

This spider is also nearly allied to *Meta quinquelineata*, Keys (from Bogota, S. Amer.), but is, I think, certainly of a distinct species. It is also distinct from another nearly allied (and undescribed) species received from Bombay from Major Julian Hobson.

Hab.—Murree, June 11th to July 14th, 1877.

Genus—*TETRAGNATHA*, Walck.

53.—TETRAGNATHA EXTENSA.

Tetragnatha extensa, Linn., Syst. Nat. Ed. 10, i, p. 621.

Several specimens, which I believe to be of this species, were contained in the collection.
Hab.—Káshghar, December 1873; Sind Valley, August 5th to 13th, 1873; Yárkand, May 21st to 27th, 1874; and route from Yárkand to Bursi, May 28th to June 17th, 1874.

Genus—*EPËIRA*, Walck.

54.—EPËIRA TARTARICA.

Epëira tartarica, Kronenberg, Reise in Turkestan v. Alexis Fedtschenko, Moscow, 1875, p. 2, pl. 1, fig. 1.

Adult and immature females were contained in the collection.
Hab.—Neighbourhood of Leh, August and September 1873.

55.—EPËIRA DIGIBBOSA, sp. n.

Immature female: rather over 1 line in length.

Although it is very young, I think this spider is of a different species from several with a protuberance or gibbosity on each side of the fore extremity of the abdomen already described by different authors.

The whole of the fore part is of a dull yellowish hue, except the sternum, which is yellow-brown; the *cephalothorax* is suffused with yellow-brown in the indentations by which the union of the caput and thorax is indicated, and has a somewhat triangular patch of a cream-white colour at the occiput.

The *legs* are rather short and moderately strong; their relative length is 1, 2, 4, 3, and they are furnished with hairs and bristles only.

The *eyes* are on slight dark tubercles; those of the hind-central pair are the largest of the eight, and considerably larger than the rest, and are divided from each other by a diameter's interval; the space which divides each from the hind-lateral on its side being double, if not more, of that diameter: those of the fore-central pair are less than half the size of the hind-centrals; they are separated from each other by two diameters' interval, but yet form a line shorter than that formed by the hind-centrals. The four central eyes form a square whose foremost side is rather shorter than the rest; those of each lateral pair are placed very obliquely and contiguously. When looked at from in front, the eye area is in the form of a triangle whose apex (at the hind-central pair) is truncated, and whose base is formed by the lateral and fore-central pairs, which, from this point of view, are in a perfectly straight line. The height of the clypeus is about one-third that of the facial space.

The *abdomen* is white; its fore part, which is broad and has a strongish, pointed, conical gibbosity on the upper side at each fore corner, projects considerably over the base of the cephalothorax; these gibbosities are tipped with dark brown. The greater part of the hinder half is occupied by a very broad dentated brown band which narrows to the spinners, and is itself mottled with white; the under side is dull brownish, with a curved white stripe on each side of the hinder part.

No doubt some variation in colours, and perhaps in markings, will be found in the adult form; but the above details will, I think, be found sufficient for the determination of the species, whether in the adult or immature state.

Hab.—Murree, June 11th to July 14th, 1873.

56.—Epëira pellax, sp. n.

Adult female: length 3⅙ lines

This spider is allied to *Epëira bigibbosa*, but may easily be distinguished by a difference in the relative size and position of the eyes, as well as by its spinous and annulated legs.

The *cephalothorax* is dull yellow-brownish on the sides, and the caput has also one or two, indistinctly defined, yellow-brown markings, and its surface is clothed with a coarse greyish-sandy pubescence.

The *eyes* are rather small, of a dull, amber yellowish-brown colour, and do not differ greatly in size; those of the hind-central pair are the largest, and, with the fore-centrals, which appear to be next in size (though not very much smaller), form very nearly a square; the fore-side of the square being, if anything, slightly longer than the hinder one: the interval between the hind-centrals is equal to a diameter, but that between each of them and the hind-lateral eye on its side is equal to at least four diameters of the hind-central eye.

The *legs* are rather short and tolerably strong; their relative length appears to be 1, 4, 2, 3, though the difference between those of the first, second, and fourth pairs is slight; their colour is yellow, clouded, and irregularly annulated, with dark yellow-brown; they are also furnished with hairs, bristles, and short spines.

The *falces* are short and moderately strong, a little prominent near their base in front; they are similar in colour to the cephalothorax, suffused a little with brownish on their outer sides.

The *maxillæ* and *labium* are normal in form, and of a pale-yellowish colour, suffused with brownish towards their bases.

The *sternum* is heart-shaped; the anterior side truncated in a hollow line.

The *abdomen* is short, and broad in front, with a short, roundish, conical protuberance on each side of the forepart; from a little way behind each of these protuberances, a sinuous, or sub-dentate, dark-brown line, edged with whitish on its outer side, runs backwards towards the spinners; these lines do not reach the spinners, nor do they meet each other, since they run nearly parallel during the latter part of their length; the space included by these lines is mottled with white, and contains some dull-brownish, curvi-angular, transverse lines, bisected through their angles (which are directed forwards) by a longitudinal median line of the same hue; the sides are alternately slashed, or broadly and obliquely striped, with white and yellowish-brown. The under side is white, with a large, central, somewhat triangular, dull-brownish area in the middle. The base of this triangle is towards the forepart, where it is joined by a pedicular patch of the same colour, to the hinder margin of the genital opening. This aperture has connected with it a long, strong, very prominent epigyne, divided into two parts, a basal part whitish, tumid, and membranous in appearance, and a terminal portion blackish-brown, corneous, curved, tapering, directed strongly forwards, and clothed with hairs on its fore, or convex, side; its hinder, or concave, side has a narrow, longitudinal slit or duct.

The *spinners* are normal, and surrounded, on the sides and behind, with a horseshoe-shaped line of six white spots with dull-brownish intervals.

Hab.—Murree to Sind Valley, July 14th to August 5th, 1873.

57.—EPËIRA GURDA, sp. n.

A number of immature specimens, both male and female, of an *Epëira* which I believe to be of an undescribed species, were contained in the collection, though, from their immaturity and the apparently faded state of their colouring, I do not venture to describe them in detail. They are very nearly allied to *Epëira cornuta*, Clerck, the pattern on the abdomen being very similar to that spider. The colours, however, are much duller, and the markings far less distinct, and the *legs* are of a uniform dull-yellowish hue. One character alone will serve to distinguish it at once from *E. cornuta*; the eyes of the hind-central pair form a line which is very nearly—indeed, as long—as that formed by those of the fore-central pair; but which in *E. cornuta* is distinctly (and indeed considerably) shorter.

Hab.—Murree, June 11th to July 14th, 1873.

58.—EPËIRA HARUSPEX, sp. n.

Adult female: length 1¾ lines.

In its general form this spider is much like *Epëira pellax*; it is, however, much smaller, and differs in many material particulars; it is also allied nearly to *E. agalena*, Walck.

The whole of the fore part, excepting the sternum, which is darker, and the normal indentations of the cephalothorax, which are dusky brown, is of a dull yellowish-brown colour.

The *eyes* are small; the four centrals form a square, but those of the hind-central pair are considerably larger than the front-centrals and are separated from each other by a diameter's distance, each being also divided from the hind-lateral on its side by more than double that which separates them from each other.

The *legs* are neither very long nor strong; their relative length is 1, 2, 4, 3, and the tibiæ are faintly annulated with reddish yellow-brown; their armature had been all rubbed off, but apparently it had consisted only of hairs and bristles.

The *falces* are moderately long, tolerably strong, and roundly prominent near their base in front.

The *maxillæ* and *labium* are of normal form, pale yellowish towards their extremities, but dark brown on the basal part.

ARANEIDEA. 47

The *abdomen* is broad in front, where it is slightly prominent (though scarcely gibbose) at the fore-corners, and the middle of the fore-margin is also blunt-angularly prominent; the upper side has almost its whole area enclosed by two longitudinal brown lines (edged outwardly with white), which run from the fore-corners obliquely backwards, and converge towards each other to the spinners; the first half of these lines is sub-angular. Along the length of the area thus enclosed, a large, but not very clearly defined, somewhat cruciform or dagger-shaped, whitish marking runs with its sharp point backwards, and is bisected longitudinally by a dull, pale-brownish line, from which several oblique lines branch off on each side. The rest of the area is dull brown, deepening to reddish brown at the hinder extremity; the sides are dull brown, mottled thickly with small white spots; the under side is also brown margined with a distinct line of white spots, which does not, however, enclose the fore extremity, as each end of the line terminates close behind the spiracular opening; a little way from the spinners, on each side, there are two white spots in a longitudinal line. The process connected with the genital aperture is prominent, and of characteristic form: it consists of two portions—the basal, which is short, broad, dark blackish-brown, and corneous; and a rather long, twisted pale epigyne, directed backwards: only a figure, however, can give any correct idea of the form and structure of this process.

Hab.—Yárkand, May 21st to 27th, 1874.

50.—EPËIRA PÆNULATA, sp. n.

Adult female: length 2¾ lines.

This spider is allied to *Epëira cornuta*, Clerck, but is much smaller, and (in the only example examined at least) wants the characteristic pattern on the abdomen.

The whole nearly of the fore part is of a dull-yellow colour; the normal indentations of the cephalothorax are suffused with dusky-brown, and from each of the hind-lateral eyes a brownish-black line runs backwards towards the occiput; and there are two strong, spinelike, curved bristles directed forwards from behind each lateral pair of eyes.

The *eyes* are small; those of the hind-central pair are the largest, and are separated from each other by no more than half a diameter's interval, if so much; and the line formed by them is distinctly shorter than that formed by the fore-central pair, and constitutes the hinder and shortest side of the square formed by the fore- and hind-central eyes; the eyes of each lateral pair are widely removed, by an interval of at least double the length of the line formed by the hind-central eyes. The clypeus is less in height than half that of the facial space.

The *legs* are rather short, but tolerably strong; their relative length is 1, 2, 4, 3, and they are furnished with hairs, bristles, and short, not very strong spines, each of which springs from a small blackish tubercle.

The *palpi* are moderately long and strong, and are furnished, chiefly towards their inner sides, with numerous bristles, and long, slender, curved spines.

The *falces* are moderately long, strong, perpendicular, and rounded in profile.

The *maxillæ* and *labium* are of normal form, and, except their extremities, which are pale, are of a yellow-brown colour.

The *abdomen* is of a broad-oval form, not excessively convex above, its upper surface being parallel to its under side, and its hinder part, as well as each side of its fore extremity, rounded, but the latter are not prominent, as in several of the species described above; the middle, however, of the fore extremity projects forwards in a strong, blunt-conical form. The whole of the upper part, including the upper half of the sides, is of a cream-yellow colour marked with two converging rows of linear black spots; these rows are almost the whole width of the abdomen apart where they begin, which is just about half way from the hinder extremity to the middle of the fore extremity, and they represent the angular or dentated lines so common on the abdomen of the genus *Epëira*. Four parallel, longitudinal, dull-brownish, venose lines (the outer ones of which curve round and almost meet at their fore extremity, and the middle pair are shorter than the outer ones) are included within the two converging rows of spots above mentioned, and are also connected by another curved line which crosses at the ends of the two middle lines. The remainder of the sides, together with the whole of the under part, is of a dull-brownish hue, indistinctly mottled with dull whitish-yellow spots; and four large blotches of white spots form a square between the spiracular plates and the spinners. The sides are also marked with some oblique, venose, yellow-brown lines. The process, or epigyne, connected with the genital opening is prominent, tapering, not very long, obtuse at its extremity, which is curved, directed backwards, and transversely rugulose throughout.

Hab.—Murree, June 11th to July 14th, 1873.

60.—EPËIRA PRÆDATA, sp. n.

Adult male: length 2½ lines.

The *cephalothorax* is broad and round-oval behind, rather produced and narrow before; the forepart of the ocular area projecting over the clypeus. The hinder part of the cephalothorax is considerably higher than the forepart, and is well rounded and convex; it is of a brownish-yellow colour, with converging paler stripes, following the direction of the normal indentations. On the sides of the caput, near the eyes, are several strong, curved, spine-like bristles, directed forwards.

The *eyes* are of tolerable size, on the fore part and sides of the extremity of the caput; those of the fore-central pair are the largest of the eight, and are seated on a somewhat tubercular prominence; the interval between them is equal to, if not a little more than, a diameter, and the line formed by them considerably longer than that formed by those of the hind-central pair. These latter are on black spots, and near together, the interval being no more than half a diameter; the figure thus formed by the four central eyes is a rectangle, with its posterior side shortest, and its longest transverse diameter less than its longitudinal one; the interval between each hind-central eye and the hind-lateral next to it is equal to rather over two diameters of the former.

The *legs* are moderately long and tolerably strong; their relative length is 1, 2, 4, 3. They are furnished with hairs, bristles, and spines; three of the longest, strongest, and darkest of the last forming a transverse row near the slightly incrassated middle part of the inner side of the tibiæ of the first pair.

The *palpi* are short, and similar in colour to the legs. The cubital joint is short, and (in profile) of a sub-angular form, with two long, strong, yellow-brown, tapering, curved, spine-line bristles, directed forwards from its fore extremity on the upper side; the radial joint is very short, but is produced considerably in an obtuse form on the outer side, the end of the produced portion being furnished thickly with strong bristles; the digital joint is of

ARANEIDEA.   49

a brown colour, large and of a long oval form, hairy, and bristly; the palpal organs are very large and complex, consisting of various yellowish-brown and dark red-brown corneous processes.

The *falces* are neither very long nor strong; they are perpendicular, but removed far back under the projecting fore part of the caput; their colour (as well as that of the *maxillæ*, *labium*, and *sternum*, whose forms are normal) is like that of the legs.

The *abdomen* is of a short, oblong-oval form, equal in size at each end, and tolerably convex above. It is of a pale dull brownish-yellow colour; the upper side is more or less thickly mottled with white, leaving a broad median dull stripe on the fore half; the hinder extremity of this stripe has four (two on each side) obliquely diverging lines issuing from it, and is itself continued by a fine line (all of the same dull hue) to the spinners. Four small brown spots form a rectangle near the middle, and close behind the foremost pair of these spots is a large, roundish patch, free of all white mottling; a little behind the middle of the sides are four or five distinct, parallel, transverse, black-brown, fine lines; the hindermost line is the strongest, and has a large spot of the same colour near its inner extremity, thus altogether forming a transverse, interrupted line, appearing to cut off the extremity of the abdomen. On the under side is a large, somewhat quadrate area of white; and immediately behind it, is a semi-circle of five distinct white spots not far in front of the spinners.

This spider apparently belongs to the *Epëira cucurbitina* group.

Hab.—Murree to Sind Valley, July 14th to August 5th, 1873.

61.—EPËIRA CUCURBITINA.

Epëira cucurbitina, Clerck, Sv. Spindl. p. 44, pl. 2, tab. 4.

An immature example of this very pretty, but common and widely-dispersed spider was found in Dr. Stoliczka's collection.

Hab.—Sind Valley, 5th to 13th August 1873.

62.—EPËIRA CORNUTA.

Epëira cornuta, Clerck, Sv. spindl.

Hab.—Immature examples, which are, I believe, *Epëira cornuta*, Clk., and are certainly not distinguishable from immature European specimens of that species, were found in those portions of the collection made at Yárkand and neighbourhood in November 1873, and *en route* from Yárkand to Bursi between May 28th and June 17th, 1874.

63.—EPËIRA PANNIFERENS, sp. n.

Adult female: length 3 lines.

The *cephalothorax* is rather strongly constricted laterally at the caput, which is tolerably produced; the normal indentations are strong, especially that at the thoracic junction, and the oblique ones which mark the union of the caput and thorax. Its colour is pale yellow, with the whole of the upper part of the caput and a broad lateral band, which runs very near the margin the whole way round the cephalothorax, of an orange yellow-brown; the band on the caput is of a rather elongated diamond shape, and is produced behind to the thoracic junction.

The *eyes* are in the usual four pairs, occupying the whole width of the fore part of the caput; the four central ones are as nearly as possible of equal size, and form a square whose posterior side is shorter than the rest; those of the hind-central pair are on largish, dark, reddish-brown spots, and are separated from each other by an eye's diameter; those of the fore-central pair are seated on a slight prominence, and are directed away from each other; the interval between them being nearly two diameters. Those of each lateral pair are seated obliquely and contiguously on a dark tubercle.

The *legs* of the third and fourth pairs (one only of each being all that remained in the example examined) are short, strong, of a pale-yellow colour, annulated with dark yellow-brown, and furnished with hairs, bristles, and a few spines.

The *palpi* are short, and similar to the legs in colour and armature.

The *falces* are moderate in length and strength, roundly prominent at their base in front, perpendicular, and of a pale-yellow colour.

The *maxillæ* are of normal form; their colour is brown, with a pale-yellowish border all round their extremity.

The *labium* is of a darker brown than the maxillæ, with a pale-yellowish apex.

The *sternum* is yellowish, suffused with dark brown; its form is heart-shaped, with the fore extremity broadly truncated and hollow.

The *abdomen* is large, oval, broadest in front, the middle of the fore margin of which is a little sub-angularly prominent; it projects considerably over the base of the cephalothorax, and is of a yellowish-brown colour mottled obscurely with whitish; the fore extremity of the upper side has a sub-angular, marginal, white stripe. Beginning at some little distance behind this is a large deep brown patch-like area, broad in front, and narrowing gradually, to about half its front width, near the spinners. This patch is bordered by a fine, deeper-brown sinuous line, outside of which is a distinct white border. On the sides, the white mottlings are gathered into broadish, though rather indistinct, oblique stripes; the under side is deep brown, bordered on each side with three large white spots, the middle one of which is the largest. The epigyne is rather short, of a pale-yellowish hue, tapering, blunt-pointed, and directed backwards.

Hab.—Murree to Sind Valley, July 14th to August 5th, 1873.

64.—EPËIRA CARNIFEX, sp. n.

Adult female : length 3¼ lines.

The *cephalothorax* is tolerably strongly constricted on the lateral margins at the caput, which is also rather produced forwards ; its colour is dull yellow, rather thickly clothed with a coarse greyish pubescence ; and the whole of the upper side of the caput is of a deep brown, the same colour being prolonged backwards to the thoracic junction. The height of the clypeus is rather less than the diameter of one of the fore-central eyes.

The *eyes* are in the usual four pairs, occupying the whole width of the fore-part of the caput; the four central eyes form apparently, as nearly as can be, a square; the hind-central pair are seated on strong black spots, on a small tubercular prominence, and they are the largest of the eight, considerably larger than those of the fore-central pair, and separated by rather more than a diameter's interval.

ARANEIDEA. 51

The *legs* are moderately long, but not very strong, and their relative length is 1, 2, 4, 3. Their colour is dull brownish-yellow, annulated with darker reddish yellow-brown, and they are furnished with hairs, bristles, and strongish, but short, spines.

The *palpi* are similar to the legs in colour, moderately long and slender.

The *falces* are moderately strong, tolerably long, arched in profile, perpendicular, of a deep-brown colour, but pale-yellowish at the base, and furnished with bristles in front.

The *maxillæ* are of normal form, and of a deep-brown colour, pale-yellowish on the margin, at their extremity. The *labium* also is similar in colour, with a pale margin at the apex.

The *sternum* is heart-shaped and of a deep-brown colour.

The *abdomen* is very large, and in its general form and appearance reminds one of that of *Cyrtophora opuntiæ*, Duf. Its shoulders have each a short, somewhat conical prominence also. It projects considerably over the base of the cephalothorax; and when looked at in profile, the hinder extremity, which projects a little over the spinners, is nearly as high as the fore-extremity. The sides are steep and sloping inwards, and the upper side rather flat. The upper side is of a dull cream-white colour marked, clouded, and mottled with brown; the lateral margins of the white area are very strongly dentated, and along the middle of it are two very distinctly defined black dentated lines, which beginning near its fore extremity, converge towards each other (but do not meet) at the hinder extremity, where they are sometimes joined by a transverse blackish line. Along the middle of the fore part of the space included by these black dentated lines, which space is frequently darker or more suffused with brown than the area outside it, is a somewhat oblong deep brown, marking with two angular points on each side of it and a row of white spots along its middle. The fore extremity of the upper side has two prominent portions of the white area near the middle; these are generally curved, and enclose a more or less well-defined brown patch bearing a white spot in the centre, and in front of this, outside the brown patch, is another larger white spot; the under side is of deep sooty-brown colour, of a quadrangular form, margined by a distinct, broad, yellowish-white border, before and on its sides, each of the posterior ends of the border being continued on either side of the spinners by two well-defined white blotches, the anterior one of which is much larger than the other; from the middle of the hinder extremity, on the upper side, to the spinners there is generally a longitudinal central yellowish-white stripe; the sides are brown, thickly mottled with dull whitish-yellow. The genital aperture is simple in form, and somewhat of a transverse, kidney shape, placed rather behind a slightly prominent process from in front of which issues a moderately long, slender, epigyne, which curves backwards and has its extremity slightly sinuous. There is considerable variety in the markings of the upper side of the abdomen in this spider according as the brown mottlings are more or less diffused, or else well defined.

Hab.—Murree, June 11th to July 24th, 1873.

65.—EPËIRA ? GIBBERA, sp. n.

Adult female: length 2½ lines.

Probably this spider will be found some day, on comparison with some other closely allied exotic species, to be of a different genus from "*Epëira*" (sensu stricto), in which event, I think, a new genus must be formed for its reception. At present I describe it as an aberrant and doubtful form of *Epëira*.

The *cephalothorax* is very short and broad, and rises gradually from the hinder extremity to the fore part of the caput; though the real convexity of the whole does not vary much in one part or another, since the basal line rises forwards with the general rise of the caput; and the middle of the fore part of the caput has a rather prominently pointed appearance, without the lateral prominences (on which the lateral pairs of eyes are placed) usual in *Epëira*; in fact, there is an approach in the form of this part to some species of *Poltys*. The colour of the cephalothorax is yellow-brown, and it is clothed with a greyish pubescence.

The *eyes* are placed as in *Epëira* generally. Those of the hind-central pair are considerably larger than the fore-centrals; the interval between them is nearly about one and half diameters; they form a longer line than the fore-central pair, and together with them they form a rectangle whose greatest transverse diameter is longer than its longitudinal one. Between the eyes of the fore-central pair are two longish, divergent, pale-grey, bent bristles directed forwards and downwards. The eyes of each lateral pair are on slight tubercles, placed obliquely, and contiguous to each other; they are very widely removed from the four central ones, and, owing to the oblique, sloping character of the sides of the caput, are placed some way back, not far (when looked at sideways) above the middle of the base of the falces.

The *legs* are short, moderately strong, their relative length being 1, 2, 4, 3; they are of a yellow-brown colour, with faint traces of darker annulation; and are furnished with hairs and slender bristles, but no spines.

The *palpi* are short, slender, and nearly similar in colour to the legs, the digital joints terminating with a curved, toothed claw.

The *falces* are moderately long, strong, and similar in colour to the cephalothorax.

The *maxillæ* and *labium* are like those of *Epëira*: they are of a dark yellow-brown colour; the extremities of the former and the apex of the latter, pale dull whitish.

The *sternum* is short, heart-shaped, the fore extremity very broad and truncated; its colour is deep yellowish-brown, and it is clothed with a prominent grey pubescence.

The *abdomen* is very large, and almost conceals the cephalothorax; it is of a sub-conical form, the upper side towards the hinder extremity being produced gradually backwards and upwards into a considerable hump, whose termination is a large round deep-brown boss. The distance from this boss to the spinners is rather less than to the fore extremity on the upper side. Its colour is a dull-brown, mottled thickly above and on the sides with dull yellowish-white, leaving a largish, irregularly-defined brown patch near the middle of either side of the upper part. The middle of the upper part has four distinct, impressed, deep black-brown spots in a quadrangular figure, whose posterior side is much longer than its anterior, and its shortest transverse diameter longer than its longitudinal one. A little way from, and on each side of, the inferior spinners is a white spot, in front of which is another, or rather a somewhat roundish, white patch. The genital aperture is of a simple transverse oval form, covered by the epigyne, which is very prominent, directed backwards, curved, flattish, and rather tapering to a broad, rounded point.

Hab.—Murree to Sind Valley, July 14th to August 5th, 1873.

Genus—*CHORIZOOPES*, Cambr.

66.—CHORIZOOPES STOLICZKÆ, sp. n.

Adult female: length rather over 2½ lines.

The *cephalothorax* is short, broadish, and massive in front; the caput elevated, especially the occipital portion of it. The colour is a dark reddish yellow-brown, and there are some coarse greyish hairs on the surface, which is also finely punctuose.

The *eyes* are small, and placed in three widely separated groups; the central group of four eyes (forming a quadrangular figure whose anterior side is the shorter, and its posterior side the longer) is placed near the lower part of the foreside of the caput, the height of the clypeus rather exceeding the diameter of one of the fore-central eyes, which are a little the largest of the eight. These of each lateral pair are placed obliquely, close above the insertion of the falces, and separated from each other by an interval of at least the diameter of the fore-lateral eyes, which are larger than the hind-lateral.

The *legs* are short and slender; there is very little difference in the length of those of the first, second, and fourth pairs, the third pair being the shortest. They are of a dull yellow colour tinged with orange, annulated with reddish yellow-brown, and furnished with hairs and slender bristles only.

The *palpi* are short, slender, and similar to the legs in colour and armature.

The *falces* are rather long, strong, prominent at their base in front, and slightly divergent, obliquely truncated at their fore extremity on the inner side, the oblique portion being armed with a single row of short strong teeth and long bristly hairs. Their colour is rather paler than that of the cephalothorax, with a darker suffusion across the middle.

The *maxillæ* are strong and considerably inclined towards the labium, which is small and of a curvilinear triangular form. The colour of the *maxillæ* is yellow-brown; that of the *labium* paler.

The *sternum* is of a similar form to that of the labium, only of course much larger, and with its apex pointed in the opposite direction; it is of a dull orange yellow-brown colour, with some red-brown marginal indentations between the points of insertion of the legs.

The *abdomen* is large, of an oval form, more pointed before than behind, where it is very bluff and rounded. The general convexity is great, though the upper surface is rather flat, and it projects greatly over the base of the cephalothorax. On each side of the upper part is a longitudinal row of small pointed (or conical) protuberances, and another longitudinal row of three similar protuberances bisects the hinder part of the abdomen; the foremost of these last is nearly in a straight, transverse line with the hinder one of each of the other row. The upper part of the abdomen is of a dull golden-yellowish colour, marked on each side of the central line with blackish-brown, including the inside half of each of the conical protuberances, and leaving a clear, broadish, longitudinal, median yellow band, from which two curved lateral stripes, edged with black, issue on each side from its hinder half, and a prominent point on each side of its fore part; the sides are obliquely rugulose, and, with the under side, are yellowish, marked with dark brown; some of the lower lateral markings are oblique, and from the genital aperture two parallel, blackish streaks, close together, run to the spinners. The epigyne connected with the genital aperture is prominent, tapering, pointed, and a little directed backwards.

This spider is remarkable from the genus having only been previously recorded as indigenous to Ceylon.

Hab.—Murree to Sind Valley, July 14th to August 5th, 1873.

67.—CHORIZOOPES CONGENER, sp. n.

Adult female : length 2 lines.

The caput is greatly elevated, broad, well rounded on all sides above, and highest at the occiput, appearing to overwhelm the thoracic portion by its disproportionate development; its sides are perpendicular, and it is divided longitudinally by a duplex, longitudinal, indented line. The colour of the cephalothorax is deep red-brown, except two largish oblique, somewhat oval patches on either side of the occiput, which are of a clear yellowish-red, and a patch on each side of the four central eyes, as well as the clypeus, which are dull reddish-yellow. The height of the clypeus is less than half that of the facial space, being equal to rather more than two diameters of one of the fore-central eyes. The surface of the cephalothorax is covered thinly with short fine hairs, and appears to be finely punctuose.

The *eyes* are small but not greatly different in size. They are placed in the usual three widely separated groups : the central one, of four eyes, is in the form of a quadrangle whose hinder side is the longer and its fore-side the shorter. The fore-centrals are slightly larger than the hind-centrals, seated on the sides of a slight prominence, and separated by rather more than a diameter's interval. Those of each lateral pair are near together, but not quite contiguous to each other, and are placed just above the hinder part of the insertion of the falces (looked at sideways).

The *legs* are short and not very strong; those of the fourth pair are distinctly the longest, and those of the third pair the shortest, the others differ but little from each other in length; those of the first pair slightly the longest. Their colour is yellow, annulated with light yellow-brown, and they are furnished with hairs and slender bristles only.

The *palpi* are short, slender, yellow, and without annuli; they are furnished with hairs, and several spine-like bristles on the inner-sides of the digital joints.

The *falces* are moderately long, very strong, roundly prominent near their base in front but retreating and directed backwards towards their extremity, where on the fore side there are two adjacent prominences, the larger and more prominent being the inner one of the two, and each is furnished thickly with strong bristles, in front of which are some strongish teeth.

The *maxillæ*, *labium*, and *sternum* are similar in form and structure to those of the preceding species; their colour is yellowish-brown, that of the sternum being the darkest; the apex of the labium and the extremities of the maxillæ being much the palest.

The *abdomen* is short, but broad and deep, the hinder extremity is broader and deeper than the fore part, and has four rounded prominences; three of these form a nearly straight transverse line along the upper margin, the middle one of the three being the largest and a little in advance of the other two, while the fourth is half way in a straight line between it and the spinners. It is clothed with short fine hairs of a greyish hue, and the upper side is yellowish and brown, with dark black-brown mixed; the most distinct of the yellow markings are in the median longitudinal line, towards the hinder extremity of which are two or three tolerably well-defined transverse angular bars or chevrons, with the angles directed forwards; and in front of them is a broad longitudinal band of yellow reaching to the fore extremity, and having a dusky brownish, ill-defined stripe along the middle; there is also a considerable patch of yellow on the lateral margins, mostly towards the hinder part of the upper side. The sides and under-side are deep brown; the former are rugulose, and the latter has some indistinct, dull orange-yellowish markings; the process (or *epigyne*) connected with the genital aperture is not very prominent, but obtuse, and directed backwards.

This spider is remarkably nearly allied to *Chorizoopes frontalis*, Cambr., from Ceylon, but is, I think, distinct, although closely resembling it in size, form, and colour.

Hab.—Murree to Sind Valley, July 14th to August 5th, 1877.

ARANEIDEA. 55

Family—*GASTERACANTHIDES.*

Genus—*CYRTARACHNE*, Thor.

68.—CYRTARACHNE PALLIDA, sp. n.

Immature female: length 1½ inch.

The whole of the fore part of this spider is of a pale straw-yellow colour: the normal grooves and indentations on the cephalothorax, as well as the occipital region, are suffused with whitish. The cephalothorax is short, broad behind, and but very slightly constricted laterally at the caput, the fore part of which is rather broad also. The occiput has some strong, erect bristles, and the height of the clypeus is equal to the diameter of one of the fore-central eyes. The eyes are in the ordinary position: they occupy the whole width of the fore part of the caput, and are of a pale dull amber colour; those of the hind-central pair are the largest of the eight, and are divided by an interval equal to an eye's diameter; those of the fore-central pair are divided by more than a diameter, and form a line very slightly shorter than that which is formed by the hind-central pair, the four central eyes thus forming very nearly a square.

The *legs* are rather short and slender, and are furnished with hairs and fine bristles only; their relative length is 1, 2, 4, 3.

The *palpi* are short and slender.

The *falces* are not very long, strong, straight, perpendicular. The *maxillæ, labium*, and *sternum* are of normal form, and similar to the legs in colour.

The *abdomen* is large, much the broadest across the middle, of a rather flattened form, and projects considerably over the base of the cephalothorax; it is of dull cretaceous-whitish hue with a longitudinal, median, dusky-brown line, which has some fine, oblique, venose lines of a similar colour issuing from its hinder part; and on either side of the fore part is a large, oblique, oblong, dull-brownish patch; the under side is sooty blackish.

Hab.—Murree to Sind Valley, July 14th to August 5th, 1873.

Family—*ULOBORIDES.*

Genus—*ULOBORUS*, Walck.

69.—ULOBORUS ALBESCENS, sp. n.

Adult female: length 2½ lines.

The *cephalothorax* is short, broad, nearly round behind, and gibbose on the thorax, on either side of the thoracic indentation: the caput, which is rather broad in front, is also constricted laterally. The colour is yellow-brown, paler on the margins along the medial line and on the outer side of the gibbous portion of the thorax.

The *eyes* are small, seated on black spots, and do not differ greatly in size; they are placed in two transverse curved rows, occupying the whole width of the fore part of the caput; the convexity of the curve of the hinder row is directed forwards, while that of the front row is directed backwards; the interval between the eyes of each lateral pair is thus greater than that between the fore and hind-central pairs. The interval between the eyes of the hind-central pair is greater than that between each of them and the hind-lateral on its side; and the interval between those of the fore-central pair, which are seated on a slight prominence, is less than that between each of them and the fore-lateral on its side, the latter interval being also less than that between the eyes of each lateral pair. The interval between the fore-centrals is also very nearly as great as that between each hind-central eye and the hind-lateral on its side. The four central eyes form a quadrangular figure whose fore-side is the shortest and its hinder side slightly the longest. The clypeus is almost obsolete.

The *legs* are short, and those of the first and fourth, and second and third pairs respectively, do not differ greatly in length; those of the first pair are much the strongest, their relative length being 1, 4, 2, 3. They are of a pale yellow-brown colour, clouded in parts with a darker hue; excepting the calamistra on the metatarsi of the fourth pair, there were scarcely any hairs on the legs, but probably some of them had been rubbed off.

The *palpi* are short, and slender, of a dull yellow colour, furnished with bristles and grey hairs, and terminate with a rather strong curved, black, toothed claw.

The *falces* are small, slightly projecting forwards, and of a pale yellow-brown colour.

The *maxillæ* and *labium* are of normal form, and similar in colour to the palpi.

The *sternum* is heart-shaped, and of a pale brownish-yellow colour.

The *abdomen* is large, very much elevated and obtuse at its anterior extremity, which projects considerably over the cephalothorax, and somewhat pointed behind on either side of the upper part; near the anterior extremity is a roundish, somewhat sub-conical protuberance. The colour of the abdomen is yellowish white; an irregular brownish venose line extends along the middle of the fore part of the upper side, and emits some other fine venose lines on each side as it runs backwards; about the middle of each side, near, and partly on, the under side, is a rather oblique brown patch; and another of the same colour extends along a portion of the middle of the under side. On each side, near the base of the spinners, are two white spots, and immediately in front of the usual spinners is the supernumerary spinning organ. An obtuse prominent pale-yellowish process, slightly indented at its extremity and pointed backwards, is connected with the genital aperture.

Hab.—Murree to Sind Valley, between July 4th and August 5th, 1873.

Family—*THOMISIDES.*

Genus—*THOMISUS*, Walck. (*ad partem*).

70.—THOMISUS ALBIDUS, sp. n.

Immature female: length 1⅜ lines.

It is with some hesitation that I describe this spider as a new species, since it is possible that in the adult state it may present some other specific characters which may either prove it to have been already described, or else render the present description quite inadequate for the determination of the species. As, however, there is no described species known to me to which I can refer it, I venture to include it here as new.

The *cephalothorax* is broadest quite at the hinder extremity, and narrows gradually to the fore extremity, which is also tolerably broad and truncated: the fore corners of the upper side of the caput are prominent and sub-angular. The margins of the cephalothorax are whitish, the sides yellow-brown with a greenish tinge; the broad median longitudinal band, to a little distance behind the eyes, is pale yellow-brown, and the remainder is suffused with white; the ocular area and the middle part of the clypeus are also suffused with white.

The *eyes* are very small, seated on strong tubercles in a crescent form. The hind-laterals are the most prominent of the tubercles, forming the fore-angles of the caput; those of the fore-central pair are slightly the largest of the eight; the intervals between those of the hinder row are equal, as are, apparently, also those between the eyes of the anterior row; the interval between those of each lateral pair is less than that between the fore and hind-central pairs. The four central eyes form a square whose posterior side is longer than the rest. The height of the clypeus is less than half that of the facial space.

The *legs* of the first pair are moderately long, slender, of a dull whitish-yellow colour, and armed with two parallel rows of short spines beneath the metatarsi. The legs of the second pair were absent; those of the third and fourth pairs are much shorter than the first—the third slightly the shortest; they are rather paler in colour than the first, and have no spines.

The *palpi* were both absent.

The *maxillæ* and *labium* are of normal form, and similar to the legs in colour.

The *sternum* is nearly round, slightly hollow at the fore extremity, and its colour is whitish yellow.

The *abdomen* is large, considerably convex above, and projects greatly over the base of the cephalothorax; its hinder extremity is the broadest and most massive, and it is of a uniform yellow-white colour above, whiter on the sides and underneath.

Hab.—On the road from Yarkand to Bursi, between May 28th and June 17th, 1874.

71.—THOMISUS ALBENS, sp. n.

Immature female: length rather over 2½ lines.

The *cephalothorax* has the slope of its sides and hinder part gradual and not very steep. The angular prominences at the fore-corners of the caput are strong; the *clypeus* projects forwards, and its height exceeds half that of the facial space. The colour of the cephalothorax is dull pale-yellowish, very slightly tinged with brown; the ocular area, all the middle portion of the clypeus, and a large arrow-head-shaped patch on the occiput (the point of the arrow running backwards to the hinder margin), are white, the sides, and part immediately behind the eyes, being also slightly veined with white.

The *eyes* are very small, seated on tubercles in two curved rows in the usual form of a crescent: those of the hind-central pair are further from each other than each is from the hind-lateral eye on its side, while the fore-centrals are considerably nearer together than each is to the fore-lateral on its side; those of each lateral pair are also much nearer together than the fore and hind-central pairs are to each other, the front row being much the more strongly curved. The four central eyes form nearly a square, the anterior side being considerably shortest, and the posterior one slightly the longest.

The *legs* of the first and second pairs are moderately long and tolerably strong; the second are, if anything, slightly the longest. They are of a pale dull yellowish colour suffused below with white, and the metatarsi are armed beneath with two longitudinal parallel rows of short spines; beneath the fore extremity of the tibiæ are one or two more spines, but besides these there appear to be no more on any of the legs. Those of the third and fourth pairs are much the shortest, the latter being a little longer than the third.

The *palpi* are short, destitute of bristles and spines, and similar to the legs in colour.

The *falces* are short, strong, sub-conical, rather projecting, and, excepting a small patch bisected with a white line at their base near the outer side, of a white colour like the clypeus.

The *maxilla* and *labium* are of normal form, and similar to the legs in colour.

The *sternum* is oval, hollow-truncate in front, and of a whitish hue.

The *abdomen* is of good size, flattened above, projecting well over the base of the cephalothorax, much broadest behind, where it is of a blunt-angular form on each side; the form of the upper side is therefore somewhat quadrangular, the fore part being a little roundly truncated; the sides, the fore part, and also the hinder extremity (which is abrupt) are rugulose and marked with rows of small impressed points; these are most apparent as a margin to the fore part and sides. The five normal impressed points are visible on the fore half of the upper side, and the whole of the abdomen is of a uniform white colour; the spinners are tolerably strong, very short, compact, and similar in colour to the legs.

This spider is allied to, but quite distinct from, *T. pugilis*, Stoliczka, found in the neighbourhood of Calcutta.

Hab.—On the route from Yárkand to Bursi, between May 28th and June 17th, 1874.

Genus—*MISUMENA*, Thor.

72.—MISUMENA EXPALLIDATA, sp. n.

Adult female: length 3¼ lines.

The whole of the fore part of this spider is a dull pale yellow. The cephalothorax slightly tinged with brown, with a pale, somewhat triangular, patch at the occiput; the falces also being similarly tinged. The abdomen is white, a little suffused on the sides with brownish-yellow, and with a narrow, median, brownish stripe on the fore half of the upper side, emitting some lateral and posterior venose lines; the usual five impressed spots are also visible on the fore half of the upper side.

The *eyes* are small, and differ but little in size, the fore-laterals being rather the largest; they are seated on white tubercles, in the form of a crescent, in two curved rows, the anterior being the shorter and more curved; the interval between those of the hind-central pair is less than that between each and the hind-lateral eye on its side, while that between the fore-centrals is slightly greater than that between each and the fore-lateral next to it. The four central eyes form a square whose posterior side is a very little longer than its anterior, and the interval between those of each lateral pair is less than that between the fore and hind-central pairs. The height of the clypeus is less than half that of the facial space.

The *legs* of the first and second pairs are long, moderately strong, and scarcely differing in length; those of the third and fourth pairs are much shorter, less strong, but also of nearly equal length. Those of the second pair appear to be slightly the longest, and the third pair slightly the shortest; all are furnished with spines, of which there are two longitudinal parallel rows beneath the metatarsi and tibiæ of the two first pairs. The metatarsi and tarsi are tinged with reddish yellow-brown.

The *palpi* are short, slender, and furnished with hairs and bristles. The genital aperture is small and simple, being of a somewhat oblong form, a little narrower at its hinder than at its fore extremity.

Hab.—Murree, between June 11th and July 14th, 1873.

73.—MISUMENA OBLONGA, sp. n.

Adult female: length nearly 4¼ lines.

The *cephalothorax* is as broad as it is long; the marginal constrictions on the sides of the caput are strong; the thorax broader than long. The colour of the cephalothorax is dull yellow-brown, with a longitudinal median white line, and a lateral, somewhat zigzag, line of the same colour along the middle of each side. The central white line has two or three lateral points on each side, and it runs from immediately behind the ocular area, to the thoracic indentation: the clypeus projects a little forwards, and its height distinctly exceeds half that of the facial space.

The *eyes* are small, seated on white tubercles in two transverse rows, in a narrow crescent form; the front row is the shorter and more strongly curved; the ocular area is comparatively rather small, and the fore-lateral eyes are but slightly larger than the fore-centrals; these last are a little further from each other than each is from the fore-lateral on its side; while those of the hind-central pair are nearer together than each is to the hind-lateral next to it. The four central eyes form very nearly a square whose longitudinal is very slightly greater than its transverse diameter. The interval between those of each lateral pair is equal to that between the fore and hind-central pairs.

The *legs* of the first and second pairs are tolerably long and rather slender: those of the first appear to be a little the longest, those of the third and fourth pairs are much shorter, the third rather the shortest; all are of a pale, dull, straw-yellow colour, and are furnished with hairs, slender bristles, and spines; of the last there are two longitudinal parallel rows of long conspicuous ones beneath the metatarsi and tibiæ of the first and second pairs; the rest of the spines on these legs, and especially those on the third and fourth pairs, are small and inconspicuous.

The *palpi* are short, slender, and similar to the legs in colour and armature; the radial joint has a long, slightly curved, pointed spine near the base on the inner side, its point directed inwards.

The *falces* are moderately long, not very strong, sub-conical, projecting a little forwards, and similar in colour to the cephalothorax.

The *maxillæ* and *labium* are of normal form and similar to the legs in colour, the sternum being heart-shaped and of a pale straw-colour.

The *abdomen* is elongated, oblong-oval in form, and more than three times the length of the cephalothorax; its fore extremity is rather roundly truncated and broader than the hinder extremity, the widest part being a little in front of the middle; its colour is a dull straw-yellow, with the sides and a broad, median, longitudinal band pretty thickly spotted with white; the median band has a largish, elongate, diamond-shaped, dull-brownish, straw-coloured marking on the fore part emitting some short venose lateral lines. The genital aperture is small and simple, consisting of two round yellow-brown openings placed side by side, and edged with dark reddish-brown.

This is in several respects a remarkable spider and aberrant from the generic type, both in the form of the abdomen, the height of the clypeus, and the small comparative size of the ocular area. I hesitate, however, at present, to form a new genus for it, though it will probably be necessary at some future time to do so.

Hab.—Murree to Sind Valley, between July 14th and August 5th, 1873.

Genus—*SYNEMA*, Simon.

74.—SYNEMA EXCULTA, sp. n.

Adult female : length 2 lines.

The *cephalothorax* is short and broad; the lateral constrictions on the margin of the caput are slight, and the caput is broadly truncated before. The height of the clypeus is rather less than half that of the facial space. It is of a dull yellow-brown colour; the clypeus, as well as the ocular area and a broad longitudinal band on each side of the upper part, being of a deep reddish-brown colour; that of the two bands being the darkest; the occiput is marked with a somewhat curvilinear, angular, pale-yellowish marking, the angle of which is directed backwards. The sides and hinder slope of the cephalothorax are steep, and its surface is thinly covered with long, curved, prominent, rather tapering bristles.

The *eyes* are on tubercles, in two transverse curved rows occupying the whole width of the broad caput; the hinder row being the longer, and, if anything, slightly the more strongly curved; thus, the eyes of each lateral pair are rather further from each other than the fore-central pair is from the hind-central one. The eyes of each row respectively are equi-distant from each other; though, if anything, the fore-centrals may be very slightly further from each other than each is from the fore-lateral on its side. The fore-laterals are the largest of the eight and considerably larger than the fore-centrals. The fore-central eyes form a square whose posterior side is longer than the other three.

The *legs* of the first and second pairs are moderately long and slender, the second pair being perhaps slightly the longest ; they are of a yellow-brown colour, the femora much the darkest; the third and fourth pairs are much shorter, of a pale-yellowish colour, and the third pair is slightly the shortest. All are furnished with hairs, slender bristles, and spines.

The *palpi* are short, slender, and similar in colour to the third and fourth pairs of legs.

The *falces* are short, strong, sub-conical, slightly projecting forwards, and of a dull reddish yellow-brown colour.

The *maxillæ* and *labium* are of normal form, and of a dull brownish-yellow colour.

The *sternum* is heart-shaped and yellow.

The *abdomen* is short, considerably convex above, and projects entirely over the hinder slope of the cephalothorax; it is much broadest, and well rounded, towards the hinder extremity, and is of a dull yellow-brown colour; the fore part of the upper side has a few deep red-brown points, and a diffused marginal border of white cretaceous spots; the hinder part is much covered with similar white spots arranged in three not very well defined transverse diffused, curved bars, the two intervals between the first three being of a deep red-brown colour; there are also some markings of the same dark, red-brown hue just above the spinners; the sides are rugulose, marked with indistinct reddish-brown streaks following the somewhat oblique course of the rugulosities. The genital opening is simple, and consists of two small, round, reddish-brown apertures in a transverse line.

This spider is allied to *Synema* (*Diæa*, Thor.) *globosa*.
Its fore-lateral eyes, however, are larger in proportion to the fore-centrals than in that species, and *S. exculta* thus diverges still more widely from the spiders of the genus *Diæa*.
Hab.—Murree, between June 11th and July 14th, 1873.

Genus—*DIÆA*, Thor.

75.—DIÆA SPINOSULA, sp. n.

Adult male: length rather less than 1½ lines.
The *cephalothorax* is as broad as, if not a little broader than, it is long, the caput short, broadly truncate in front, and constricted laterally at the lower margins; it is of a bright reddish orange-yellow colour, with a largish patch of a paler hue on the occiput; the surface is covered thinly with strong, prominent, dark-coloured spine-like bristles, and the margins are armed with minute but distinct teeth.
The *eyes* are small and seated on whitish tubercles, the fore-laterals being rather the largest; they are in two transverse, concentric, curved rows, the curve directed forward; and they occupy the whole width of the fore extremity of the caput, the front row being the shorter: the eyes of the hinder row are equidistant from each other, but those of the fore-central pair are nearer to each other than each is to the fore-lateral eye on its side. The tubercles on which the eyes of each lateral pair are placed are large and round, the interval between the eyes themselves being equal to that between the fore and hind-central pairs. The four central eyes form a quadrangular figure whose posterior side is the longest and anterior the shortest. The height of the clypeus is considerably less than half that of the facial space.
The *legs* are exceedingly slender; those of the first and second pairs are very long, and appear scarcely to differ in length; these two pairs are of a rather a paler duller colour than the cephalothorax, the metatarsi, and the two-thirds of the tibiæ next to them, being of a deep reddish chocolate-brown; those of the third and fourth pairs are yellow, the third pairs being rather the shorter; excepting two or three small spines on the femora of the first and second pairs, the armature of the legs consists of hairs and slender bristles only.
The *palpi* are short and not strong; the radial joint is rather shorter than the cubital, and has a small, short, tapering, pointed apophysis at its outer extremity, with several longish bristles on its upper side: the digital joint is small and of a rather narrow, oval form (its length being about equal to that of the radial and cubital joints together), and it is a little suffused with brown. The palpal organs are very simple and not prominent.
The *falces* are short, but moderately strong, perpendicular, subconical, and similar in colour to the cephalothorax.
The *maxillæ*, *labium*, and *sternum* are yellow.
The *abdomen* is rather narrow, oval, and of a somewhat flattened form; its colour on the upper side, which is of a somewhat coriaceous nature, is a slightly brownish yellow, and is covered, like the cephalothorax, with erect, strong, tapering, spine-like, dark bristles; and there are five impressed yellow-brown spots on the fore half of the upper side, enclosing an acute angle directed forwards. The sides, and the hinder extremity of the upper side, are rugulose, and, with the under part, are of a pale straw-yellow colour.
Hab.—Murree, between June 11th and July 14th, 1873.

76.—DIÆA SUBDOLA, sp. n.

Adult male: length rather more than 1¼ lines.

The cephalothorax is round-oval behind, broad and truncated in front, longer than it is broad, and the lateral constrictions of the caput are slight; its colour is dull brownish orange-yellow, the hinder part of the caput, and some short lateral converging stripes, being pale yellow; its surface is smooth and glossy, but covered very thinly with long, nearly erect, curved black bristles; the height of the clypeus is a little less than half that of the facial space.

The *eyes* are seated on rather strong, greenish-white tubercles in the form of a crescent; they do not differ greatly in size; the fore-laterals are, however, distinctly the largest of the eight, and the tubercles on which they are seated are also the largest; the other eyes differ very slightly in size; the fore-centrals, however, appear to be rather larger than those of the hinder row: the front row being the shorter and more curved, a more strongly crescent form than usual is given to the ocular area, and the interval between the eyes of each lateral pair is consequently less than that between the fore and hind-central pairs: the intervals between the eyes of the hinder row are as nearly as possible equal, while that between the fore-centrals is distinctly greater than that between each and the fore-central on its side. The four central eyes from a quadrangular figure whose longitudinal is slightly greater than its transverse diameter at the hinder part, and its fore-side the shortest.

The *legs* are not very slender; those of the first and second pairs are long, the latter slightly the longer; the third pair is the shortest, but that and the fourth pair, in proportion to the first and second, are not so short as usual; they are very nearly of the same colour as the cephalothorax, and are furnished with bristles and longish slender spines.

The *palpi* are short and similar to the legs in colour; the radial and cubital joints are short and of nearly equal length; the former is, if anything, rather the shorter, but a little stronger; it has a few strong spine-like bristles, and its extremity on the outer side is prolonged into a longish projection, bent a little downwards and backwards, rather broadest near its extremity, which is rather bifid or slightly furcate; and there is another strong, curved obtusely-pointed process beneath the joint. The digital joint is large, broad, and rounded behind, pointed in front, and is somewhat angularly prominent on the outer margin; the palpal organs are simple but encircled by a long, strongish, black spine which issues from their base on the inner side.

The *falces* are neither long nor very strong; they are nearly perpendicular, and similar in colour to the cephalothorax; the *maxillæ* and *labium* are of the ordinary form and rather duller and paler than the falces.

The *sternum* is heart-shaped and of a brightish yellow colour.

The *abdomen* is round and broadest behind, narrower and more pointed before; it is of a dull brownish-yellow colour, marked with cretaceous white spots on either side of the upper part, defining indistinctly the normal dentated central band so conspicuous generally in *xysticus*; there are also several deep red-brown spots on each side, and a large patch suffused with red-brown at the hinder extremity surrounding the spinners, but chiefly placed on each side of them; the under side is paler than the upper; the upper side is furnished with a few scattered, long, strong bristle; and an oblong-oval patch between the spiracular plates is similar in colour to the *sternum*. It is probable that there may be, in a series of examples,

ARANEIDEA. 63

such considerable variety in the extent and nature of the abdominal markings, as is found be to in some others of this group.

Hab.—Murree, between June 11th and July 14th, 1873.

77.—DIÆA SUFFLAVA, sp. n.

Adult male: length rather more than 2¼ lines.

The *cephalothorax* is round behind, and constricted laterally at the caput; its colour is yellow, and the upper surface of the caput has a few strong, blackish, prominent bristles. The height of the clypeus is less than half that of the facial space.

The *eyes* are seated on round tubercles, in two curved transverse rows, in the form of a crescent; the curves of the rows are directed forwards, and the front row is the shorter and more strongly curved. The fore-lateral eyes are slightly the largest, and are nearer to the hind-laterals than the fore-central pair are to the hind-central; each is also nearer to the fore-central eye on its side than the fore-centrals are to each other; the hind-centrals are slightly nearer to each other than each is to the hind-lateral on its side; the four central eyes describe very nearly a square, its fore side being slightly shorter than its hinder one, and its longitudinal very slightly longer than its transverse diameter.

The *legs* of the first and second pairs are long and tolerably strong; they are similar in colour to the cephalothorax, and, with those of the third and fourth pairs, are furnished pretty freely with spines, besides hairs and bristles. The difference in length between the first and second pairs is very slight; if anything, those of the first pair are a little the longer: the third and fourth pairs are short, the fourth slightly the longer; they are rather paler than those of the first and second pairs.

The *palpi* are short and of a pale yellow colour. The cubital and radial joints are short, but about equal in length; the former has a long, strong, tapering, curved bristle at the middle of its fore extremity on the upper side, and the latter has two or three upon it, but less strong than that on the cubital joint. The radial joint also has its fore extremity on the outer side, produced into a not very large, slightly tapering, sharp-pointed apophysis, the point being of a corneous claw-like nature, and directed slightly outwards and downwards. There is also another apophysis, at the extremity, underneath this joint, stronger, curved, and obtuse at the extremity; the digital joint is tolerably long, equal in length to the radial and cubital joints together, oval, and pointed at its anterior extremity; the palpal organs are small, simple, and apparently without any marked spines or processes.

The *falces* are short, strong, straight, perpendicular, not greatly broader at their base in front than at their extremity, and their colour is similar to that of the legs. The *maxillæ* and *labium* are of normal form; the former are rather paler in colour than the legs, the latter is yellow-brown.

The *abdomen* is rather large, of an elongate-oval form, decreasing gradually in breadth from its fore to its hinder extremity; its convexity on the upper side is not great, but tolerably uniform. It is of a pale dull yellow-brown above, and pale dull straw-yellow on the sides and underneath; the upper side is margined by a belt of whitish cretaceous spots, on the inner side of which is an irregular row of dark red-brown spots which increase in size towards the hinder extremity, and evidently represent the ends of a series of broken transverse angular bars; the surface is also thinly covered with a few prominent dark-coloured bristles, and the spinners are short and of a yellow-brown colour.

Hab.—Murree, between June 11th and July 14th, 1873.

78.—DIÆA SUSPICIOSA, sp. n.

Adult male: length nearly 2⅜ lines.

This spider is very nearly allied to *Diæa dorsata*, Fabr. (*Thomisus floricolens*, Blackw.), but may be distinguished by its generally lighter hue and less distinct markings, as well as by a quite different structure of the palpi and palpal organs.

The *cephalothorax* is yellow; the sides, the fore part of the upper side of the caput, and the normal indentations are strongly suffused with yellow-brown; and there are a few strongish bristles on upper margins of the caput. The ocular region has none of the deep rusty red-brown suffusion characteristic of *Diæa dorsata*. The height of the clypeus is less than half of that of the facial space.

The *eyes* are seated on round, whitish tubercles, in two nearly concentric curved rows: the front row being a little the more strongly curved, and thus the eyes of each lateral pair are brought rather nearer together than the fore- and hind-central pairs are to each other. The fore-laterals are largest of the eight, and seated on the strongest tubercles; the interval between those of the hind-central pair is rather less than that between each and the hind-lateral on the same side; and the interval between the fore-centrals is very slightly, if anything, greater than that between each and the fore-lateral on its side. The four central eyes form a square whose anterior side is the shortest.

The *legs* of the first and second pairs are very long; those of the first the longer, slender, and of a yellow colour, suffused with reddish yellow-brown at the fore extremity of the femora and genua, and at both extremities of the tibiæ, but the colouring scarcely amounts to annulation; and the under sides of the femora are speckled with red-brown; those of the third and fourth pairs are much shorter than the others; the third pair rather the shorter, and paler in colour than the rest; all are furnished with hairs and spines.

The *palpi* are short, and pale yellow; the digital joints suffused with brown. The radial joint is shorter than the cubital, and has its outer side, at the fore extremity, produced into a tolerably strong and long, tapering, sharp-pointed apophysis, with a distinct angular point about the middle underneath. In *Diæa dorsata* this point is replaced by a larger and rounded protuberance close at the end of the apophysis, which gives it a more bifid form. The digital joint is of tolerable size, broad-oval behind, and with a slightly constricted, narrow extremity, and the outer margin near the base is somewhat sub-angularly prominent; the palpal organs are simple and encircled by a black filiform spine. The radial and cubital joints are furnished with two or three strong tapering bristles, and the digital joint is also hairy and bristly. This joint is smaller in *Diæa dorsata*, and the palpal organs in that species have no encircling black spine.

The *falces* are short, strong, straight, sub-conical, perpendicular, and similar in colour to the cephalothorax.

The *maxillæ*, *labium*, and *sternum* are of normal form, and of a pale-yellow colour.

The *abdomen* is oval, of a rather flattened form; its upper side is of a dull pale-yellowish hue, thinly pencilled with whitish, and deep brownish, rusty-red spots: the sides of the upper part are a little suffused with brownish rusty-red, and its margins have a tolerably distinct white border; the sides have a longitudinal brownish rusty-red band, which runs round the fore extremity, including the spinners, and joining in with the rusty-red colouring at the hinder extremity of the upper side. The under side is of a uniform pale dull yellow.

Hab.—Route from Yárkand to Bursi, between May 29th and June 17th, 1874.

79.—DIÆA SUBARGENTATA, sp. n.

Adult male: length rather under 2 lines.

This spider is nearly allied to *Diæa* (*Xysticus*) *Pavesii*, Cambr., Journ. Linn. Soc., vol. xi, p. 540, pl. 15, fig. 8, but it may be distinguished without difficulty, if the descriptions of the markings on the abdomen, and the structure of the palpi in the two species are carefully compared.

The *cephalothorax* is round behind, slightly constricted on the lateral margins at the caput, the fore extremity of which is broad and slightly roundly truncated. It is of a brownish-yellow colour, with a not very strongly defined, longitudinal, darker reddish yellow-brown band on each side of the upper part; the caput is also rather suffused with lighter reddish yellow-brown, and there is a somewhat arrow-head-shaped yellow marking on the occiput, with the point directed backwards. The height of the clypeus is just half that of the facial space.

The *eyes* are on round, whitish tubercles in the usual two-curved rows, which are very nearly concentric, making the interval between the eyes of each lateral pair nearly equal to that between the fore and hind-central pairs. The interval between the eyes of the hind-central pair is distinctly greater than that between each and the hind-lateral eye on its side; while that between the fore-centrals is less than that between each and the fore-lateral eye on its side. The four central eyes form a rectangle whose posterior side is the longest and anterior the shortest. The fore-laterals are but slightly the largest of the eight.

The *legs* of the first and second pairs are long and moderately strong; those of the second pair are, if anything, slightly the longer; they are of a brownish yellow colour, the genua, as well as the two extremities of the tibiæ, and the fore extremity of the metatarsi, being of a darker reddish yellow-brown, giving them an annulated appearance. Those of the third and fourth pairs are much the shortest, the third pair being the shorter of the two; these are of a plain pale yellowish hue; and all the legs are furnished with hairs, slender bristles, and spines.

The *palpi* are short, slender, and of a brownish-yellow colour. The cubital and radial joints are short, and are furnished with several longish, tapering, dark bristles; the radial is the shortest and has a not very long nor strong tapering apophysis at its extremity near the outer side, terminating with a sharp, somewhat corneous-looking point. There is also another apophysis on the under side, apparently rather stronger, and obtusely pointed. The digital joint is as long as the radial and cubital joints together, and is of a narrow-oval form, sharpish pointed at its anterior extremity. The palpal organs are small and simple in form, apparently encircled, or nearly so, with a very slender filiform spine.

The *falces* are moderate in length and strength, sub-conical, and directed a little backwards. Their colour is like that of the cephalothorax.

The *maxillæ* and *labium* are of normal form, and of a light yellowish-brown colour.

The *sternum* is heart-shaped and yellow.

The *abdomen* is oval, moderately convex above, though of a somewhat flattened form on the upper side; it is broadest towards the hinder extremity, which is obtuse-pointed, and its fore extremity is roundly truncated. The upper part and sides are of a dull yellowish hue, thickly covered with somewhat scale-like spots of a silvery whitish colour. Five impressed spots form a triangle on the fore half, whose apex is directed forwards. The apical spot is surrounded with dull reddish yellow-brown; and immediately following the last impressed spot on each side is a row of three or four reddish yellow-brown blotches, decreasing in size as they run backwards, the two rows of blotches converging to the spinners; these last are short and yellow-brown in colour; the superior and inferior pairs are of equal length, but the latter are the strongest; and at the extremity of the abdomen on each side of the spinners is an oblong patch of red-brown.

The female is altogether lighter coloured than the male; the abdomen has no markings, excepting the normal five impressed spots on the upper side, and the oblong patch (which, however, is very indistinct) on each side of the spinners; the legs also are of a uniform lape yellow, and those of the third and fourth pairs are destitute of spines, or at any rate they are no stronger than an ordinary bristle.

Hab.—Murree, between June 11th and July 14th, 1873.

Genus—*XYSTICUS*, C. L. Koch.

80.—XYSTICUS CRISTATUS.

Xysticus cristatus, Clerck (sub *Araneus*), Sv. Spindl., p. 136, pl. 6, tab. 6.

Hab.—Examples of a spider, which I believe to be of this species, were contained in a portion of the collection labelled 'Road across the Pamir from Sirikol to Panja and back, April 22nd to May 7th, 1874;' and 'Yárkand and neighbourhood, November 1873'.

81.—XYSTICUS PINI, Hahn.

Hab.—Young examples of this spider were contained in a part of the collection from the Sind Valley, 5th to 13th August 1873, and Hills between Sirkol and Aktalla, 8th to 13th May 1874.

82.—XYSTICUS MACULOSUS, sp. n.

Adult female: length 2 lines.

In form and structure this spider closely resembles *Xysticus audax*, Bl.; its colours are a speckled mixture (both above and below) of white, yellow-white, yellow-brown, dark-brown, and red-brown.

The sides of the *cephalothorax* are dark-brown, marbled and marked with pale yellow-brown: the upper part, consisting of a broad longitudinal band, is yellow-white, suffused with pale yellow-brown forwards, and spotted all over with small spots of a darker hue, two rather distinct parallel lines of the darker yellow-brown running close together from between the hind-central pair of eyes to the occiput. The lower margin of the clypeus has a row of strong prominent bristles directed forwards.

The *legs* of the first and second pairs are rather strong and moderately long, those of the first pair a little the longer; they are distinctly spotted and blotched with yellow-brown, dark-brown, and white on a pale-yellowish ground, the outer sides of the femoral and tibial joints being marked, rather distinctly, with a longitudinal white stripe, on each side of which is a dark-brown one; the inner sides of the tibial and metatarsal joints are armed with two longitudinal rows of strong spines springing from tubercular eminences. The legs of the third and fourth pairs are much shorter than the rest, and marked with similar colours, but presenting a more annulated appearance.

The *palpi* are short, pale-yellow, roughly annulated with deep-brown, and armed with bristles and short spines.

The *falces* are short, tolerably strong, sub-conical, perpendicular, marbled with pale yellow-brown, white, and deep brown, and furnished with some strong prominent black bristles.

The *maxillæ* and *labium* are dark dull brown; and the *sternum* is yellowish-white, distinctly speckled with small, deep black-brown points.

The *abdomen* is oval, broadest behind, where it is rounded, the fore extremity being rather truncate, and projecting over the whole hinder slope of the cephalothorax. The upper side is flattish, of a dull pale yellow-brown colour, thickly and minutely speckled with darker yellow-brown and whitish, with a few deep reddish-brown spots round the margins, and some smaller ones of the same colour thinly dispersed over the whole; the sides are rugulose and whitish, speckled thinly with yellow-brown and deep red-brown, the under side being dull yellow-brown, thickly and minutely speckled with small white and red-brown points.

This spider is nearly allied to *Xysticus græcus*, C. L. Koch, from which, as also from another nearly allied Egyptian species, *X. promiscuus*, Cambr., it is certainly distinct; from the latter it may at once be distinguished by the almost total absence of the characteristic dentated pattern on the upper side of the abdomen. This is quite distinct in *X. promiscuus*, while in the present spider it can scarcely be traced excepting by a very slightly paler tone in the general hue.

Hab.—Murree, between 11th June and 14th July 1873.

83.—XYSTICUS SETIGER, sp. n.

Adult female: length nearly 3 lines.

The whole of the fore part of this spider is of a reddish, orange-yellow-brown colour. The *cephalothorax* is of ordinary form, and has two longitudinal, darker yellow red-brown bands running backwards, one from each hind-lateral eye, the fore part of the median band being rather darker than the rest; and it is covered thinly with long, strong, dark, prominent bristles, directed a little forwards.

The *eyes* are on small yellowish tubercles, and differ a little from the typical position of *Xysticus*. The fore-laterals being placed farther back, give a stronger curve to the front row, and bring the eyes of each lateral pair nearer together; the interval between them in the present spider being distinctly less than that between the fore and hind-central pairs, while in the typical *Xysticus* it is equal, if not greater. The position of the eyes is thus more like that of *Philodromus*. The four central eyes form very nearly a square, whose fore side is rather longer than the hinder one, and its sides slightly longer than its fore side. The height of the clypeus is nearly equal to half that of the facial space.

The *legs* are tolerably long and strong; those of the first pair are slightly longer than those of the second; these latter are thinly speckled with red-brown, and a little clouded, on the femora of the first pair, with a darker hue than that of the ground-colour. They are furnished with hairs, bristles, and spines; the last form two longitudinal parallel rows beneath the tibiæ (6—6) and metatarsi (5—5), and issue from tubercular eminences; the legs of the fourth pair are distinctly longer than those of the third. Each tarsus has a small claw-tuft beneath the two terminal claws.

The *palpi* are short, not very strong, and are furnished with hairs, bristles, and spines.

The *falces* are short, strong, sub-conical, perpendicular, and furnished with strong prominent bristles in front.

The *sternum* is oval-pointed behind, truncated in a hollow line in front, and of a pale orange-yellow colour, destitute of bristles and (apparently) of hairs also.

The *abdomen* is broadest towards its hinder extremity, which is obtusely pointed, the fore extremity being truncated; it is of a deep yellow-brown on the upper side, mottled with reddish yellow-brown along the middle, indistinctly indicating the normal dentated band, and some transverse, slightly curved line towards the hinder part; the upper side is also covered with long, strong, slightly curved, nearly erect blackish bristles: the sides are rugulose, paler than the upper part, slightly suffused with white, and thinly speckled with few dark black-brown points; the under side is yellow-brown, and has a large quadrate, central area thickly mottled with small, whitish-yellow spots.

Hab.—Murree, between 11th June and 14th July 1873.

84.—XYSTICUS BREVICEPS, sp. n.

Adult female: length 3½ lines.

The *cephalothorax* is short, its breadth at least equalling its length; the caput, constricted laterally, is broad and particularly short; when looked at in profile, the hinder slope is very abrupt, and the depth of the cephalothorax is greatest there, sloping, in a slight curve, very gradually thence to the eyes. The colour is pale-yellow, irregularly streaked and marked with whitish-yellow; it is margined laterally with a distinct whitish, narrow border, and a broad, reddish yellow-brown, longitudinal band occupies the upper part of each side; the normal spade-shaped marking on the upper side is indicated by a reddish-yellow suffusion, and a posterior limit, formed by a curvi-angular, whitish-yellow distinct stripe. The space enclosed by this stripe is also marked with whitish-yellow striæ, bearing short erect bristles; some stronger bristles occur in the ocular region, and on the lower margin of the clypeus, which is less in height than half that of the facial space.

The *eyes* of each lateral pair are perceptibly nearer together than the fore- and hind-central pairs are to each other, owing to the fore-lateral eyes (which are the largest of the eight) being placed farther back than usual, giving the front row a stronger curve than that of the hinder one. The four central eyes form very nearly a square, the longitudinal being rather less than the transverse diameter.

The *legs* are rather short, and strong: those of the first and second pairs scarcely differ in length, those of the third pair being distinctly shorter than those of the fourth. They are of the same colour as the cephalothorax, striped with whitish-yellow, and furnished with hairs, bristles, and spines; the last are, principally, in two parallel rows beneath the tibiæ

ARANEIDEA. 69

and metatarsi of the first and second pairs; those on the metatarsi are much the strongest and most numerous. The femora of the first pair have three smaller erect spines in a longitudinal line on the upper side.

The *palpi* are short, and similar to the legs in colour and armature.

The *falces* are strong, moderately long, subconical, and a little projecting forwards; they are of a reddish yellow-brown colour, marked and suffused with whitish-yellow, and furnished with bristles in front.

The *maxillæ* and *labium* are normal in form, and similar in colour to the legs.

The *sternum* is oval, blunt-pointed behind, and broadly truncated in front; it is of a pale whitish-yellow colour, thinly clothed with slender, erect, bristly hairs.

The *abdomen* is oval, rounded in front and obtusely pointed behind, tolerably convex above, and thinly clothed with hairs. The upper side is of a whitish-yellow or dull cream-colour, thickly speckled with minute red-brown specks; the sides are rugulose, and pale yellow-brown; the rugulosities yellow-white, minutely spotted with red-brown; the under side is pale whitish-yellow, like the sternum. The ordinary longitudinal, dentated band on the abdomen is imperceptible; probably, however, some variety exists in this respect in different examples.

Hab.—Yárkand to Bursi, between May 28th and June 17th, 1874.

85.—XYSTICUS MUNDULUS, sp. n.

Immature male: length just over 2 lines.

The *cephalothorax* is of ordinary form, and has a whitish, narrow marginal border. The sides are of a dull reddish yellow-brown colour, irregularly but distinctly marked with short whitish streaks and markings, leaving a broad, median, longitudinal, nearly white band, slightly narrowest at its hinder extremity; the fore part of this band contains the normal spade-shaped marking, which is of a dull pale-brownish hue, rather peculiar in form, and marked with some red-brown lines and markings; its posterior extremity being also continued, by a red-brown line, to the thoracic indentation.

The *eyes* are in the normal position; the four central eyes form very nearly a square, the longitudinal being slightly greater than the transverse diameter, and the fore side slightly shorter than the hinder one; the interval between those of the hind-central pair is distinctly less than that between each and the hind-lateral eye on its side, and the interval between those of each lateral pair is equal to that between the fore- and hind-central pairs. The height of the clypeus is scarcely more than one-third of that of the facial space.

The *legs* are tolerably long and strong; those of the second pair are slightly longer than those of the first, and the third pair are a little the shortest. They are of a yellowish colour, more or less suffused and striped longitudinally with white, especially on the femora of the first and second pairs, which are also prettily spotted with reddish yellow-brown. The other legs are also spotted, though more faintly; the tarsi and metatarsi of all being of an almost unmarked pale-yellow colour. The tibiæ and metatarsi of the first and second pairs are armed with a few longish, not very strong, spines, in two parallel longitudinal rows on the under sides.

The *palpi* are similar in colour to the legs.

The *falces* are short, strong, subconical, perpendicular, furnished with a few strong bristles; they are of a whitish colour, excepting at the base on the upper side, where they are yellow-brown.

The *maxillæ*, *labium*, and *sternum* are of normal form, and their colour is nearly white; the *sternum* spotted thinly with small, deep reddish-brown points.

The *abdomen* is oval, of a rather flattish form, and not much broader at any part than it is before and behind, at both which points it is rounded. The sides of the upper part are of the same colour as the sides of the cephalothorax; the normal longitudinal, median, dentated band is of a paler hue, bordered with white, and marked with a few red-brown points; the sides are whitish, rugulose, and thinly spotted with red-brown; the outer side is also similarly coloured.

Hab.—Sind Valley, between August the 5th and 13th, 1873.

Genus—*MONASTES*, Luc.

86.—MONASTES DEJECTUS, sp. n.

Adult female: length nearly 2¼ lines.

The *cephalothorax* of this spider is nearly round, excepting the clypeus, which is broad, square at the fore extremity, and projecting; the hinder extremity also is rather flattened; the sides are sloping, and the upper surface flattish. It is of a reddish yellow-brown colour, mottled and marked with yellowish-white, showing a broad, pale, longitudinal, median band of the latter hue (including the eyes and clypeus), with two short, yellow-white streaks and red-brown spots, on either side, near its hinder extremity, indicating some of the usual converging furrows. On each side of the median band (also near the eyes) is another short, yellow-white longitudinal streak, terminating posteriorly in a red-brown spot; the lower part of the sides is more mottled with white than the rest. A few strongish bristles are dispersed over the cephalothorax, but most of them had apparently been broken off.

The *eyes* are in two concentric, curved, rather widely-separated rows; the convexity of the curve is directed forwards, and the front row is much the shorter. The fore-central pair are the smallest of the eight, and the fore-laterals slightly the largest, being rather larger than the hind-laterals. The eyes of the front row are separated by nearly equal intervals, that between the central pair being perhaps rather greater than that between each and the lateral on its side. The four central eyes form a quadrangular figure, whose fore side is considerably the shortest, and whose longitudinal diameter is much greater than its widest transverse diameter; the interval between the hind-centrals is less than that between each and the hind-lateral on its side. The four lateral eyes are seated on large, roundish, tubercular eminences; and the height of the clypeus equals half that of the facial space.

The *legs* are slender: those of the first and second pairs are long, and very nearly equal in length; the second, if anything, slightly surpass the first; those of the third and fourth pairs are short, and scarcely differ in length; the third, if anything, being slightly the longer; they are of a pale brownish-yellow colour, mottled, chiefly beneath, with white, and spotted thinly with small red-brown tubercles, each of which is surmounted by a short slender spine.

The *palpi* of the male are short, of a dull-yellow colour, slightly mottled with white; the radial joint is shorter than the cubital, and both have some bristles springing from dark red-brown spots; also, besides some lesser projections on the under side, the radial joint has, at the extremity of the outer side, a rather long tapering one, with a curved, obtusely-pointed dark-brown termination; the digital joint is of moderate size and almost wholly white, of an oval form, with its fore extremity pointed and rather elongated: the palpal organs are not prominent, but of simple form, with a curved, sharp-pointed, dark red-brown spiny process at their fore extremity.

The *falces* are moderately long, but not very strong; they are of a subconical form, and project in a continuous line with the clypeus; their colour is a pale yellow-brown, mottled (chiefly at their fore extremity) with white.

The *maxillæ* and *labium* are of normal form, and of a pale dull-yellowish colour.

The *sternum* is oval, pointed behind, of a pale whitish-yellow colour, sprinkled with dull yellow-brown points.

The *abdomen* is of a somewhat pentagonal form, broadest and subangular at the hinder part; the hinder extremity is blunt-pointed below, but has a slightly angular prominence at the middle of its upper part; it is of a dull brownish-yellow colour, mottled, suffused, and marked with white, chiefly along the middle line of the upper side, and the lower part of the sides; the upper side is also thinly and symmetrically sprinkled with small, red-brown, tubercular spots, each of which bears a strongish, slightly curved bristle, directed backwards.

Hab.—Murree to Sind Valley, between July 14th and August 5th, 1873.

Genus—*SAROTES*, Sund.

87.—SAROTES REGIUS.

Aranea regia, Fabr., Entom. system. t. iii, p. 408, No. 4.
Olios leucosius, Walck., Ins. Apt. i. p. 566.

Hab.—Two or three immature females, found at Murree between June 11th and July 14th, 1873, are, I believe, of this species; but in the immature state it is impossible to be quite certain of their specific identity.

88.—SAROTES PROMPTUS, sp. n.

Adult female: length 6¼ lines.

The *cephalothorax* is rather longer than broad, a little constricted on the lateral margins near the fore extremity of the caput, and broadly truncated at the lower margin in front. The colour is a dark reddish yellow-brown, marked with still deeper stripes following the course of the normal indentations, and converging to the thoracic junctional one; it is thinly clothed with greyish-sandy pubescence, and the clypeus (which is of a paler yellowish colour and considerably less in height than half that of the facial space, or about equal to, or a little more than, the diameter of a fore-lateral eye) is furnished with a few prominent, black bristles.

The *eyes* are in two transverse, nearly parallel rows; the fore-laterals are the largest of the eight, and considerably larger than the fore-centrals; these last are further from each other than each is from the fore-lateral on its side, the interval between each fore-central and the fore-lateral eye next to it being equal to the diameter of the latter. The eyes of the hind-central pair are nearer to each other than each is to the hind-lateral eye nearest to it; the interval between the eyes of each lateral pair is rather less than that between the fore- and hind-central pairs, owing to the small size of the fore-central eyes.

The *legs* are moderate in length and strength; in respect to the former, they do not differ greatly; relatively this appears to be 2, 4, 1, 3. Their colour is reddish yellow-brown, growing darker gradually to the tarsi; the femora are much the palest, and are obscurely spotted with small, red-brown spots; all are armed with long spines, and the tarsi and metatarsi are furnished with claw-tuft and scopula.

The *palpi* are moderately long, the digital is equal in length to the radial and cubital joints together, while in colour and armature they resemble the legs.

The *falces* are tolerably long, powerful, straight, perpendicular, and rounded in the profile line; their colour is reddish yellow-brown, somewhat longitudinally striped with a darker hue.

The *maxillæ* are rather long, straight, slightly inclined towards the labium, and rounded at their extremities, which are of a yellowish colour, the rest being dark red-brown.

The *labium* is small, of a somewhat semi-circular form, and its height is not half the length of the maxillæ. Its colour is dark red-brown, with a pale apical margin.

The *sternum* is heart-shaped, and of a yellow colour, like that of the basal joints of the legs.

The *abdomen* is oblong-oval, rather truncate before, and rounded behind, and moderately convex above; it is clothed with somewhat silky, sandy-grey pubescence, and is of a dark red-brown and reddish yellow-brown colour, mixed in variously mingled spottings and linear markings. An indistinct, longitudinal, narrow, dark red-brown, tapering marking occupies the middle of the fore part of the upper side; and towards the hinder extremity is a slightly sinuous, transverse, dark blackish line, edged posteriorly with pale-yellowish, and rendered conspicuous by short white hairs. Along the middle of the under side, from the genital aperture to the spinners, is a broad, black-brown band, laterally margined with a pale stripe. The genital aperture, which is large, conspicuous, and of a somewhat triangular form, has two large, nearly round, prominent lobes or processes connected with its posterior margins. The spinners are small, short, and compact; those of the superior pair are deep blackish red-brown, the inferior pair yellow-brown.

Hab.—Murree, between June 11th and July 14th, 1873.

Genus—*SPARASSUS*, Walck.

89.—SPARASSUS TIMIDUS, sp. n.

Immature female: length nearly 3½ lines.

This spider is nearly allied to *Sparassus suavis*, Cambr. (Spid. of Egypt, P. Z. S., 1876, p. 588), resembling it very closely in its colours and markings; the femora of the fore-legs, however, have no trace of the reddish-brown spots found on those of that species, and the eyes are closer together.

The whole of the fore part is yellow, the *cephalothorax* having a slightly radiate appearance owing to the rather darker hue of the normal converging indentations; and the *maxillæ* have a central, dark, reddish-brown patch.

The *eyes* are of almost uniform size, seated on distinct black spots. The interval between the fore-centrals is considerably less than a diameter, and each is very close, but not quite contiguous, to the fore-lateral eye on its side. The interval between those of each lateral pair is rather less than a diameter of the hind-lateral eye; the eyes of the hinder row are equidistant from each other, and the four central eyes form a square whose anterior side is shorter than the other three.

The *legs* are long, slender, furnished with hairs and a few longish fine spines; their relative length is apparently 2, 4, 1, 3. The tarsi and metatarsi have some divergent hairs of uniform length underneath, but scarcely amounting to a scopula; and there is a strong claw-tuft beneath the two terminal claws of the tarsi.

The *abdomen* is of a dull straw-yellow colour. The upper part and sides are marked with red-brown spots and markings; two broken longitudinal lines of those spots on the fore half of the upper side enclose a long wedge-shaped marking, which is followed by a series of somewhat angular spots of the same hue reaching to the spinners. A few whitish cretaceous spots are scattered along the middle longitudinal line of the upper side as well as on the under side.

Hab.—Neighbourhood of Leh, August or September, 1873.

90.—SPARASSUS FUGAX, sp. n.

Immature female: length 2¾ lines.

This spider is closely allied to the foregoing, but is of a much duller hue, the yellow portions being suffused with dull brownish. The abdomen is shorter and more convex above; the red-brown spots and markings are more thinly scattered, while the white cretaceous spots are larger and more numerous, and spread over the whole abdomen. The femora, genua, and tibiæ are speckled with small red-brown spots, and the spines are longer. The maxillæ also have no central brown patch. With these differences the general character of the markings is similar to that of *Sparassus timidus*.

Hab.—Murree to Sind Valley, between July 14th and August 5th, 1873.

91.—SPARASSUS FLAVIDUS, sp. n.

Adult female: length 10 lines.

The *cephalothorax* is nearly as broad as long, truncated before and constricted laterally at the caput; the height of the clypeus is nearly equal to twice the diameter of one of the fore-central eyes. Its colour is yellow, tinged with brownish orange, deepening to red-brown on the fore part of the caput; and it is thickly clothed with sandy-grey pubescence.

The *eyes* are in a somewhat crescent form, in two transverse rows, the hinder one the longer and straight, or very nearly so; the front row curved, the convexity of the curve directed forwards. They are of moderate size, and relatively differ but little, those of the fore-central pair being a little the largest; the intervals between those of the hinder row are equal; that between the fore-centrals is more than double that between each and the fore-lateral on its side, being near about one diameter. The four central eyes form very nearly a square whose longitudinal is a little greater than its transverse diameter.

The *legs* are long, moderately strong; their relative length appears to be 4, 2, 1, 3. Their colour is yellow, with the tarsi and metatarsi reddish-brown; they are clothed with light sandy hairs and red-brown spines, and there is a rather dense, dark, mouse-coloured scopula beneath the metatarsal and tarsal joints, with a strong claw-tuft beneath the terminal tarsal claws.

The *palpi* are moderately long, yellow, with the under side of the radial and digital joints dark, blackish red-brown; and they are armed with spines, bristles, and hairs.

The *falces* are moderately long, powerful, straight, perpendicular, of a deep, blackish red-brown colour reflecting somewhat of a violet tint, and clothed with sandy hairs and strong dark bristles.

The *maxillæ* are of normal form, their colour is dark red-brown, the inner side at the extremity pale yellow.

The *labium* is similar to the maxillæ in colour, with a pale-yellow apex.

The *sternum* is yellow.

The *abdomen* is of a dull straw-yellow hue, clothed with sandy-grey and darker hairs. The genital aperture is red-brown and of characteristic form, and has two round corneous lobes or eminences at its hinder extremity.

Hab.—Yárkand, between the 21st and 27th of May, 1874.

Genus—*PHILODROMUS*, Walck.

92.—PHILODROMUS CINERASCENS, sp. n.

Adult male: length rather over 2½ lines.

This spider is nearly allied to *Philodromus fallax*, Westr.; its general hue, however, is of a far more ashy-grey, especially that of the abdomen, whereas *P. fallax* is of a sandy colour, and the characteristic median marking on the fore half of the upper side is truncated at its hinder extremity instead of pointed, as in the present spider; besides which the details of the other abdominal markings are different.

The *cephalothorax* is roundish oval, narrower before than behind, decreasing in width gradually, the lateral marginal constrictions of the caput being slight. The upper convexity is moderate, the sides roundly sloping, and the median part flattish. This part, forming a broad, longitudinal, median band, is of a greyish sandy colour, the sides being suffused with brown, most deeply and distinctly on each side towards the hinder part of the median band. The lateral margins of the cephalothorax are greyish white, and the height of the clypeus is very nearly equal to half that of the facial space.

The *eyes* are small and do not differ greatly in size. The fore-laterals, however, are distinctly the largest. The hinder row is straight, the fore one much the shorter and curved, the curve directed forwards. The interval between the eyes of the hind-central pair is rather greater than that between each and the hind-lateral eye on its side; and that between the fore-centrals is also greater than that between each and the fore-lateral next to it, this latter interval being equal to the diameter of one of the fore-central eyes; the interval between the eyes of each lateral pair is considerably less than that between the fore- and hind-central pairs, and as nearly as possible equal to that between the eyes of the hind-central pair.

The *legs* are long and moderately strong, but do not differ greatly in length; their relative length is 2, 4, 1, 3. They are of a greyish-sandy colour tinged with brown, minutely speckled with darker brown, and furnished with hairs, bristles, and spines; the tarsi and a small portion of the metatarsi have a thin scopula on their under sides.

The *palpi* are moderately long, similar to the legs in colour, except the radial and digital joints, which are strongly tinged with brown. The radial and cubital joints are short (the former being the shorter), and are armed with a few strong spine-like, tapering bristles. The digital joint is large, of an elongate oval form, rather pointed before and equal in length to the humeral joint, exceeding that of the radial and digital joints together. The palpal organs are simple, rather the most prominent at their base, with a long, contorted, dark-brown, narrow stripe (probably indicative of an internal duct) on their surface and a strongish, curved, prominent tooth-like spine at their anterior extremity; the radial joint has a very small angular prominence at its extremity on the outer side, and a short, broadish, truncated apophysis underneath.

The *falces* are rather long and slender, straight, and a little directed backwards; their colour is like that of the cephalothorax.

The *maxillæ* and *labium* are of normal form and similar to the falces in colour, the extremities of the maxillæ, however, being of a pale-whitish hue.

The *sternum* is heart-shaped, granulose, and of a brownish-yellow colour.

The *abdomen* is oval, of a stone-white colour, speckled thickly with small punctures and minute black specks. The normal longitudinal marking on the fore half of the upper side is of a dark-grey hue, and has a prominent, obtuse point at the middle of each side, and its posterior extremity is pointed; its outer margins and extreme hinder point are also indicated by a few black, and mostly linear, spots. The sides of the upper part are clouded with dark-grey, leaving a pale, median tapering band on the hinder half, and several oblique, white indistinct stripes on the outer margins, where there is also a line of three or four black spots on each side; these lines converge in the direction of the spinners. The sides are rugulose and the spinners short, compact, and tinged with a sandy colour.

The *female* is rather larger than the male, but in colours and markings resembles it. The oblique white stripes on the lateral margins of the upper part are better defined, and consist of more or less confluent spots and elongate blotches. The form of the genital aperture is characteristic.

Hab.—On the road from Tanktze to Chagra and Pankong Valley, between the 15th and 21st of September, 1873; and from Yárkand to Bursi, between May 28th and June 17th, 1874.

93.—PHILODROMUS MEDIUS.

Philodromus medius, Cambr., Spid. Palest. and Syria, Proc. Zool. Soc., 1872, p. 311.

Hab.—One or two immature examples of this spider found at Murree (June 11th to July 14th, 1873) exactly resemble the types found in Palestine.

Genus—*TIBELLUS*, Sim.

Thanatus, C. L. Koch, *ad partem*.

94.—TIBELLUS PROPINQUUS, sp. n.

Immature female: length rather more than 2½ lines.

This spider is very nearly allied to *Tibellus oblongus* (Walck.), which it resembles closely in form and colour. In the present species, however, the tibiæ and metatarsi of the legs, together with the upper sides of the femora of the first and second pairs, are speckled with minute, dark red-brown spots, while, among a large number of examples of the European species (*T. oblongus*), I can find no trace of this speckling. It is possible that the discovery of the adult males may show that this spotting of the legs, as well as a less definite abdominal marking, is merely a local variation not amounting to a specific distinction.

Hab.—Káshghar, December 1873.

Genus—*THANATUS*, C. L. Koch.

95.—THANATUS THORELLII.

Thanatus thorellii, Cambr., Spid. Pal. and Syria, Proc. Zool. Soc., 1872, p. 309.

Hab.—Immature examples were found in the collections made at Yárkand in November 1873, and on the road thence to Bursi, between May 28th and June 17th, 1874.

96.—THANATUS ALBESCENS, sp. n.

Adult female: length 2¼ lines.

The *cephalothorax* is of a very flattened form; it is as broad as, or broader than, long, truncated behind, and somewhat obtusely pointed at its anterior extremity in the ocular region; the lateral marginal constrictions of the caput are exceedingly slight; it is of a pale dull yellow-brown hue, and has a narrow lateral white margin with a little white venose suffusion above it; the occipital region is also paler than the surrounding surface.

The *falces* are small, straight, nearly perpendicular, like the cephalothorax in colour, at their bases, and paler at their extremities.

The *legs*, *palpi*, *maxillæ*, and *labium* are of a pale dull straw-colour. The legs are rather long and slender. Those of the second pair are distinctly the longest, and the third rather the shortest.

The *eyes* are very small, scarcely differing in size, and seated on round white tubercles in two curved rows, of which the anterior is much the shorter and more strongly curved. The interval between those of each lateral pair is distinctly greater than that between the fore- and hind-central pairs; those of the hind-central pair are a little further from each other than each is from the hind-lateral on its side, while the interval between the four centrals is more than double that between each and the fore-lateral next to it, and just equal to that between each and the hind-central opposite to it. The fore-centrals appear to be very slightly larger than the fore-laterals, and the interval between the fore-central and its nearest fore-lateral eye is but a little more than the diameter of the former.

ARANEIDEA. 77

The *abdomen* is oval, moderately convex above, but a little flattened on its upper side; its colour is stone-white, speckled with very minute blackish points, and with a dull brownish somewhat emarginate lanceolate marking along the middle of the fore half of the upper side, followed by a series of obscure, and almost confluent, diminishing, angular bars of the same hue.

Hab.—On the road from Murree to the Sind Valley, July 14th to August 5th, 1873.

Family—*LYCOSIDES.*

STOLICZKA, genus novum.

Eyes unequal in size; in two transverse, rather widely separated, slightly curved, and nearly parallel rows, the front row much the shorter, and the convexity of the curves directed forwards, the fore-lateral eyes considerably larger than the fore-centrals.

Cephalothorax longer than broad, strongly constricted at the caput on the lateral margins, the fore extremity being truncated and a little broader than the constricted part.

Maxillæ moderately long, strong, broader at their extremity than just above the insertion of the palpi; their outer extremity rounded, the inner one obliquely truncated.

Labium short, convex in front, of a somewhat oval form, truncated at its apex.

Legs moderately long, strong, relative length 4, 1, 2, 3, spinous; and the tarsi are furnished with three claws.

The *abdomen* is rather small, but broader behind than before.

This genus is allied closely to *Nilus* (Cambr., Spiders of Egypt in P. Z. S., 1876, p. 596, pl. ix, fig. 13), but, among other differences, the great disproportion in size between the fore-central and fore-lateral eyes is an essential one.

97.—STOLICZKA INSIGNIS, sp. n.

Adult female: length rather over 5 lines.

The *cephalothorax* is clothed with a short sandy-grey pubescence; its colour is deep brown, with a broad longitudinal band and a narrow irregular lateral one, on each side, a little way from the margin, of a much paler, yellow-brown hue. The median band has, on each side, a little way behind the ocular area, a slight enlargement in the form of a small, angular point; this is most conspicuous in young examples, but is traceable in adults as well, and is a strong specific character. The height of the clypeus is equal to the diameter of a fore-lateral eye.

The *eyes* of the hind-central pair are very much nearer to each other than each is to the hind-lateral eye on its side, being separated by no more, or even by less, than a diameter's interval; those of the fore-central pair are rather further from each other than each is from the fore-lateral eye next to it; the length of the front row is as nearly as possible equal to the length of the line formed by either three of the eyes, adjacent to each other, of the hinder row: the hind-lateral eye on each side is equally distant from the hind-central and fore-central eye next to it, forming the apex of an isosceles triangle; and the four central eyes form a quadrangular figure whose longitudinal is much greater than its transverse diameter, and whose anterior side is slightly shorter than its posterior one.

The *legs* do not differ greatly in length; they are of a yellow-brown colour, deepening gradually to deep red-brown on the tarsi. They are indistinctly annulated with a deeper hue; but this annulation is generally lost more or less in adults, being pretty distinct in young examples. The tarsi and metatarsi are furnished beneath with a thin scopula; all the legs are tolerably thickly furnished with hairs, bristles, and spines, and the inferior tarsal claw is very small and sharply bent downwards, being not easy to distinguish in the tuft of hairs which surrounds it; the two upper claws are strong, curved, and armed with about five denticulations.

The *palpi* are short, of a deep red-brown colour, similar to the legs in their armature, and terminate with a curved claw.

The *falces* are tolerably long, strong, perpendicular; their basal half in front is roundly protuberant, smooth, strong, and of a very dark rich red-brown colour, yellowish red-brown at the extremity.

The *maxillæ* and *labium* are rather less deep and rich in colours than the falces.

The *sternum* is roundish-oval, pointed behind and truncate before, and of a reddish yellow-brown colour.

The *abdomen* fits pretty close up to the steepish hinder slope of the cephalothorax; it is broader behind than before, this form becoming intensified in adults that have deposited their eggs. In adults, the abdomen is of a deep-brown colour, palest underneath, and clothed with a short, somewhat sandy-grey pubescence, besides longer prominent hairs; and on the fore-half of the upper side is a yellow, longitudinal, median, somewhat tapering stripe. In immature specimens, the abdomen is yellow-brown, marked with dark-brown, shewing the yellow stripe on the fore-half of the upper side, as well as some angular bars of the same colour between it and the spinners. These are short, compact, the inferior stronger than, but of equal length with, the superior pair. The genital aperture consists of two somewhat roundish openings, one on each side, at the hinder part of an oval prominence.

This spider is an extremely interesting form, and appears to be an abundant species. Some of the examples had large, round, dark-brown lycosiform bags of eggs attached by silken fastenings to their spinners.

Hab.—Murree, June 11th to July 14th, 1873.

Genus—*OCYALE*, Sav.

98.—OCYALE RECTIFASCIATA, sp. n.

Immature male: length nearly 6 lines.

The *cephalothorax*, legs, falces, and other fore parts of this spider are of a dull yellow-brown colour. A broad, dark yellow-brown, median band, edged with a marginal border of white hairs, runs throughout, and includes the ocular area. This band is very distinct and its margins are parallel to each other.

The *eyes* are in the ordinary position; the anterior row is equal in length to the interval between the two eyes of the posterior row; it is curved, the curve directed backwards, and its four eyes are small and do not differ greatly in size; the two lateral ones are smallest, and, being each seated in front of a dark tubercle (the tubercle itself being in a straight line with the two central eyes), give to the row the appearance at first sight of being straight, but, as above stated, the row is in reality curved, its eyes being equidistant from each other, and separated by an interval of less than the diameter of one of the centrals. Those of the middle row are rather larger than the fore-centrals, and form a line nearly equal in length to the intervals between the lateral eyes of the anterior row, and are separated by more than a diameter's interval. Each of them is also equally distant from the fore-lateral and hind-lateral eyes on its side, the interval between these two last being equal to that between the two fore-lateral eyes. The height of the clypeus is just equal to half that of the facial space.

The *legs* are long and slender, but do not differ very greatly in length. Their relative length is 4, 2, 1, 3, and they are furnished with hairs and spines.

The *palpi* are short; the digital joint large, and, not being yet fully developed, tumid; its length is equal to that of the humerus, its fore extremity being considerably drawn out. The radial joint is rather longer than the cubital, and has a not very large, sharp-pointed, tapering apophysis at its fore extremity on the outer side. This apophysis, as well as some strong bristles on the upper side of the joint, were plainly visible beneath the cuticle, the moulting of which would have brought the example to the adult state, in which the palpal organs would be fully developed.

The *maxillæ*, *labium*, and *sternum* are of normal form; the two first of a more yellow-brown than the sternum.

The *falces* are moderately long but not particularly strong, straight but slightly divergent, perpendicular, and a little roundly prominent in profile at their base.

The *abdomen* is of an elongate-oval form, tapering pretty gradually from the fore to the hinder extremity. It is of a dull yellowish whitey-brown colour; the sides are marked with a few scattered, indistinct, brown spots, and a broad, darkish yellow-brown, tapering band runs along the middle of the upper side from end to end, and is edged with a marginal border of white hairs; the edges of this band towards the narrowest (or hinder) extremity are slightly sinuous; the band itself has the appearance of a continuation of that on the cephalothorax. The under side of the abdomen has a broad, median, longitudinal, slightly tapering, whitish band reaching from the spiracular plates to the spinners. These are short, but those of the superior pair are rather longer, though less strong, than those of the inferior pair.

The example above described being immature, the abdomen is very much larger than it would be in the adult state, in which it is probable that the total length of the spider would not exceed 4½ lines.

This spider is very nearly allied to one (not yet described) of a larger size, but almost exactly similar in colours and markings, though of quite a distinct species, received from Ceylon and also from Bombay.

Hab.—Murree to Sind Valley, between July 14th and August 5th, 1873.

99.—OCYALE DENTIFASCIATA, sp. n.

Adult female : length rather more than 4½ lines.

The spider is nearly allied to the foregoing species; it is, however, not only smaller (which may not be a constant character), but the abdominal band is very deeply dentated on its margins, and has a series of brownish-yellow, somewhat angular markings along its middle. The sides of the abdomen are irregularly, but extensively, covered with almost confluent brown markings, leaving, however, next to the upper side, a tolerably distinct dentated, pale dull yellowish-brown band slightly spotted with brown. The *cephalothorax* has a median longitudinal band edged with white hairs like that of *O. rectifasciata*.

The *eyes* are in a similar position to those of that species, but at the same time are rather more separated from each other.

The *legs* are rather long, of a dull, darkish yellow-brown colour, armed with spines; and their relative length appears to be 4, 2, 1, 3.

The *falces* are similar to the last species in form and size, and are of a dark shining yellow-brown, deeper in hue than the legs.

The *maxillæ* are yellow-brown, palest at the extremities, and the *labium* is of the same colour, with a pale apical margin.

The *sternum* is yellowish, with a distinct, broadish, yellow-brown marginal border, and is clothed with coarse grey, and a few dark-brown, hairs.

Hab.—Murree to Sind Valley, between July 14th and August 5th, 1873.

Genus—*TROCHOSA*, C. L. Koch.

100.—TROCHOSA RUBIGINEA, sp. n.

Adult female: length 4¼ lines.

It is not without considerable hesitation that I have included this very interesting spider in the genus *Trochosa*. It is probable that future collectors will discover other species presenting similar special peculiarities in the position of the eyes, joined to the rather short, but strong, unattenuated legs of the present spider; in which case it might become necessary to form a separate genus, or ub-genus, for their reception.

The *cephalothorax* is oval, broad, and truncate at its fore extremity; the marginal lateral constrictions of the caput are slight, and the height of the clypeus is at least equal to, or even exceeds, double the diameter of one of the central eyes of the front row. Its colour is yellowish, with a narrow marginal band, and two broad longitudinal lateral bands, of a rusty red-brown hue, leaving a rather indistinct, median, tapering, yellowish band strongly constricted near the occiput, and having a large part of its surface along the middle line suffused with rusty red-brown lines, all radiating or converging to the thoracic indentation, which is marked by a fine, deep red-brown line: the middle of each side is occupied by a longitudinal, well-defined, but not very broad, yellow band. The fore part of the area enclosed by the middle and posterior rows of eyes is of a dark reddish-brown colour; the hinder part of this patch contains two oval, parallel, yellowish markings. Sometimes the slender red-brown lines defining the outer sides of these oval markings are obsolete, leaving a short, dark red-brown stripe, ending a little way behind the posterior row of eyes, its termination, more or less, laterally dilated. The broad lateral rusty-brown bands are traversed by numerous deep red-brown lines, all radiating or converging to the thoracic indentation. The surface of the cephalothorax is covered with yellow-grey pubescence, and there are numerous blackish bristles on the upper part and sides of the caput.

The *eyes* are in the usual three rows—4, 2, 2; the central pair of the first row are larger than the laterals, and are divided by an interval exceeding a diameter, and each is very near, but not quite contiguous, to the lateral on its side; the front row is very slightly, if anything, shorter than the second; the eyes of the second row are, if anything, slightly smaller than those of the third row, those of both the second and third rows being very considerably larger than the eyes of the fore-central pair. The distance between each eye of the second row and the lateral of the first row opposite to it is equal to the diameter of the former. The length of the third row is double that of the second, and the interval between these two rows is double that between the first and second.

The *legs* are rather short, strong, tapering, but not attenuated at the extremities; they are of a yellowish colour, pretty densely clothed with hairs, armed with a few not very strong spines, and annulated with rusty red-brown, most distinctly on the upper side of the femora; their relative length is 4, 1, 2, 3, but the difference is not great.

The *palpi* are tolerably long, and similar in colour and armature to the legs.

The *falces* are moderate in length and strength, straight and perpendicular; they are of a yellow colour, with two longitudinal red-brown lines at their base, and are furnished with numerous bristles in front.

The *maxillæ* and *labium* are of ordinary form, hairy, and similar in colour to the falces.

The *sternum* is oval, truncated before, of a deep rusty red-brown colour, bordered with a broad yellow margin, and with a median, longitudinal, sharp-pointed, yellow stripe at its fore extremity.

The *abdomen* is of an oblong-oval form, broadly, but rather roundly, truncated at its fore extremity, and pretty densely clothed with greyish-yellow and other hairs. The upper part and sides are of a dark rusty-reddish colour, freckled with small, pale-yellowish spots. The fore part of the upper side has the normal longitudinal marking of an orange-yellowish colour, slightly margined with deep red-brown and rather bluntly pointed at its hinder extremity: this marking is rather broadest just behind the middle, and on each side of the broadest (or subangularly prominent) part is a short, orange-yellow, oblique stripe: and following the hinder extremity of the marking is a series of oblique, rather elongate, opposed, oval markings of a similar colour in pairs, each oval marking containing a small but distinct central red-brown spot: the two lines of these oval markings converge towards the spinners, but become obsolete before they reach these parts. They evidently represent the normal angular bars or chevrons. The under side is dull orange-yellow, with a longitudinal median rusty band.

Immature males resembled the females in colour and size.

Hab.—This spider appears to be pretty common. Its localities are Yárkand and neighbourhood, November 1873; Káshghar, December 1873; and route from Yárkand to Bursi, between May 28th and June 17th, 1874.

101.—Trochosa hebes, sp. n.

Adult male: length 2¾ lines.

This spider, which is of the *Trochosa picta* group, is very closely allied to *Arctosa amylacea*, C. L. Koch, which it resembles in size and in the general character of its markings, but it is not nearly of so bright a hue; and the form of the genital aperture of the female is quite distinct. I have only been able to compare the females of the two species, not possessing a male of *A. amylacea*.

The *cephalothorax* is broad-oval behind, and somewhat drawn out forwards, though with but slight lateral constrictions on the margins of the caput; its colour is brownish-yellow, with a dentated marginal band, and a broader lateral strongly dentated one, of a dark-brown colour on each side, leaving a large central star-shaped, or radiated, brownish-yellow marking. The ocular area is dark-brown, and the whole surface of the cephalothorax is pretty thickly clothed with hairs, many among which are prominent, dark brown, and of a bristly nature.

The *eyes* are grouped as in *T. picta*. Those of the hinder row are smaller than those of the middle row, but considerably larger than the central pair of the front row; the eyes of the middle and hinder rows form a quadrangular figure whose posterior side is not greatly longer than the anterior one, the length of the sides being apparently equal to that of the posterior side. The anterior row of eyes is, if anything, slightly shorter than the middle row, and the interval between the eyes of its central pair is larger than that between each and the lateral eye next to it, to which last it is very close, though not quite contiguous. The height of the clypeus is at least equal to twice the diameter of one of the central eyes of the front row.

The *legs* are moderately long and tolerably strong, particularly the femoral joints; they are of a dark-yellowish colour with dark-brown annuli, and are thickly clothed with hairs and long prominent slender bristles, those of the third and fourth pairs being armed with spines.

The *palpi* are rather short, hairy, and similar in colour and markings to the legs. The radial joint is a little shorter, but of equal strength with the cubital; the digital joint is dark brown at its base, paler at the extremity; it is long and narrow, being only a little broader at its basal part than the radial joint; its length is equal to that of the radial and cubital joints together; the palpal organs are small and simple, being very like those of *T. picta*.

The *falces* are long, moderately strong, straight, perpendicular, and of a deep brown colour.

The *maxillæ* and *labium* are of normal form; their colour is yellowish-brown; the extremities of the former and the apex of the latter being of a paler hue.

The *sternum* is oval, hairy, and of a dark yellow-brown colour.

The *abdomen* is rather broader behind than in front; it is hairy and of a brownish-yellow colour; the markings, which are of the general *Lycosa* type, and almost exactly similar to those of *T. picta*, are delineated by dark blackish-brown lines and spots. The under side is also more or less marked with the same.

Hab.—Yárkand and neighbourhood, November 1873; Yangihissár, April 1874; Yárkand, between 21st and 27th May 1874; hills between Sirikol and Aktalla, between 8th and 18th May 1874; route from Yárkand to Bursi, between May 28th and June 17th, 1874.

102.—TROCHOSA PROPINQUA, sp. n.

Adult female: length just over 5 lines.

This spider is very closely allied to *T. ruricola*, De Geer, but is, I think, certainly of a distinct species.

The *cephalothorax* is broader behind and narrower before than in *T. ruricola*. The broad, lateral, brown bands, instead of stopping behind the hinder row of eyes, run through and include the laterals of both the middle and hinder rows. The median longitudinal yellow band is similarly constricted at the occiput; but is broader behind that point, and more radiated than in *T. ruricola;* and the two longitudinal brown stripes on the fore part of this band are confluent with the sides of the brown lateral bands.

The *eyes* occupy a larger area, and are of a pale, dull, yellowish-brown hue, being much paler than in the other species mentioned.

The *legs* are rather shorter in proportion, and are pretty distinctly annulated with brown; whereas they have rarely any trace of annulation in *T ruricola*.

The *abdomen* is of a much darker hue, being of a blackish yellow-brown colour, the normal median longitudinal marking on the fore half of the upper side is of a brighter orange-yellow, and is margined by a much more distinct black border. The form of the genital aperture differs but very slightly. The under side of the abdomen is suffused with dark brown, and on each side is a marginal border of a darker black-brown.

Hab.—Sind Valley, between 5th and 13th August, 1883.

103.—TROCHOSA ADJACENS, sp. n.

Adult female: length just over 5 lines.

This spider is very closely allied to *T. terricola*, Thor., differing from it in about the same degree as *T. propinqua* does from *T. ruricola*, De Geer. It is rather a smaller spider, and the cephalothorax appears also to be of a rather broader form, and the whole spider is of a much duller hue and less distinctly marked; the bands on the cephalothorax are scarcely discernible.; the whole being of a dull yellowish-brown colour, pretty densely clothed with short, greyish-sandy pubescence, and with some indistinct, darker brown, radiating stripes indicating the normal indentations.

The *eyes* of the front row are much larger than in *T. terricola*, and are very nearly equal in size to those of the hinder row; these last, however, being much smaller than the corresponding ones in that species.

The *legs* in the present spider have no trace whatever of annulation, while those of *T. terricola* are frequently annulated with brown, though never very distinctly, and, in general, chiefly on the femora.

The markings on the abdomen are very similar, as also is the form of the genital aperture, though a slight difference in this respect is observable.

Hab.—Yangihissár, April 1874.

104.—TROCHOSA SABULOSA, sp. n.

Adult female: length 10 lines.

The *cephalothorax* of this fine spider is strongly compressed laterally on the margins of the caput, and the lateral slopes are much depressed; the caput, however, is broad at its lower margin and tolerably massive above. Its colour is reddish yellow-brown, totally obscured by a dense clothing of short, pale sandy-grey pubescence, leaving but very slight and broken traces of the ordinary radiating indentations of a darkish brown colour. The height of the clypeus is equal to the diameter of one of the fore-central eyes.

The *eyes* are in the ordinary position, and occupy an area whose length and breadth are, as near as possible, equal; the front row is distinctly longer than the middle one; its central pair of eyes are much larger than the laterals, though distinctly smaller than those of the hinder row, and are separated by an interval less than a diameter, but double as great as that which divides each from the lateral next to it; those of the middle row are divided by, as nearly as possible, a diameter's interval; and each is separated from the eye of the posterior row on its side by an interval of about one and a half diameters.

The *legs* are moderately long, strong, and tapering, but not attenuated; their relative length is 4, 1, 2, 3, and their colour is of a pale-yellowish hue, deepening to brownish-red; the metatarsi and tarsi are pretty densely clothed with sandy-grey pubescence, mixed with other darker hairs and bristles, and armed (chiefly on the third and fourth pairs) with spines; the under sides of the genua and the fore extremities of the tibiæ are black-brown; the under sides of the metatarsi and tarsi of the first and second pairs, and of the metatarsi of the third and fourth pairs, as well as of the digital joint of the palpi, are furnished with a dense scopula of black-brown hairs.

The *palpi* are similar in colour to the legs.

The *falces* are long and powerful, straight, perpendicular, the profile-line convexly curved, of a deep black-brown colour, thickly clothed with sandy-grey and brown hairs and bristles.

The *maxillæ* and *labium* are of normal form, and bristly; their colour is red-brown, the extremities of the former and the apex of the latter having a pale-yellowish tinge.

The *sternum* is oval, truncate at its fore extremity, of a dark brownish-black colour, clothed with sandy-grey pubescence.

The *abdomen* is oval; it projects well over the base of the cephalothorax, and is considerably convex above; it is densely clothed with sandy-grey, black, whitish, and brown hairs. On the upper side, the ordinary *Lycosa* pattern is indistinctly visible, being indicated by the scattered markings formed by the darker and whitish hairs. The normal elongate marking on the fore half is truncate at its posterior extremity, and a prominent subangular point on each side, about the middle, is indicated by a strong blackish spot; on the hinder half, the only markings traceable (besides a generally thin sprinkling of small blackish spots over the whole of the upper part and sides) are two rows of obscure spots of whitish hairs, converging towards the spinners; the whole of the under side, including the spiracular plates and the genital aperture, is black.

The male is smaller, but resembles the female in colours and markings. The radial joint of the palpus is considerably longer than the cubital; the digital joint is red-brown, and a little longer than the radial; the palpal organs present no very marked peculiarity of structure.

Hab.—Yangihissár, April 1874; between Yangihissár and Sirikol, March 1874; road across the Pamir from Sirikol to Panja and back, between April 22nd and May 7th, 1874; and Yárkand, between 21st and 27th May 1874.

105.—TROCHOSA APPROXIMATA, sp. n.

Adult female : length 5¼ lines (nearly).

This spider is almost exactly like *T. sabulosa* in colour and markings, the grey hue, however, being less marked; but it may readily be distinguished by its comparatively small size, and by the under side of the abdomen being of a dull sandy hue, instead of black, as in *T. sabulosa*. The sides of the cephalothorax also appear to be more depressed, and the fore-central eyes of the front row are more nearly equal in size to those of the hinder row,

The genital aperture is very minute, being of a transverse narrow-oval form divided longitudinally by a septum.

Hab.—Yárkand, November 1873.

106.—TROCHOSA RUBROMANDIBULATA, sp. n.

Immature male : length 5¼ lines.

This spider is nearly allied to both the foregoing species, but may easily be distinguished by the following characters. The general hue is less grey than in *T. sabulosa*, and the darker markings on the abdomen are more distinct ; the normal longitudinal marking on the fore half of the upper side is of a dark brown hue, with some black spots and markings on its outer margins : there are also some black spots alternating with the pale spots on the hinder half (these latter spots not being so white as in *T. sabulosa*). The under side of the abdomen is jet-black, distinctly and abruptly enlarged laterally from near the middle to the spinners, and there is a distinct short black bar on each side near the base of the spinners.

The *legs* are unicolorous, having no trace of the black suffusion underneath the fore extremity of the tibiæ, except very slightly beneath those of the fourth pair.

The *eyes* of the fore-central pair are smaller than in either *T. sabulosa* or *T. propinqua*; and a striking character, which distinguishes it at a glance from both, is the dense clothing of scarlet (somewhat squamose) hairs on the front of the falces.

It is probably also a smaller spider than *T. sabulosa*, though this is not certain, as the only example examined was not adult.

Hab.—Murree to Sind Valley, between July 14th and August 15th, 1873.

107.—TROCHOSA LUGUBRIS, sp. n.

Adult male : length nearly 5 lines.

The *cephalothorax* is of a dark, rich red-brown colour, thickly clothed with silky, light grey hairs disposed in a broad longitudinal and narrower marginal bands,—the sides being clothed with black hairs, forming thus alternate bands of white and black hairs ; the caput is considerably produced, and constricted on the lateral margins. The height of the clypeus is no more than, if quite so much as, the diameter of one of the fore-central eyes.

The *eyes* are in the ordinary position ; the length of the front row is perceptibly longer than that of the middle row, whose central eyes are larger than the laterals, though much smaller than those of the hinder row ; those last are rather smaller than those of the middle row, and form a line very nearly indeed equal to that formed by each of them, and that one of the middle row on its side; the interval between those of the middle row a little exceeds a diameter; the eyes of the hinder and middle rows thus form very nearly a square whose anterior side is the shortest.

The *legs* are tolerably long and strong, though rather attenuated at their extremities. They are of a yellowish, dark red-brown colour; the femora being much the darkest, and clothed with grey hairs, not only of a pubescent nature, but also with numerous long, slender, prominent ones like those on the legs of *Tegenaria* and *Argyroneta*. They are also armed with strong spines; the tarsi are furnished underneath with a thin scopula. Their relative length is 4, 1, 2, 3, but the difference between those of the first and fourth, and of the second and third pairs, respectively, is not great.

The *palpi* are tolerably long and strong, similar to the legs in colour and hairy clothing; the humeral joint has three spines of equal length close together in a transverse line on the upper side at the fore extremity. The radial joint is longer than the cubital, and the digital joint, which is darker than the rest, slightly exceeds in length the radial joint, whose width it considerably exceeds at the base, its fore extremity being rather attenuated. The palpal organs are rather simple, with a prominent subconical hook-pointed process, about the middle of their outer side.

The *falces* are long, powerful, straight, perpendicular, of a deep black-brown colour; clothed with grey pubescence and long dark bristly hairs.

The *maxillæ* and *labium* are of normal form, and of a deep blackish red-brown colour; the *sternum* is of the same colour, oval and truncated before; these parts are furnished with strong dark bristles.

The *abdomen* is of moderate size and convexity above; the upper part and sides are dark brown, thickly clothed with grey hairs, shewing some curved transverse lines, formed by these hairs on the hinder half. The whole of the under part, extending also a little way up the sides, is jet-black.

Hab.—On the road across the Pamir from Sirikol to Panja and back, between April 22nd and May 7th, 1874.

Genus—TARENTULA, Sund.

108.—TARENTULA IRASCIBILIS, sp. n.

Immature female: length 3¼ lines.

The *cephalothorax* is oval, the caput a little produced and rather strongly constricted on the lateral margins; the fore margin is broad and truncated, and the lower part of the sides rather gibbous; it is of a yellow colour tinged with orange-brown; on the upper part of each side is a broad longitudinal darkish yellow-brown band traversed by still darker converging lines showing the normal indentations; the lateral margins are also marked with some broken irregular brown spots and markings. The ocular area is blackish-brown, and the height of the clypeus is nearly about equal to the diameter of one of the fore-central eyes; the surface of the cephalothorax is thinly clothed with a greyish silky pubescence.

The *eyes* are in the ordinary position, forming an area as long as it is broad, though narrower in front than behind; the eyes of the middle and posterior rows are very large, and appear to be very nearly, if not quite, equal in size; the interval between the middle ones is equal to, or a little more than, a diameter, being less than that between each and that of the hinder row opposite to it; the length of the hinder row is greater, though not very much, than that of the middle row, which is also, if anything, a very little longer than the front row; the eyes of this last are small and equally separated; those of the central pair being but little larger than the laterals.

The *legs* are tolerably long and strong; their relative length being 4, 1, 2, 3; they are yellow, annulated, though not very distinctly, with broken and angular brown annulations; they are furnished with hairs and spines, but have no scopula beneath the tarsi and metatarsi.

The *palpi* are similar in colour and markings to the legs.

The *maxillæ*, *labium*, and *sternum* are of ordinary form, and of a yellow-brown colour. The *falces* are also of a similar colour, rather long, powerful, and perpendicular, and furnished with a few bristles in front.

The *abdomen* is a little wider behind than in front, its colour is yellowish, clothed, but not very densely, with a few greyish, and a few longer, coarser brown hairs; there is, along each lateral margin of the upper side, a broad dentated brown band, from the lower side of which two or three oblique, but very regular, rows of brown spots traverse the sides; along the middle of the fore half is the normal marking of a deep brown colour edged with black, with a prominent angular point on each side, and truncate at its posterior extremity, which merges in the first of a series of broadish, angular, brown chevrons; these decrease in size as they approach the spinners; the point of each chevron, which is (as usual) directed forwards, touching the inside of the angle of the chevron in front of it. The under side is immaculate.

Hab.—Neighbourhood of Leh, August or September, 1873.

109.—TARENTULA INIMICA, sp. n.

Adult female: length rather more than 6 lines.

The *cephalothorax* is yellow-brown, with a broad longitudinal band, on each side, of a darker hue; the whole covered with a short sandy-grey pubescence. The clypeus is low, not much exceeding in height the diameter of one of the fore-central eyes. The eyes of this row are placed on somewhat of a ridge, making this part look prominent when seen in profile. The facies is low.

The *eyes* occupy an area about equal in length and breadth. The front row is distinctly shorter than the middle one; its eyes are very small; the centrals are but slightly, if at all, larger than the laterals, and the interval between them is greater than that between each and the lateral eye on its side. The eyes of the middle row are much larger than those of the posterior one, and are separated by slightly over a diameter's interval; the hinder row is considerably longer than the middle one.

The *legs* are tolerably strong, but not very long; those of the fourth pair are the longest, the rest not varying very much; they are of a yellow-brown colour, and are furnished with hairs and spines; the tarsi of the first and second pairs have a very thin scopula on their under sides.

The *palpi* are short, but similar in colour to the legs.

The *maxillæ* and *labium* are of a rich deep red-brown colour; the former have their extremity, and the latter has its apex, pale yellow.

The *sternum* is oval, somewhat truncated at its anterior extremity, and similar in colour to the maxillæ.

The *abdomen* is of a short-oval form considerably convex above; it is of a reddish-brown colour mottled with much clearer reddish spots; the normal longitudinal macula on the fore half of the upper side is large, considerably prominent past the middle on each side, and truncated at its posterior extremity; it is of an obscure brown hue, indistinctly margined with darker brown; on the hinder half is a median longitudinal series of strongish, but not very conspicuous, yellowish-red, angular bars or chevrons. The under side is dark, of a rather sooty-brown hue; the form of the genital aperture is distinct and characteristic; the hairy clothing of the abdomen had been entirely denuded.

Hab.—On the road across the Pamir from Sirikol to Panja and back, between April 22nd and May 7th, 1874.

Genus—*LYCOSA*, Latr. ad partem=*LYCOSA*, Thor.

110.—LYCOSA CONDOLENS, sp. n.

Adult male : length 2⅜ lines.

The general form and appearance of this spider are like those of *Lycosa agricola*, Thor. and some other closely-allied European species. The *cephalothorax* is deep brown, in some cases approaching to black, with a narrow median, and, on each side, a submarginal brownish-yellow stripe; the median stripe is often very indistinct, and seldom runs (towards the eyes) beyond the occipital region, certainly not reaching nearly to the ocular area, and the marginal stripes are irregular, or somewhat dentated, on their edges. These stripes are clothed with pale hairs. The ocular area is black, and the clypeus, which rather exceeds in height the diameter of one of the fore-central eyes, is yellow.

The *eyes* are in the ordinary position; the foremost row is distinctly shorter than the middle one, and its eyes are very small; those of the central pair being scarcely larger than the laterals, and the interval between them is double that between each and the lateral eye on its side. The eyes of the middle row are considerably farther apart than a diameter's interval,—in fact, nearly equalling two diameters; being equal to the interval between each and the lower margin of the clypeus at its nearest point. The hinder row is longer than the middle one, and its eyes are smaller than those of that row.

The *legs* are long, rather attenuated, furnished with hairs, bristles, and spines; they are of a brownish-yellow colour, the femora and tibiæ annulated and marked with black-brown and yellow-brown; the femora are often more or less completely suffused with black-brown. This is only, however, the case with some adult males, and is probably owing to their having been longer in the adult state; the legs of the fourth pair are the longest, and those of the third pair slightly the shortest.

The *palpi* are moderately long; the humeral joint is nearly black, the cubital and radial joints yellow; the latter is the longest, and black on the whole (more or less) of the under side; the upper side is furnished with white hairs, mostly close to the fore extremity; the digital joint is of tolerable size, round, oval, and black behind, pointed and of a paler brownish hue in front. The palpal organs are characteristic in their structure, though they do not present anything very remarkable in form; there is, about their middle, a not very prominent, somewhat crescent-shaped, process, one of whose limbs is truncated, and the other, the shorter, is blunt-pointed.

The *maxillæ* are dark reddish brown, yellowish at their fore extremities.

The *labium* is also of a similar colour,—yellowish at the apex.

The *falces* are moderately long, not particularly strong, straight, perpendicular, of a brownish-yellow colour, more or less clouded with deep brown.

The *sternum* is oval and nearly black.

The *abdomen* is black-brown on the upper side; the normal longitudinal marking on the fore part is of a reddish yellow-brown hue, blunt-pointed at its posterior extremity, and followed towards the spinners by a series of short, angular bars of the same colour; these bars (often broken at the angle) thus consist of two oblique, opposed, oblong-oval markings, each of which has a black spot in the middle; there is also on each side of this series, towards the margin of the upper side, a longitudinal series of pale spots formed by small tufts of whitish hair; the sides are mottled with yellow-brown, and the under side is yellow-brown marked with a median, and two (lateral) longitudinal dark blackish stripes, rendered more or less indistinct by the pale (among other) hairs with which the surface of the abdomen is generally covered.

The female is paler-coloured, and the markings are more distinct than in the male, preserving, however, the same essential characters. It seems to be an abundant species.

Hab.—Yárkand and neighbourhood, November 1873; Káshghar, December 1873; between Yangihissár and Sirikol, March 1874; Yangihissár, April 1874; on the road across the Pamir, from Sirikol to Panja and back, between April 22nd and May 7th, 1874; hills between Sirikol and Aktalla, between 8th and 18th May 1874; road from Yárkand to Bursi, between May 28th and June 17th, 1874.

111.—LYCOSA FORTUNATA, sp. n.

Adult male: length 3 lines.

This spider is very nearly allied to *Lycosa condolens*; but it is rather smaller, and generally lighter-coloured. The following points of distinction will serve to distinguish it readily.

The central yellow band on the *cephalothorax* is much broader, more distinct, reaches more nearly to the eyes, behind which it is strongly constricted, being broader and somewhat radiated at the thoracic junction, immediately behind which it is again constricted; the lateral yellow stripes are broken, and scarcely extend more than half way to the fore extremity. The height of the clypeus is a little greater, and the two central eyes of the front row are larger in proportion to the laterals.

The *legs* have the femora and tibiæ in general obscurely annulated, but the former are not black as in *L. condolens*. The radial joints of the palpi are longer in proportion to the length of the cubital than in that species, and are a little clouded with brown towards their fore extremities, which are furnished thickly with long, black, bristly hairs, particularly underneath and on the inner sides. The fore part of the digital joint is less attenuated, and it is clothed thickly with black hairs, and terminates with a strongish curved claw; the palpal organs differ also in structure; they are more prominent at their hinder extremity, and the process corresponding to that described in reference to *L. condolens* as somewhat "crescent-shaped" is much larger and more prominent; its larger limb being strongly curved. The abdominal markings are very similar, but the usual one on the upper side, at the middle of the fore part, is distinctly margined with black.

The female resembles the male in colours and markings, but the annulations of the legs are darker and more distinct.

This spider appears to be equally abundant with *L. condolens*.

Hab.—Neighbourhood of Leh, August and September 1873; Tanktze to Chagra and Pankong Valley, 15th to 21st September 1873; Yárkand and neighbourhood, November 1873;

Káshghar, December 1873; between Yangihissár and Sirikol, March 1874; Yangihissár, April 1874; on the road across the Pamir from Sirikol to Panja and back, April 22nd to May 7th, 1874; hills between Sirikol and Aktalla, 8th to 18th May 1874; Yárkand, 21st to 27th of May 1874; road from Yárkand to Bursi, May 28th to June 17th, 1874.

112. LYCOSA STELLATA, sp. n.

Adult female: length from 4 to 5¼ lines.

The *cephalothorax* of this distinct spider is of a brown colour, and clothed with a short sandy-grey pubescence; there is a large, very distinct, star-shaped or radiate yellowish marking at the occiput, divided longitudinally by a dusky red-brown line, strongish at each end, and produced before into an obtuse, somewhat transverse, oblong marking a little way behind the ocular area, notched at its fore-margin, and often marked with the bifid continuation of the bisecting line on the stellate portion; there is also a pale yellowish submarginal, and generally broken, band on each side; these markings, seen very distinctly on immature examples, are more or less obscured by the pubescence in adult specimens.

The *eyes* of the foremost row form a line distinctly shorter than the middle row, and its central pair are distinctly larger than the laterals, and are divided by an interval larger than that which separates each from the lateral eye on its side; the height of the clypeus very little, if at all, exceeds the diameter of one of the fore-central eyes. The ocular area appears to be broader behind than it is long, and the eyes of the middle row are considerably larger than those of the hinder one, forming a line nearly about equal to that formed by the laterals of these rows.

The *legs* are rather long, tolerably strong, of a yellowish hue, and pretty distinctly annulated with dark brown; they are furnished with hairs and spines, and the colour and markings are liable (in adults) to be obscured more or less by a rather dense, short, sandy-grey pubescence; their relative length is 4, 1, 2, 3.

The *palpi* are similar in colour and armature to the legs.

The *falces* are long, strong, straight, and perpendicular: they are of a reddish-yellow-brown colour—red-brown at the base and extremities in front, and furnished with numerous long, prominent bristles.

The *maxillæ* are yellow-brown, and the *labium* deep brown with a pale-yellowish apex.

The *sternum* is deep brown, clothed with grey pubescence.

The *abdomen* is of a blackish-brown colour on the upper side. The normal longitudinal marking on the fore half is indicated by broken, surrounding, submarginal, reddish-yellow markings, and its posterior extremity is truncated: following it is a series of opposed, oblique, yellowish markings, these being the broken portions of the normal angular bars, which are, however, sometimes perfect; and each bar contains a black spot: outside these angular bars is, on each side, a longitudinal row of yellowish spots. The sides are brownish-yellow, spotted and marked with black-brown; and the under side is also brownish-yellow, without any markings. The genital aperture is of a characteristic form, and its colour is red-brown.

The male resembles the female in colours and markings, but is rather smaller.

Hab.—Yárkand and neighbourhood, November 1873; Káshghar, December 1873; Yangi-hissár, April 1874; on road across the Pamir from Sirikol to Panja and back, April 22nd to May 7th, 1874; hills between Sirikol and Aktalla, 8th to 13th of May 1874; Yárkand, 21st to 27th May 1874; Yárkand to Bursi, May 28th to June 17th, 1874.

From the localities recorded, this spider, though perhaps less numerous, appears to be distributed nearly equally with the two foregoing species, and all three are probably found together. The present species is very nearly allied to *L. injucunda*, Cambr., found in Egypt, but quite distinct.

113. LYCOSA CREDULA, sp. n.

Adult female: length nearly 2¾ lines.

This spider is very nearly allied to *Lycosa nigriceps*, Thor., which it resembles closely in form and general appearance, but may be distinguished by the absence of any constriction of the median, longitudinal yellow band on the *cephalothorax*, and by the normal longitudinal marking on the fore half of the abdomen being sharp-pointed instead of truncated at the posterior extremity.

The *cephalothorax* is of a bright yellow colour, somewhat obscured by a greyish pubescence, which probably soon becomes more or less denuded; the sides are narrowly edged with black, a very little way above which edging is a narrow, dark, yellow-brown, submarginal stripe, with a broad lateral band of the same colour along the upper part of each side, leaving a median, longitudinal yellow band of equal width throughout, and scarcely wider than the lateral bands of the same hue. The ocular area is black.

The *eyes* are in the usual position; the front row is distinctly shorter than the middle one, and its two central eyes are placed on a small prominence; these two are larger than the laterals of the same row, and the interval between them is greater than that between each and the lateral on its side; the eyes of the middle row are much the largest, and form a line shorter than those of the third row, though this latter is not so long, proportionately, as in some other groups of *Lycosa*. The four eyes of the middle and hinder rows form a square whose posterior side is longer than the rest. The height of the clypeus is more than double the diameter of one of the fore central eyes.

The *legs* are moderately long, and rather slender; they are of a yellow colour, indistinctly marked and annulated on the femora with yellowish-brown, and are clothed with hairs, spines, and grey pubescence. Their relative length is 4, 1, 2, 3.

The *palpi* are yellow, marked with brown.

The *falces* are rather long, slender, straight, and directed backwards; their colour is yellow, slightly marked longitudinally with brown.

The *maxillæ* and *labium* are yellowish, tinged with brown.

The *sternum* is oval, rounded before, and pointed behind; its colour is black-brown, irregularly margined with yellow, and a median longitudinal stripe of the same colour extends from the fore extremity rather more than half-way to the hinder one.

The *abdomen* is dark-brown above, spotted minutely and striated with yellow; a tapering dentated yellowish median band runs throughout the upper side to the spinners; the fore part of this band contains the normal marking, distinctly defined by a dark-brown line, and sharp-pointed at its hinder extremity; in the hinder half of the dentated band may be indistinctly traced the usual series of angular bars or chevrons, each of which is charged with two small brown spots in a transverse line. The sides of the abdomen are irregularly striated with dark-brown on a yellow ground, and the under side is paler, with still fewer brown markings.

The genital aperture is not large, but is, as usual, of characteristic form.

Hab.—Hills between Sirikol and Aktalla, 8th to 13th of May 1874; road from Yárkand to Bursi, May 28th to June 17th, 1874.

114. LYCOSA VINDEX, sp. n.

Adult female: length 2½ lines.

This spider is very closely allied to *Lycosa credula*, but it is of a shorter, stouter form, and the colours are of a generally duller hue. The median longitudinal yellow band on the *cephalothorax* is broader, distinctly constricted at the occiput, and enlarged at the thoracic indentation. The ocular area is black, with a geminated reddish-yellow spot between the eyes of the hinder row; the lateral brown bands are more distinctly traversed by darker lines radiating towards the thoracic indentation. The clypeus is lower, not exceeding in height two fore central eyes' diameter; the legs are entirely annulated (though not very distinctly excepting the tarsi), and they are also longer than those of *L. credula*.

The *sternum* is very similar in its markings; but the normal marking on the fore half of the upper side of the abdomen is less distinctly marked, and is blunt-pointed at its hinder extremity; the usual angular bars which succeed it are longer, and, with the marking on the fore part, are of a reddish yellow-brown hue; this is also the prevailing tint of the upper side, of which the rest is marked and striated with dark-brown; the under side is of a dull-yellowish hue, without any markings; and the genital aperture is of a distinct and characteristic form.

Hab.—Yárkand, November 1873.

115. LYCOSA VINDICATA, sp. n.

Adult female: length 2 lines.

This spider is exceedingly closely allied to *L. vindex*, but I am induced to record it as a distinct species, not only on account of its smaller size, but because the median longitudinal yellowish band on the *cephalothorax* is narrower, and has no constriction at the occiput, nor any lateral enlargement at the thoracic indentation; the submarginal lateral brown stripe is also more distinct and continuous, and the genital aperture differs a little in its form. In most other respects it resembles *L. vindex*, though the legs are more distinctly annulated. Its smaller size, shorter, stouter form, and reddish-brown hue of the paler markings on the abdomen, as well as the far more distinctly and completely annulated legs, and lower clypeus distinguish it readily from *L. credula*.

Hab.—Murree, June 11th to July 14th, 1873, and between Yangihissár and Sirikol, March 1874.

116. LYCOSA PASSIBILIS, sp. n.

Adult male: slightly over 3 lines.

The *cephalothorax* is of a deep brown colour, with a broadish longitudinal median band, and two lateral, narrower, sub-marginal ones slightly paler, and clothed with greyish hairs, with which, indeed, the rest of the cephalothorax is, though more thinly, covered.

The *eyes* occupy an area longer than broad; the front row is shorter than the middle one, and its two central eyes are placed on a slight prominence, and are larger than the lateral ones; the four being very nearly, if not quite, equally separated from each other, and the height of the clypeus is greater than the diameter of one of the fore central eyes, but not as much as two diameters. The eyes of the middle row are much larger than those of the hinder one, and, with them, form a square whose posterior side is the shortest.

The *legs* are long, attenuated at the extremities, of a deep reddish-brown colour, furnished with hairs and spines, and clothed with greyish pubescence.

The *palpi* are rather long and strong, and similar in colour to the legs; the radial and cubital joints are of equal length; the digital joint is of tolerable size, and nearly equal in length to the radial and cubital joints together; it is oval behind and rather attenuate before. The palpal organs are not very complex, but from the middle there projects a short but prominent process with three prominent divergent points, the middle one being the longest and strongest. Like the legs, the palpi are covered more or less with greyish hairs.

The *falces* are moderately long, but not very strong, slightly divergent and directed backwards: their colour is deep rich reddish black-brown, with a reddish-yellow broad longitudinal stripe on the inner side at the fore extremity.

The *maxillæ* are reddish-brown, pale-yellowish at the extremities and on the inner side.

The *labium* is similar to the maxillæ in colour, with a pale-yellowish apex.

The *sternum* is oval and of a deep shining reddish-brown colour.

The *abdomen* is of a blackish-brown hue, with the normal longitudinal marking on the fore half of the upper side, and a series of succeeding angular bars of an obscure brownish red; the whole is thickly clothed (especially along the median portion of the upper side) with greyish hairs, forming there a broad, longitudinal grey band, emitting some short lateral prominent lines on each side of the hinder part.

Hab.—Hills between Sirikol and Aktalla, between the 6th and 18th of May 1874.

117. LYCOSA FLAVIDA, sp. n.

Adult female: length 2½ lines.

The *cephalothorax* is of a pale-yellow colour, with two broad, longitudinal, lateral yellow-brown bands reaching from the fore to the hinder margin, and thinly clothed with greyish hairs; the height of the clypeus is no more than equal to the diameter of one of the fore central eyes.

The *eyes* are in the ordinary position, on large black spots; the whole of the fore part however, of the ocular area is more or less black. The front row is shorter than the middle one; its two central eyes are larger than the laterals, and the interval between them is greater than that between each and the lateral eyes next to it. The eyes of the middle row are very large, and separated from each other by no more than one diameter; this interval being but little less than that which separates each from the posterior eye opposite to it; the four hinder eyes form a square whose posterior side is longest and anterior one slightly the shortest.

The *legs* are moderately long and not very strong; their relative length is 4, 1, 2, 3, and they are of a pale yellow colour, rather paler than the cephalothorax, armed with spines and furnished thinly with hairs.

The *palpi* are moderately long, and similar to the legs in colour, deepening to a brownish hue on the last two joints.

The *falces* are moderate in length and strength, slightly divergent, perpendicular, of a yellow-brown colour, and clothed with bristly hairs.

The *maxillæ* and *labium* are of a paler hue than the falces, and the *sternum* is heart-shaped and similar in colour to the legs.

The *abdomen* is of a rather shortish oval form. On the upper side is a broad, longitudinal, pale-yellow band sharply dentated on its hinder half; the fore part of this band contains the normal longitudinal marking, of a slightly clearer colour, and faintly defined by a broken, brownish, indistinct line, and its hinder extremity is truncated. Some other indistinct, fine, brown, broken, angular lines on the hinder part, indicate the ordinary chevrons. On each side of the. median dentated band, and, in fact, defining it, is a broad brown band diffused in scattered spots a little over the sides; the under side is immaculate. The genital aperture is small, but of a characteristic form.

An immature male exactly resembled the female.

Hab.—Yárkand and neighbourhood, November 1873; Káshghar, December 1873; between Yangihissár and Sirikol, March 1874; Yangihissár, April 1874; road from Yárkand to Bursi, May 28th to June 17th, 1874.

BORDE, Genus Nov.

I am induced to form this new genus for the reception of four remarkable *Lycosids*, one received from Sinai, and described (P. Z. S., 1870, p. 822, pl. 1., fig. 3) as a *Lycosa* (*L. prælongipes*, Cambr.), another from the present collection, a third, *L. ungulata*, Cambr. Spiders of Egypt, Proc. Zool. Soc., 1876, p. 603, and a fourth, *L. arenaria*, Sav., Egypt. These (or at least three of them, for *L. arenaria*, Sav. is unknown to me, except from Audouin's figure and description, which do not detail the special points under consideration, though I have but little doubt of its possessing them), though exactly agreeing in several peculiar points of structure, are quite distinct species. The points in which they differ from *Lycosa*, *Trochosa*, and *Tarentula* may be seen from the following diagnosis of generic characters.

Cephalothorax oval, truncate before, and strongly constricted on the lateral margins of the caput; the normal indentations, especially the one dividing the caput from the thorax, are strong, and the upper side of the thorax on each side of the normal longitudinal indentation is gibbous, so that there is, when the spider is looked at in profile, a strong angular depression between the caput and thorax, the lateral thoracic margins being much depressed.

The *eyes*, as regards their general position, are like those of *Lycosa*, &c., but those of the second row have their vertical axes directed very nearly straight forwards, that is to say, scarcely at all upwards, though a little outwards; in this respect there is a marked approach to *Dinopis*, the facies being very vertical.

The *legs* are long and attenuated, especially those of the fourth pair. Two parallel rows of spines run throughout the under side of the tibiæ, metatarsi, and tarsi; at the fore end of each tarsus there is the appearance of a kind of short obsolete, or fixed, joint. It has apparently no movable articulation, but there is both a visible constriction and a kind of suture as though of a joint either consolidated by disuse, or in process of development towards a perfect supernumerary joint such as we find in *Hersilia*. The superior terminal claws are unusually long, slightly curved, and have four or five denticulations at their posterior extremity.

The *maxillæ* are not very long, but rather enlarged at their extremities, where they are rounded; and, instead of forming a straight line with the labium, they are turned distinctly sideways, thus in another point resembling *Dinopis*.

The *labium* is short, broad, and truncated in a slightly curved convex line at the apex.

The *palpi* of the female are truncated at the extremity, and the terminal claw, which is nearly straight and finely pectinated, issues from the middle of the truncation.

118. BOEBE BENEVOLA, sp. n.

Adult female: length 4½ lines.

The *cephalothorax* is clothed with short, sandy-grey pubescence, and is of a yellow-brown colour, with a broad yellow longitudinal median band, strongly constricted at the occiput, and enlarged at the thoracic indentation, the portion in front of the constriction forming a very distinct, transverse, oblong-oval area. There is also a broken, narrow, submarginal yellowish band on each side. The height of the clypeus is equal to twice the diameter of one of the fore central eyes, and the colour of the ocular area is black.

The *eyes* of the front row form a straight line shorter than that of the middle row; the central pair of the front row are larger than the laterals, and are separated by an interval wider than that which divides each from the lateral eye on its side; those of the middle row are very large, and are separated by nearly about one and a half diameter's interval, forming a line not far from equal to that formed by each and the posterior eye opposite to it; the eyes of the hinder row are large, but smaller than those of the middle one, and form a much longer line.

The *legs* of the fourth pair are considerably the longest; and those of the second pair apparently the shortest; while there is not so much difference between those of the first and third pairs, the first being the longer of the two. They are yellow in colour; the femora marked distinctly with brown spots, patches, and some other linear markings of the same hue; the spines beneath the tarsi are numerous, of equal length, much shorter than those on the other joints, and give the joint a comb-like appearance.

The *palpi* are like the legs in colours, and in the markings on the humeral joints, and are furnished with hairs and a few spine-like bristles.

The *falces* are powerful, of moderate length, rounded in profile, clothed with sandy-grey hairs and long bristles, and of a dark reddish yellow-brown colour.

The *maxillæ* are yellow, strongly tinged with yellow-brown, particularly on their inner sides, and pale yellowish at their inner extremities.

The *labium* is dark yellowish-brown with a pale apex.

The *sternum* is of a short heart-shape and dark yellow-brown colour, thinly clothed with sandy-grey pubescence.

The *abdomen* is oval and moderately convex above; its colour is a dull brownish-yellow, marked with dark brown, occasionally approaching to black; the intersecting portions of the yellow ground-colour are spotted more or less with cretaceous-white spots. The normal longitudinal median marking (of a deep brown colour) on the fore half of the upper side is large, somewhat wedge-shaped, and roughly dentated on its margins, and its posterior extremity is prolonged into a more or less distinct median line to the spinners, and gives off on each side various oblique lines and markings, forming some tolerably distinct, angular, yellow bars of different sizes, and some of which have a dark-brown spot at their extremities. There is a tolerably clear, marginal yellow space round the normal marking on the fore half. The under side is almost all occupied by a broad longitudinal light-brown band.

The genital aperture is small, but of characteristic form. The spinners are very short, but those of the superior pair are stronger and rather longer than those of the inferior. An immature male resembled the female in colours and markings.

Hab.—Yárkand and neighbourhood, November 1873; Káshgbar, December 1873; between Yangihissár and Sirikol, March 1874; Yangihissár, April 1874; Yárkand, 21st to 27th May 1874, and Yárkand to Bursi, May 28th to June 17th, 1874.

Family—*SPHASIDES*.

Genus—*OXYOPES*, Latr.

119. OXYOPES JUBILANS, sp. n.

Adult male: length rather more than 3½ lines.

This spider is nearly allied to *Oxyopes* (*Sphasus*) *lepidus*, Blackw., of which the female only has yet been described; the latter differs, however, from the female of the present species in being of a more robust form and in having shorter legs, as well as in the abdominal markings.

The general form and appearance are similar to those of most others of the genus; the *cephalothorax* is of a brownish-yellow colour, and the normal indentations are distinctly marked. The ocular area, and the middle of the clypeus are clothed with grey hairs; a fine brown line runs obliquely along the margins of the upper side, and so downward to the lower corners of the clypeus; two others run, one from each of the two foremost eyes, nearly perpendicularly to the falces (to the extremity of which they are continued), bisecting them in front. There are also two parallel brown lines along the middle of the cephalothorax, not reaching further forward than the occiput, and less distinct in the male than in the female; the eyes are on black spots and in the usual position, six posterior ones forming a transverse hexagonal figure whose sides scarcely differ in length; they may be also taken as in four transverse rows of two each. Those of the foremost row are very minute and separated from those of the next row by an eye's diameter. Those of the second row are the largest of the eight or nearly so, and are separated by an interval of one diameter, both from each other, and from the eyes of the third row; this row is considerably the longest, and the fourth row is slightly longer than the second, its eyes being rather further from each other than each is from the lateral of the third row on its side.

The *legs* are long and slender, their relative length seems to be 4, 1, 2, 3; they are of a yellow colour, and are armed with numerous long spines. The femora of the first and second pairs have longitudinal brown lines on the under side, a faint trace of two only of these existing on the femora of the third and fourth pairs.

The *palpi* are short, similar in colour to the legs; the cubital joint is very short with but a very slight angular prominence at its fore extremity on the upper side; the radial joint is much stronger than the cubital; it is strongly tinged with yellow-brown, much enlarged

In a blunt angular form on the outer side, with a short red-brown irregular projection rather underneath; the digital joint is round-oval, brownish-coloured behind, and of a narrow, slightly tapering, pointed, beak-like form in front; the beak portion is yellowish, and less in length than the oval part; this part has a small angular prominence at its base on the outer side. The palpal organs are prominent and rather complex, but do not present any remarkable processes. The radial and cubital joints are furnished with several long, curved, spine-like bristles.

The *falces* are not very long nor strong; they are of a subconical form, straight, perpendicular, similar in colour to the cephalothorax, and bisected in front by a longitudinal brown line.

The *maxillæ* and *labium* are of normal form, the colour of the former is yellow, and of the latter yellow-brown.

The *abdomen* is long and narrow, being of an elongate, tapering, or pyramidal form; the fore part is the largest, and it gradually narrows to the spinners: it is of a yellowish hue, somewhat freckled with white cretaceous spots of small size; the upper side is margined on each side with a double longitudinal brown line, and a faint, narrow, tapering, dusky band along the middle; on the under side is a broadish, tapering, dusky, longitudinal band, margined with reddish-brown.

The female is more distinctly marked than the male, and the process connected with the genital aperture is blackish and prominent. The colours of this spider, as above described from examples for several years immersed in spirits, cannot be considered altogether reliable, inasmuch as the yellow tints may have possibly been more or less green when the spider was living.

Hab.—Tinali; route from Murree to Sind Valley, July 19th, 1873.

I come to the conclusion that this is the locality, because Dr. Stoliczka, in his diary of July 19th, 1873, mentions having found that evening a good number of spiders, "chiefly *Thomisus* and *Sphasus*" (=*Oxyopes*), and in the one unlabelled bottle I find the only examples of *Sphasus* contained in the whole collection. These are of three species, the present and the next one very nearly allied to each other, the third quite distinct both in form and markings; all three are, I believe, of undescribed species, though Dr. Stoliczka says of those he found "among the latter (*Sphasus*)" he recognized *Sphasus viridanus*. This is a Calcutta species described by Dr. Stoliczka in Journ. Asiat. Soc., Bengal, vol. xxxviii, p. 220, pl. xx, fig. 1, and is undoubtedly a species of *Pasithea*, Bl. (*Peucetia*, Thor.), which, though generically nearly allied to *Sphasus*, is yet easily recognized by the difference in the position of the eyes.

120. OXYOPES PRÆDICTA, sp. n.

Adult male: length 4 lines.

This spider is very closely allied to *Oxyopes jubilans*; it is, however, rather larger, and this, I think, may probably prove to be a constant character. In general colouring, form, and appearance, the two species are strikingly similar, but the following distinctions will serve to separate them without difficulty.

The *cephalothorax* has no lateral brown lines running to the fore corners of the clypeus, and the two parallel median ones are here replaced by a not very distinct, median, longitudinal rusty-reddish band, which runs quite to the hinder row of eyes.

The *palpi* present an easily observed difference from those of *Oxyopes jubilans* both in the radial and digital joints. The former is not enlarged on the outer side, but is, on the contrary, rather excavated there, with a somewhat corneous, red-brown ridge just behind the excavated part: the radial joint is also somewhat angularly prominent underneath towards the inner side. The digital joint has its short-oval, posterior portion of a darker hue, and more angularly prominent at its base on the outer side. The anterior, or beak-like, portion is also distinctly longer than the oval part, and terminates in a sharpish and somewhat corneous point.

The *legs* have the femoral joints of the third and fourth pairs as distinctly marked with one (if not two) longitudinal blackish-brown lines as the whole of the first and second pairs.

The *abdomen* has on the upper side a longitudinal, median, tapering, rusty-reddish band, at the fore part of which the normal elongate marking, of a somewhat spear-headed form and yellow colour, is visible.

Hab.—Found at the same time and in the same locality as *O. jubilans*. Tinali; route from Murree to Sind Valley, July 19th, 1873.

121. OXYOPES REJECTA, sp. n.

Adult female: length rather less than 3¾ lines.

This spider is nearly allied to *Oxyopes* (*Sphasus*) *gentilis*, C. L. Koch. It may easily be distinguished from the two foregoing species by its shorter legs, as well as by its shorter, stouter form, and by the short cephalothorax, which has the sides and hinder slope very steep, and the normal indentations very slightly marked, so that the divisional line between the thorax and caput is scarcely visible. The colour of the cephalothorax is brownish-yellow, paler in the ocular area and at the occiput; it is indistinctly marked in the median longitudinal line, as well as on the sides, with blackish-brown; there are also two slightly curved lines, of the same colour, running down from the two foremost eyes over to the middle of each of the falces, and continued over them in a slightly sinuous form, but stopping somewhat considerably short of their extremity.

The *eyes* are on conspicuous black blotches, those of the third row are considerably nearer to those of the second than to those of the fourth (or posterior) row. The length of the two last (2nd and 4th) rows are exactly equal, though, owing to the difference in the size of the eyes, the interval between those of each row is different. The height of the clypeus is rather less than half that of the facial space.

The *legs* are rather short, and their relative length is 4, 1, 2, 3. They are armed with long spines, and their colour is yellow, the femora being marked underneath with two longitudinal parallel, blackish-brown lines.

The *palpi* are similar to the legs in colour, rather long, slender, and armed with a few strong spine-like bristles.

The *falces* are not very long nor strong; they are of a subconical form, straight, and perpendicular; their length is less than the height of the facial space, and their colour is like that of the cephalothorax, with a longitudinal blackish-brown line from the base to two-thirds of the distance towards their extremity.

The *maxillæ* and *labium* are of normal form, and of a light brownish-yellow hue.

The *abdomen* is oval, pointed behind; on the upper side is a broad, longitudinal central slightly tapering yellowish band, spotted with small cretaceous-white spots, and showing the normal marking on the fore part of a clearer, though slightly brown, colour, and of an elongate diamond-shape; the marginal portions of the upper part are marked with blackish-brown oblique linear markings, which extend more or less over the sides. These parts, as well as the under side, are similar in colour to the middle of the upper side; the under side having a broad, well-defined, longitudinal, median, black-brown band, marked along the middle with pale yellowish. The genital aperture is small but of characteristic form.

Hab.—Found at the same time and place as the two foregoing species. Tinali; route from Murree to Sind Valley, July 19th, 1873.

Family—*SALTICIDES*.

Genus—*HELIOPHANUS*, C. L. Koch.

122. HELIOPHANUS DUBIUS.

Heliophanus dubius, E. Simon, Arachnides de France, tom. iii, p. 146, pl. x., fig. 4.

I have not been able yet to compare this spider (♂ adult) with a type of *H. dubius*, Sim., but I believe it to be identical with that species, as it agrees well with the figures and description given *l. c.*

Hab.—Hills between Sirikol and Aktalla, 8th to 18th May 1874.

Genus—*PLEXIPPUS*, C. L. Koch.

123. PLEXIPPUS ADANSONII.

Attus adansonii, Sav., Egypte, p. 169, pl. 7, fig. 8.

Hab.—Both sexes of this spider, differing in no respect from examples found in Egypt, and received from Bombay, were contained in the portion of the collection without date or locality but probably (as before observed) made between Murree and Sind Valley about the end of July 1873.

Genus—*MENEMERUS*, E. Simon.

124. MENEMERUS CINCTUS, sp. n.

Adult male : length rather over 2½ lines.

The *cephalothorax* is of a flattened form with a strong transverse depression indicating the junction of the caput and thorax. The upper area of the caput is black, the rest of the cephalothorax is dark yellowish-brown, paler towards the margins.

It is clothed with hairs mostly of a golden hue, a longitudinal median stripe and a marginal one on each side being furnished with white hairs; the marginal stripe is formed of two narrow parallel ones. There are also some prominent bristly hairs on the cephalothorax, strongest on the sides of the ocular area, below which three of them form a longitudinal line.

The *eyes* form an area broader than long; those of the anterior row are separated from each other by a small and equal interval, and those of the middle row appear to be as nearly as possible half-way between the first.and third rows, and slightly within the straight line formed on each side by the laterals of those rows, of which the first is shorter than the third.

The *legs* are moderately strong and not very long, those of the first pair are the longest, considerably the strongest, and of a dark yellow-brown colour, the femora being the lightest in hue; the rest are yellow, tinged with brown, and all are furnished with hairs and long prominent bristles, but no spines except some short ones beneath the tibiæ and metatarsi of the first pair: there is a compact claw-tuft beneath the terminal tarsal claws. The third pair appear to be slightly the shortest.

The *palpi* are short, of a yellow-brown colour, and furnished with hairs and bristles; the cubital joint is short and strong, the radial is shorter and less strong, but is considerably produced on its outer side, the produced portion ending in a tapering, pointed, slightly blunt apophysis. The digital joint is long, of a slightly bent oblong-oval form. The palpal organs have a large, nearly globular lobe at their base, extending beneath and rather on the inner side of the radial and cubital joints.

The *falces* are short, strong, straight, projecting strongly forward, and but very slightly divergent; their anterior extremity is as broad almost as the posterior, truncated, and with a strongish tooth at the inner corner; their colour is dark yellow-brown tinged with red.

The *maxillæ* are short, strong, broadest, and rounded at their extremities, and inclined towards the *labium*, which is of a somewhat oblong-oval form; these parts are of a deep yellow-brown hue, the extremities of the former, and the apex of the latter being of a paler colour.

The *sternum* is oval and of a palish yellow-brown colour.

The *abdomen* is oval and of a somewhat flattish form; it is banded transversely on the upper part and sides with alternate broad whitish and dark-brown bands, the first band encircling the fore margin, is white, and the second and third are divided in the middle by a narrow brown patch; the brown bands are considerably the broadest, and all become more or less tapering when they reach the sides: the surface is clothed with hairs; a broad longitudinal median band, and a lateral one on each side, are formed by hairs of a rusty scarlet hue, those on the intermediate spaces being whitish grey; on the hinder half of the upper side is a longitudinal, median series of whitish, angular bars, of course visible only when they occur upon the brown transverse bands, the under side is dull whitish, with a broad median longitudinal tapering yellowish-brown band.

Hab.—Yárkand, May 1874.

125. MENEMERUS INCERTUS, sp. n.

Adult female: length 2¾ lines.

The *cephalothorax* of this spider is short, of a rather flattened form, and the profile of the ocular area slopes, but very slightly, downwards; this part is of a brownish black hue, the rest of the cephalothorax being brownish-yellow, and the whole clothed with appressed grey hairs; some erect bristly ones being dispersed thinly over the surface. The margins are black.

The *eyes* of the anterior row are separated from each other by distinct intervals, that between the centrals being less than that between each and the fore lateral next to it; the posterior row is a little longer than the anterior one. A single row of strong bristles runs longitudinally just below the lateral eyes of the three rows, and the eyes of the middle row are nearer to the anterior than to the posterior row.

The *legs* are short, those of the fourth pair are the longest but less strong than those of the first pair, and those of the second pair appear to be a little the shortest: they are furnished with hairs, bristles, and a few spines, with a small claw-tuft beneath the terminal tarsal claws. The colour of the legs is yellow.

The *palpi* are similar to the legs in colour, short, slender, and furnished with white hairs.

The *falces* are short, not very strong, straight, projecting, and of a yellow-brown colour.

The *maxillæ* and *labium* are lighter-coloured than the falces.

The *sternum* is small, oblong-oval, and similar to the legs in colour.

The *abdomen* is of a rather elongate oval form, and of a dull yellow colour, somewhat clouded with reddish yellow-brown on the upper side, and clothed with fine yellowish and grey hairs, with a few dark, slender bristly ones intermixed; on the fore part is a brown marking consisting of two short parallel lines looped in front, and near the hinder extremity are three confluent bright red-brown patches. The genital aperture is of a distinctive form.

It is possible that this may be the female of *Menemerus cinctus*, but, as the colours and pattern of that species are different, it is best to describe it at present as distinct, until we have other evidence of their identity; dissimilarity of colours and pattern, as well as of structure, are often found in the sexes of spiders, though *primâ facie* such dissimilarity is proof of specific difference.

Hab.—Yárkand, end of May 1874.

126. MENEMERUS DELETUS, sp. n.

Adult female: length 2½ lines.

The form of the *cephalothorax* is flattish; it is of a deep yellow-brown colour, darkest on the caput, with an indistinct, ill-defined, brownish-yellow, marginal border, and a still less distinct, longitudinal, median stripe on the thorax. The caput and sides of the cephalothorax are clothed with light-grey hairs; those on the other parts had probably been rubbed off.

The *ocular* area is broader than long; the length of the anterior row of eyes is slightly shorter than that of the posterior one, and the eyes of the middle row are almost exactly intermediate between them. The fore central eyes are of a dull mother-of-pearl colour.

The *legs* are moderately long and strong; those of the first pair are the strongest, but not quite so long as the fourth pair, and the second pair are slightly the shortest. They are of a yellow colour, those of the first pair light yellow-brown, and with some short, strong spines in pairs beneath the tibiæ and metatarsi; beneath the terminal tarsal claws is a black claw-tuft.

The *palpi* are slender, not very long, and clothed with white hairs.

The *falces* are yellow-brown, the *maxillæ* and *labium* a little paler, and the *sternum* darker; the last clothed with coarse, whitish hairs.

The *abdomen* is oval, truncate before, pointed behind, and of a pale-yellow colour mottled thickly with whitish cretaceous spots; on the fore part of the upper side is a small, median, longitudinal, dull yellowish-brown, somewhat arrow-headed marking, continued in an attenuated line of the same colour to the spinners; a little way from this, on each side, is an indistinct, longitudinal broad band clothed with coppery-red-hairs; the spinners are moderately long and strong, and of a pale yellow-brown colour. The form of the genital aperture is characteristic.

Hab.—Route from Yárkand to Bursi, May 28th to June 17th, 1874.

127. MENEMERUS FRIGIDUS, sp. n.

Adult female: length 2½ miles.

The *cephalothorax* is short, though distinctly longer than broad, and of the usual flattish form; the upper side is dark-brown, tinged with yellowish, the caput being the darkest, and there is a larger longitudinal patch of brownish-yellow on the middle; the sides are yellow, with a narrow white marginal border; the whole is thinly clothed with sandy-grey and whitish hairs.

The *eyes* of the posterior row form a line scarcely, if at all, longer than the anterior row, and the middle row is nearer to the anterior than to the posterior one. Those of the anterior row are of a dull mother-of-pearl colour, and are divided by distinct intervals; that which separates the central pair is less than that which divides each from the lateral eye next to it.

The *legs* are rather short, and not very strong; those of the first pair are a little stronger but distinctly shorter than the fourth pair, if, indeed, they be not also slightly shorter than the third, the second pair being the shortest; they are of a brownish-yellow colour, and are armed with a few spines, as well as with bristles and hairs.

The *palpi* are short, slender, yellow, and clothed with whitish hairs.

The *falces* are short, straight, projecting, and of a yellow-brown colour.

The *maxillæ* and *labium* are rather paler than the falces, and the *sternum* is similar to the legs in colour.

The *abdomen* is oval, and of a dull yellowish hue, marked irregularly with brown along the sides of the upper part, and with a longitudinal brown stripe along the middle of the fore part; this stripe is bifid at its hinder extremity, and followed by a series of brown, sharply-angular bars, some of which run into the brown markings on the sides. The form of the genital aperture is characteristic.

Hab.—Murree, June 11th to July 14th, 1873.

Genus—*ATTUS*, E. Simon.

128. ATTUS DEVOTUS, sp. n.

Adult female: length 1¾ lines.

The *cephalothorax* is of a slightly flattened form; the hinder slope is short, and at about an angle of 45°; the forward slope of the caput being slight, and but very little convex; its colour is yellow-brown, darkest on the upper part of the caput, and it is clothed with yellowish and grey hairs intermixed; the lateral margins are black, immediately above which is a not very broad band of white hairs, and a similar band or stripe runs along the middle of the hinder slope to the occiput. The clypeus is very low and retreating.

ARANEIDEA.   103

The *eyes* form an area broader than long; those of the anterior row (which is of equal length with the posterior one) are very near to each other, if not quite contiguous. The fore centrals are of very large size and of a yellowish-brown mother-of-pearl hue; those of the middle row are equi-distant between the anterior and posterior lateral eyes.

The *legs* are neither very long nor strong; their relative length appears to be 4, 1, 3, 2; those of the first pair are the strongest, and those of the fourth pair are the most attenuated; their colour is yellow, and they are furnished with hairs and spines, the latter on the tibiæ and metatarsi, but only underneath these in the first and second pairs; beneath the terminal tarsal claws is a compact, blackish claw-tuft.

The *palpi* are moderately long, hairy, and yellow, the digital joint tinged with yellow-brown.

The *falces* are short, strongish, straight, directed forwards, though placed rather far back, and of a dark yellow-brown colour.

The *maxillæ* and *labium* are yellow-brown; the *sternum* being of a dark brownish-yellow, and of a rather elongate-oval form.

The *abdomen* is oval, truncated before and rounded behind, and projects over the hinder slope of the cephalothorax; it is clothed with grey, brassy-yellowish, and white hairs. The upper side is of a dull yellowish-brown colour, with an elongate, whitish marking along the middle of the fore part, followed by some not very distinctly defined, small, angular bars, on each side of which (as well as of the elongate marking) is a series of short transverse whitish markings, giving an appearance, when taken in connection with the markings along the middle, of irregular transverse stripes across the upper side; the sides and under side are dull yellowish, the upper part of the former slightly marked with faint brownish spots and markings, and the latter clothed with short, greyish hairs.

Hab.—Murree, June 11th to July 14th, 1873.

129. ATTUS BENEFICUS, sp. n.

Adult female: length nearly 2¼ lines.

Cephalothorax short and broad, the hinder slope steep, at an angle of 45°; the ocular area slopes a little forwards in a convex line and there is a distinct, though not unusually strong, transverse depression at the occiput. The clypeus is very low, being almost obsolete.

The upper part, with a portion of the sides, is black-brown, the ocular area quite black, with an oblong yellow stripe on the upper part of the hinder slope; the remainder of the sides is yellow, clothed with fine, white hairs.

The *eyes* form an area much broader than long; the posterior and anterior rows are equal in length; the fore centrals are very large and of a mother-of-pearl hue; they are separated by a small interval, less than that which divides each from the fore lateral on its side; the lateral eye, on each side, of the middle row is equi-distant from the laterals of the posterior and anterior rows.

The *legs* are short and strong; their relative length is apparently 4, 1, 2, 3, but the difference between 4 and 1, and 2 and 3 respectively, is very slight. Their colour is yellow, those of the first pair being clouded in parts with brown; the tibiæ and metatarsi of the third and fourth pairs, and the under sides of those of the first and second pairs, are armed with spines, and there is a compact claw-tuft beneath the terminal claws of each tarsus.

The *palpi* are short, yellow, and furnished with coarse hairs, principally on the digital joints.

The *falces* are short, strong, straight, nearly perpendicular, but removed rather far backwards, and of a dark yellow-brown colour.

The *maxillæ* and *labium* are yellow-brown; the *sternum* is yellow, with dusky margins, and of a rather elongate-oval form.

The *abdomen* is of a short-oval form, rather broader behind, where it is rounded, the fore extremity being rather truncated, and projecting over the base of the cephalothorax; the upper side is black-brown, thinly speckled with yellowish points; on the middle of the fore part is a small, somewhat triangular, pale-yellow patch, produced backwards in a short stalk-like form with a prominent blunt point or patch on each side, and followed towards the spinners by a series of large, angular lines, or chevrons, of the same colour; the first of these chevrons is of a rather sinuous form, and they all vary in strength and distinctness of definition, and have, here and there, a black-brown spot upon them; the sides are pale-yellow, spotted, chiefly on the hinder half, with black-brown, and the under side is also pale-yellowish, with a broad, longitudinal, median, dusky-brownish band. The genital aperture is small, and of characteristic form, its colour being yellow-brown, edged with red-brown; the spinners are short; the superior pair are of a dark-blackish hue; the inferior pair yellow-brown, slightly shorter, but a little stronger, than the superior pair.

Hab.—Sind Valley, August 1873.

130. ATTUS DIDUCTUS, sp. n.

Adult female: length rather over 2⅛ lines.

This spider is nearly allied to *Attus beneficus*, which it resembles in general colours and markings, but may be distinguished at once by the less convex cephalothorax and the flatter ocular area. The sides of the cephalothorax also, instead of constituting a broad, well-defined yellow band along almost its whole width, have only an irregular and not very well-defined brownish-yellow, narrow, marginal border, the margin itself being black; the fore central pair of eyes are also much darker-coloured, and the legs are rather less strong, those of the fourth pair being distinctly, though not greatly, longer than the first, which last are rather the stoutest and are marked along each side with deep brown.

The colour of the *sternum* is dark yellow-brown, and the *abdomen* has a very similar pattern to that of *A. beneficus*, though less distinct, and the form of the genital aperture is quite distinct.

Hab.—Murree, June 11th to July 14th, 1873.

131. ATTUS AUSPEX, sp. n.

Adult male: length 2⅛ lines.

The *cephalothorax* is broader behind than in front; looked at in profile the hinder slope is long, gradual, and but very slightly convex, running to the third posterior row of eyes, from which the caput slopes rapidly downwards to the anterior row; its colour is yellow-brown, deepening gradually to the caput, which is black-brown; there is a narrow blackish marginal line, and the whole is clothed pretty thickly with mixed yellowish, coppery-golden, and grey squamose appressed hairs, those immediately round the eyes of the front row being very bright and forming, probably in most cases, scarlet 'irides.'

The *eyes* form an area broader than long, and the posterior row is larger than the anterior one; the central pair of the anterior row are very large and close together, but not contiguous, being separated by an interval a very little less than that which divides each from the lateral of the same row on its side. These laterals are rather larger than the eyes of the posterior row, and the small eye (on each side) of the middle row is in a straight line with the inner edges of the fore lateral and hind lateral eyes, being also nearer to the hind lateral than to the fore lateral eye. The height of the clypeus is equal to the diameter of one of the fore central eyes.

The *legs* are strong and moderately long. Their relative length is 4, 1, 2, 3; they are of a pale-yellowish colour, furnished thickly with hairs, bristles, and spines. Some of the hairs are squamose and appressed, others long and prominent, especially on the first pair; those beneath the tarsi and metatarsi are the most numerous, and black, the rest being mostly grey or sandy-coloured. The terminal tarsal claws have a claw-tuft beneath them, and are long and slender, especially those of the fourth pair; these have only 1—3 minute teeth about the middle of the under side; on some, if not all, of the other legs, even these denticulations appear to be wanting. The legs of the first pair are considerably the strongest, while those of the fourth pair are much the longest.

The *palpi* are short and strong, similar in colour to the legs, and furnished with long (as well as some shorter squamose) grey hairs; the radial joint is shorter and less strong than the cubital, and its fore extremity on the outer side is produced into a not very long, tapering, sharp-pointed, curved projection whose extremity is of a deep reddish-brown colour; the digital joint is of great length, the base is of a somewhat angular shape, and the fore part is produced into a long cylindrical curved form; the palpal organs are bulbous, tumid, placed chiefly beneath the hinder part of the digital joint, and encircled at their base and round the inner side by a long, strongish, tapering spine, which runs more or less closely alongside the inner margin of the digital joint, and forms a very conspicuous and characteristic feature of the species.

The *falces* are short and straight, placed considerably backwards, and of a dark yellow-brown colour.

The *sternum* is small, oval, yellow-brown, and clothed with coarse grey hairs.

The *maxillæ* are short and almost touch, at their extremities, over the *labium*; these parts are yellow-brown, paler at the extremities of the former and the apex of the latter.

The *abdomen* is oval, of a yellowish-brown colour with an indistinct dark brown stripe along the middle of the fore part of the upper side, and clothed pretty densely with short squamose, mixed yellowish, grey, sandy, and shining coppery hairs; the under side is of a pale dull brownish-yellow hue, clothed with grey, squamose hairs.

The female is larger than the male, but resembles that sex in colours and other general characters. It is probable that a series of examples would show, in some instances, a more or less distinct pattern on the upper side of the abdomen, depending on the distribution of the colours of the hairs, which are subject to much variation in different individuals of the same species in this group. Traces of this pattern in brown blotches and markings are visible in the female. The *palpi*, however, are so characteristic in the adult male that the species can hardly be mistaken for any other.

Hab.—Yárkand and neighbourhood, November 1873; hills between Sirikol and Aktalla, 8th to 13th May 1874.

132. ATTUS AVOCATOR, sp. n.

Adult male : length slightly over 1½ lines.

The *cephalothorax* of this small species is less high at the hinder row of eyes than in *Attus auspex*, the hinder slope is (when looked at in profile) a little convex, as also is that of the upper part of the caput, or ocular area. Its colour is yellow-brown, the ocular area being the darkest; it is thickly clothed with grey and yellowish mixed, appressed hairs, showing, however, three longitudinal stripes of white hairs, one on each side, just below the margin of the upper part, and the third along the middle of the hinder slope.

The *eyes* form an area considerably broader than long; the anterior row is equal in length to the posterior; in other respects the eyes are like those of *Attus auspex*.

The *legs* are rather short and moderately strong; their relative length being 4, 1, 3, 2. They are of a brownish-yellow colour, indistinctly annulated with darker brown, and the extremities of the tarsi of the first pair are blackish. They are clothed with hairs, bristles, and spines; the terminal tarsal claws are long and slender, and are apparently devoid of denticulations beneath; underneath them, on each tarsus, is a compact claw-tuft.

The *palpi* are short, similar in colour to the legs, and clothed with coarse (and principally grey) hairs; the radial is shorter than the cubital joint, and has a small pointed apophysis at the outer extremity; the digital joint is long but not very broad, being of a somewhat oblong form; and the palpal organs are simple and of a blackish-brown colour.

The *falces* are moderately long, not very strong, straight, perpendicular, but placed considerably backwards, and of a dark yellow-brown colour.

The *maxillæ* and *labium* are also yellow-brown, the former are pale at their extremities which do not nearly meet over the latter; the labium also has the apex of a pale hue.

The *sternum* is small, oval, of a dark yellow-brown colour clothed with coarse grey hairs.

The *abdomen* is oval, rather truncated in front. The upper side is dark brown, mottled obscurely with yellowish, the margin being a little notched and bordered with white hairs, a short streak along the middle of the fore part, of a dull yellowish hue, is followed to the spinners by a series of short, but pretty distinct, angular bars of the same colour; these markings are clothed with white hairs: the sides are longitudinally striated with brown, and the under side is of a dull brownish yellow.

Hab.—Yángihissár, April 1874.

SYSTEMATIC LIST OF SPIDERS ABOVE DESCRIBED AND RECORDED.

N. B.—The figures denote those districts in which the Spiders were found, (*vide* Introductory Remarks and the Separate Lists *postea*).

Fam. THERAPHOSIDES.

Gen. *Idiops*, (Perty).

Idiops designatus, sp. n., 1.

ARANEIDEA. 107

Fam. FILISTATIDES.

Gen. *Filistata*, (Latr.).

Filistata reclusa, sp. n., 2.

Fam. DYSDERIDES.

Gen. *Dysdera*, (Latr.).

Dysdera cylindrica, sp. n., 1.

Fam. DRASSIDES.

Gen. *Drassus*, (Walck. *ad partem*).

Drassus troglodytes, (C. L. Koch), 2, 4, 5, 3.
„ infletus, sp. n., 5.
„ interruptor, sp. n., 2.
„ invisus, sp. n., 5.
„ interpolator, sp. n., 5.
„ dispulsus, sp. n., 5, 2, 4, 3, 1.
„ interlisus, sp. n., 5, 4, 2, 3.
„ involutus, sp. n., 1.
„ lapsus, sp. n., 5.

Gen. *Gnaphosa*, (Latr.).

Gnaphosa stoliczkæ, sp. n., 5, 3.
„ plumalis, (Cambr.), 3.
„ mœrens, sp. n., 5, 3.

Gen. *Prosthesima*, (L. Koch).

Prosthesima cingara, (Cambr.), 4, 5.

Gen. *Micaria*, (Westr.).

Micaria convexa, sp. n., 5.
„ pallida, sp. n. 5.

Gen. *Clubiona*, (Latr.).

Clubiona deletrix, sp. n., 1.
„ laticeps, sp. n., 1.
„ laudata, sp. n., 3.

Gen. *Cheiracanthium*, (C. L. Koch).

Cheiracanthium adjacens, sp., n., 1.
„ approximatum, sp. n., 1.

Gen. *Agröeca*, (Westr.).

Agröeca debilis, sp. n., 5.
„ flaveus, sp. n., 4.
„ molesta, sp. n.

Gen. *Trachelas*, (L. Koch).
Trachelas costata, sp. n., 1, 2.

Fam. DICTYNIDES.
Gen. *Dictyna*, (Sund.).
Dictyna albida, sp. n., 5.

Fam. AGELENIDES.
Gen. *Argyroneta*, (Latr.).
Argyroneta aquatica, (Walck.), 4.

Gen. *Cælotes*, (Blackw.).
Cælotes tegenarioides, sp. n., 1.
 „ simplex, sp. n., 1.

Gen. *Tegenaria*, (Latr.).
Tegenaria ? 3.

Fam. THERIDIIDES.
Gen. *Episinus*, (Walck.).
Episinus algiricus, (Luc.) 1.

Gen. *Theridion*, (Walck.).
Theridion saxatile, (C. L. Koch), 1.
 „ lepidum, sp. n., 1.
 „ subitum, sp. n., 1.
 „ confusum, sp. n., 1.
 „ expallidatum, sp. n., 1.
 „ tuberculatum, (Kronenberg), 1.
 „ incertum, sp. n., 1.

Gen. *Steatoda*, (Sund.).
Steatoda nigrocincta, sp. n., 1, 3.
 „ mandibularis, (Lucas), 3.
 „ sordidata, sp. n., 5.

Gen. *Phycus*, (Cambr.).
Phycus sagittatus, sp. n., 1.

Gen. *Erigone*, (Sav.).
Erigone atra, (Blackw.), 3.
 „ dentipalpis, (Westr.), 1.

Gen. *Pachygnatha*, (Sund.)
Pachygnatha clerckii, (Sund.), 5.

ARANEIDEA. 109

Gen. *Linyphia*, (Latr.).

Linyphia consanguinea, sp. n., 1.
,, albopunctata, sp. n., 1.
,, straminea, sp. n., 1.
,, perampla, sp. n., 1.
,, pusilla, (Sund.), 4,3.

Fam. EPĒIRIDES.
Gen. *Meta*, (C. L. Koch).
Meta mixta, sp. n.

Gen. *Tetragnatha*, (Latr.).
Tetragnatha extensa, (Linn.), 3.

Gen. *Epëira*, (Walck.).
Epëira tartarica, (Kronenberg), 2.
,, bigibbosa, sp. n., 1.
,, pellax, sp. n., 1.
,, gurda, sp. n., 1.
,, haruspex, sp. n., 4.
,, pœnulata, sp. n., 1.
,, prædata, sp. n., 1.
,, cucurbitina, (Clerck) 1.
,, cornuta, (Clerck) 1, 3, 4.
,, panniferens, sp. n., 1.
,, carnifex, sp. n., 1.
,, gibbera, sp. n., 1.

Gen. *Chorizoopes*, (Cambr.).
Chorizoopes stoliczkæ, sp. n., 1.
,, congener, sp. n., 1.

Fam. GASTRACANTHIDES.
Gen. *Cyrtarachne*, (Thor.).
Cyrtarachne pallida, sp. n., 1.

Fam. ULOBORIDES.
Gen. *Uloborus*, (Latr.).
Uloborus albescens, sp. n., 1.

Fam. THOMISIDES.
Gen. *Thomisus*, (Walck. *ad partem*).
Thomisus albidus, sp. n., 3.
,, albens, sp. n., 3.

c 1

Gen. *Misumena*, (Thor.).
Misumena expallidata, sp. n., 1.
" ? oblonga, sp. n., 1.

Gen. *Synema*, (Sim.).
Synema exculta, sp. n., 1.

Gen. *Diæa*, (Thor.).
Diæa (?) spinosula, sp. n., 1.
" subdola, sp. n., 1.
" sufflava, sp. n., 1
" suspiciosa, sp. n., 3.
" subargentata, sp. n., 1.

Gen. *Xysticus*, (C. L. Koch).
Xysticus cristatus, (Clerck), 5.
" audax (?), (C. L. Koch), 5.
" maculosus, sp. n., 1.
" setiger, sp. n., 3.
" breviceps, sp. n., 3
" mundulus, sp. n., 1.

Gen. *Monastes*, (Luc.).
Monastes dejectus, sp. n., 1.

Gen. *Sarotes*, (Sund.).
Sarotes regius (Fabr.), 1.
" promptus, sp. n., 1.

Gen. *Sparassus*, (Walck.).
Sparassus timidus, sp. n., 2.
" fujax, sp. n., 1.
" flavidus, sp. n., 4.

Gen. *Philodromus*, (Walck.).
Philodromus cinerascens, sp. n., 2; 3.
" medius, (Cambr.), 1.

Gen. *Tibellus*, (Sim.).
Tibellus propinquus, sp. n., 5.

Gen. *Thanatus*, (C. L. Koch).
Thanatus thorellii, (Cambr.), 3.
" albescens, sp. n., 1.

Fam. LYCOSIDES.
Stoliczka, gen. nov.
Stoliczka insignis, sp. r. 1.

ARANEIDEA. 111

Gen. *Ocyale*, (Sav.)
Ocyale rectifasciata, sp. n., 1.
„ dentifasciata, sp. n., 1.

Gen. *Trochosa*, (C. L. Koch).
Trochosa rubiginea, sp. n., 4, 3.
„ hebes, sp. n., 4, 5, 3.
„ propinqua, sp. n., 1.
„ adjaceus, sp. n., 5.
„ sabulosa, sp. n., 5, 4.
„ approximata, sp. n., 1.
„ rubromandibulata, sp. n., 4.
„ lugubris, sp. n., 5.

Gen. *Tarentula*, (Sund.).
Tarentula irascibilis, sp. n., 5.
„ inimica, sp. n., 2.

Gen. *Lycosa*, (Latr. *ad partem*).
Lycosa condolens, sp. n., 4, 3.
„ fortunata, sp. n., 2, 4, 5, 3.
„ stellata, sp. n., 5, 4, 3.
„ eredula, sp. n., 5, 3.
„ vindex, sp. n., 4.
„ vindicata, sp. n., 1, 5.
„ passibilis, sp. n., 5.
„ flavida, sp. n., 4, 5, 3.

Bœbe, gen. nov.
Bœbe benevola, sp. n., 4, 5. 3.

Fam. SPHASIDES.
Gen. *Oxyopes*, (Latr.).
Oxyopes jubilans, sp. n., 1.
„ prædicta, sp. n., 1.
„ rejecta, sp. n., 1.

Fam. SALTICIDES.
Gen. *Heliophanus*, (C. L. Koch).
Heliophanus dubius, Sim., 5.

Gen. *Plexippus*, (C. L. Koch.)
Plexippus adansonii, Sav., 1.

Gen. *Menemerus*, (Lin.).
Menemerus cinctus, sp. n., 4.
„ incertus, sp. n., 4.
„ deletus, sp. n., 3.
„ frigidus, sp. n., 1.

Gen. *Attus*, (Sim.).

Attus devotus, sp. n., 1.
,, beneficus, sp. n., 1.
,, diductus, sp. n., 1.
,, auspex, sp. n., 4, 5.
,, avocator, sp. n., 5.

SEPARATE LISTS OF SPECIES FOUND IN THE SEVERAL DISTRICTS.

N. B.—Where no figure is added, the spider was found only in the district under consideration.

DISTRICT 1.

Murree, Murree to Sind Valley, and Sind Valley.

Idiops designatus.
Dysdera cylindrica.
Drassus dispulsus, 2, 3, 4, 5.
,, involutus.
Clubiona deletrix.
,, laticeps.
Cheiracanthium adjacens.
,, approximatum.
Trachelas costata, 2.
Cœlotes tegenarioides.
,, simplex.
Episinus algiricus, (Luc.).
Theridion riparium, (Blackw.).
,, lepidum.
,, subitum.
,, confusum.
,, expallidatum.
,, tuberculatum, (Kron.).
,, incertum.
Steatoda nigrocincta, 3.
Phycus sagittatus.
Erigone dentipalpis.
Linyphia consanguinea.
,, albopunctata.
,, straminea.
Meta mixta.
Epëira bigibbosa.
,, pellax.
,, gurda.
,, punctata.
,, prædata.
,, cucurbitina.
,, panniferens.
,, carnifex.
,, gibbera.

Chorizoopes stoliczkæ.
,, congener.
Cyrtarachne pallida.
Uloborus albescens.
Misumena expallidata.
,, (?) oblonga.
Synema exculta.
Dicæa spinosula.
,, subdola.
,, sufflava.
,, subargentata.
Xysticus maculosus.
,, setiger.
,, mundulus.
Monastes dejectus.
Sarotes regius, (Fabr.).
,, promptus.
Sparassus fugax.
Philodromus medius, (Cambr.).
Thanatus albescens.
Stoliczka insignis.
Ocyale rectifasciata.
,, dentifasciata.
Trochosa propinqua.
,, rubromandibulata.
Lycosa vindicata, 5.
Oxyopes jubilans.
,, prædicta.
,, rejecta.
Plexippus adansonii.
Menemerus frigidus.
Attus (?) devotus.
,, beneficus.
,, diductus.

ARANEIDEA. 113

DISTRICT 2.

Neighbourhood of Leh, and Tanktze to Chagra and Pankong Valley.

Filistata reclusa.
Drassus troglodytes, (C. L. Koch.), 3, 4, 5.
„ interemptor.
„ dispulsus, 1, 3, 4, 5.
„ interlisus, 3, 4, 5.
Trachelas costata.

Epëira tartarica.
Sparassus timidus.
Philodromus cinerascens, 3.
Tarentula irascibilis.
Lycosa fortunata, 3, 4, 5.

DISTRICT 3.
Yárkand to Bursi.

Drassus troglodytes, (C. L. Koch.), 2, 4, 5.
„ dispulsus, 1, 2, 4, 5.
„ interlisus, 2, 4, 5.
Gnaphosa stoliczkæ, 5.
„ plumalis, (Cambr.).
„ mœrens, 5.
Clubiona laudata.
Tegenaria (?).
Steatoda nigrocincta, 1.
Drepanodus mandibularis, (Luc.).
Erigone atra, (Blackw.).
Linyphia pusilla, (Sund.), 4.
Tetragnatha extensa, (Linn.).
Epëira cornuta, (Clerck), 4.
Thomisus albidus.

Thomisus albens.
Dicæa suspiciosa.
Xysticus breviceps.
Philodromus cinerascens, 2.
Thanatus thorellii, (Cambr.).
Trochosa rubiginea, 4.
„ hebes, 4, 5.
Lycosa condolens, 4.
„ fortunata, 2, 4, 5.
„ stellata.
„ credula, 5.
„ flavida, 4, 5.
Bœbe benevola, 4, 5.
Menemerus deletus.

DISTRICT 4.
Yárkand and neighbourhood, and Yárkand.

Drassus troglodytes, (C. L. Koch), 2, 3, 5.
„ dispulsus, 1, 2, 3, 5.
„ interlisus, 2, 3, 5.
Prosthesima cingara, Cambr., 4.
Agröeca flavens.
Argyroneta aquatica, (Walck.).
Linyphia pusilla, (Sund.), 3.
Epëira haruspex.
„ cornuta, (Clerck), 3.
Sparassus flavidus.
Trochosa rubiginea, 3.
„ hebes, 3, 5.

Trochosa sabulosa, 5.
„ approximata.
Lycosa condolens, 3.
„ fortunata, 2, 3, 5.
„ stellata, 3, 5.
„ vindex.
„ flavida, 3, 5.
Bœbe benevola, 3, 5.
Menemerus cinctus.
„ incertus.
Attus auspex, 5.

DISTRICT 5.

Káshghar; between Yángihissár and Sirikol; Yángihissár; road across the Pamir from Sirikol to Punjah and back; and hills between Sirikol and Aktalla.

Drassus troglodytes, (C. L. Koch.), 2, 3, 4.
„ infletus.
„ invisus.
„ interpolator.

Drassus dispulsus, 1, 2, 3, 4.
„ interlisus, 2, 3, 4.
„ lapsus.
Gnaphosa stoliczkæ, 3.

1 D

Gnaphosa plumalis, (Cambr.), 3.
Prosthesima cingara, (Cambr.), 4.
Micaria connexa.
„ pallida.
Agröeca debilis.
Dictyna albida.
Steatoda sordidata.
Pachygnatha clerckii, (Sund.).
Xysticus cristatus, (C. L. Koch.).
„ audax, (C. L. Koch.).
Tibellus propinquus.
Trochosa hebes, 3, 4.
„ adjacens.

Trochosa sabulosa, 4.
„ lugubris.
Tarentula inimica.
Lycosa fortunata, 2, 3, 4.
„ stellata, 3, 4.
„ credula, 3.
„ vindicata, 1.
„ passibilis.
„ flavida, 3, 4.
Bœbo benevola, 3, 4.
Heliophanus dubius.
Attus auspex, 4.
„ avocator.

EXPLANATION OF THE PLATES.
Pl. I.

Fig. 1. *Idiops designatus*, sp. n., ♂.
    a. spider in profile with legs and palpi truncated ; *b.* eyes from above and behind ; *c.* palpus; *d., e.* portion of leg of first pair in different positions ; *f.* natural length of spider.

„ 2. *Filistata reclusa*, sp. n., ♀.
    a. spider in profile without legs or palpi ; *b.* eyes and falces from in front; *c.* natural length of spider.

„ 3. *Dysdera cylindrica*, sp. n., ♂.
    a. spider in profile without legs or palpi ; *b., c.* palpus in two different positions; *d.* natural length of spider.

„ 4. *Drassus infletus*, sp. n., ♀.
    a. spider in profile without legs or palpi ; *b.* eyes from behind ; *c.* genital aperture ; *d.* natural length of spider.

„ 5. *Drassus interemptor*, sp. n., ♂.
    a. spider in profile without legs or palpi ; *b.* eyes from above and behind ; *c.* palpus ; *d.* natural length of spider.

„ 6. *Drassus invisus*, sp. n., ♀.
    a. spider in profile without legs or palpi ; *b.* eyes from above and behind ; *c.* genital aperture ; *d.* natural length of spider.

„ 7. *Drassus interpolator*, sp. n., ♂.
    a. spider in profile without legs or palpi ; *b.* eyes from above and behind ; *c.* palpus ; *d.* natural length of spider.

„ 8. *Drassus dispulsus*, sp. n., ♂.
    a. spider in profile without legs or palpi ; *b.* eyes from above and behind ; *c.* palpus of ♂ ; *d.* genital aperture of ♀ ; *e.* natural length of spider.

„ 9. *Drassus interlinus*, sp. n., ♂.
    a. spider in profile ; *b.* eyes from above and behind ; *c.* palpus; *d.* natural length of spider.

„ 10. *Drassus involutus*, sp. n., ♀.
    a. spider in profile ; *b.** eyes from above and behind ; *c.* genital aperture; *d.* natural length of spider.

    * In this figure (10 *b.*), the eyes of the hind-central pair are placed too obliquely.

Pl. II.

Fig. 11. *Drassus lapsus*, sp. n., ♀.
    a. spider in profile ; *b.* eyes from above and behind ; *c.* genital aperture ; *d.* natural length of spider.

ARANEIDEA. 115

Fig. 12. *Gnaphosa stoliczkæ*, sp. n., ♂.
  a. spider in profile; *b.* eyes from above and behind; *c.* palpus of ♂; *d.* genital aperture of ♀; *e.* natural length of spider.
,, 13. *Gnaphosa moerens*, sp. n., ♂.
  a. spider in profile; *b.* eyes from above and behind; *c.* palpus of ♂; *d.* portion of palpus showing form of radial apophysis; *e.* genital aperture of ♀; *f.* natural length of spider.
,, 14. *Clubiona deletrix*, sp. n., ♂.
  a. spider in profile; *b.* eyes from above and behind; *c.* palpus of ♂; *d.* genital aperture of ♀; *e.* natural length of spider.
,, 15. *Clubiona laticeps*, sp. n., ♀.
  a. spider in profile; *b.* eyes from above and behind; *c.* genital aperture; *d.* natural length of spider.
,, 16. *Clubiona laudata*, sp. n., ♂.
  a. spider in profile; *b.* eyes from above and behind; *c.* palpus; *d.* natural length of spider.
,, 17. *Chiracanthium adjacens*, sp. n., ♂.
  a. spider in profile; *b.* eyes from above and behind; *c.* palpus of ♀; *d.* genital aperture of ♂; *e.* natural length of spider.
,, 18. *Chiracanthium approximatum*, sp. n., ♀.
  a. spider in profile; *b.* eyes from above and behind; *c.* genital aperture; *d.* natural length of spider.
,, 19. *Agroeca debilis*, sp. n., ♀.
  a. spider in profile; *b.* eyes from above and behind; *c.* maxillæ and labium; *d.* genital aperture; *e.* natural length of spider.
,, 20. *Agroeca flavens*, sp. n., ♀.
  a. spider in profile; *b.* eyes from above and behind; *c.* maxillæ and labium; *d.* genital aperture; *e.* natural length of spider.
,, 21. *Trachelas costata*, sp. n., ♀.
  a. spider in profile; *b.* eyes from above and behind; *c.* maxillæ and labium; *d.* genital aperture; *e.* natural length of spider.

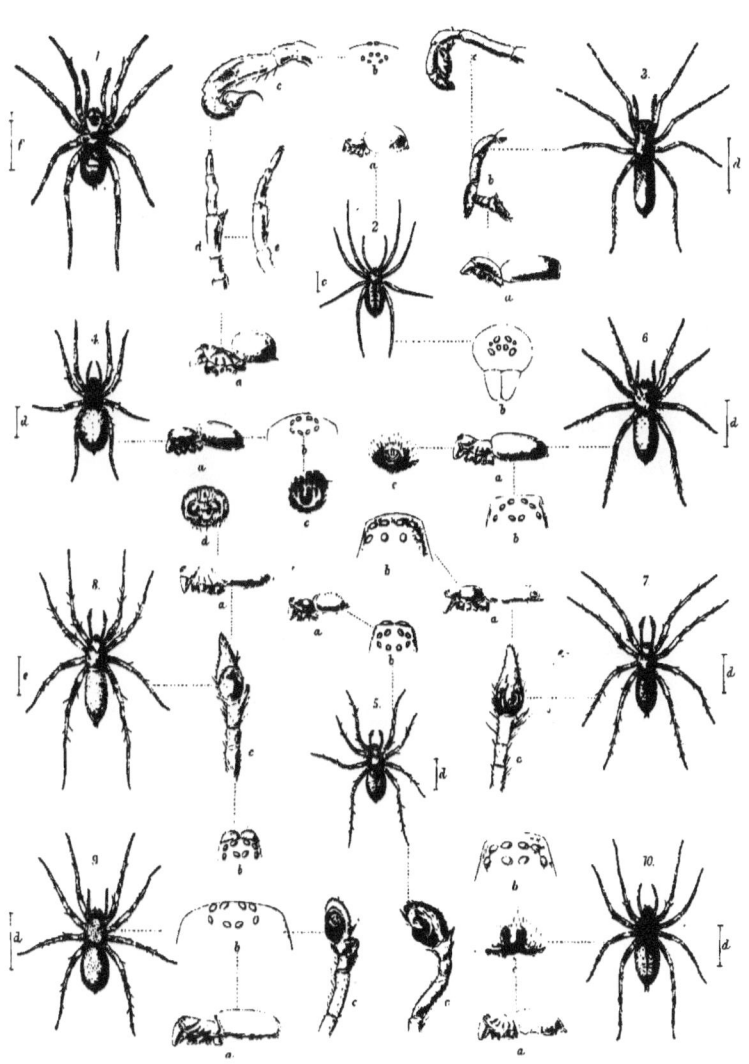

Plate II

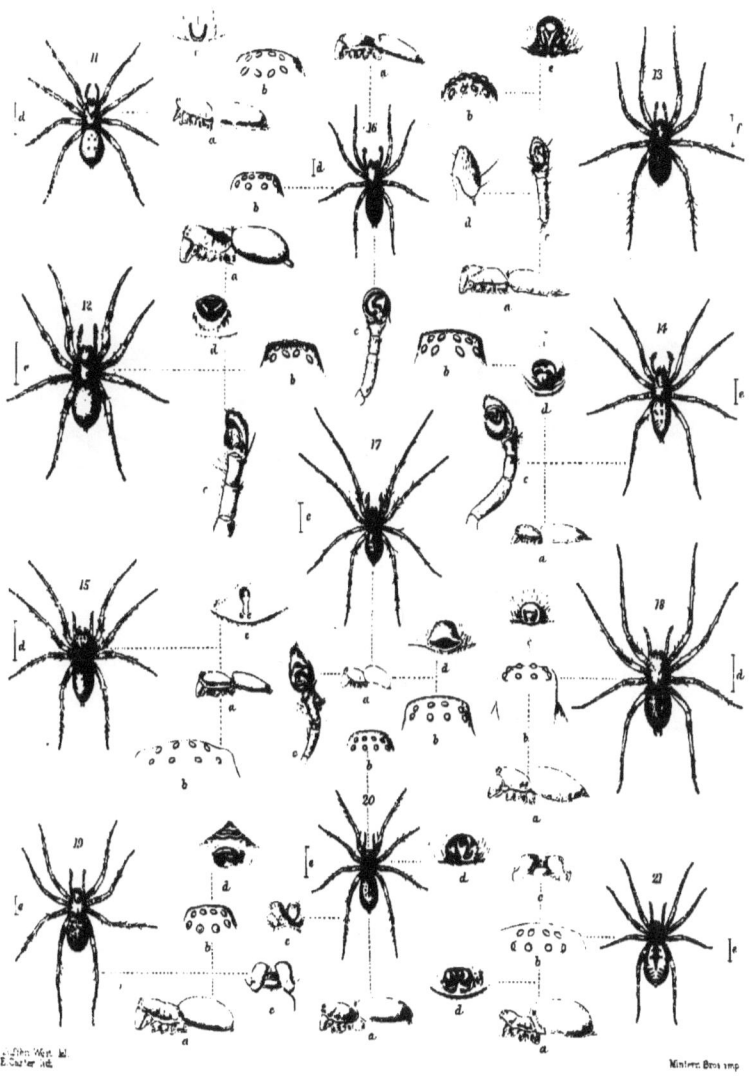

SCIENTIFIC RESULTS

OF

THE SECOND YARKAND MISSION;

BASED UPON THE COLLECTIONS AND NOTES

OF THE LATE

FERDINAND STOLICZKA Ph.D.

MOLLUSCA.

BY

GEOFFREY NEVILL, C.M.Z.S.

Published by order of the Government of India.

CALCUTTA:
OFFICE OF THE SUPERINTENDENT OF GOVERNMENT PRINTING.
1878.

SCIENTIFIC RESULTS

OF

THE SECOND YARKAND MISSION;

BASED UPON THE COLLECTIONS AND NOTES

OF THE LATE

FERDINAND STOLICZKA Ph.D.

MOLLUSCA.

BY

GEOFFREY NEVILL, C.M.Z.S.

Published by order of the Government of India.

CALCUTTA:
OFFICE OF THE SUPERINTENDENT OF GOVERNMENT PRINTING.
1878.

CALCUTTA:
PRINTED BY THE SUPERINTENDENT OF GOVERNMENT PRINTING,
8, HASTINGS STREET.

SCIENTIFIC RESULTS

OF

THE SECOND YARKAND MISSION.

MOLLUSCA.

BY GEOFFREY NEVILL, C.M.Z.S.

I.—MOLLUSCA FROM EASTERN TURKESTAN AND LADÁK.

THE following is a list of the mollusca obtained by the late Dr. Stoliczka in Central Asia and Ladák, while attached as naturalist to the second embassy to Yárkand; Dr. Stoliczka also collected a considerable number of shells in Kashmir and its neighbourhood; as, however, nearly, if not all, the land mollusca from those parts belong to our Indian fauna proper, I have thought it best to give a separate list of them. As was to be expected, the molluscous fauna of Yárkand proves to be exceedingly poor and entirely European in its affinities; the freshwater shells, indeed, are either identical with, or most closely allied to, well-known European forms; very nearly all the species are recorded from Turkestan in the account of the Mollusca of Fedschenko's 'Reise.' I take this opportunity of acknowledging the great obligation I am under to Dr. E. von Martens, not only for a copy of the above work, of which he is the author, but also for a critical opinion on the species here recorded, of which I have availed myself in several instances. The only striking novelty is the new *Succinea martensiana*: its thickness and opaqueness of texture and its vivid orange-coloured aperture ma ce it one of the most interesting and peculiar forms of the genus. It is interesting to find such characteristic shells as *Helix phæozona* and *H. plectotropis* extending southwards from Kokand and the Tian Shan Range as far as Sásak Taka; even more remarkable are the new localities for *Pupa cristata*, originally found in the Sarafshan Valley; the absence of the genus *Hydrobia* from Dr. Stoliczka's collection strikes me as noteworthy, especially as no species of *Valvata*, on the other hand, is recorded by von Martens from Turkestan. The most interesting fact, however, seems to me to be the entire disappearance, on leaving Sonamarg on the confines of Kashmir, of the characteristic Indo-Malayan genus *Nanina*, which re-appears again (with two species of the sub-genus *Macrochlamys*) in the Sarafshan Valley; the same is also the case with species of *Buliminus* (*Napæus*), *Parmacella*, and *Limax* (?); the two last, however, belong to the European fauna and species of them are mere stragglers on the extreme north-west confines of India. Stoliczka remarks that the shells recorded as found in the Pankong Lake were taken from a "stratified shaly and sandy deposit on the west side of the Pankong plain, about 50 feet above the level of the present edge of the water and about two miles distant from it;

some of the specimens of *Valvata* still have the epidermis, and it is possible that where the water of the lake is fresh, the shells may live."

The re-appearance of two of M. Issel's new species of *Limnæa* (originally described from Persia) is important, as proving the constancy of these respective forms. The same remark holds good with regard to one of my new Yunnan species.

1. VITRINA PELLUCIDA, Müll.

Shell perfectly undistinguishable from European specimens from Mennighüfen and other localities. Dr. Stoliczka had previously collected some twenty specimens of a similar form at Lahoul. Von Martens does not record the species from Turkestan, but describes a new species as *V. rugulosa*, Koch, the Latin description and measurements of which seem to agree fairly with the Mataian form; unfortunately I am unable to understand the Russian description, in which he compares his new species with *V. pellucida*. Dr. Stoliczka describes the animal of this Mataian shell as "blackish, with the tentacles very short."

Sixteen specimens from Mataian, near Drás, Upper Indus Valley: diam. 6, alt. 3¼ mm.; apert. diam. 3½, alt. 3¼.

2. HYALINA (CONULUS) FULVA, Drap.

Perfectly undistinguishable, as far at least as regards the shell, from the typical European form. Stoliczka had previously found the species in abundance at Spiti and Lahoul. Mr. Blanford also found the species at Mazendaran in Persia. Species from Pekin are well represented by Deshayes (Nouv. Archiv. Museum, vol. x, pl. 1).

Three specimens from Wakhan and three from Mataian.

3. HELIX (FRUTICICOLA) PHÆOZONA, v. Mart., Figs. 1—3.

E. v. Martens, Fedsch. Moll., pl. i, fig. 8 (Kokand).

Shell of solid texture, about the size of *H. similaris*, which indeed it somewhat resembles; umbilicate, conoidly globose, irregularly and roughly striate, decussated with almost microscopical spiral lines; straw-white, with a single, very broad brown band, just above the periphery; in a single specimen only is this band altogether absent; spire conoidal, varying in being more or less raised; whorls six, the last more or less subangulate, convex at base; aperture lunately rounded, with the peristome much thickened, and the columella exceedingly broadly reflected.

Diam. 16½, alt. 12½; apert. diam. 9, alt. 8 mm.
Depressed variety from Pasrobat; diam. 16, alt. 10½.
I ought to note that I include the margins in recording measurements of the aperture.
Twenty specimens from Sásak Taka (6,500 ft.) and five from Pasrobat, west of Yárkand.

4. HELIX (FRUTICICOLA) PLECTOTROPIS, v. Mart., Figs. 4—6.

E. v. Martens, Malakozoologische Blätter, XI, 1864, and Fedsch. Moll., pl. i, fig. 11 (Tinusohang).

Shell about the same size as the preceding; openly umbilicate, depressedly conoidal, with a raised keel which is distinctly visible to nearly the apex, sutures not excavated; beautifully and somewhat regularly sculptured, with sharp and raised oblique ribs, about half the breadth of their interstices, above of a light brown, with the keel and ribs of a straw colour, about one-fourth of the base nearest the periphery pale brown, the rest straw colour; spire depressed, convex, with brown apex, whorls six, the last one sharply and prominently keeled and more or less convex at base, aperture diagonal (produced laterally), peristome reflected, angled at the periphery, the columella, as in the preceding, exceedingly broadly expanded; the apertures of several specimens were closed with a calcareous epiphragm.

Diam. 18, alt. 10; apert. diam. 11, alt. 8¼ mm.

Twenty-five specimens from Sásak Taka found living with the preceding.

5. HELIX (FRUTICICOLA) MATAIANENSIS, n. sp., Figs. 7—9.

Shell a little smaller than *H. plectotropis*, in many respects a good deal resembling it, but of much thinner and more delicate texture; openly umbilicate, depressedly conoidal, whorls five and a half, with excavated suture and without a raised keel, in both of which respects it materially differs from the preceding, last whorls with a medium-sized keel, base convex, above sculptured irregularly, with more or less strongly developed ribs, beneath sculpture obsolete, almost smooth; white, irregularly mottled with pale horn colour, apex horn brown; aperture ovate, subangulate at periphery, almost as high as broad; peristome lightly reflected, columella expanded.

Diam. 13⅜, alt. 7 mm.; apert. diam. 6¾, alt. 6¾ mm.

Nine specimens from Mataian, in the Drás Valley, at 11,200 feet. Unfortunately most are quite young shells, only one or two being sufficiently full grown to show the reflected outer lip. Stoliczka describes the animal in his journal as "uniform greenish dusky, no trace of a tail gland, the body very short, the posterior part of the foot shorter than the anterior."

6. HELIX (XEROPHILA) STOLICZKANA, n. sp., Figs. 10—12.

Shell rather thin, about the size of *H. ericetorum* and closely resembling it, but more depressed, umbilicus slightly less open, colouration different and aperture differently shaped; openly umbilicate, flatly depressed; above irregularly striate, below sculpture obsolete; white; invariably ornamented with two striking brown bands, one of which in most, but not all the specimens, can be traced as far as the apex, the two bands are, of course, near the periphery; the space between them is about the width of the two bands together; apex bright brown; whorls five and a half with distinct suture, not keeled, convex at base; aperture as high as broad, dilated above, considerably higher than the periphery in fine full-grown specimens; peristome slightly thickened, columella moderately reflected; the aperture in many of the specimens was closed with a thin epiphragm.

SECOND YARKAND MISSION.

Diam. 16⅜, alt. 7¼; apert. diam. 8, alt. 7½ mm.
About a hundred specimens from Sásak Taka and Pasrobat, west of Yárkand.
Twelve specimens, in poor and weathered condition, from north of Tangitar on carboniferous limestone; they are a remarkably small variety, about half the typical size, the two bands are scarcely discernible, and they are not quite so flat.

7. HELIX (VALLONIA) COSTATA, Müll., var. ASIATICA, nov.

This is probably the variety recorded by von Martens from Turkestan, measuring 3 mm. in diam. and 1¼ in height; it only differs from the typical European form by its larger size and slightly stronger subangulation at base near the umbilicus. More than a hundred and fifty specimens were collected by Dr. Stoliczka, all of approximately the same size, at Pasrobat, Sásak Taka and Wakhan; one of the specimens from the last locality I have taken as my type of var. *asiatica*.

I take this opportunity of noting that Mr. W. T. Blanford brought back from Mazendaran in Persia numerous specimens of a variety, the same size as the European form and with similar sculpture, but having the umbilicus a shade more open.

8. HELIX (VALLONIA) LADACENSIS, n. sp. (*an H.* COSTATA, var.?)

I have long separated this form, which can be distinguished from all the varieties of *H. costata* at a glance by its much more open umbilicus, at least half as open again; it is a much larger shell than typical *H. costata*, about the same size (a trifle larger) as the above described var. *asiatica*; the sculpture is finer, closer together and more beautifully regular; the spire is flatter, the suture more excavated; the base is scarcely, if at all, subangulate near the umbilicus, as it is in so marked a way in the preceding; one of the best characteristics of *H. ladacensis* is the considerably higher and more expanded aperture with a corresponding less oblique columella; the umbilicus is so much more open that the whorls within can be clearly traced up to the apex itself.

Diam. 3¼, alt. 1⅜ mm.
Type from Mataian in the Drás Valley (Ladák), where Dr. Stoliczka found about sixty specimens. One of the specimens I sent Dr. von Martens from this locality possesses, he informs me, a "little plait on the wall of the mouth." Unfortunately I have not been able myself to detect this plait in any other specimens. Ten specimens were brought from Leh (chief town of Ladák); twenty from "Narka" (?) in West Tibet, slightly smaller and with more raised spire than Mataian specimens.

9. PUPA (PUPILLA) MUSCORUM, L.

Fourteen specimens from Pasrobat, 3¼ mm. in length; fifty from Kaskasu, 3¾ mm. in length; fifty from shores of Lake Pankong, a form remarkable for its produced whorls, 3¾ to 4¼ mm. in length; twenty from Spiti,[1] agreeing with the preceding form, in the great difference in the length of the spire in different individuals, the whorls being sometimes much produced, at other times curiously shortened and compressed; four specimens from Mataian, one only perfect unfortunately. I have considerable doubts in referring this Mataian form to

[1] Procured by Dr. Stoliczka on a previous visit to the Himalayas.

P. muscorum at all, the spire is less produced, striation less developed, form of aperture simpler and less angular; length 2¾ mm.; no tooth.

Not a single one of the Ladák specimens possesses even a rudiment of a tooth on the wall of the aperture, nor have I been able to detect any in the Kaskasu form; in one or two of the Pasrobat shells only is a very slight tooth just discernible; as far as I have seen, this absence of the tooth appears to be characteristic of our Asiatic forms.

10. PUPA (PUPILLA) CRISTATA, v. Mart.

E. v. Martens, Fedsch. Reise, Moll. pl. ii, fig. 19 (Sarafshan Valley).

The specimens of this very distinct and interesting form agree exactly with typical figures 19C. and E. Figure B, on the contrary, has the spire a little more produced, with the whorls a trifle more convex, and the aperture slightly more contracted, the margins of which, in our Museum specimens, are somewhat considerably more delicately dilated; I can only detect, after a most careful search under the lens, a single tooth on the outer margin, as in the above figure C, not two, as in the description and figure B.

Shell ovate, rimate, of horny brown colour, obliquely slightly striated, apex obtuse; seven whorls, the 4th, 5th and 6th of equal width, the last one somewhat compressed at the base, with an obtuse keel round the umbilicus continued more prominently in a raised ridge, parallel with the outer margin of the aperture; aperture small and rounded, with the peristome broadly reflected; a prominent tooth on the wall of the mouth, a single fold on the columella (lying rather far back) and a single obtuse tooth within the outer margin. Long. 3½, diam. 2 mm.

Eleven specimens from Sásak Taka, where it is by no means common; sixteen from Pasrobat, where it occurs more abundantly in company with *P. muscorum*.

11. SUCCINEA MARTENSIANA, n. sp., Figs. 30-31.

Shell unusually thick, about the size of *S. girnarica*, Theob., (Conchologia Indica, pl. lxvii, fig. 6,) which it at first sight much resembles; it is, however, quite half as thick again, of much intenser colouring and of more convexly shaped whorls; whorls four, convex, produced and separated: in *S. girnarica* there are only three, which increase less rapidly and are less obliquely inclined; the last whorl of the Yárkand species is shorter and not nearly so ovately oblong; the texture is more rugose, the irregular longitudinal furrows being unusually strongly developed; the colouration is peculiar, being of an opaque milky white, more or less purple near the apex; the aperture is internally of a brilliant orange colour and more laterally expanded than in *S. girnarica*: the columella varies, but is always straighter than is the case in its ally; the callosity joining the columella and outer lip is strongly marked.

S. martensiana (type), long. 17, diam. 11; apert. long. 12, diam. 8¼ mm.

S. girnarica, long. 18½, diam. 11¾; apert. long. 14, diam. 9¼ mm.

This species is very variable in shape; the Museum possesses a very fine series of it, all from Kathiawad (Kattywar).

I have named this handsome species after Dr. E. von Martens of Berlin, to whose great kindness, in sending me a critical opinion of these Yárkand shells, I am so much indebted.

Of the Yárkand species, Dr. Stoliczka found about fifty specimens, in all stages of growth, at Sásak Taka, many of them alive; also about twenty at Pasrobat.

12. SUCCINEA PFEIFFERI, Rossm., var.

This Yárkand variety is only distinguishable from typical European specimens by its smaller proportions, slightly stouter texture, and deeper amber colour.
Long. 11, diam 6; apert.long. 7½, diam. 4 mm.
Ten specimens from Yárkand and nine from Sásak Taka.

SUCCINEA PFEIFFERI, var. SUBINTERMEDIA, nov., Figs. 32-33.

From near Yárkand, Dr. Stoliczka also brought back about twenty specimens of a small form, easily distinguishable from the preceding by its less everted last whorl, thinner texture and lighter colour; it is in some respects intermediate between $S.$ $putris$ and $S.$ $pfeifferi$, but its more produced spire seems to me to compel its classification with the latter; the nearest European form we possess in the Museum is a Transylvanian shell sent from Germany as $S.$ $amphibia$ ($putris$) var. $intermedia$. The Museum also possesses three specimens from Candahar, presented by the late Captain Hutton, in no way to be distinguished from the Yárkand form, except in being about half as large again; the columella is less rounded and decidedly more subangulate at the base, than in German and French specimens. I found a variety, however, from England agreeing in this respect with our Asiatic forms, though the spire is less produced in the latter; it seems to me that the transition as regards the shell itself from $S.$ $putris$ to $S.$ $pfeifferi$ is almost, if not quite, imperceptible ?
Long. 11, diam. 5¾; apert. long. 7¾, diam. 4½ mm.

13. SUCCINEA PUTRIS, L. var.

About forty specimens were found living on grass in a marsh near Yárkand city; it is a small, thin and glassy variety, resembling in miniature a form from Wales sent me by the late Mr. F. Layard as $S.$ $putris$, L., var. $vitrea$; its more swollen shape, less produced spire and more everted last whorl distinguish it from the form I have described above as $S.$ $pfeifferi$, var. sub-$intermedia$; its more globose shape, less produced spire, thinner and more vitreous texture from my var. $yarkandensis$.
Long. 10; diam. 6; apert. long. 7, diam. 4½ mm.

14. LIMNÆA AURICULARIA, L., var.

This form agrees fairly with Kobelt's figure (Mal. Bl., 1870, pl. lii, fig. 8, $L.$ $auricularia$, var. $ventricosa$; London); the principal difference is the apparently constantly more broadly reflected columella, which is also more rounded at the base; the great tendency to deformity in the Sirikul specimens is very striking; it appears to me that this form would be almost as well classified as an extreme variety of $L.$ $lagotis$, allied to var. $obliquata$.

Long. 23, diam. 10¾; apert. long., 18½ diam. 14 mm. columella, at junction with body whorl, 2 to 2¾ mm. in breadth.

About 20 specimens (dead) on the shore of Lake Sirikul or Victoria, Pámír.

Another variety is smaller and more delicate than the above, but with the same remarkably thickened and rounded columella, as is well represented on pl. ii, fig. 20, "Fedsch. Moll."; the spire, however, in the Aktásh specimens is more prominent and the broadly reflected columella even more marked.

Long. 16¾, diam. 13; apert. long. 13, diam. 10 mm.

A deformed specimen measures long. 12½, diam. 12 mm.

About 30 specimens were taken alive in a stream at Aktásh (Sarikol).

15. LIMNÆA DEFILIPPII, Iss., var. SIRIKULENSIS, nov.

Issel, Moll. Persia, 1865, pl. iii, figs. 26 & 63 (Lake Gokcha, 5,500 feet).

This is perhaps the most remarkable of the Yárkand species of *Limnæa* and the furthest removed from the typical forms of *L. auricularia* and *L. lagotis*, even more so than typical *L. defilippii*. As justly pointed out by Issel, it is intermediate between the above group and that of *L. stagnalis*. It differs from Issel's figure by the much more swollen, subangulate whorls, and by the shorter, not twisted and evenly rounded columella; the produced spire and malleated texture are very characteristic of both.

I had already written the following description before I read that of Issel.

Shell in size intermediate between *L. stagnalis* and *L. lagotis*; of moderately thin texture, the same as in *L. stagnalis*; spire much more produced than in *L. lagotis*; whorls six, remarkably subangulate; aperture expanded as in fig. 10, pl. ii, "Mal. Bl.," 1870; columella broadly reflected, almost completely covering the umbilicus, not twisted in the least, evenly rounded at base as in fig. 9 (*loc. cit.*); very young specimens present a remarkably close resemblance to those of *L. stagnalis*, the subangulation of the whorls and short, straight columella being naturally less distinctive than in full-grown specimens; the surface of most specimens is more or less roughly decussately malleated; under the lens a very fine and close longitudinal striation can be seen.

Type of var. *sirikulensis*: long. 30¼, diam. 21; apert. long. 20, diam. 14¼ mm.; the ante-penultimate whorl measured from the outer lip 6¾ mm.; a young specimen measured long. 24¾, diam. 14; apert. lat. 14, alt. 9¼.

Fourteen dead specimens found, on the shores of Lake Sirikul, in company with *L. auricularia*, var.

16. LIMNÆA LAGOTIS, Schr.

Limnæa lagotis, Schr., Fauna Boica, iii, 1803.

L. lagotis, var. *solidissima*, Kobelt, Malakozoologische Blätter, 1872, pl. ii, figs. 17 & 18.

L. obliquata, v. Mart., Mal. Bl., 1864, pl. iii, figs. 9 & 10 (Lake Issik-kul).

A fine series of this remarkable variety was procured by Dr. Stoliczka in all stages of growth; it varies greatly in the more or less produced spire, though never, even in its most elongated form, approaching the preceding form; there is little, if any, trace of the malleated sculpture, often so characteristic of the preceding: the same fine longitudinal striation however exists; all the specimens, young and old, are without exception of the peculiar thickness which suggested its excellent name of *solidissima*; the five whorls agree with those of Kobelt's original figure, which I suspect was taken from a Lake Pankong specimen, and do not show the subangulation described in the preceding; the aperture is much more expanded than in Kobelt's typical figure, which was evidently taken from a rather young shell, the very thick columella in most specimens agrees with that of the type, but in some few it is abruptly twisted back, as in pl. ii, fig. 21 of "Fedsch. Moll." (*L. obliquata*, v. Mart.)

These specimens are interesting as removing one of the few slight differences between *L. obliquata* and *L. solidissima*; my Pankong specimens clearly show the more expanded aperture to be merely a question of age and condition, as is also the gradual slope of the outer lip; nearly all my specimens agree in this latter respect with typical *obliquata*, only very few showing the angular outer lip of typical *solidissima*; Kobelt in his description pointed out the close affinity of the two forms, and also that *L. obliquata* must be classed rather with *L. lagotis*, than *L. auricularia*; the shortened columella seems to me the best characteristic of the latter group, as shown in the form I have already described as a variety of that species; the difference is also excellently portrayed in von Marten's figures, pl. ii, figs. 20 and 21, "Fedsch. Moll." The Pankong shell, though always preserving its chief characteristics, varies most remarkably, as will be seen from the accompanying measurements.

The ordinary form:—long. 22, diam. 17½; apert. long. 18, diam. 12½ mm.

A form with more produced spire and contracted aperture, agreeing with Kobelt's figure: —long. 22, diam. 15½; apert. long. 15¾, diam. 10¼ mm.

A unique form, with quite depressed spire:—long. 19, diam. 15½; apert. long. 17¼, diam. 12 mm.

A form (represented by six or seven specimens), with unusually expanded and more rounded aperture:—long. 18, diam. 17; apert. long. 14½, diam. 12 mm.

About a hundred specimens from the shores of the Pankong Lake: both young and old specimens show the same peculiar thickness of shell.

LIMNÆA LAGOTIS, var. COSTULATA.

Limnæa lagotis, var. *costulata*, v. Martens, Fedsch., Reise, Moll., pl. ii, fig. 24.

More than a hundred specimens were collected by Dr. Stoliczka at Leh, agreeing exactly with figs. 22 and 24 (*loc. cit.*). I cannot consider the forms there figured as belonging to even different varieties; there are numerous individuals amongst the Leh specimens of all the forms and of every conceivable connecting link; the variety, as I understand it, appears to be fairly constant as regards size and colour; the spire, too, does not appear to vary much more than in the figures quoted; the columella, however, graduates from even a more rounded shape than in figure 22 B to the straight (or slightly bent back) form of figure 24 A.

Long. max. 18¾, diam. 12; apert. long. 13¾, diam. 8¾ mm.

LIMNÆA LAGOTIS, var. YARKANDENSIS, nov.

This is a striking and handsome form, close to the preceding, but half as large again, with more produced spire, of five less convex whorls, much stouter texture and straighter, more evenly rounded columella, which is very broadly reflected; these characters of the columella appear to be its only marked difference from the European form figured by Kobelt, "Mal. Bl.," 1870, pl. iii, fig. 9.

About forty specimens from Yárkand and from near Sásak Taka, on the road to Sarikol; fourteen specimens from North Tangitar, of even stouter texture than the preceding; twenty specimens from a marsh, 5 miles west of Panjah, in Badakshán; this is a shorter, dwarf form.

Type of var. *yarkandensis* (from near Sásak Taka): long. 22, diam. 15¼; apert. long. 16, diam. 10½ mm.

LIMNÆA LAGOTIS, var. SUBDISJUNCTA, nov.

More than a hundred specimens from the neighbourhood of Leh; shell smaller even than var. *costulata*, of a peculiarly dark horn colour; whorls four to five, more convex and generally a little more produced, though varying in this respect, than fig. 22 B (*loc. cit.*); aperture unusually narrow, especially above; columella sharp, scarcely reflected, almost or altogether detached from the body whorl, and continuous with the outer lip, in consequence of this peculiar character the variety is always more or less openly umbilicate.

Typical and ordinary form of the variety: long. 11, diam. 7; apert. long. 7¾, diam. 5 mm.
An extremely elongate form: long. 12, diam. 6½; apert. long. 6¾, diam. 4½ mm.
A depressed form: long. 10¼, diam. 7; apert. long. 8, diam. 5 mm.

17. LIMNÆA ANDERSONIANA, Nov.

This interesting small species, which I have described in my paper on the mollusca brought back by Dr. Anderson from Yunnan and Upper Burma, is probably the form mentioned in the systematic list of the "Conchologia Indica" as *L. marginata*, Mich., from the Shan Provinces; at least Mr. Theobald gave me a single specimen from the Shan States agreeing exactly with typical specimens of *L. andersoniana* from Nantin (Yunnan). Dr. von Martens by letter informs me that my Yárkand specimens belong to his "*L. pervia*, which again is the *L. davidi* of Deshayes from Tibet." I cannot, however, accept this identification as the original description throughout makes a great point of the open umbilicus, which it compares with that of *L. truncatula*, also stating that it is only half covered by the dilated columella. Out of several hundred specimens from Yárkand and Yunnan I am unable to discover a single specimen with what could be called an open umbilicus; they all have it almost, and generally quite covered with the very broadly reflected columella.

More than a hundred specimens, of a rather distinct variety, from North Tangitar and Káshghar; with distinctly rimate aperture and spire more produced, whorls more convex than in the typical Yunnan form, columella not so short or straight, and less thickened. This must be the form I suppose nearest *L. pervia* ?

Long. 11¼, diam. 7; apert. long. 7⅛, diam. 5 mm.

About a hundred specimens from Yárkand; after a most careful examination quite undistinguishable from the Yunnan type specimens: the umbilicus is completely covered.

18. LIMNÆA TRUNCATULA, Müll.

About thirty specimens from Leh, agreeing fairly with pl. ii, fig. 26 of 'Fedsch. Reise Moll.' Dr. Stoliczka on a former visit to the Himalayas, found a still more produced form abundant at Spiti; also a shorter form at Kulu, Kotegarh, &c.

19. LIMNÆA LESSONÆ, Iss.

Issel, Moll.. Persia, 1865, pl. iii, figs. 64—66.

I cannot separate this form, even as a variety, from Issel's Persian shell, for specimens of which, from Karmán (Persia), I am indebted to Mr. W. T. Blanford. Dr. Stoliczka collected some fifty specimens of an almost perfectly similar form in a stream east of the Pamír-kul; they are like the type form imperforate, with similar short spire and rather expanded aperture. The Pamír specimens are of rather thicker substance; the characteristic orange colour is also more marked.

Long. 8, diam. 5⅞; apert. alt. 5¾, lat. 3¾ mm.

20. PLANORBIS (GYRAULUS) ALBUS, Müll., var.

More than a hundred specimens were found on the shores of Lake Pankong; they consist mainly of two forms, apparently equally plentiful, one with a more narrow umbilicus than in any European specimens I have seen, in this respect agreeing with some varieties of *P. convexiusculus*, Hutt., and with pl. iv., fig. 35, "Mal. Bl.," 1875 (*P. riparius*); in other respects, however, resembling figs. 1—3, *loc. cit.*, of typical *P. albus*: diam. 4⅜, alt. 1½ mm.

The other, with more open umbilicus, agreeing with figures 4—6 and 10—12, *loc. cit.*, intermediate between the two: diam. 5, alt. 1½ mm.

There are also two specimens with very open umbilicus, more so than in fig. 14, in other respects more like *P. lævis*: diam. 6¼, alt. 1½ mm.

Two or three deformities were also found, in which the last whorl is completely detached and the spire curiously raised, presenting some analogy to specimens of *Valvata*.

From Leh, also, some hundred specimens were brought of a form agreeing exactly in colour and every other respect with figs. 1—3. Mixed up with them equally abundantly was another allied form, which however, I have classed separately as *P. lævis*, var.

More than a hundred specimens were collected at Yárkand; the majority fairly represented by figs. 4—6, *loc. cit.* Some few however, have the last whorl near the aperture considerably deflected, as in figs. 15 and 21; the umbilicus varies in being a little more or less open. Nine specimens from 5 miles west of Panjah (Badakshan); they agree fairly with the preceding Yárkand form.

21. PLANORBIS (GYRAULUS) LÆVIS, Ald., var. LADACENSIS nov.

Planorbis lævis, Alder, Trans. Nat. Hist. Northumb., 1830.
———— *glaber*, Jeffr., Trans. Linn. Soc. Lond., 1830.

I confess I am unable to distinguish quite satisfactorily the differences between this species and the preceding. This Leh form, in any case, seems fairly separable from all the others brought back by Dr. Stoliczka; it differs mainly in two respects, colour and shape of the aperture, in the latter respect agreeing with pl. iv, figs. 10—12, "Mal. Bl." xxii, (*P. lævis*, Ald.)—shell resembling the above figures, but of a rich chestnut brown, and with the umbilicus a little more open; the aperture is considerably more laterally expanded than in the forms I have grouped under *P. albus*, and consequently relatively not so high.

Diam. 6, alt. $1\frac{1}{2}$ mm.
About a hundred specimens from Leh.

22. PLANORBIS (TROPIDISCUS) SUBANGULATUS, Phil., var.

Planorbis subangulata, Phil., "Moll. Sicil." 1844, pl. xxi, fig. 6 (Sicily).

Four specimens only were found at North Tangitar; the form is a very remarkable one, and may, I think, prove to be new; it is very different from Persian specimens of *P. subangulatus*, as also from European *P. marginatus*; the angulation is less distinct than in the former, the whole shell more compressed and flattened out, the spire showing distinctly all five whorls; the aperture is more contracted, and the under side less deeply sunk.

Pl. iii, figs. 23-24, "Malakozoologische Blätter," 1875, gives an almost exact representation of the form; the shape of the aperture is quite different from that of fig. 22, being higher than the body whorl and not bent down; of course these figures are magnified views of a minute and quite different species; a fair idea of the shell may, however, be obtained from them.

Diam. 8, alt. $1\frac{3}{4}$ mm.
Persian specimens of *P. subangulatus* measure—diam. $7\frac{1}{2}$, alt. 2 mm.

23. PLANORBIS (SEGMENTINA) NITIDUS, Müll.

Planorbis nitidus, Müller, Hist. Vermium, p. 163.

Twelve specimens of a small form from Yárkand.

24. PLANORBIS (HIPPEUTIS) COMPLANATUS, Lin.

Planorbis fontanus, Lightf. (England).

Ten specimens were found with the preceding at Yárkand; they are also a small variety.

25. PLANORBIS (ARMIGER) NAUTILEUS, Lin.

(*Fide* Westerl., Mal. Bl., 1875, p. 115 = *P. crista*, Lin., var.)

I detected seven specimens of this interesting minute form inside the apertures of the Yárkand specimens of *Limnæa*; the margins of the aperture are continuous; I can detect no signs of transverse ribs, and the form is most certainly specifically distinct from my English specimens of *L. crista*, L., as represented in "Malakozoologische Blätter," pl. iv, figs. 25—27; the Yárkand shells agree very fairly with figs. 28-30, *loc. cit.*

Diam. $2\frac{1}{4}$ mm.

26. VALVATA PISCINALIS, Müll.

Nerita piscinalis, Müller, Hist. Verm., p. 172.

About thirty specimens from the Pankong Lake, quite undistinguishable from European specimens.

27. VALVATA STOLICZKANA, n. sp. Figs. 34–36.

This is a distinct and interesting new species; in its size and depressed form it resembles *V. depressa*, C. Pfr., Küster, pl. xiv, figs. 20 & 21; it can be at once distinguished from it by the remarkably deep and narrow umbilicus, only half as open as that of Pfeiffer's shell. There are four whorls, which are slightly subangulate, forming a faint depression near the suture; under the lens it is distinctly, closely and regularly striated; the colour is a light glossy green, the aperture is not perfectly circular and is not quite so broad as high.

Diam 4, axis 1¾ mm.

Abundant at Yárkand.

28. PISIDIUM, n. sp.

It is a great pity that the figures in Clessin's new monograph of *Pisidium*, in Küster's edition of the "Conchylien-Cabinet," are so bad as to be almost without exception perfectly unrecognizable; a glance at Baudon's figures, "Monog. Pisidies Francaises," published in 1857, will show the great inferiority of the former; the shell described by Clessin as *Corbicula* (?) *minima* in "Fedsch. Moll.," pl. iii., fig. 30, is a most remarkable form, and I hope Dr. von Martens will give us further and more correct information as to its proper classification.

The present species bears a close resemblance to European forms of *P. pulchellum*; it is certainly not allied even to the species represented in Fedschenko's Mollusca; the form is well characterized by its obtuse and tumid umbones, by its extreme shortness, by its distinct concentric sculpture, and by its light-grey (cineraceous) colour; it somewhat resembles Baudon's pl. i, fig. E (*P. obtusale*), but is less extremely tumid, and not so high, compared with its breadth; compared with pl. iii, fig. D, *loc. cit.*, it is not so high, more tumid at the umbones, which are less central, and Baudon's shell is apparently smooth; the position of the umbones is exactly represented by pl. ii, fig. H. (*P. limosum*), *loc. cit.*, from which indeed the Yárkand shell would seem to be scarcely separable.

Diam. 3, alt. 2½, crass. 2⅛ mm.

Abundant at Yárkand.

29. PISIDIUM, n. sp.

This is a very small, almost circular species, flatter than the last when of the same size and with the umbones less tumid and more central; the sculpture is the same; it is more tumid and less polished than the next form, with the sides less produced and more rounded, the umbones more central; it has more the shape of Baudon's pl. III, fig. D, than the last species has.

Diam. 2¼, alt. 2, crass. 1¾ mm.

About a dozen specimens from Yárkand.

30. PISIDIUM, n. sp.

This small form is quite distinct from the two preceding; it can be at once distinguished by its great flatness, by being more broadly truncate anteriorly, more produced posteriorly, by its very flatly appressed umbones and by its polished glabrous surface; it resembles Baudon's pl. ii., fig. E, (*P. thermale*, Dup.), and also somewhat "Fedsch. Moll.," pl. iii., fig. 33, though apparently the latter does not possess the characteristic appressed umbones.

Diam. 3, alt. 2½, crass 1½ mm.

Only two or three specimens from Yárkand.

31. PISIDIUM OBTUSALE, Pfr.

Agrees fairly with Clessin's figure of *P. obtusale*, *loc. cit.*, pl. ii., fig. 22.

Diam. 4½, alt. 3¾, crass. 2⅔ mm.

About twenty specimens from Pankong Lake.

II.—MOLLUSCA FROM KASHMIR AND THE NEIGHBOURHOOD OF MARI (MURREE) IN THE PUNJAB.

THE change from the Indo-Malayan to the so-called European molluscous fauna at the northern watershed of the Kashmir Valley is most abrupt and distinct; every species found at Sonamarg belonging to the former, while, at only two days' march from thence at Mataian, every shell belongs to the latter, as already above recorded. Major Godwin-Austen, who has personally visited the locality, has been kind enough to inform me that it is on crossing the pass called the "Zoji-la" into Drás, that the change becomes at once very great, the aspect of the country entirely changing, the forest-clad hills of Kashmir disappear, and, instead, one enters a sterile, dry country of higher elevation, altogether Tibetan in character; Sonamarg is within the drainage of the River Jhelum, whilst Mataian, on the other hand, is within that of the River Indus.

1. HELICARION AUSTENIANUS, n. sp., Figs. 22—24.

This is a very distinct and peculiar form, well distinguished from all other Indian species; it is most like a dwarf *H. flemingi*, from which it is distinguished by its short, almost globose form, &c.

Shell much smaller than that of *H. flemingi*, more globose, suture more excavated, and the spire more raised, apex more distinct; more rudely and regularly concentrically plicated; whorls five, more convex, the last one not nearly so much dilated; texture thinner and more membranaceous, of an equally dark, but brighter and more glossy colour; aperture about as high as broad; base a shade more convex, imperforate; columella less oblique, very short and abruptly triangularly reflected.

Diam. $15\frac{1}{2}$, axis $7\frac{1}{2}$; apert. lat. $9\frac{1}{2}$, alt. $9\frac{1}{2}$ mm.

Some dozen specimens, several of which are preserved with the animal in spirit, were brought back from Sonamarg.

2. HELICARION FLEMINGI, Pfr.

Vitrina flemingi, Pfr., P. Z. S., 1856, p. 324 (Sind).

Young specimen, of approximately same size as full grown *H. austenianum* (for comparison): diam. $14\frac{1}{2}$, axis $5\frac{3}{4}$, alt. max. 9; apert. lat. $8\frac{3}{4}$, alt. $8\frac{3}{4}$ mm.

Dr. Stoliczka found this fine species tolerably abundant at Murree and Tinali. There are several specimens with the animal in spirit.

Diam. 40, axis 12, alt. max. 23·5; apert lat. 25, alt. 20 mm.

MOLLUSCA. 15

3. HELICARION STOLICZKANUS, n. sp., Figs. 19—21.

Vitrina monticola of Reeve and Conchologia Indica, not Pfr.
(?) *Vitrina* sp., from Almora, Bens., J. A. S. B., VII, p. 214.
(?) *Vitrina monticola* of Benson in MSS., not of Pfr.

This shell is a close ally of *H. cassida*, and might indeed be ranked as a smaller variety, with less exserted whorls and with a rather differently coloured epidermis; the close relationship was noted as above by Benson, and is well shown by Reeve, figs. 10 and 11, and by Hanley, pl. clii, figs. 1—4, who represent both species side by side, no doubt purposely. A comparison of these figures with Pfeiffer's original description, as detailed here under the next species, at once shows that the two belong to totally different sections of the genus. I have discovered a very similar misunderstanding with *Nanina petrosa*, Hutton, originally described from Mirzapur. On Benson informing Hutton that his Mirzapur *N. petrosa* was only the Calcutta *N. vitrinoides*, the latter transferred his name of *N. petrosa* to an undescribed Himalayan allied smaller form, the animal of which he knew to be distinct. Benson was wrong; Hutton's species from the Rájmahál Hills (Bhágalpur, Mirzapur, &c.), proves quite different, both as regards shell and animal, from the Calcutta form, and of course retains its name *N. petrosa*. It is well and correctly figured in the "Conchologia Indica," pl. lxxxviii, figs. 7 and 10, where our common Calcutta *N. vitrinoides* is not represented at all. I think it very likely something similar may have happened, causing the confusion of this *Helicarion* and the next species; some one may have pointed out that Pfeiffer's flat and depressed shell was only a variety of Benson's *H. scutella* from Teria Ghát, whereupon the name of *monticola* was transferred to the other North-West form, which had previously not been distinguished by a separate name from *H. cassida*, though probably the allied form from Almorah referred to by Benson in the original description (J. A. S. B., VII, p. 214). Indeed from this passage I conclude Benson's manuscript name of *monticola* really referred to this shell, and not to the species described as such by Pfeiffer. This would account for this form being named *monticola* in Cuming's collection, and hence figured for it by Reeve and Hanley; Pfeiffer's actual type of *monticola* should be looked for in the Cumingian collection, amongst the variety of *Vitrina scutella* from the North-West Himalayas. Benson probably, when describing his *Vitrina scutella*, did not compare it with Pfeiffer's *monticola*, because he assumed the latter to be his own true manuscript *monticola*, and not the flat-whorled, depressed shell Pfeiffer really described for it, and which Benson considered (possibly correctly) to be a variety of his Teria Ghát *scutella*.

Dr. Stoliczka found a single specimen at Tinali. I have not taken this specimen as my type, but one of the common Naini Tál specimens, represented in most collections. Type from Naini Tál: diam. 22, axis 8, alt. 13; apert. lat. 14¼, alt. 12 mm.

4. HELICARION MONTICOLA, Pfr.

Vitrina monticola, Pfr., P. Z. S., 1848 (Landour, Almorah, &c.)
Vitrina scutella (pars), Bens., Ann. & Mag. Nat. Hist., 1859, ser. 3, vol. iii, p. 188 (Khási Hills and Kashmir).

Unfortunately, in his original description of *H. scutella*, Benson does not say whether he takes the Khási or Kashmir form for his type; the two must, I believe, be specifically separated. If, however, they should prove identical, the *scutella* of Benson will be a synonym of *monticola*. According to the "Conchologia Indica," the type form of *H. scutella* is from the Khási Hills, and the variety from Kashmir; after a careful consideration of the original description, I think Mr. Hanley is correct in this view. Instead of $3\frac{1}{4}$, *H. monticola* has $4\frac{1}{2}$ whorls, which increase more regularly than in *H. scutella*; the colour is of a greenish-brown, instead of bright green; the apex less acute; the aperture much higher in proportion to its breadth; the columella not oblique at all, almost straight and rounded at the base. This species is found abundantly everywhere throughout the North-West Himalayas in company with the preceding.

Specimen from Murree: diam. $16\frac{1}{2}$, axis $5\frac{3}{4}$, alt. $8\frac{1}{2}$; apert. alt. $10\frac{1}{2}$, alt. $10\frac{1}{4}$ mm.

Pfeiffer's original measurements of *H. monticola* are:—diam. maj. 18, alt. $7\frac{1}{2}$ mm. This is evidently an even more depressed form than the one here recorded from Murree, and does not at all agree with the preceding species, which possesses moderately exserted whorls and has been figured by both Reeve and Hanley for *H. monticola*; the latter author's figure measures:—diam. $20\frac{1}{2}$, alt. 13 mm. Pfeiffer's description, too, suits this shell, and not the preceding, when he says, "*Depressa, &c., spira plana; anfract. 4, celeriter accrescentes planiusculi, ultimus depressus, non descendens*, &c."

5. NANINA (ROTULA) CHLOROPLAX, Bens.

Helix chloroplax, Benson, Ann. & Mag. Nat. Hist., 1865, ser. 3, vol. xv, p. 14 (near Simla).

Found abundantly near Murree, agreeing exactly with the original description and the figure in "Conchologia Indica," pl. xxxii, figs. 1 and 4.

A few of the specimens found were larger than the type, which was only 8 mm. in diameter. Diam. max. 11, axis 5, alt. 6; apert. lat. 6, alt. 4 mm.

6. NANINA (ROTULA) KASHMIRENSIS, n. sp., Figs. 13—15.

Shell small, closely resembling the preceding, from which it can, however, be easily distinguished by its smaller size, less depressed shape, much more closely wound whorls, higher spire and less acute keel; by the more convex base, which does not possess the excavated depression round the umbilicus so characteristic of its ally; the umbilicus itself also is smaller; the sculpture is apparently the same, above subplicately striate, below the same but less developed than above. I think both should rather be described as most minutely punctuate, rather than "*tenuissime decussata*" as in the original description of *N. chloroplax*. The aperture is quite different, being much less dilated in the present species, with scarcely any trace of the acute angulation in the middle of the outer margin, and with the columella less oblique and more rounded at the base. Full-grown type of *N. kashmirensis*, diam $7\frac{1}{2}$, axis $3\frac{3}{4}$, alt. $4\frac{1}{2}$; apert. lat. $3\frac{1}{4}$, alt. 3 mm. Young specimen of *N. chloroplax* (for comparison) : diam $7\frac{1}{2}$, axis $3\frac{1}{4}$, alt. $4\frac{1}{4}$; apert. lat. 4, alt. 3 mm.

Abundant at Sonamarg.

7. NANINA (MICROCYSTIS ?) SONAMURGENSIS, n. sp., Figs. 16—18.

Shell small, depressed, thin, horny-brown, with the suture distinct; roughly, regularly and closely ribbed above; sculpture of a similar kind, but almost obsolete, can be traced on the

MOLLUSCA. 17

base; whorls seven, closely wound; the last scarcely, if at all, broader than the previous one, more or less subangulate at the periphery: base convex, distinctly excavated round a deep narrow umbilicus; aperture very shallow, the outer margin distinctly thickened, slightly subangulate in the middle; columella very slightly reflected, oblique, evenly rounded, without any angulation at the base, in this character resembling *N. splendens* and differing from *N. prona*. I know of no Indian species like this interesting little shell; in shape it somewhat resembles the smooth *N. woodiana*. Diam. 11½, alt. 5⅔, axis 4½; apert. lat. 5¼ mm.

Dr. Stoliczka found a few specimens alive at Sonamarg; he notes that the animal is provided with a mucous pore.

8. NANINA (MACROCHLAMYS) PRONA, n. sp.

Shell small, of the same group as *N. petrosa*, Hutt., &c., but with closer wound whorls; it is a form which apparently is widely spread throughout the North-Western Himalayas, as the Museum possesses numerous specimens from Simla, Masuri, Naini Tál and Sahúranpur; two specimens, found by Colonel Godwin-Austen in the Daffla Hills, also apparently belong here. A very similar small form, but I think specifically distinct, is also found in the Bombay Presidency. Dr. Stoliczka's specimens from Murree are all young, or in bad preservation; I have therefore determined on not naming the species from his Murree specimens, but take as my type the common North-West Himalayan form, the animal of which is known and which is usually recorded in collections as *N. petrosa*. Colonel Godwin-Austen informs me that Hutton himself transferred his own name *petrosa* from the Mirzapur shell to the Masuri one, on the strength of Benson's statement that the former was identical with the Calcutta *N. vitrinoides*, in which, as already stated, Benson was quite wrong. This species is not figured in the "Conchologia Indica," as far as I can see. Whorls six, closely wound, the last only slightly deflected, sometimes not at all, in which case, of course, the aperture is quite vertical; spire almost or quite flat; periphery rounded; umbilicus resembling that of *N. petrosa*, more open than in all the other allied species; horny-brown colour, smooth and polished above and below; margins of aperture distinctly, but slightly thickened. Type from Naini Tál: diam. 12, axis 4¼, alt. 5¾; apert. lat. 6, alt. 4¾ mm.

9. NANINA (BENSONIA) MONTICOLA, Hutt., var. MURRIENSIS, nov.

Nanina monticola, Hutt., J. A. S. B., vii, 1838, p. 215 (North-Western Himalayas).
Helix labiata, Pfr., P. Z. S., 1845, p. 65 (Loc.—?—)

Both species are recorded and figured in the "Conchologia Indica" as distinct, and I think very possibly the two forms there given may prove separable. Unfortunately, typical *N. monticola* is typical *N. labiata*, as figured l. c., pl. xxvii, fig. 5. This I am able to prove by a fine series of typical *N. monticola*, presented years ago by Captain Hutton to the Asiatic Society, and now in the Indian Museum. Theobald correctly unites the two species in his catalogue, though I consider him mistaken in also uniting Reeve's *H. convexa*. The form found by Dr. Stoliczka is near the much rarer one figured in the "Conchologia Indica," pl. lii, fig. 3, as *H. monticola*, and may prove distinct; the Murree specimen differs indeed, even more markedly than the one there figured, in the characters which separate it from the type

3

form, namely, open umbilicus, compressed whorls, more vertical aperture and peculiar, abruptly raised apical whorls.

A single specimen only was found at Changligali near Murree.

10. NANINA (BENSONIA) SPLENDENS, Hutt.

Nanina splendens, Hutton, J. A. S. B., 1839, p. 216 (North-Western Himalayas); "Conchologia Indica," pl. li, figs. 7 and 10.

This is one of the puzzling species, apparently intermediate between *Macrochlamys* and *Xesta*. The question of its correct generic rank can only be settled by a careful examination of its anatomy. In the excellent original description, the animal is described as of " a dark verdigris green, living under fallen timber at 9,000 to 11,000 feet above the sea," &c. Dr. Stoliczka found a few specimens at Tinali.

11. NANINA (BENSONIA) ANGELICA, Pfr.

Helix angelica, Pfr., P. Z. S., 1856, p. 33 (Punjab).

Dr. Stoliczka found several living specimens, all unfortunately young, at Uri (between Tinali and Srinagar). The form is distinguished from the preceding by the almost closed umbilicus, more closely wound whorls, &c.; the rounded periphery and numerous varices appear to be characteristic.

12. NANINA (BENSONIA) JACQUEMONTI, v. Mart.

Nanina jacquemonti, v. Mart., Mal. Bl., xvi, 1869, p. 75 (Himalayas).

A single specimen of this well-marked species was found at Murree; it is a common shell in the Punjab Salt Range. I give below the measurements of the Murree specimen, as they differ somewhat considerably from those of the type.

Diam. 20, axis 7¼; alt. 10¼, apert. lat. 10¾, alt. 8¼ mm.

13. HELIX (PATULA) HUMILIS, Hutt.

Helix humilis, Hutt., J. A. S. B., 1838, p. 217 (Simla).

Found tolerably abundant near Murree. Hutton records the animal " as that of a true *Helix*, of a dark grey or blackish colour, abundant during the rains on moist rocks, under dead leaves, &c., and at the roots of shrubs."

14. SUCCINEA PFEIFFERI, Rossm.

A few specimens from near Srinagar.

15. CLAUSILIA WAAGENI, Stol.

Clausilia waageni, Stoliczka, J. A. S. B., 1872, pl. ix, fig. 10 (Changligali).

About a dozen specimens of this species were found near Murree, under the bark of trees.

16. CLAUSILIA CYLINDRICA, Gray.

Clausilia cylindrica, Gray, Pfr., Symb. III, p. 93 (India).

Found in great abundance, under the bark of oak trees, near Murree.

17. BULIMINUS (PETRÆUS) STOLICZKANUS, n. sp., Figs. 25—27.

Shell in shape resembling *B. rufistrigatus*; deeply and narrowly rimate, oblong, for a species of *Petræus* of rather thin and diaphanous texture; obliquely, very irregularly striated, the striæ often very broad, more or less crowded together, with gaps between the "fasciculi." The ground colour is dark horny brown, with the striæ pure white, having the appearance (owing to the epidermis) in a fresh state of being a bright yellow; spire oblong, conical, apex obtuse; whorls seven, scarcely convex; aperture oblique and oblong, peristome white, outer margin scarcely reflected, columella moderately broad. It can be easily distinguished from its next ally *B. rufistrigatus*, by the less convex whorls, the more produced spire, less obtuse apex, by the considerably broader last whorl (in proportion to the others) and by the more dilated aperture; the sculpture also is peculiar and characteristic: it is nearer pl. xxiii, fig. 10, of the "Conchologia Indica" than pl. xx, fig. 4.

Long. 16, diam. 7 (last whorl to base of aperture 9); apert. 5¾, lat. 4¼ mm.

Found fairly abundant living on currant-bushes at Sonamarg.

18. BULIMINUS (PETRÆUS) MAINWARINGIANUS, n. sp., Fig. 28.

There is no Indian species with which I can compare this species. As to shape, the nearest I know of are some small dwarf forms of *Cylindrus insularis*; the species is, however, next allied to *B. pretiosus* and *B. rufistrigatus*.

Narrowly and superficially rimate, subcylindrically conical, of stout, smooth and polished substance; striated, striæ less oblique than in the preceding, fewer and more regular, not crowded together in the same way, here and there one more developed than the others, with intermediate ones more or less obsolete; light horny-brown, variegated with opaque white markings, as in *B. pretiosus*; these markings are fewer, of a more zigzag, broader and more irregular nature than those of the preceding; spire produced, apex scarcely obtuse; whorls 7, the three apical ones unusually short compared with the others, last whorl compressed; aperture very small, almost as broad as high, peristome pure white, outer margin considerably thickened, columella very broadly reflected, straighter than in the preceding, slightly subangulate, instead of rounded, at base.

Long. 10, diam. 4½ (last whorl to base of aperture, 5¼); apert. alt. 3⅞, lat. 3 mm.

Fairly abundant, near Murree.

I have named this pretty little shell after my friend Colonel Mainwaring, B.S.C., who has lately discovered very many interesting, rare and new forms round Calcutta, in Behar, and near Darjiling.

19. BULIMINUS (PETRÆUS) BEDDOMEANUS, n. sp., Fig. 29.

This is a very interesting species, resembling somewhat, in shape of the whorls and aperture, *B. smithei*, "Conchologia Indica," pl. xx, fig. 3, but it is still nearer *B. eremita*, Bens., *l. c.*, fig. 8, from which its produced spire, narrower whorls, and aperture easily distinguish it. Narrowly rimate, subcylindrically turreted, of solid, scarcely polished substance; closely, obliquely striate, striæ more regular and crowded together than in the two preceding forms; of a very pale horn colour, only here and there discernible, on account of the crowded striæ, which are of a chalk white colour; spire much produced, apex obtuse; whorls 10, increasing very gradually and regularly, last whorl compressed; aperture very small, peristome white, outer margin broadly reflected, very slightly arcuate (much as in pl. xx, fig. 3, *l. c.*), columella dilated, obliquely rounded at base.

Long. $13\frac{3}{4}$, diam. $4\frac{3}{4}$ (last whorl to base of aperture, 5); apert. alt. $3\frac{1}{2}$, lat. $2\frac{1}{4}$ mm.

Rather scarce near Murree.

I have named this shell after Colonel Beddome, who has contributed so extensively to our knowledge of the plants, reptiles and mollusks of South India.

20. BULIMINUS (PETÆUS) PRETIOSUS, Cantor.

Four specimens were found at Tinali, and a single one, of a slightly different form, near Murree.

21. BULIMINUS (PETRÆUS) RUFISTRIGATUS, Bens.

A single specimen of the typical form from the Jhelum Valley, and two specimens from Kashmir of the var. *gracilis* of the "Conchologia Indica."

22. BULIMINUS (PETRÆUS) DOMINA, Bens.

A few specimens were found alive near Murree.

23. BULIMINUS (PETRÆUS) CANDELARIS, Pfr., var.

A peculiarly shortened form found very abundantly near Tinali; the dextral form appears to have been found more abundant than the sinistral. Mr. Lydekker, of the Geological Survey of India, informs me he has noticed that the two forms are not usually found absolutely together.

MOLLUSCA.

24. ANADENUS ALTIVAGUS, Theob.

Limax altivagus, Theob., J. A. S. B., 1862, p. 489.

A few specimens were found at Changligali, under a log of wood. I am by no means sure that my friend Mr. Theobald is correct in uniting with this species the *A. giganteus*, Heyn.; the latter seems to me to agree better with a still larger slug of which the Indian Museum possess several fine specimens in spirit, found at Katmandu in Nipal.

25. ANADENUS MODESTUS, Theob.

Limax modestus, Theob., J. A. S. B., 1862, p. 489 (Simla Hills).

A few specimens of this small form, as far as I can see, only differing in external aspect by their smaller size and finer texture, were found with the preceding.

26. ANADENUS, sp.

I should not have ventured on separating this single specimen, found with the two preceding, but for a note of Dr. Stoliczka, which says—" I also found near here four specimens of an *Arion*, and specimens of two other *Arion*-like slugs." It is slightly larger than the preceding, and of a black, instead of light liver colour; otherwise I can see no difference.

27. ANADENUS, sp.

Described by Stoliczka in his notes as "a slug like the one I found at Changligali, but with the foot sharply crested."

EXPLANATION OF THE PLATE.

Fig. 1—3. *Helix* (*Fruticicola*) *phæozona*, v. Mart., p. 2.
 „ 4—6. „   „ *plectotropis*, v. Mart., p. 3.
 „ 7—9. „   „ *mataianensis*, Nevill, p. 3.
 „ 10—12. „ (*Xerophila*) *stoliczkana*, Nevill, p. 3.
 „ 13—15. *Nanina* (*Rotula*) *kashmirensis*, Nevill, p. 16.
 „ 16—18. „ (*Microcystis*) *souamurgensis*, Nevill, p. 16.
 „ 19—21. *Helicarion stoliczkanus*, Nevill, p. 15.
 „ 22—24. „ *austenianus*, Nevill, p. 14.
 „ 25—27. *Buliminus* (*Petræus*) *stoliczkanus*, Nevill, p. 19.
     28. „   „ *mainwaringianus*, Nevill, p. 19.
     29. „   „ *beddomeanus*, Nevill, p. 20.
 „ 30—31. *Succinea martensiana*, Nevill, p. 5.
 „ 32—33. „ *pfeifferi*, var. *subintermedia*, Nevill, p. 6.
 „ 34—36. *Valvata stoliczkana*, Nevill, p. 12.

ERRATUM.

In names at foot of plate *for* " var. intermedia," *read* " var. subintermedia."

MOLLUSCA.

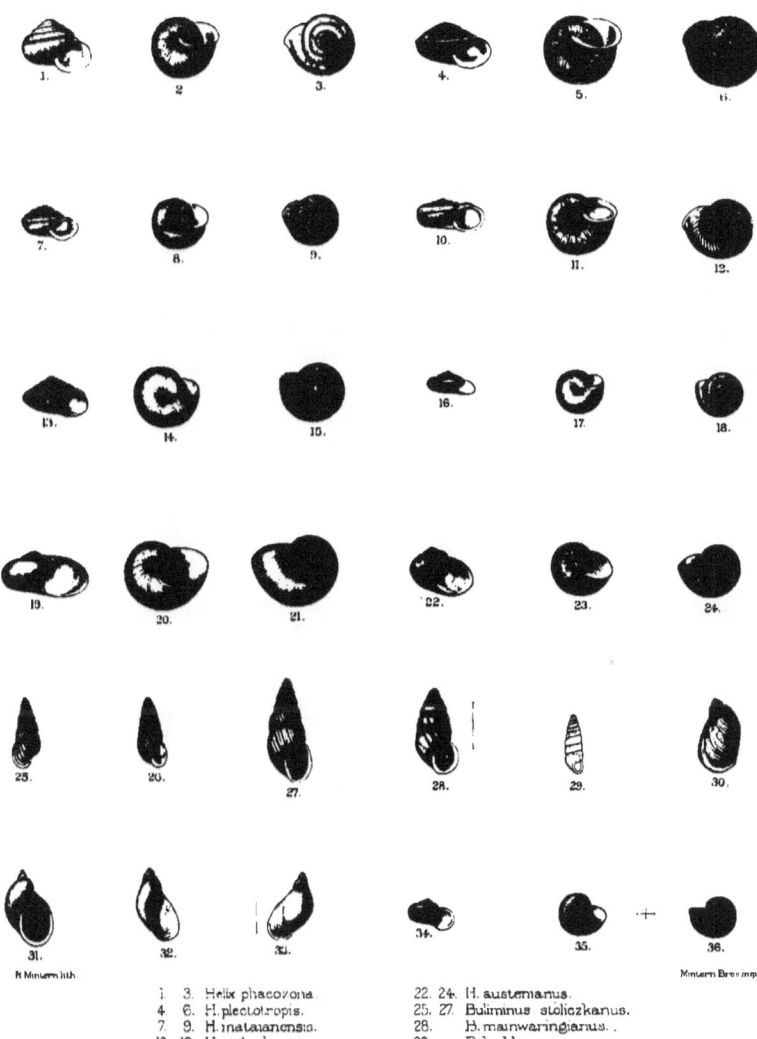

1. 3. Helix phaeozona.
4. 6. H. plectotropis.
7. 9. H. nataiancnsis.
10. 12. H. stoliczkana.
13. 15. Nanina kashmirensis.
16. 18. N. sonamurgensis.
19. 21. Helicarion stoliczkanus.
22. 24. H. austenianus.
25. 27. Buliminus stoliczkanus.
28. B. mainwaringianus.
29. B. beddomeanus.
30. 31. Succinea martensiana.
32. 33. Succinea pfeifferi (var intermedia)
34. 36. Valvata stoliczkana.

SCIENTIFIC RESULTS

OF

THE SECOND YARKAND MISSION;

BASED UPON THE COLLECTIONS AND NOTES

OF THE LATE

FERDINAND STOLICZKA, Ph.D.

ICHTHYOLOGY.

BY

FRANCIS DAY, F.L.S., F.Z.S.

Published by order of the Government of India.

CALCUTTA:
OFFICE OF THE SUPERINTENDENT OF GOVERNMENT PRINTING.
1878.

SCIENTIFIC RESULTS

OF

THE SECOND YARKAND MISSION;

BASED UPON THE COLLECTIONS AND NOTES

OF THE LATE

FERDINAND STOLICZKA, Ph.D.

ICHTHYOLOGY.

BY

FRANCIS DAY, F.L.S., F.Z.S.

Published by order of the Government of India.

CALCUTTA:
OFFICE OF THE SUPERINTENDENT OF GOVERNMENT PRINTING.
1878.

CALCUTTA:
PRINTED BY THE SUPERINTENDENT OF GOVERNMENT PRINTING,
8, HASTINGS STREET.

SCIENTIFIC RESULTS

OF

THE SECOND YARKAND MISSION.

ICHTHYOLOGY.

By FRANCIS DAY, F.L.S., F.Z.S.

THE following descriptions refer to the entire collection of fishes obtained during the expedition (except, so far as I know, two specimens[1]). With them I have compared some types of Steindachner's excellent paper on Dr. Stoliczka's "Fishes of Tibet" (Verh. z.-b. Ges. Wien, 1866), which specimens were given me by Dr. Stoliczka.

Mr. Hume, C.B., has since then obtained a few more skins of fishes from those regions through the exertions of Dr. Scully. These have likewise been forwarded to me; and one appears to be a very aberrant form of *Ptychobarbus*.

Order PHYSOSTOMI.

Family—*SILURIDÆ*

1. EXOSTOMA STOLICZKÆ. Plate I, fig. 1.

Day, Proc. Zool. Soc. 1876, p. 782.

D. $\frac{1}{6}$, P. $\frac{1}{17}$, V. $\frac{1}{6}$, A. 6, C. 15.

Length of head from 4 in the young[2] to 5⅔, of caudal 8, height of body 7¼ in the total length. *Eyes* minute, situated in the middle of the length of the head; the width of the interorbital space equals half that of the snout, or the distance between the eye and hind

[1] These two specimens are in the British Museum.
[2] The remarkable difference in the comparative length of the head to that of the total length is shown in the following figures:—

3 specimens	4	inches in length.	Head 4	to 4¼ in the total length.
4	,,	4·2 to 4·5	,,	4½ to 5¼ ,,
5	,,	5·0 to 5·7	,,	5 to 5½ ,,
3	,,	6·0 to 6·6	,,	5¼ to 5¾ ,,
2	,,	7	,,	5¼ to 5¾ ,,

nostril. *Head* depressed, as broad as long, and obtusely rounded. Mouth inferior; lips thick, and studded with small tubercular elevations; the upper and lower lips continuous at the angle of the mouth; but the transverse fold across the lower jaw is interrupted in the middle. Nostrils close together, the anterior round and patent, the posterior tubular: a barbel divides the two nostrils; it is situated on a bridge of skin, below which the two nostrils are continuous. *Barbels:* the nasal ones reach the hind edge of the eye; the maxillary ones have a broad basal attachment, and reach the root of the pectoral. Of the mandibular barbels the anterior are situated just behind the inner end of the lower labial fold: they are shorter than the outer pair, which latter extend to the gill-opening. *Gill-opening* situated on the side of the head in front and above the base of the pectoral fin. *Teeth:* several rows of pointed ones in each jaw, of which the outer is slightly the larger, rather wide apart, and with rather obtuse summits. *Fins:* the dorsal arises midway between the snout and the commencement of the adipose fin; its greatest height is one-third more than the length of its base; its spine is rudimentary and enveloped in skin. Adipose dorsal very long and low. Pectoral nearly as long as the head, having its outer half horizontal and its inner vertical; its spine is rudimentary, with a broad, striated, cutaneous covering. Ventral of a similar form to the pectoral: its first and a portion of its second ray also with a striated cutaneous covering; the fin commences on a vertical line falling just behind the base of the dorsal fin; it is rather nearer the snout than the posterior end of the adipose dorsal, and commences midway between the bases of the ventral and caudal fins; it is half higher than long. Caudal cut almost square. Free portion of the tail half higher than long. *Skin* tuberculated from the head, along the lower surface of the body, to nearly as far as the base of the ventrals. *Colours:* of a dull yellowish green, becoming lightest along the abdomen. Fins yellowish, with dark edges or bands.

Hab. Basgo, Snima, and Leh on the Upper Indus. The longest specimen 7 inches in length.

I propose here to shortly remark upon the distinction between the six species of *Exostoma* at present known—

A.—*Teeth in jaws pointed.*

1. Exostoma labiatum.—Lower labial fold uninterrupted. The interspace between the first and adipose dorsal fins equals two-thirds the length of the latter. Anal commences much nearer the base of the caudal than the base of the ventral. Mishmi Mountains, East Assam.

2. E. blythii.—Lower labial fold interrupted. Interspace between dorsal fins very slight. Anal commences in last third of distance between ventral and base of caudal. Head-waters or affluents of Ganges.

3. E. berdmorei.—Snout more pointed. Caudal forked. Tenasserim.

4. E. davidi[1].—The interspace between the first and adipose dorsal fins equals the length of the latter. Pectoral reaches the ventral. Eastern Tibet.

5. E. stoliczkæ.—Lower labial fold interrupted. Anal commences nearer the base of the ventral than that of the caudal. Pectoral does not extend to the ventral. Upper waters of Indus.

B. *Outer row of teeth flattened.*

6. E. andersonii.—Lower labial fold interrupted. Bhamo, Burmah.

[1] *Chimarrichthys davidi*, Sauvage.

ICHTHYOLOGY.

Family—*CYPRINIDÆ*

The majority of the fishes in the collection consist of carps, those from the more elevated regions being confined to such as have the vent and base of the anal fin bounded by a row of tiled scales, or the ubiquitous Loaches.

2. OREINUS SINUATUS.

Only one species of *Oreinus* exists in this collection, the *O. sinuatus*, Heckel, from Leh in Ladák, and which has likewise been captured in Kashmir. Although some of the fish were obtained in Kashmir, where the genus *Oreinus* has representatives, there was no example from that locality.

Having observed upon the great variation in proportions existing in a species of *Exostoma* captured on the hills, it may be worth while drawing attention to the same fact as occurring in specimens of this genus. Thus, in examining the following ten examples of *O. richardsonii*, Gray, in the British Museum, I found them as follows:—

4 specimens, in spirit, from 3·3 to 3·8 inches in length. Head from 4 to 4⅜ in the total.
1 specimen, in spirit, 4 inches in length. Head 4¼ in the total.
1　　„　　„　5¼　　„　　„　4¼　„
1　　„　　„　9　　„　　„　5⅓　„
1　　„　stuffed, 10　„　　„　5　„
1　　„　　„　15　„　　„　5½　„
1　　„　　„　18　„　　„　6　„

3. SCHIZOTHORAX CHRYSOCHLORUS. Plate I, fig. 2.

Racoma chrysochlorus, M'Clelland, Cal. Journ. Nat. Hist., ii., p. 577., t. xv., f. 3.
Schizothorax biddulphi, Günther, Ann. & Mag. Nat. Hist., 1876, xvii., p. 400.
Schizothorax chrysochlorus, Day, Proc. Zool. Soc., 1876, p. 784.

B. iii. D. $\frac{4}{7-1}$, P. 18, V. 10, A. $\frac{2}{5}$, C. 20, L. l. 110 to 120.

Length of head 4¼ to 5¼, of caudal 6 to 6¼, height of body 6¼ in the total length. *Eyes*: Diameter 5¼ (in a fish 7 inches long), 7 to 9 in the length of head, 2 to 2¼ diameters from the end of snout, and the same apart. Upper surface of the head nearly flat; its width rather exceeds its height, and equals half its length. Snout rather compressed, and overhanging the upper jaw. Mouth directed forwards, horseshoe-shaped, the lower labial fold interrupted in the middle. The maxilla reaches to below the front nostril. The depth of the cleft of the mouth equals the width of its gape. A very thin horny covering to the inside of the lower jaw. Posterior edge of opercle cut square. *Barbels*: the rostral ones as long as the eye, the maxillary rather longer, sometimes twice as long, and reaching to beneath the middle or hind edge of the orbit. *Teeth* pharyngeal 5, 3, 2, 2, 3, 5 pointed, and with rather compressed summits. *Fins*: the dorsal, which is as high as the body, arises midway between the end of the snout and the base of the caudal, its last undivided ray osseous, strong, finely serrated posteriorly, from a little longer than the head, in a specimen 11·9 inches in length, to ⅔ the length in the adult. Pectoral as long as the head excluding the snout; it reaches halfway to the base of the anal. Anal, when laid flat, reaches about halfway to the base of the caudal, which latter fin is forked. *Scales:* the row which bears the lateral line consists of larger scales than those above or below it; those forming the anal sheath are equal to half a diameter of the eye. *Colours* greyish along the back, becoming yellowish-white on the sides and beneath; a black mark over the eye, and a few dull spots on the back.

Hab. Káshghar, Yangihissár, and Yárkand, up to 20 inches in length: also Afghanistan. Dr. Scully collected four specimens in Káshghar (4,043 feet above the sea), which are 13, 16, 17, and 18 inches respectively in length.

4. SCHIZOTHORAX PUNCTATUS. Plate I, fig. 3.

Day, Proc. Zool. Soc., 1876, p. 785.

B. iii., D. $\frac{4}{8}$, P. 20, V. 11, A. $\frac{3}{5}$, C. 20.

Length of head $3\frac{1}{4}$ to 4, caudal $5\frac{1}{4}$, height of body 6 to 7 in the total length. *Eyes:* diameter $6\frac{1}{3}$ in the length of head, $2\frac{1}{4}$ diameters from end of snout, and 2 apart. Interorbital space flat. The greatest width of the head exceeds its height by one-fourth, and is $\frac{4}{9}$ of its length. Mouth anterior, with the upper jaw somewhat the longer; the cleft commencing opposite the middle of the eyes, whilst the maxilla reaches to below the front edge of the orbit. Lower labial fold interrupted in the middle. A thin striated horny covering to the lower jaw. *Barbels:* the maxillary ones equal the diameter of the eye; the rostral ones are slightly longer. *Fins:* dorsal rather higher than the body; it commences midway between the front edge of the eye and the base of the caudal fin; its last undivided ray is strong, coarsely serrated posteriorly, and as long as the postorbital portion of the head. Pectoral does not quite reach the ventral, which latter arises on a vertical line below the first articulated dorsal ray, and extends two-thirds of the distance to the anal. Anal rather above twice as deep as its base is long; when laid flat it does not extend to the commencement of the caudal. Free portion of the tail one-half longer than deep at its highest part. *Scales:* those along the lateral line larger than those above or below it. The tiled row along the base of the anal fin small, and equalling one-third of the diameter of the orbit. *Colours:* silvery, covered with largish black spots.

Racoma nobilis, M'Clelland, has more fleshy lips, whilst the mouth appears more transverse, as in *Oreinus*, and the under jaw much the shorter.

Hab. Kashmir Lake.

5. SCHIZOTHORAX ESOCINUS. Plate I, fig. 4.

Schizothorax esocinus, Heckel, Fische Kasch, p. 48, t. ix.; M'Clelland, Cal. Journ. Nat. His., ii., p. 579; Günther, Cat. vii., p. 166. Day, Proc. Zool. Soc., 1876, p. 785.

B. iii., D. $\frac{4}{8}$, P. 20, V. 10, A. 7, C. 20.

Length of head $4\frac{1}{4}$ to $4\frac{1}{2}$, of caudal $5\frac{1}{4}$; height of body $7\frac{1}{4}$ in the total length. *Eyes:* diameter $6\frac{1}{4}$ in the length of head, 2 diameters from end of snout and also apart. Interorbital space flat. The greatest width of the head equals its height or its postorbital length. Mouth very slightly oblique, horse-shoeshaped, the upper jaw longer than the lower, the maxilla reaching to nearly below the front edge of the eye. Lower labial fold interrupted

ICHTHYOLOGY.

in the middle. A horny covering to inside of lower jaw. *Barbels:* the rostral ones more than half longer than the eye, reaching to below its first third; the maxillary ones are slightly shorter. *Fins:* the dorsal as high as the body; it commences midway between the nostrils and the base of the caudal; its last undivided ray osseous, coarsely serrated posteriorly, and its bony portion being as long as the head, excluding the snout. Pectoral does not quite reach the ventral, which latter fin commences on a vertical line slightly behind the origin of the dorsal, and extends two-thirds of the distance to the anal. Length of base of anal $\frac{1}{3}$ of its height; it reaches, when laid flat, to the base of the caudal, which latter fin is deeply forked. Free portion of the tail as high at its base as it is long. *Colours* silvery, with numerous black spots, most distinct in the upper half of the body.
Hab. Leh, on the Upper Indus, Kashmir, and Afghanistan.

6. SCHIZOTHORAX INTERMEDIUS. Plate II, fig. 1.

Schizothorax intermedius, M'Clell., Cal. Journ. Nat. Hist. 1842, ii, p. 579; Günther, Cat. vii, p. 165.

B. iii., D. $\frac{4}{4-8}$, P. 19, V. 10, A. $\frac{2}{5}$, C. 20, L. l. 105.

Length of head $4\frac{1}{2}$, of caudal 5 to 6, height of body 6 in the total length. *Eyes:* diameter $5\frac{1}{4}$ in the length of head, $1\frac{1}{3}$ diameter from the end of snout and also apart. Upper surface of the *head* flat; its greatest width equals its postorbital length, whilst its height equals its length excluding the snout. Upper jaw rather longer than the lower, and not overhung by the snout. Mouth horseshoe-shaped, the depth of the cleft equalling the width of its gape. The maxilla reaches to below the hind nostril. Lower labial fold interrupted in the middle. A thin, smooth, deciduous horny covering to the lower jaw. *Barbels* four, as long as the eye in the young, longer in the adult. *Teeth:* pharyngeal, 5, 3, 2, 2, 3, 5, pointed and rather crooked at their summits. *Fins:* dorsal as high as the body in the young, not quite so high in the adult; it commences midway between the end of the snout or front nostril and base of the caudal; its last undivided ray strong, rather coarsely serrated posteriorly, one-half to two-thirds as long as the head in the immature, four-fifths of its length in the adult. Pectoral as long as the head excluding the snout, and reaching more than half-way to the base of the ventral, which latter fin arises below the first dorsal ray and extends more than half-way to the anal. The length of the base of the anal equals half its height, which latter equals the length of the pectoral; if laid flat it almost reaches the base of the caudal, which is forked. *Scales:* depth of those in tiled row equals half a diameter of the eye. Free portion of the *tail* about as high at its commencement as it is long. *Colours* silvery, usually without spots; but in some specimens from Yangihissár there are minute black spots on the upper half of the body.
Hab. Káshghar, Yangihissár, and Sarikol. M'Clelland likewise obtained it (through Griffith) from Afghanistan, the Cabul River at Jellalabad, and Tarnuck River. He sent three specimens to the East India Museum.

7. SCHIZOTHORAX MICROCEPHALUS. Plate III, fig. 2.

Day, Proc. Zool. Soc., 1876, p. 787.

B. iii., D. $\frac{4}{5}$, P. 18, V. 11, A. $\frac{2}{5}$, C. 18, L. l. 105, L. tr. 25/.

Length of head 5 to $5\frac{1}{2}$, of caudal 6, height of body $5\frac{3}{4}$ to 6 in the total length. *Eyes:* diameter 7 in the length of head, $2\frac{1}{4}$ diameters from end of snout, and $2\frac{1}{4}$ apart. Interorbital,

B

space flat. The greatest width of the *head* equals its length behind the middle of the eyes; its height equals its length excluding the snout. *Mouth* broad, anterior, with the upper jaw the longer, and overhung by the snout; the cleft of the mouth nearly horizontal; it extends to below the hind nostril, and is scarcely above half the extent of its gape; lower labial fold interrupted in the middle. A thin horny covering to the lower jaw. *Barbels*: the rostral ones reach to below the hind edge of the eye, the maxillary ones to the hind edge of the preopercle. *Fins*: dorsal anteriorly nearly as high as the body, commencing slightly nearer the snout than the base of the caudal fin, or midway between the two; its last undivided ray weak, articulated, and with some very small obsolete denticulations posteriorly about its centre (absent in some specimens). Pectoral as long as the head behind the front nostril, and reaching rather above half-way to the ventral, which latter is shorter than the pectoral, reaching about half-way to the base of the anal. Anal almost reaching base of caudal when laid flat, the length of its base being only one-third of its height. Caudal with rounded lobes. Free portion of the tail rather longer than high. *Scales*: in the first third of the body those along the lateral line are larger than those above or below them, but posteriorly they are of the same size; the tiled row equal about half the diameter of the eye. *Colour* silvery.

M'Clelland says of *S. edeniana* that its spine is slender, soft, and denticulated at its base, but the reflected fold of the lower lip is uninterrupted. *Racoma gobioides*, M'Clell., from the Bamean River, shows the head almost as short as in this species; but it has a strong serrated dorsal spine, whilst that fin is on an elevated base. The anal does not appear to reach above half-way to the base of the caudal.

Hab. The specimens are from Panjah (9,000 feet) in Wakhán, the waters going to the Oxus. The dorsal spine approaches that of *Ptychobarbus*.

8. SCHIZOTHORAX IRREGULARIS. Plate IV, fig. 1.

Day, Proc. Zool. Soc., 1876, p. 787.

? *Schizothorax edeniana*, M'Clell., Cal. Journ. Nat. Hist. ii, p. 579.

B. iii, D. $\frac{3}{8}$, P. 18, V. 9, A. $\frac{2}{7}$, C. 20, L. l. 98, L. tr. 26/.

Length of head 5, of caudal 6, height of body 6 in the total length. *Eyes*: diameter $6\frac{1}{4}$ in the length of head, $2\frac{1}{2}$ diameters from the end of snout, and about 2 apart. Interorbital space nearly flat. The greatest width of the *head* equals its height or its length behind the orbit. *Mouth* narrow; the upper jaw slightly the longer, and only slightly overhung by the snout. Cleft of mouth a little oblique, its width equal to its length, and the maxilla reaching to beneath the front nostril. Lips very thick, lobed in the centre, and with an interrupted labial fold. *Barbels*: the rostral ones reach to below the front edge of the eye; the maxillary ones are one-half longer than the diameter of the eye. *Fins*: dorsal anteriorly about two-thirds as high as the body below it: its last undivided ray weak, very feebly serrated posteriorly, whilst the extent of its osseous portion does not exceed one-third of the length of the head; the fin commences midway between the front edge of the eye and the base of the caudal fin. Pectoral as long as the head excluding the snout, and reaching half-way to the ventral, which latter is rather shorter and extends rather more than half-way to the base of the anal. Anal two-fifths as long at its base as it is high; when laid flat it almost reaches the caudal, which latter is slightly forked. Free portion of the tail rather longer than high at its base. *Scales*: those behind the pectoral region to as far as the end of the anal, and below the lateral line, are much smaller than those above the lateral line. The tiled row small, not above

ICHTHYOLOGY. 7

half the diameter of the eye. *Colours* silvery, becoming lightest and glossed with gold below the lateral line.

Hab. The specimen described is stuffed, and 20·5 inches in length. It was obtained at Tash-kurgan in Sarikol. If this is identical with *S. edeniana*, M'Clell., it is also found in the Cabul River, in the Mydan Valley, and Sir-i-chusmah.

9. SCHIZOTHORAX NASUS. Plate IV, fig 3.

Schizothorax nasus, Heckel, Fische Kasch., p. 33, t. vi.; Günther, Cat. vii., p. 166.

B. iii, D. $\frac{4}{8}$, P. 18, V. 10, A. $\frac{2}{7}$, C. 19, L. l. 90-100.

Length of head $4\frac{2}{3}$, of caudal $5\frac{1}{4}$, height of body 5 in the total length. *Eyes*: diameter $5\frac{1}{2}$ in the length of head, $1\frac{1}{4}$ diameter from the end of snout, and also apart. Dorsal profile more convex than that of the abdomen. Upper surface of the *head* nearly flat; its greatest width equals its postorbital length, while its height equals its length excluding the snout. Upper jaw rather longer than the lower and overhung by the snout. *Mouth:* horseshoe-shaped, its gape equalling its cleft. The maxilla reaches to below the hind nostril. Lower labial fold interrupted. *Barbels:* four; the maxillary ones two-thirds as long as the eye; the rostral ones slightly shorter. *Fins:* dorsal as high as the body below it; it commences midway between the middle of the eye and the base of the caudal fin; its last undivided ray is strong, rather coarsely serrated, and nearly as long as the head. Pectoral about as long as the head excluding the snout, and reaching above half-way to the base of the ventral, which latter fin arises below the last undivided dorsal ray, reaching half-way to the base of the anal, which is above twice as high as wide at its base, and nearly reaches the caudal when laid flat. *Scales:* depth of those in the tiled row scarcely one-third of the diameter of the eye. Free portion of the tail not quite as high at its commencement as it is long. *Colours:* silvery, with black spots on the upper half of the body.

This species has a more elevated dorsal profile and shorter barbels than *S. intermedius*.

Hab. Kashmir Lake.

10. PTYCHOBARBUS CONIROSTRIS. Plate III, fig. 3.

Ptychobarbus conirostris, Steindachner, Verh. z.-b. Ges. Wien., 1866, p. 789, t. xvii, f. 4; Günther, Cat. vii., p. 169.

B. iii, D. $\frac{4}{8}$, P. 22, V. 10, A. 7-8, C. 19, L. l. 95, L. tr. 24/.

Length of head $4\frac{3}{4}$ to 5, of caudal $7\frac{1}{4}$, height of body $6\frac{1}{4}$ to $6\frac{2}{3}$ in the total length. *Eyes*: diameter from $4\frac{1}{4}$ to $5\frac{1}{2}$ in the length of the head, 2 diameters from the end of snout, and $1\frac{1}{4}$ apart. The greatest width of the *head* equals its postorbital length, but is slightly less than its height. *Mouth:* horseshoe-shaped, with the upper jaw a little the longer, and rather overhung by the snout; the maxilla reaches to below the front edge of the eye. Lower labial fold very broad, uninterrupted, and with a cleft in the median line posteriorly. *Barbels:* a pair at the angle of the mouth, which reach the posterior edge of the preopercle; in a small specimen, 3·1 inches long, they only equal half a diameter of the eye in extent. *Teeth:* pharyngeal ones in two rows. *Fins:* the dorsal commences much nearer the snout than the base of the caudal, its entire base being equidistant from these two points; it has no osseous ray, and is as high as the body below it. Pectoral as long as the head behind the nostrils, and does not reach quite so far as the ventral, which latter fin arises under the last few dorsal rays and reaches two-thirds of the distance to the base of the anal. The anal, when laid flat, reaches the base of the caudal, its base is 2½ in its height. *Scales:* the tiled row small, not one-third of the diameter of the eye. *Colours:* silvery, darkest along the back and upper half of body, where most of the scales have black margins, thus causing small reticulations in the colour. Upper surface of the head spotted with black; some dark spots on the dorsal fin, and sometimes a few light ones on the caudal.

Hab. Head-waters of Indus, Hanle in Tibet, and Chiliscomo, near Drás.

11. PTYCHOBARBUS LATICEPS. Plate III, fig. 1.

Day, Proc. Zool. Soc. 1876. p. 789.

B. iii, D. ⅗, P. 18, V. 9, A. ?, C. 20, L. l. 145.

Length of head 4¼, of caudal 9⅔, height of body 7 in the total length. *Eyes:* diameter 12 in the length of head, 2½ diameters from the end of snout, and also apart. *Mouth* anterior, with the lower jaw somewhat the longer; the depth of the cleft of the mouth equals half the width of the gape. Upper surface of the *head* broad, its width being nearly twice its height. No lower labial fold under the mandible. *Barbels:* a maxillary pair as long as the eyes. *Fins:* dorsal arises slightly nearer the base of the caudal than the end of the snout; its last undivided ray weak, articulated at its extremity, and not serrated. Pectoral two-fifths as long as the head. Ventral arises below the anterior dorsal rays. Caudal forked. *Scales* are scarcely imbricated, but cover the entire body; those forming the tiled sheath along the base of the anal fin are two-thirds of the diameter of the eye. *Colours* silvery superiorly, becoming dull white beneath; a few blackish spots along the back.

This interesting skin has unfortunately had its anal fin removed, whilst the pharyngeal teeth have not been preserved. The specimen is 52 inches in length.

It may be considered that as this fish differs from *P. conirostris* in the form of its mouth and snout, also in the position of the ventral fin, it might form a new genus; but we have yet much to learn of the mountain barbels; perhaps a more extensive acquaintance will diminish the number of genera into which they are at present subdivided.

Hab. Káshghar (4,043 feet elevation), the river from which place eventually joins the Yárkand River.

12. PTYCHOBARBUS LONGICEPS. Plate IV, fig. 2.

Day, Proc. Zool. Soc., 1876, p. 790.

B. iii, D. ⅖, P. 19, V. 12, A. ⅗, C. 20, L. l, 112, L. tr. 31.

Length of head 3⅔ to 4, of caudal 7 to 7¼, height of body 5¼ to 6 in the total length. *Eyes:* diameter 7 to 9 in the length of head, 1½ diameter from the end of snout, and 2 apart. *Mouth* anterior, cleft oblique, commencing superiorly opposite the upper margin of the eye. Lower jaw somewhat the longer; the maxilla reaches to below the middle of the eye. The greatest width of the *head* rather exceeds its height, and equals half its length. Interorbital space flat. No lower labial fold under the mandibles. *Barbels:* a maxillary pair half as long as the eye. *Fins:* the dorsal commences midway between the hind edge of the preopercle and the base of the caudal fin. Its last undivided ray is osseous, of moderate strength, and very finely serrated posteriorly; its osseous portion equals a little above one-fourth of the length of the head. Pectoral half as long as the head, and reaches half-way to the ventral; the latter fin commences under the first divided dorsal ray, and does not extend quite half-way to the root of the anal. Anal twice as high as its base is long; it does not reach the caudal when laid flat; the latter fin forked. *Scales* oval, nearly as wide as high and slightly imbricate; the tiled row half the diameter of the eye. Free portion of the tail rather longer than high. *Colours* bluish on the back, lightest below, dorsal and caudal spotted.

Hab. Yárkand, whence the stuffed specimen described was brought. It is 31 inches in length. This species scarcely accords with the definition of *Ptychobarbus*, the last undivided dorsal ray being osseous and finely serrated. The specimen, however, is large, whilst *P. laticeps* forms the intermediate form between it and *P. conirostris*.

13. SCHIZOPYGOPSIS STOLICZKÆ. Plate II, fig. 2.

Schizopygopsis stoliczkæ, Steind. Verh. z.-b. Ges. Wien., 1866, p. 785; Günther. Cat. vii, p. 170.

B. iii, D. $\frac{1-3}{7-8}$, P. 13, V. 11, A. $\frac{1}{5-6}$, C. 19.

Length of head 5 to 5¾, of caudal 5¼ to 5¾, height of body 7 to 8 in the total length. *Eyes:* diameter 4 to 5 in the length of head, 1 to 1⅛ diameters from end of snout, and 1½ to 2 apart. The greatest width of the *head* equals its length behind the middle of the eyes; and its height equals its length excluding the snout. *Mouth* inferior, overhung by the snout; the maxilla reaches to below the front edge of the eye. A sharp, anterior, horny edge to the mandible. *Barbels* absent. *Fins:* the dorsal commences about midway between the end of the snout and the root of the caudal; its upper edge is nearly straight, oblique; the fin is as high as the body below it, and one-third higher than its base is long; its last undivided ray osseous and finely serrated posteriorly. Pectoral not quite so long as the head, and reaching rather above half-way to the ventral, which latter, arising below the middle of the dorsal, is slightly the shorter, and does not reach the anal. Anal, when laid flat, reaches the base of the caudal; it is rather above twice as high as its base is long. Caudal deeply forked. Free portion of the tail as high as long. *Lateral line* at first descends gently, and then reascending, attains the middle of the body opposite the posterior extremity of the dorsal fin. *Colours* olive superiorly, becoming white on the sides and beneath; the whole covered with irregular blackish spots.

The ova are comparatively large. The serrated dorsal spine is strongest in specimens from Leh.

These fishes appear to be much attacked by parasites, which occasion yellowish elevated tubercles, not only on the head and body, but also on the dorsal fin.

One specimen, from Balakchi, had a shot (No. 2) imbedded in the isthmus, where the parts around it had healed.

Hab. Leh, Tánkse, and fry or small fish from Lukong and Chagra (15,090 feet), all from waters directly or indirectly going to the Indus. Some fry from Sarikol, the waters of which go to the Yarkand River[1], Aktash, Upper Kara-kul and Panjah, tributaries of the Oxus or Amu River. This fish has also been taken at Gnari Khorsum by Schlagintweit.

[1] I am very dubious of these specimens, and hardly think they can have been obtained from waters that flow into the Yárkand River, as the adults have not been obtained thence. The adult, however, has been taken in the Oxus; and I find by the diary that on the day the specimens in question were captured the camp was at Sarikol, a few miles from a valley where a stream enters the Aksu River, a tributary of the Oxus.

Largest specimen 8·5 inches in length. There is also a specimen from Balakchi, the streams there apparently flowing towards the Yárkand River, which goes to the east.

14. DIPTYCHUS MACULATUS. Plate II, fig. 3.

Diptychus maculatus, Steindachner, Verh. z.-b. Ges. Wien., 1866, p. 787; Günther, Cat. vii., p. 171. Day, Proc. Zool. Soc., 1878, p. 792.

Diptychus severzowi, Kessler, Fish. Turkestan, p. 17, t. iv, f. 12.

B. iii, D. $\frac{12}{3-4}$, P. 19, V. 9, A. $\frac{7}{5}$, C. 19, L. l. 80–90.

Length of head 5 to 6, of caudal 5 to 6; height of body $7\frac{1}{2}$ to 8 in the total length. *Eyes*: diameter $4\frac{1}{2}$ in the young to 6 in the adult in the length of the head, $1\frac{1}{4}$ to 2 diameters from the end of snout, and $1\frac{1}{2}$ apart. The greatest width of the *head* equals its height, or its length behind the front edge or middle of the eyes. *Mouth* transverse, inferior, having an anterior sharp horny covering on the lower jaw. Lower labial fold interrupted in the middle. *Barbels*: one at each maxilla, having thick bases, and hardly so long as the eye. *Teeth* pharyngeal, 4, 3, 3, 4, curved at the outer extremity and pointed. *Fins*: the dorsal commences rather nearer the snout than the base of the caudal, its upper edge is straight; it is as high as the body below it, its last undivided ray articulated. Pectoral not quite so long as the head; it reaches rather above half-way to the ventral, which latter commences on a vertical line below the last dorsal ray; it reaches rather above half-way to the base of the anal. Anal when laid flat reaches the base of the caudal; its height is nearly three times the length of its base. *Scales* not imbricated, but scattered over the upper two-thirds of the body and pectoral region, in which latter locality the skin is often rugose: the tiled row well developed. Free portion of the tail one-half longer than high at its base. *Colours* bluish, lightest inferiorly, indistinctly blotched and spotted along the upper half of the body; often a narrow, dull band along the lateral line, and a second below it. The dorsal and caudal fins much spotted in some specimens.

The very young are destitute of scales; they first appear along the lateral line. One specimen from Basgo, 1·1 inch long, has no barbel on the left side. There are two specimens from the west of Sarikol: one has an adipose lid, covering the anterior half of the left eye; the other has a similar lid covering the lower half of the left eye. Brown tubercles are common on some of the specimens, and do not appear to be normal. Some specimens from Leh have the eye small.

Diptychus severzowi, from the Rivers Aksai and Ottuck appears to be the above species.

Hab. Specimens were brought from Kharbu, Basgo, Snima, Leh, Tánkse, and Chagra, from waters going directly or indirectly to the Indus; from Pasrobat (9,370 feet), and Tarbashi (11,515 feet), whence the waters go to the Yárkand River; also from west of Sarikol, which goes to the same river. Some specimens are also labelled as from Chiliscomo. This fish has also been captured in other parts of Tibet, and likewise in Nepal.

15. LABEO SINDENSIS. Plate II, fig. 4.

Cirrhina sindensis, Day, Proc. As. Soc., Beng., 1872, p. 319.

B. iii, D. 12-13, P. 18, V. 9, A. 7, C. 19, L. l. 43, L. tr. 8-9.

Length of head $6\frac{1}{2}$, of caudal $4\frac{1}{4}$; height of body $5\frac{1}{4}$ in the total length. *Eyes*: diameter $5\frac{1}{4}$ in the length of the head, 2 diameters from the end of the snout, and $2\frac{1}{4}$ apart. Snout

ICHTHYOLOGY. 11

rather overhanging the mouth, without any lateral lobe. Lips continuous at the angle of the mouth, and having a thin cartilaginous covering. *Barbels:* a short maxillary, but no rostral pair.

Hab.—Sind, Punjab, and the Deccan. The specimen figured came from Murree.

16. CIRRHINA GOHAMA.

Cyprinus latius and *gohama*, Hamilton Buchanan, Fishes of Ganges, pp. 346, 393.
Barbus diplochilus, Heckel, Fisch. Kasch., p. 53, t. x, f. 1.
Tylognathus barbatulus, Heckel, Hügel's Reise, iv., p. 376.
Chondrostoma wattanah, Sykes, Trans. Zool. Soc., ii., p.   t. 62, f. 4: Bleeker, Beng., p. 25.
Gonorhynchus brevis, M'Clell *and* Ind. Cypr. p. 373, t. 43, f. 6.
Crassocheilus latius and *gohama*, Bleeker, Prod. Cypr., p. 110; Günther, Cat. vii., p. 72.
Crassocheilus rostratus, Günther, *loc. cit.*
Crossocheilus barbatulus, Günther, *loc. cit.*

B. iii., D. $\frac{3}{14}$, P. 15, V. 9, A. $\frac{2}{5}$, C. 19, L. l. 38–40.

There are several specimens of this fish from the lake in Kashmir; and, curiously enough, they show the links between Hamilton Buchanan's and Heckel's species. All have a pair of rostral barbels and minute mandibular ones (*C. barbatula*). Some have $5\frac{1}{2}$, some $4\frac{1}{2}$ rows between the lateral line and base of first dorsal ray. Others possess 3, $3\frac{1}{2}$, and $4\frac{1}{2}$ rows between the lateral line and base of ventral fin. The proportions, likewise, vary with age and other causes.

The localities this fish inhabits, and its mode of frequenting stones, very much resemble those of *Discognathus lamta*, Hamilton Buchanan, whilst its jaws are wide (not deep); and its under surface is similarly flattened, but it has no labial sucker.

17. BARBUS TOR. Plate III, fig. 4.

Cyprinus tor, Hamilton Buchanan, Fishes of Ganges, pp. 305, 388.
Barbus (Labeobarbus) hamiltonii, Gray and Hardwicke, Ind. Zool., pl.; Jerdon, Mad. Journ. Lit. and Sci., 1849, p. 311.
Barbus progeneius, M'Clell and, Ind. Cyp., pp. 270, 334, pl. lvi, f. 3; Cuv. and Val., xvi, p. 208.
Labeobarbus macrolepis, Heckel, Fisch. Kashmir, p. 60, pl. x, f. 2, Cuv. and Val., xvi, p. 209.
Labeobarbus tor, Bleeker, Cobit. *et* Cyp. Ceylon, in Nat. Verh. Holl. Maat. Haar., 1864, p. 10, f. 2. Day, Proc. Zool. Soc., 1867, p. 290; 1870, p. 372.
Barbus khudree, Sykes, T. Z. S. ii, p. 57.
Barbus tor, Cuv. and Val., xvi, p. 199.
Barbus (Barbodes) tor, Day, Proc. Zool. Soc., 1869, pp. 270, 334.
„   *mosal*, Günther, Cat. vii, p. 130.
„   *macrolepis*, Günther, Cat. vii, p. 131.
„   *longispinis*, Günther, Cat. vii, p. 132.

B. iii, D. $\frac{4}{9}$, P. 18, V. 9, A. $\frac{2}{5}$, C. 19, L. l. 23-24, L. tr. 4/4.

This fish, the Mahaseer of India, is too well known to need describing.

Habitat.—From Sind throughout India and Ceylon, and generally ascending mountain rivers for the purpose of breeding. Should such rivers be snow-fed, it deposits its ova in the side streams.

Before describing the Loaches, I will give my reasons why it appears to me that the genus *Diplophysa*, Kessler, may probably be a synonym of *Nemacheilus*.

It is said to consist of "elongated fishes, strongly compressed posteriorly," which we perceive in *Nemacheilus stoliczkæ* and *N. yarkandensis*; but in an equally elongated species *N. tenuis*, the free portion of the tail is not compressed, but is as wide as deep.

"The eyes are surrounded with a fold of skin forming a lid." This is also perceived in specimens amongst the species I have enumerated from Yárkand; and I have likewise noted that some of the other fishes from the same cold region have folds of skin more or less covering the eyes.

"Lips fleshy, the upper more or less denticulated, the inferior bilobed, and more or less papillated." I have figured the inferior surface of the head of all the Loaches; and although some, as *N. stoliczkæ* and *N. tenuis*, have the lips as described by Kessler, the *N. yarkandensis* has not, whilst the three certainly cannot be separated into distinct genera.

"Air-vessel in two parts, the anterior enclosed in a bony capsule, the posterior elongated and free in the abdominal cavity." This is the only portion of Kessler's definition not perceived in these fishes in which the air-vessel is enclosed in bone; and I cannot resist suggesting a re-examination of Western Turkestan specimens. It would be very remarkable were the *Nemacheili* found in Europe, in fact throughout Asia, even in the Oxus, to have their air-vessels enclosed in bone, whereas in the river Ili going to Lake Balkash, and the river Urdjar falling into Lake Ala (Ala-kul), they have the same organ partially free in the abdomen, as is seen in genus *Botia*. But granting Kessler's description to be accurate, I cannot think that such a fact alone would justify instituting a new genus for the reception of his species.

The reason for air-vessels being more or less enclosed in bone in some fishes is obscure; and I some time since adverted, in the 'Proceedings of the Zoological Society,' to the circumstance of such not being infrequent in Indian *Siluridæ*.

I found amongst the Indian genera of Siluroids of the fresh waters, or those which entered fresh waters, as follows :—

    A.—Air-vessel, when present, free in the abdominal cavity—

        1. *Rita* ; 2. *Erethistes* ; 3. *Pseudeutropius* ; 4. *Silurus* ; 5. *Olyra* ; 6. *Macrones* ; 7. *Callichrous* ; 8. *Wallago* ; 9. *Arius* ; 10. *Hemipimelodus* ;[1] 11. *Osteogeniosus*; 12. *Batrachocephalus* ; 13. *Pangasius* ; 14. *Plotosus*. Of these, five (Nos. 9, 10, 11, 12, and 14) are marine forms, entering fresh waters for predaceous purposes.

    B.—Air vessel more or less enclosed in bone—

        1. *Ailia* ; 2. *Ailiichthys* ; 3. *Sisor* ; 4. *Bagarius* ; 5. *Amblyceps* ; 6. *Saccobranchus* ; 7. *Silundia* ; 8. *Eutropiichthys* ; 9. *Gagata* ; 10. *Nangra* ; 11. *Pseudocheneis* ; 12. *Exostoma* ; 13. *Clarias* ; 14. *Glyptosternum*. All of these are fresh water genera.

[1] *Hemipimelodus* appears to be *Arius* destitute of teeth on the palate.

These fourteen fresh water genera having the air-vessel enclosed in bone are divisible as follows:—
1.—Waters of plains—
　(α.)—Large rivers. No suckers on the chest:—*Ailia, Ailiichthys, Sisor, Bagarius, Silundia, Eutropiichthys, Gagata, Nangra.*
　(β.)—Large rivers: descending to the sea. An accessory air-breathing apparatus:— *Clarias.*
　(γ.)—Smaller rivers, tanks, &c. An accessory air-breathing sac:—*Saccobranchus.*
2.—Waters of the plains or hills—
　No sucker on chest:—*Amblyceps.*
　Sucker on chest:—*Glyptosternum.*
3.—Waters of hills—
　Sucker on chest:—*Pseudecheneis.*
　Chest adhesive:—*Exostoma.*

As we find genera with the air-vessel enclosed in bone decrease in number the further we are from Hindustan Proper, it is but natural to conclude that the necessity for this bony capsule is greater in India than in other tropical countries, and also that it is only useful for freshwater forms.

When we see that all fishes (except the *Nemacheili*) from Yárkand have the air-vessel free in the abdominal cavity, it stands to reason that heat or cold can scarcely be that which involves the necessity of this form of organization.

It appears most probable that the air vessel being more or less enclosed in bone is for the purpose of developing some function specially required or to an abnormal extent, and that whatever this may be it is most necessary in a mountain torrent, but unnecessary in a marine existence.

We find in fishes that the air-vessel has two distinct functions—

(1).—In the *Acanthopterygii*, where it is free in the abdominal cavity, its use is more or less a mechanical one, and by contracting or expanding the fish is enabled to maintain itself at a desired level.

(2).—In the *Physostomi* we find a very different formation, as in all there is a duct opening from the air-vessel into the upper portion of the alimentary canal. In some of these fishes the mechanical function appears to be alone served by it. In others, that of hearing seems to entirely supersede that for flotation, for being more or less enclosed in bone contraction and expansion would be impeded. These bones or auditory ossicles lead to the internal ear, and it is evident that in some way the air vessel serves for auditory purposes to an extent for which we, at present, are hardly in a position to account.

It is remarkable that *Siluroid* forms do not appear to thrive in cold climates. The *Cyprininæ* of this collection have all small scales, or are more or less destitute of any; whilst the Loaches of Yarkand and Tibet have none at all; neither have those recorded from the Oxus or the Jaxartes.

There is one characteristic of the hill Loaches which seems almost invariable: the pectoral fins are stiff at their bases, as if employed for adhesive purposes. I have observed the outer ray in some of the Loaches of the plains forming a distinct bony ray with an enlarged and flattened outer extremity: but this is used for the purpose of assisting them to dig into the sand, in which they will bury themselves with great rapidity on the approach of danger.

18. NEMACHEILUS STOLICZKÆ Plate V, fig. 2.

Cobitis stoliczkæ, Steindachner, Verh. z.-b., Ges. Wien., 1866, p. 703, t. xiv, f. 2.
Cobitis tenuicauda, Steindachner *loc. cit.* p. 792, t. xvii., f. 3.
Nemacheilus stoliczkæ, Günther, Cat. vii, p. 360.
Nemacheilus tennicauda, Günther, *loc. cit.*, p. 357.

B. iii, D. $\frac{3}{7}$, P. 13, V. 8, A. $\frac{2}{5}$, C. 15.

Length of head 6, of caudal 6; height of body 8 in the total length. *Eyes*: diameter 8 in the length of head, 3 diameters from the end of snout, and 2 apart. *Snout* rounded, slightly projecting over the mouth. *Lips* rugose; and in some specimens from Yárkand the edges are fimbriated: lower lip with a lobe on either side, but the lower labial fold interrupted in the middle. The greatest width of the *head* equals its height, or its length excluding the snout. In some specimens the preorbital has a free lower edge. *Barbels* six; the maxillary ones reach beyond the hind edge of the eye; the rostral ones are shorter. *Fins*: the dorsal commences midway between the eye and the base of the caudal, it is one-third higher than its base is long, and equals the greatest depth of the body; its last ray is divided to its base; its upper edge is oblique, with a rounded anterior angle. Pectoral nearly as long as the head, and reaching rather above half-way to the ventral; the latter fin arises on a vertical line below the anterior dorsal rays, is almost as long as the pectoral, and reaches above half-way to the anal. Anal with a very narrow base: caudal slightly emarginate. Free portion of the tail from twice to two-and-a-half times as long as high at its base. *Scales*: absent. *Air-vessel*: in two portions, enclosed in bone. *Colours*: greyish along the back, becoming lighter beneath, marbled all over with dark green or black spots or bands. Dorsal, caudal, and sometimes outer pectoral rays barred.

In specimens from Sarikol the snout is rather more pointed than described above.

Hab.—Leh (11,518 feet); Snima; Lukong stream (14,130 feet); and Chagra (15,000 feet), all being waters directly or indirectly going to the Indus. Also Yárkand (3,923 feet) and Sarikol, where the waters go to the easterly or Yárkand River; and Aktásh (12,600 feet), which is on the Aksu or Oxus.

I have a specimen in my collection given me by Dr. Stoliczka: he procured it, along with those sent to Steindachner, from the Tso-Morari in Rupshu (Tibet), on his first visit to that country.

19. NEMACHEILUS YARKANDENSIS. Plate V, fig. 3.

Day, Proc. Zool. Soc., 1876, page 796.

B. iii, D. $\frac{3}{7}$, P. 17, V. 8, A. $\frac{2}{5}$, C. 15.

Length of head 4½, of caudal 6, height of body 6¾ in the total length. *Eyes*: diameter 6 to 7 in the length of the head, 2½ diameters from the end of snout, and 2 to 3 apart. *Snout*: rather elevated in the adult. Upper surface of the *head* nearly flat; its greatest width equals its height or its length excluding the snout. *Mouth* inferior, horseshoe-shaped; lips smooth, lower labial fold interrupted in the middle and destitute of lobes. *Barbels* six; the maxillary ones reach (in adults) the angle of the preopercle. *Fins*: the dorsal commences midway between the front edge of the eye and the base of the caudal fin; its upper edge is straight and oblique; its height rather exceeds that of the body below it, and is one-fourth more than the extent of its base. Pectoral as long as the head excluding the snout, and reaching two-thirds of the distance to the ventral. Ventral commences below the first dorsal ray, is shorter than the pectoral, and reaches two-thirds of the distance to the anal. Anal twice as high as wide at its base. Caudal emarginate, its outer rays being a little produced. Free portion of the *tail* at its commencement nearly equals its length in the adult, but is less in the young. *Scales* absent. *Air-vessel* in two portions, enclosed in bone. *Colours:* greyish, having in some specimens numerous fine blackish or dark spots on the body. In some there is a silvery lateral band.

Hab.—Yárkand, Pasrobat, Yangihissár, and Káshghar, all from waters in connection with the Yárkand and Yangibissár or Great Easterly River.

20. NEMACHEILUS TENUIS. Plate V, fig. 4.

Day, Proc. Zool. Soc., 1876, page 796.

B. iii, D. $\frac{3}{8-9}$, P. 13, V. 8, A. $\frac{2}{5}$, C. 17.

Length of head $5\frac{1}{2}$ to $5\frac{3}{4}$, of caudal $7\frac{1}{4}$, height of body 9 to 10 in the total length. *Eyes:* diameter $5\frac{1}{2}$ in the length of head, $2\frac{1}{4}$ diameters from the end of snout, and 1 apart. *Snout* rather compressed and overhanging the mouth; the greatest width of the *head* equals its height or its length excluding the snout. In some specimens the lower edge of the preorbital is free. *Lips* thickened and fimbriated in the adult; lower labial fold interrupted in the middle, and rather lobed on either side. *Barbels* six; the outer rostral pair extend to below the hind edge of the eye, the maxillary ones to the opercle in the adult. *Fins:* dorsal commences midway between the end of the snout and the base of the caudal fin; its upper edge is slightly concave, with a rounded upper angle; it is rather more than one-half higher than the extent of its base or than the body below it. Pectoral nearly as long as the head, and reaches rather above half-way to the ventral, which latter commences under the third dorsal ray; is as long as the pectoral, and reaches the base of the anal. Anal twice as high as wide at its base. Caudal slightly emarginate. Free portion of the tail one-third as high at its base as it is long, while its breadth equals its height. *Scales* absent. *Air-vessel* in two portions, enclosed in bone. *Colours:* yellowish white, the surface and sides sometimes with dark blotches and spots: dorsal and caudal fins with dull spots.

This fish is allied to *N. ladacensis*, Günther, but is distinguished by a more elongated body and longer barbels, &c.

Hab. Aktásh (12,600 feet elevation), whence the waters pass to the Oxus; and Yangihissár (4,320 feet elevation), where the rivers go to the Yárkand River.

21. NEMACHEILUS LADACENSIS. Plate IV, fig. 4.

Nemacheilus ladacensis, Günther, Cat. vii, p. 356.

B. iii., D. $\frac{3}{9}$, P. 13, V. 9, A. $\frac{2}{5}$, C. 19.

Length of head 5, of caudal $5\frac{3}{4}$; height of body $5\frac{1}{2}$ in the total length. *Eyes:* diameter 5 to $5\frac{1}{4}$ in the length of head, $2\frac{1}{4}$ diameters from end of snout, and 2 apart. Greatest width of *head* equals its height or its length excluding the snout. *Lips* moderately thick and rugose; lower labial fold interrupted in the middle. *Barbels* 6; the maxillary ones scarcely reach to below the front edge of the eye, the longest rostral ones to below the front nostril. *Fins*: dorsal commences midway between the front edge of the eye and the base of the caudal fin: it is as high as the body below it and half higher than its base is long; its upper anterior corner rounded. Pectoral as long as the head behind the angle of the mouth, and reaching nearly to the ventral, which latter fin arises below the commencement of the dorsal fin: it is shorter than the pectoral, but extends to the base of the anal. Anal twice as high as long, and reaches above half-way to the base of the caudal which is emarginate. Free portion of the tail twice as long as high at its base. *Scales* absent. *Colours*: of a light fawn, with sixteen or eighteen interrupted darker and sinuous bands passing from the back down the sides; a silvery lateral band. Dorsal and caudal finely spotted in lines: a darkish band on pectoral, ventral and anal.

Hab. Gnari Khorsum, Tibet. The specimen described is the largest of two obtained by Messrs. von Schlagintweit, and deposited in the Indian Museum. The size of the British Museum specimen, and the broken state of its caudal fin, must be accepted as the reason why my proportion of the free portion of the tail does not agree with Dr. Günther's (nearly $\frac{1}{4}$); whilst I find the caudal fin "emarginate," and not "rounded."

22. NEMACHEILUS GRACILIS. Plate IV, fig. 5.

Day, Proc. Zool. Soc., 1876, p. 798.

B. iii, D. $\frac{2}{7}$, P. 13, V. 8, A. $\frac{2}{5}$, C. 17.

Length of head $5\frac{1}{4}$, of caudal $6\frac{1}{4}$, height of body $6\frac{1}{2}$; in the total length. *Eyes*: diameter 11 in length of head, 4 diameters from end of snout, and $2\frac{1}{6}$ apart. *Snout* overhanging the mouth. The greatest width of the head equals its height or its length excluding the snout. *Lips* thickened; lower labial fold interrupted in the middle and rather lobed on either side. *Barbels* six; the maxillary ones nearly twice as long as the eye; the external rostral ones reach the hind nostril; the other pair are shorter. *Fins*: dorsal commences midway between the eye and vertical border of the preopercle; its upper edge is nearly straight; it is not quite so high as the body below it, and one-fourth less than the extent of its base. Pectoral as long as the head behind the angle of the mouth; it reaches rather above half-way to the base of the ventral, which latter fin arises somewhat in advance of the commencement of the dorsal; it is of about the same length as the pectoral, and extends half-way to the anal. Anal twice as high as wide at its base: it reaches, when laid flat, a little more than half-way to the base of the caudal, which is slightly emarginate. Free portion of the tail half as high at its base as it is long. *Scales* absent. *Colours* brownish along the back, becoming yellowish beneath: dorsal and caudal with dull spots.

Hab. Basgo, on the head waters of the Indus.

ICHTHYOLOGY. 17

23. NEMACHEILUS MARMORATUS. Plate V, fig. 1.

Cobitis marmorata, Heckel, Fische Kasch., p. 76, t. xii., figs. 1 and 2: Hügel, Kaschm. iv., p. 380.

Cobitis vittata, Heckel, *loc. cit.* p. 80, t. xii., figs. 3 and 4; Hügel, *loc. cit.* p. 382.

Nemacheilus marmoratus, Günther, Cat. vii., p. 356; Day, Proc. Zool. Soc., 1876, p. 798.

B. iii, D. $\frac{2}{8}$, P. 11, V. 7, A. $\frac{2}{6}$, C. 17.

Length of head $4\frac{3}{4}$ to 5, of caudal 7, height of body 7 in the total length. *Eyes*: diameter 5 in length of head, 2 diameters from end of snout, and $1\frac{1}{3}$ apart. *Snout* somewhat pointed; and in some the preorbital is slightly projecting, *Lips* wrinkled; the lower labial fold interrupted. The greatest width of the head equals its height or its length excluding the snout. *Barbels*: the maxillary ones reach to below the hind edge of the eye; the rostral ones are nearly as long, *Fins*: dorsal commences midway between the end of the snout and the base of the caudal; its upper edge is nearly straight, oblique, and with rounded angles; its height rather exceeds that of the body below it; and it is nearly twice as high as its base is long. Pectoral as long as the head excluding the snout, and extending half-way to the ventral. Ventral one-third shorter than the pectoral, and reaching half-way to the anal. Anal twice as high as long at its base. Caudal cut square, with rounded angles or slightly emarginate. Free portion of the tail from one-and-a-half to twice as long as high at its base. *Scales* absent. *Colours* marbled or irregularly blotched and spotted with brown; fins also more or less spotted.

Hab. Kashmir Lake.

24. NEMACHEILUS RUPICOLA.

Schistura rupicola, M'Clelland, Journ. A. Soc. Bengal, vii., pl. lv, fig. 3, and Ind. Cypr., p. 309, pl. lvii., f. 3.

The Kashmir species are almost or quite destitute of scales, and otherwise agree with M'Clelland's fish.

25. NEMACHEILUS MICROPS.[1]

Cobitis microps, Steindachner, Verh. z.-b. Ges. Wien., 1866, p. 794, t. xiii., f. 3.

Nemacheilus microps, Günther, Cat. vii., p. 357.

This species is entirely destitute of scales. The head is as wide as it is long. It was obtained by Dr. Stoliczka in Tibet on his first journey, but no specimens exist amongst the Yárkand collection.

[1] *Oreias Dabryi*, Sauvage, Rev. et Mag. Zool., 1874, p. 3, is closely allied to this species.

E

If we examine the localities whence the fishes which form this collection were procured, omitting the Murree and Kashmir examples, we find as follows:—

Name of species.	Head waters of Indus.	Yárkand river, or its branches.	Oxus, or its tributaries.
Exostoma stolickza	1	...	...
Oreinus sinuatus	1	...	...
Schizothorax esocinus	1	...	...
———— *chrysochlorus*	...	1	...
———— *intermedius*	...	1	1
———— *irregularis*	...	...	1
Ptychobarbus conirostris	1	...	...
———— *laticeps*	...	...	1
———— *longiceps*	...	1	...
Schizopygopsis stolickza	1	...	1
Diptychus maculatus	1	1	...
Nemacheilus stolickza	1	1	1
———— *gracilis*	1	...	...
———— *yarkandensis*	...	1	...
———— *tenuis*	...	1	1
Total	8	9	4

Thus, we have eight species from the head-waters of the Indus, two of which extend to the great easterly, or Yárkand, River of Eastern Turkestan, and one to the Oxus of Western Turkestan; nine species from the Yárkand River, two common to the Indus and three to the Oxus; and four species from the Oxus, three of which are also found in the Yárkand River, and one in the head waters of the Indus.

If these species are examined in accordance with the districts traversed by this Mission and mapped out by Mr. Hume, we obtain the following results :—

(*1st*).—From the hilly region between Murree and the Zoji-la Pass, there exists one species of *Schizothorax* showing an affinity to the Turkestan fauna: one *Oreinus*, a Himalayan genus: and two species of *Nemacheilus*, a genus common to Turkestan and Hindustan.

(*2nd*).—From Zoji-la to the head of the Pankong there are;—one Siluroid, *Exostoma*, evidently a Himalayan and hilly form. Of carps, the Himalayan *Oreinus* and four genera which may be considered as common to Turkestan, and mostly to the upper hilly regions, *viz., Schizothorax, Schizopygopsis, Ptycobarbus* and *Diptychus*: lastly, a *Nemacheilus*, an almost universally distributed genus.

(*3rd*).—From the plains of Yárkand, two species of *Schizothorax* and two of *Ptycobarbus*, evidently the most typical forms of the fishes in these elevated regions: the genus *Nemacheilus* is likewise represented.

(*4th*).—From the west of Yárkand to the Pámir *Schizothorax, Schizopygopsis* and *Diptychus*, all forms found in Turkestan or adjacent regions, and likewise *Nemacheilus* were obtained.

The foregoing species constitute the fish-collection made in the cold and inhospitable regions traversed by the Mission; and they are of interest for the purpose of ascertaining what are the chief characteristics of the fish-fauna, and what relationship it bears to those of contiguous Asiatic regions, so far as such have been ascertained.

In this inquiry it will be necessary to take a survey of the fishes of Afghanistan, Western Turkestan, and Hindustan, before proceeding further respecting those of Tibet and Yárkand or Eastern Turkestan.

Most of our knowledge of the fishes of Afghanistan is due to the labours of Griffith, who remarked:—"The characteristic forms of Afghan fish are doubtless the small-scaled *Barbi* and *Oreini*; and these far exceed the others in number . . . . The fish are as distinct from the Indian forms as the plants are . . . . By characteristic I do not mean that these forms are limited to Afghanistan, because they occur perhaps to an equal extent in the Himalayas, to the streams of which those of Afghanistan approximate more or less in the common features of rapids and bouldery beds."

Having crossed the high range of mountains separating Afghanistan from the plains of Western Turkestan, he found "a great change in the fish to occur, and *Salmonidæ*[1] seem to take the precedence of the *Cyprinidæ*. A species of trout abounds in the Bamean River and up its small tributaries, derived from the Koh-i-Baba, to an altitude of about 11,000 feet. A species of *Barbus* with small scales is likewise common in the Bamean River"[2] (Cal. Journ. Nat., Hist., ii. p. 565).

He observes that Indian species were in the majority in the Cabul river (a tributary of the Indus) at Peshawur; and in accordance with the facility or the reverse of access from the plains did he find a predominance of Indian or Afghan forms.[3]

The nature of the fishes of Afghanistan appears to be much as follows:—Absence of Acanthopterygian or spiny-rayed families, except the spineless and widely distributed *Ophiocephalus gachua*, Ham. Buch., and the spiny eel, *Mastacembelus armatus*, Lacép., so common in the East from the plains to the summits of mountains. Few Siluroids, but perhaps a *Callichorus* and *Amblyceps*. Numerous Cyprinoids which appear to belong to the following genera—*Oreinus, Schizothorax, Bungia*, from near Herat, *Barilius*, and a Loach (*? Nemacheilus*), perhaps *Discognathus* and *Barbus*.

The fullest account we possess of the fishes of Western Turkestan is that lately given by Kessler, from which I have extracted the following:—

ACANTHOPTERYGII. *Perca fluviatilis*, Linn., obtained exclusively from the Jaxartes and some of its tributaries. *P. schrenckii*, Kess., from Lake Balkash. *Lucioperca sandra*, Cuv., from the Jaxartes. *Cottus spinulosus*, Kess., very rare in Turkestan, two specimens from Khojend.

None of these spiny-rayed fishes were captured at so south a latitude as Káshghar. Out of the four species three came from the Jaxartes or its tributaries, the other from Lake Balkash.

[1] This remark appears to have been a little too strong, as he only found one species of *Salmo*; probably it was very abundant.
[2] The stuffed type presented to the British Museum from the Indian Museum seems to have been lost or destroyed.
[3] Griffith states that the Cabul River at Jellalabad presents us with two or three small-scaled *Barbi* (*? Schizothorax*) and *Oreini* together with certain tropical forms, as the Mahasir (*Barbus*) and a *Silurus* very like, if not identical with, the Poftah (*? Silurus afgana*). Also the same river at Lalpur possesses a fish, I believe, identical with the Nepoorn of Assam (*Labeo*) and a *Gonorhynchus* (=*Discognathus*). Griffith also mentions a Loach-like *Silurus* from near Jubraiz (*? Amblyceps*).

SILURIDÆ. *Siluris glanis*, Linn. Generally spread throughout Western Turkestan, having been received from the Jaxartes, Oxus, and Sarekshan or Tarafshan Rivers.

CYPRINIDÆ. *Cyprinus carpio*, Linn., from the Jaxartes, Oxus, Sarekshan Rivers. *Barbus conocephalus*, Kess., from Sarekshan. *B. platyrostris*, Kess., from the River Aksu falling into Lake Balkash. *B. lacertoides*, Kess., from Jaxartes and its tributaries. *B. brachycephalus*, Kess., from Jaxartes and Oxus. *Schizothorax aksaiensis*, from the River Aksai. *S. fedtschenkoi*, Kess., *S. affinis*, Kess., and *S. eurystomus*, Kess., from the Sarekshan River. *S. orientalis*, Kess., from a lake on the Alatau Mountains, the waters on the Western Turkestan side of which drain to Lake Balkash. *Diptychus severzowi*, Kess., Aksai and Ottuk Rivers to 10,000 feet. *D. dybowskii*, Kess., River Aksu. *Gobio fluviatilis*, Cuv., widely distributed in Western Turkestan, specimens received from near the towns of Tashkend, Khojend, Djisak, and from the Ak Daria. *Abramis brama*, Linn., Jaxartes and its tributaries. *A. sapa*, Pallas, rare, from the Jaxartes. *Acanthobrama kuschakewitschi*, Kess., Jaxartes. *Pelecus cultratus*, Linn., Sea of Aral. *Abramis chalcoides*, Güld., rather rare, obtained in the Ak Daria and Durman Kul. *A. iblioides*, Kess., creeks near Janikurjan. *A. fasciatus*, Nord., Sarekshan. *A. tæniatus*, Kess., Jaxartes. *Aspius rapax*, Pallas, Jaxartes and its tributaries. *A. esocinus*, Kess., Jaxartes and Oxus. *Leuciscus erythrophthalmus*, Linn., Jaxartes. *L. squaliusculus*, Kess., from near Khojend on the Jaxartes and Janikurjan. *L. rutilus*, Linn., Jaxartes and Aigus Lake.

CONITIDINÆ. *Cobitis longicauda*, Kess. (scaled), one specimen from the Jaxartes. *C. uranoscopus*, Kess., from near Magian, Tashkend, Ilhodjaduk, and Lake Iskander, the waters of which appear to drain to the Sarekshan River. *C. dorsalis*, Kess., creeks near Janikurjan. *C. elegans*, Kess., and *C. tænia*, Kess., river near Tashkend, a tributary of the Jaxartes. *Diplophysa strauchii*, Kess., river Ili, falling into Lake Balkash. *D. labiata*, Kess., River Urdjar, falling into Lake Ala.

SALMONIDÆ. *Salmo oxianus*, Kess., river Darant, falling into the Kisil-su, one of the upper tributaries of the Oxus.

ESOCIDÆ. *Esox lucius*, Linn., Jaxartes and its tributaries.

CHONDROPTERYGII. *Acipenser schipa*, Lovetsky, Jaxartes, Casalius River. *Scaphirhynchus fedtschenkoi*, Kess., Oxus.

The foregoing fishes of Western Turkestan [1] mainly consist of—

(*1st*).—Those descending from the north or spreading from the east or west, such as

Perca, Lucioperca, Cottus, Gobio, Abramis, Acanthobrama, Pelecus, Alburnus, Aspius, Squalius, Leuciscus, Acipenser, and *Scaphirhynchus*.

(*2nd*).—Those common to Afghanistan and Yárkand, as *Schizothorax, Barbus,* Loaches (? genus).

(*3rd*).—Those found also in Yárkand, as *Schizothorax* and *Diptychus*.

(*4th*).—*Silurus*, (which will be alluded to).

Lastly, Salmo, on the slopes of the mountains where the rivers descend to the Oxus.

The existence of one of the *Salmonidæ*, termed *Salmo orientalis* by M'Clelland, was well known to Dr. Stoliczka; and a special object of his search (as he informed me previous to starting) would be to try and ascertain its distribution. Griffith found this fish "in the Bamean River, a stream that falls from the northern declivities of the Hindoo Koosh into the Oxus."

[1] I have to thank Mr. F. Carl Craemers for kindly translating some Russian localities, which I should not otherwise have been able to give.

ICHTHYOLOGY. 21

Kessler does not record any of this family from the Jaxartes, or, in fact, from the rivers immediately descending from the Tian Shan or the Alatau Mountains. We are, therefore, left to surmise that in the hills whence these fishes were taken is the abrupt termination of members of the family *Salmonidæ*, which does not possess a solitary representative in Hindustan, except the *S. levenensis* (introduced on the Nilgiris in Madras).

If we now take a short review of the *Fresh Water Fishes of India* we find much as follows:—

ACANTHOPTERYGII.

Genera *Ambassis, Badis, Nandus, Pristolepis, Sciæna, Gobius* and some allied genera, *Rhynchobdella, Mugil, Anabas, Polyacanthus, Osphromenus, Trichogaster, Etroplus* exist in India, but are absent from the fresh waters of Afghanistan, Turkestan, and Yárkand. Whether existing only in large rivers or distributed more generally over India, none pass the boundary of the Himalayas.

Mastacembelus and *Ophiocephalus* are found in India and in Afghanistan; both ascend for some height the Himalayas and other hill ranges.

PHYSOSTOMI.

SILURIDÆ. Genera *Erethistes, Macrones, Rita, Pangasius, Pseudeutropius, Wallago, Olyra, Chaca, Clarias, Saccobranchus, Silundia, Ailia, Ailiichthys, Eutropiichthys, Sisor, Gagata, Nangra, Bagarius, Pseudecheneis, Glyptosternum* exist in India, but not in Afganistan, Turkestan or Yárkand.

Callichrous and *Amblyceps*, which are found in India, appear to be present in Afganistan, and the former also in Kashmir.

Exostoma is found along the Himalayas; *Silurus* in Turkestan and India.

CYPRINODONTIDÆ. *Cyprinodon* and *Haplocheilus* are found in India.

CYPRINIDÆ. Genera *Homaloptera, Psilorhynchus, Cirrhina, Osteochelus, Scaphiodon, Semiplotus, Catla, Amblypharyngodon, Nuria, Rasbora, Aspidoparia, Rohtee, Danio, Perilampus, Chela,* and various genera of *Cobitidinæ* exist in India.

Discognathus, Labeo, and *Barilius* are common to India and Afghanistan, but are evidently Indian forms.

Oreinus, Schizothorax, and *Barbus,* are found in India, also in Afghanistan, and the two last in Turkestan, whilst *Schizothorax* is common in Yárkand. *Cobitis* or *Nemacheilus* seem to extend everywhere.

CLUPEIDÆ and NOTOPTERIDÆ. Of the genera belonging to these families, and which exist in the fresh waters of India, none go beyond the base of the Himalayas.

The *Fishes of Yárkand*[1] or *Eastern Turkestan* consist of species of the following genera:— *Schizothorax,* found also in Afghanistan and Western Turkestan; one species on the slopes

[1] I here omit the genera *Exostoma* from the Himalayas, and *Oreinus* from the Himalayas and Afghanistan.

F

of the Himalayas, and sometimes even descending to the plains. *Diptychus*, Tibet, Yárkand and Western Turkestan. *Schizopygopsis*, Tibet and Yárkand. *Ptychobarbus*, Tibet and Yárkand. The remainder are Loaches.

Diptychus Dybowskii, Kess., would almost seem to be a *Schizopygopsis* with an articulated dorsal ray and a pair of maxillary barbels. Perhaps several of these hill-genera will, at some future date, be properly amalgamated, as has been done with the low-country Barbels (*Barbus*).

An examination of the genera of spiny-rayed or Acanthopterygian fishes clearly shows that as we proceed inland in India they diminish; at the Himalayas they cease. Two Indian species[1] only have been observed to exist in Afghanistan; and they are amongst the most widely distributed of their respective genera. Neither of these extends in the north-east, either to Western Turkestan or Yárkand. In Western Turkestan, it is true, three genera of this order are represented; but they have evidently extended southwards. Yárkand and Tibet appear to be unsuited for this order of fishes: and thence none have been brought.

The Physostomi include all the Yárkand and Tibet fishes. Among Siluroids the Indian genera *Callichrous* and ? *Amblyceps* have been doubtfully recorded from Afghanistan; but neither have spread to Western Turkestan, where, however, the *Silurus glanis* is found, evidently a wanderer from its more northern home.

It is clear that in India there is a gradual diminution of Siluroids as we proceed inland until we arrive at the Himalayas. On the slopes of these mountains we at first obtain a few peculiar genera and species organized for a mountain-torrent life; but as we rise, eventually (as was the case in this Mission), an elevation is attained which, taken in connection with the latitude and paucity of food, seems to be beyond the limit of the Indian Siluroids.

The Siluroids along the slopes of the Himalayas appear to be mostly confined to the following :—A few, as *Macrones* and *Callichrous*, ascend a short distance, which may be considered accidental. *Pseudecheneis* is a more distinct hill-form, possessing a sucker formed of transverse folds between its pectorals on the chest, and by the aid of which it prevents itself being carried away by the torrents. *Glyptosternum* has also an adhesive sucker, but of longitudinal folds, and likewise placed on the chest. These fishes, however, appear to be more intended for rapid rivers in the plains, but some ascend the slopes of the Himalayas. I have taken *large specimens* from the rivers at the base of the hills in which the suckers were scarcely visible: whether they had outgrown them, or, owing to the suckers not having been primarily well developed, they had been unable to maintain their footing in the hill-streams, of course, one cannot decide. *Amblyceps* is a Loach-like form found in the waters of the plains and also of the hills; it is abundant near Kangra. *Exostoma*, an example of which exists in the Yárkand-Mission collection, is also a remarkable form. It has a broad and depressed head and chest, the latter forming a species of sucker to enable it to sustain a mountain-torrent life.

This fish (*Exostoma stoliczkæ*) belongs to a genus which has only been recorded from hilly regions, neither extending to the waters of the comparatively level plateaus of the high lands, nor descending any distance towards the plains. The following six species are known :—
(1) *E. stoliczkæ*, from the head-waters of the Indus; (2) *E. blythii*, from near Darjeeling, where the waters descend to the Ganges; (3) *E. labiatum*, from the Mishmi Mountains and Eastern Assam ; (4) *E. andersonii*, from near Bhamo on the confines of China; (5) *E. davidi*,

---

[1] *Ophiocephalus gachua* and *Mastacembelus armatus*.

from the most easterly portion of Tibet near the head waters of the Yang-se-kiang; (6) *E. berdmorei*, from Tenasserim.

The distribution of the foregoing six species of this genus is interesting, because it is suggestive of whether, at some remote period, the Himalayan range, the mountains between Tibet and China, and the spur or continuation southwards through Burma and Siam, may not have been connected one with another.

Whilst adverting to this point, I would mention another circumstance: the only Siluroid stated to be found in Western Turkestan is the *Silurus glanis*, Linn. Three other species of the same genus have been captured on the hill-ranges of India; and their distribution somewhat accords with that of *Exostoma*—

(1).—*Silurus cochinchinensis*, Cuv. & Val. = *Silurichthys berdmorei*, Blyth, and

(2).—*Silurus wynaadensis*, Day. These fishes, found in hills up to about 2,500 feet, have been obtained in the Western Ghâts, Akyab Hills, Tenasserim and Cochin China. They would appear to be restricted to those mountains which are not far removed from the seacoast. How it is that several species of fishes are common to Malabar and Siam, or the countries contiguous to it, whilst they are entirely absent from the intermediate districts of India, is a question which I do not propose entering upon.

(3).—*S. afghana*, Günther, from Afghanistan, is identical with *S. dukai*, Day, from Darjeeling.

Cyprinidæ form the entire collection of the Yarkand Mission, after its arrival beyond the upper waters of the Indus. If we examine the members of this family found on the Himalayas in the same manner as we have the Siluroids, we find as follows:—*Discognathus*, so easily recognizable by the sucker on the lower lip, is found some distance up the mountains, but is rare above 5,000 feet. *Oreinus*, with its small scales, broad mouth, and likewise a sucker behind the lower jaw, becomes more and more common the higher we ascend. The Expedition obtained one species at Leh, in the Upper Indus; and it has been found as a genus extending from Afghanistan along the Himalayan Range, and near Bhamo by the last Yunnan Mission, or the same district as the Siluroid genera *Exostoma* and *Silurus*. It appears to essentially prefer the sides of hills and impetuous torrents.

Some of the stronger *Labeos*, Barbels (*Barbus*), and a *Barilius* are found here and there on the slopes and in the side streams of the Himalayas up to very considerable heights. They, however, are Indian forms which, if able to do so, appear to migrate during the breeding-season to the mountains to deposit their ova in the side streams which are unreplenished by snow-water. Here the fry are often compelled to remain until the succeeding year's rains swell the waters, washing food into their retreats to enable them to grow, or else to permit them to descend to the plains.

Once near the summit of these mountains, and beyond districts where adhesive suckers are a necessity for moderate-sized fishes to possess to prevent their being washed away, we come upon genera as rare in the plains of India as are the Indian forms at the summit of the Himalayas.

Kashmir is a locality traversed by this Mission, a hilly Himalayan district, and one to which it is necessary to refer. In Hügel and Heckel's "Fische aus Kaschmir" we find the following species recorded:—

Oreinus plagiostomus, Heckel; *O. sinuatus*, Heck,; *Schizothorax curvifrons*, Heck.; *S. longipinnis* Heck.; *S. niger*, Heck.; *S. nasus*, Heck.; *S. huegelii*, Heck.;

S. micropogon, Heck.; *S. planifrons*, Heck.; *S. esocinus*, Heck.; *Cirrhina gohama*, Ham. Buch.; *Barbus tor*, Ham. Buch.; *Labeo varicorhinus*, Heck.; *Nemacheilus marmoratus*, Heck.; *Callichrous pabda*, Ham. Buch.

These fishes demonstrate relationship with three districts :—

Schizothorax with Afghanistan and East and West Turkestan;
Oreinus with the slopes of the Himalayas in their whole extent;
Cirrhina, *Barbus*, and *Callichrous* with the neighbouring fauna of Hindustan.

Having examined what are the ingredient parts of the fish fauna of Western Turkestan, Afghanistan, Hindustan, Yárkand or Eastern Turkestan, Tibet, and Kashmir, it will be interesting to endeavour to discover if these localities are possessed of any indigenous forms, and, if so, how far they extend into contiguous countries.

I do not propose inquiring into whether the great desert region of Central Asia can or cannot be included in one Tartarian subregion; but, as the zoology of this portion of the globe is at present rather obscure, I think it will be more useful to limit oneself strictly to ascertained facts.

Sir D. Forsyth's Mission has led naturalists into the fringe of an ichthyological region of which Yárkand may be the centre; certainly it is richer in forms of *Schizothoracinæ* than Western Turkestan appears to be.

In the cold and hilly districts of Tibet and Yárkand we observe an absence of spiny-rayed and Siluroid fishes; whilst amongst Carps we see the genera *Schizothorax*, *Ptychobarbus*, *Schizopygopsis*, and *Diptychus*—fishes belonging to a peculiar division *Schizothoracinæ*, (or Hill-Barbels of M'Clelland), which may be thus defined :—

Carps *more or less covered with minute scales, or destitute of any. A membranous sac or slit anterior to the anal fin, which is laterally bounded by a row of vertically placed scales, like eave-tiles, and which are continued along the base of the anal fin.*

The fishes composing this are mostly of an elongated form, and are divisible into :—

a. Those with transverse mouths, as *Oreinus*, *Ptychobarbus*, *Schizopygopsis*, *Diptychus*.

b. Those with compressed mouths, as *Schizothorax*.

The genus *Oreinus* is spread from the Helmund River and Jellalabad in Afghanistan, along the whole Himalayan and contiguous ranges of hills to at least the confines of China. So far as I know, these fishes appear to be strictly residents of rivers in hilly regions, neither descending far into those of the plains nor found on the level plateaus on the summits of the mountains. This accounts for their absence from the Yárkand collection; and from the foregoing extracts it appears probable that they are not found to the north of the Oxus. This genus appears to be on the outskirts of the rest of its group; and its mouth armed with a sucker, to resist its being washed away, makes it well able to sustain a mountain-torrent life.

The other genera are more or less spread in the following districts. From the Helmund River and the eastern portion of Afghanistan, the upper part of the Oxus, and the eastern portion of Western Turkestan, the Tian Shan or Celestial Mountains, and also the Alatau mountains more to the south, they extend along the Himalayan region, certainly as far as the most easterly part of Assam.

These fishes (*Schizothoracinæ*) are confined to cold regions, as a rule, or at least to localities possessing snow-fed rivers, many of which rivers end in lakes and do not go to the sea.

They extend from Eastern Afghanistan and Western Turkestan through Tibet, and the most westerly portion of China, along the Himalayas to the hills in the Yunnan direction.

Loaches (*Nemacheilus*) are likewise generally distributed; and it is remarkable, as I have already observed, that all are scaleless. The same appears the rule in Western Turkestan.

The conclusion, I think, we may fairly arrive at, after examining the fishes of Yárkand and the adjoining countries, is that we find a peculiar group of Carps (*Schizothoracinæ*) which has spread almost due east and west from the cold and elevated regions of Eastern Turkestan, but of which the southern progress has been barred by the Himalayas.

If we look to the south, we see, as it were, that a wave of tropical forms of fishes has, at a prehistoric period, expanded over that portion of the globe where the Nicobars, Andamans, and the most southern portions of the continent of Asia and the islands of the Malay Archipelago now are, that this fish fauna has its northward progress arrested by some cause at or near where the Himalayas now exist and mark the division between the fish-fauna of India and that of Turkestan.

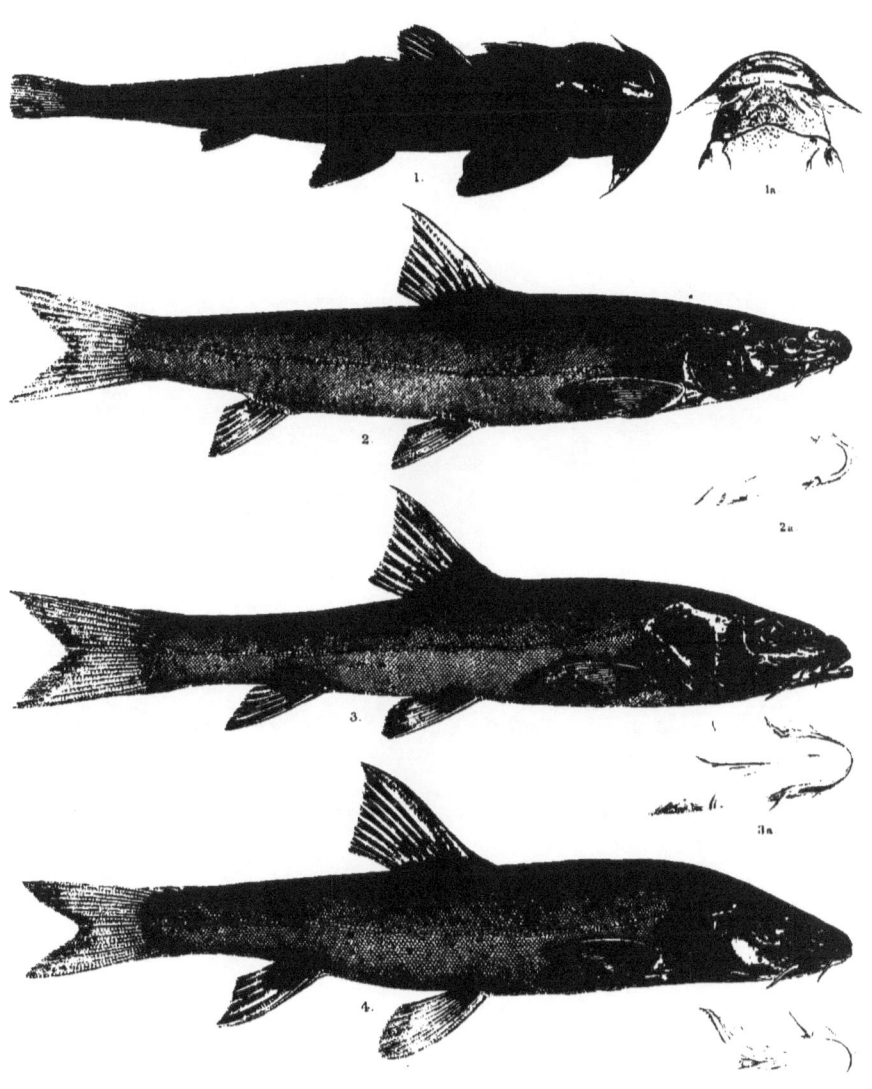

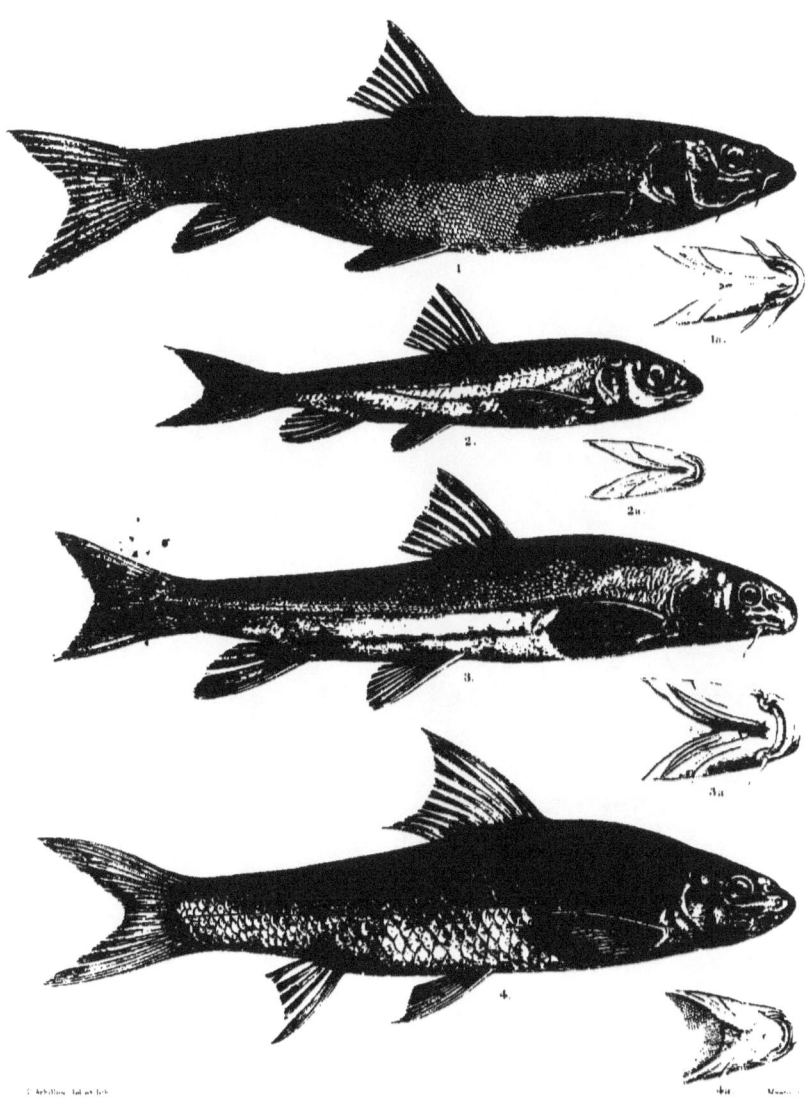

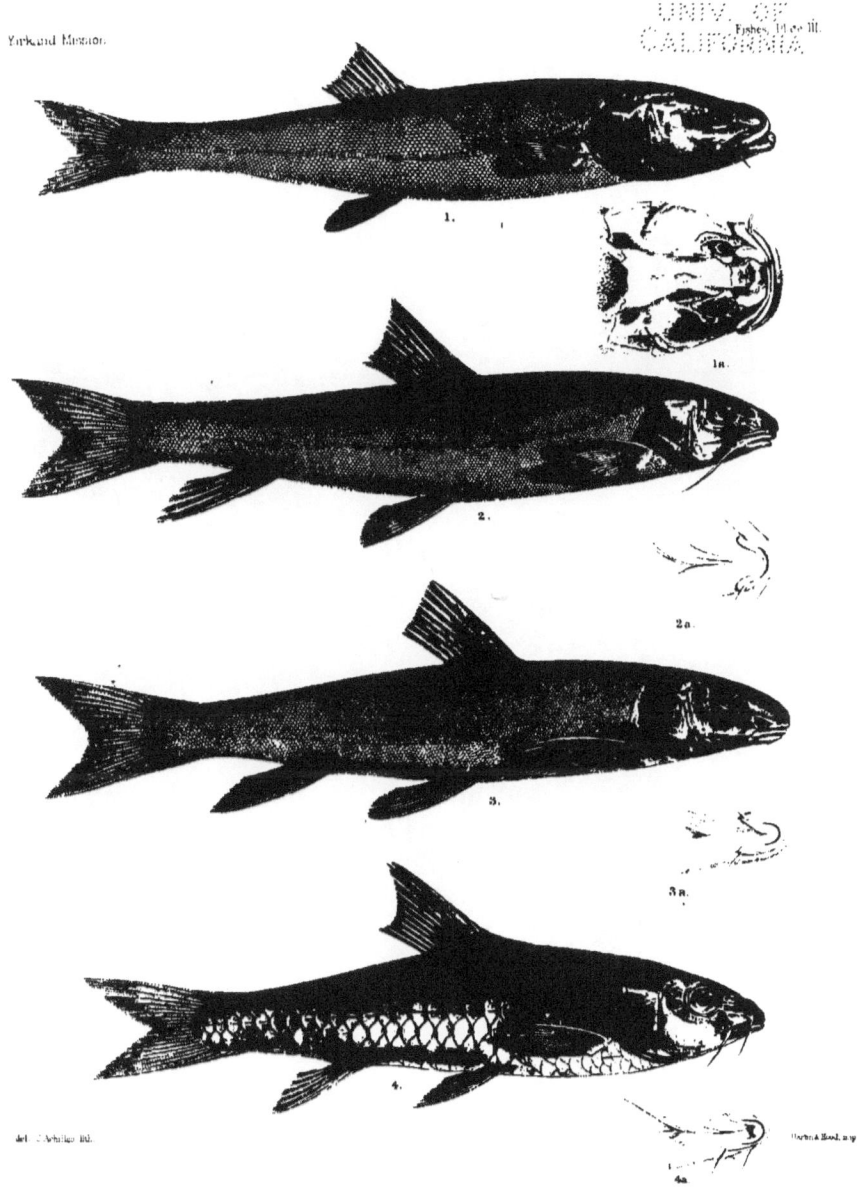

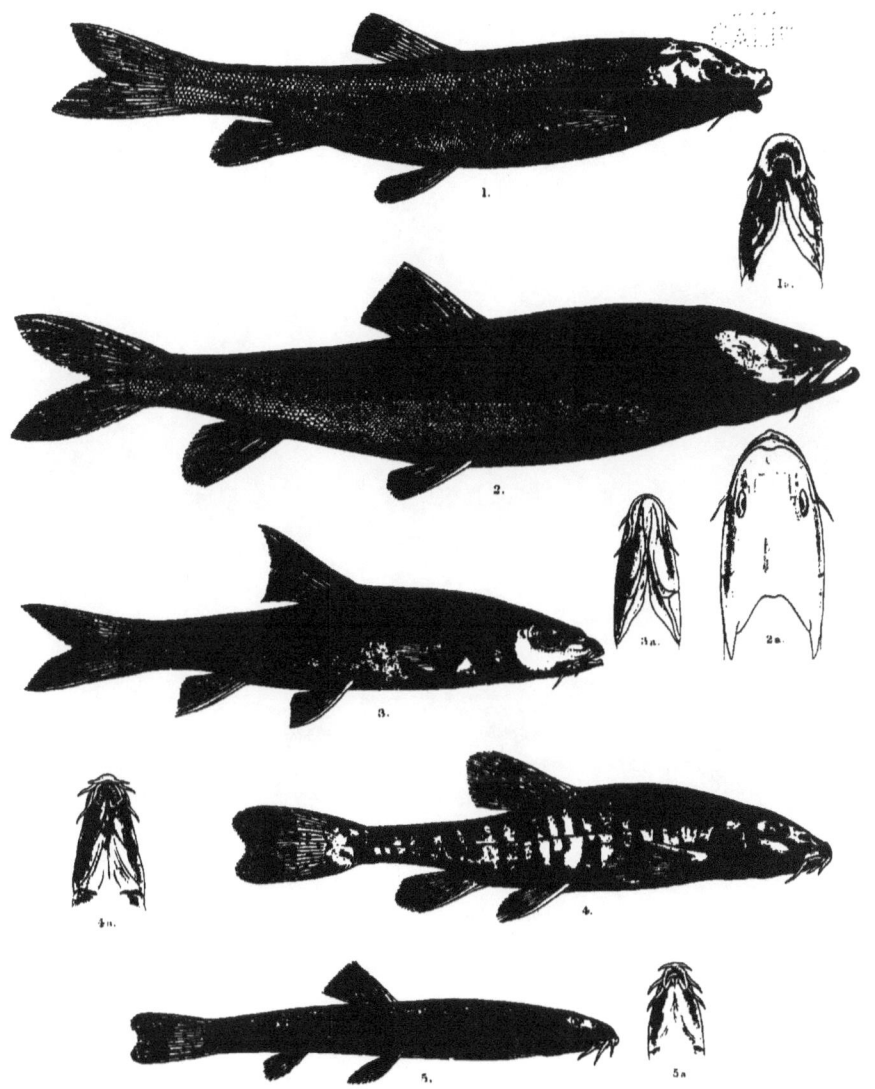

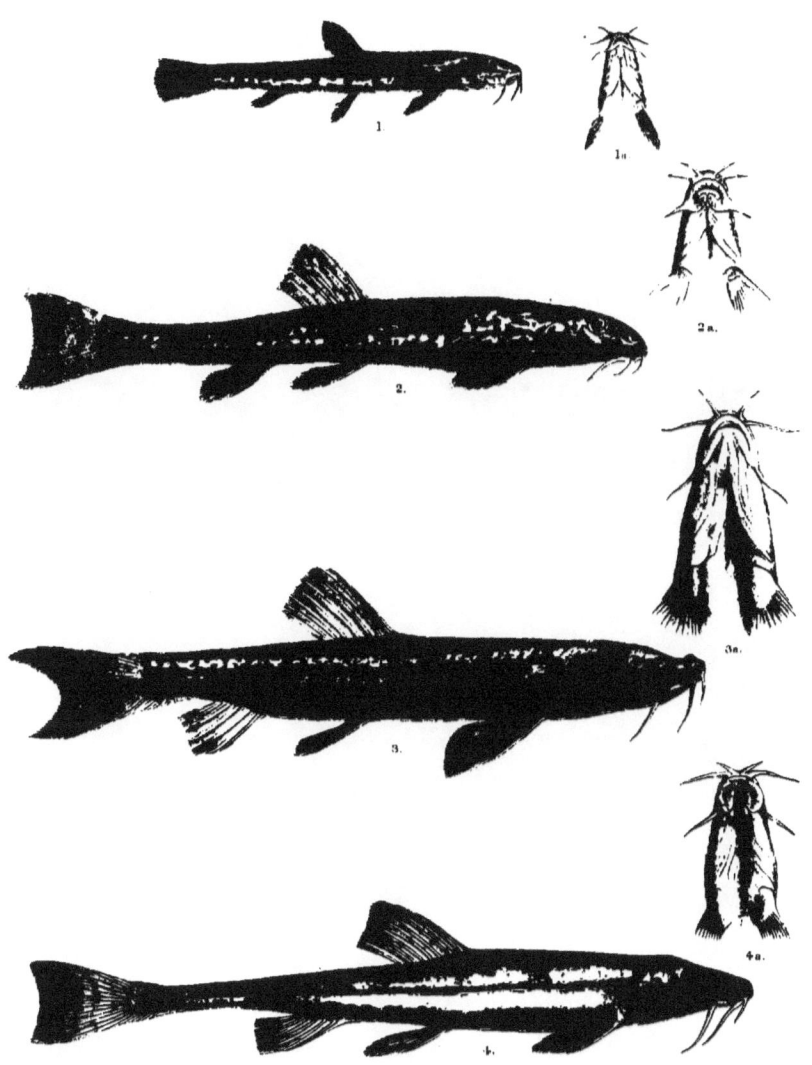

1. NEMACHEILUS MARMORATUS. 2. N. STOLICZKÆ. 3. N. YARKANDENSIS. 4. N. TENUIS.

SCIENTIFIC RESULTS

OF

THE SECOND YARKAND MISSION;

BASED UPON THE COLLECTIONS AND NOTES

OF THE LATE

FERDINAND STOLICZKA, Ph.D.

REPTILIA AND AMPHIBIA.

BY

W. T. BLANFORD, F.R.S.

Published by order of the Government of India.

CALCUTTA:
OFFICE OF THE SUPERINTENDENT OF GOVERNMENT PRINTING.
1878.

SCIENTIFIC RESULTS

OF

THE SECOND YARKAND MISSION;

BASED UPON THE COLLECTIONS AND NOTES

OF THE LATE

FERDINAND STOLICZKA, Ph.D.

REPTILIA AND AMPHIBIA.

BY

W. T. BLANFORD, F.R.S.

Published by order of the Government of India.

CALCUTTA:
OFFICE OF THE SUPERINTENDENT OF GOVERNMENT PRINTING.
1878.

CALCUTTA:
PRINTED BY THE SUPERINTENDENT OF GOVERNMENT PRINTING,
8, HASTINGS STREET.

SCIENTIFIC RESULTS

OF

THE SECOND YARKAND MISSION.

REPTILIA.

BY W. T. BLANFORD.

THE collection of reptiles made by Dr. Stoliczka during his travels with the second expedition to Eastern Turkestan was small, owing partly to the country traversed not being rich in forms of animal life, but still more because of the unfavourable season at which many of his journeys were made. The Thian Shan was visited in the depth of winter, and the Pámir steppes and Wakhán long before the snow had melted, and, under these circumstances, no snakes, lizards, or other forms of reptilian life could be found. The bulk of the collection consists of specimens procured on the journey from India to Káshghar, in the Punjab hills beyond Marí (Murree), in Kashmir and in Ladák, and those obtained on the return journey between Yárkand and the Karakoram. Of several of the species, fine series have been obtained.

The only reptiles previously collected in the districts traversed beyond Kashmir were (1) those procured by the Messrs. von Schlagintweit, who, in 1857, obtained one species of lizard, which was described by Dr. Günther in the Reptiles of British India; (2) by Dr. Stoliczka himself, who, when in Ladák in 1865, collected several reptiles, of which Dr. F. Steindachner gave an account, together with the Reptiles of the Novara Expedition; and (3) a few specimens obtained by the first expedition to Yárkand in 1870, which were examined and described by Dr. Anderson in the Proceedings of the Zoological Society for 1872. The last was the only collection which included specimens from Turkestan, but, unfortunately, the localities had apparently, in some cases, not been correctly marked on the labels. It is well known that there is much confusion in the localities of the specimens collected by the Messrs. von Schlagintweit. Nearly the whole of Dr. Stoliczka's collections are carefully labelled, and in the very few cases in which, from the labels having been omitted or lost, there is doubt as to the original locality of a specimen, this is noted in the subsequent pages in the list of the specimens collected.

The following is a list of the species of Reptiles hitherto procured from Ladák and the Upper Indus valley:—

LACERTILIA:

Stellio himalayanus.
Phrynocephalus theobaldi.

Gymnodactylus stoliczkæ.
Mocoa stoliczkæ (? =M. ladacensis).

OPHIDIA:

Zamenis ventrimaculatus (Z. ladacensis, Anderson).

The last-named is the only species not obtained by Dr. Stoliczka in his last journey. It had, however, previously been procured by him in Spiti (Steindachner, Rept. Nov. Exp., p. 65). All the other species named had also been obtained previously, no additions having been made to the fauna by the present collection.

The very moderate list of species as yet procured from Eastern Turkestan comprises the following forms:—

LACERTILIA :

Stellio stoliczkanus.
Phrynocephalus theobaldi, var. (P. forsythi).
P. axillaris.
Teratoscincus keyserlingii.

Gymnodactylus elongatus.
G. microtis.
Eremias yarkandensis.
E. vermiculata.

OPHIDIA :

Zamenis ravergieri.

Tropidonotus hydrus.

Taphrometopon lineolatum.

Of these species, only *Phrynocephalus theobaldi* and *Eremias yarkandensis* had been obtained before the country was visited by Dr. Stoliczka; another species, *Cyrtodactylus yarkandensis*, recorded as having been brought from Yárkand, having really, I believe, been collected in Ladák, and wrongly labelled.

In the present account the following species are also mentioned, specimens of them having been collected by Dr. Stoliczka in the Punjab hills or in Kashmir :—

LACERTILIA :

Stellio tuberculatus.
S. agrorensis.

Eumeces tæniolatus.
Mocoa himalayana.

OPHIDIA :

Typhlops porrectus, var.
Compsosoma hodgsoni.
Ptyas mucosus.

Tropidonotus platyceps.
Vipera obtusa.
Halys himalayanus.

With the possible exception of the last, none of these species appears to be found in the dry region of Ladák, north of the dividing range between Kashmir proper and the Indus valley.

It is thus evident that, so far as the Reptiles are considered, the countries traversed by Dr. Stoliczka between the plains of India and Káshghar yield three entirely distinct faunas: (1) that of the Punjab hills and Kashmir, comprising a majority of Himalayan forms, with a few species common to the plains of India and some types belonging to palæarctic genera; (2) that of Western Tibet; and (3) that of Eastern Turkestan, both the latter belonging to the palæarctic region, but to distinct sub-divisions, only one species having hitherto been found in both areas, and even that is represented by well-marked varieties.[1]

[1] Since the present account was first written, I have received, through the kindness of Dr. Strauch, a copy of his description of the reptiles collected by Colonel Przevalski in Central Asia. The work was published in 1876, and is, therefore, later in date than my preliminary account of Dr. Stoliczka's collections in the "Journal of the Asiatic Society of Bengal" for 1875 (vol. xliv, p. 191). The greater portion of Dr. Strauch's paper is unfortunately in Russian, but the descriptions are in Latin, and excellent lithographs of all the new species are given. One form of *Eremias*, *Podarces (E.) pylzowi*, appears to me possibly the same as *E. vermiculata* from Yárkand; but of this I am not certain, and I am unable to identify any of the other forms described, including five species of *Phrynocephalus*, and five (besides *E. pylzowi*) of *Eremias*, with the species inhabiting Eastern Turkestan.

REPTILIA. 3

Order LACERTILIA.

Family—*AGAMIDÆ*.

1. STELLIO HIMALAYANUS.

Steindachner: Novara Reise, Reptilien, p. 22, Pl. i, fig. 8.
Stoliczka: Jour. As. Soc. Bengal, 1872, xli, Pt. 2, p. 113.

1, 2, Dras valley; 3, 4, Tashgaon, near Dras; 5-7, Chiliscomo; 8, Shargol; 9, Kharbu; 10, 11, Snemo near Leh; 12-22, Leh;—all in the Upper Indus valley, north of Kashmir.

These specimens are from the original locality and its neighbourhood. *Stellio himalayanus* has hitherto only been found in the Upper Indus valley in Ladák, where it was originally discovered by Dr. Stoliczka.

In his diary Dr. Stoliczka remarks that the male of this lizard is smaller, and has the whole head, breast, and shoulders tinged with yellow, and the sides of the neck umber red. These colours are probably assumed in the breeding season; the date when they were noticed was August 17th.

2. STELLIO TUBERCULATUS.

Gray *apud* Günther: Reptiles of British India, p. 157.
Stoliczka: Jour. As. Soc. Bengal, 1872, Pt. 2, xli, p. 115, Pl. iii, fig. 3.

1, 2, Kashmir.

Though labelled Kashmir, the specimens were probably obtained on the road from Mari (Murree) to Srinagar. The species is common about Mari.

3. STELLIO AGRORENSIS. Pl. I, fig. 3.

Stoliczka: Proc. As. Soc. Bengal, 1872, p. 128.

1-6, Kashmir.

The specimens agree well with the types from the Agror valley in the Punjab hills. In his diary, Dr. Stoliczka records obtaining this species near Chatarkailas in the Jhilam valley, north-east of Mari.

As no figure of this species has ever appeared, one is published herewith. A full description was given by Dr. Stoliczka.

4. STELLIO STOLICZKANUS. Pl. I, figs. 1 & 2.

W. Blanf.: Jour. As. Soc. Bengal, 1875, xliv, Pt. 2, p. 191.

S. squamis dorsalibus mediis majoribus, haud in lineas regulares ordinatis, obtuse carinatis, lateralibus minoribus, acute carinatis, postice subæqualibus ; nonnullis mucronatis circum tympanum, et in fasciculos ad latera colli et supra humeros dispositis; caudalibus carinatis, mucronatis, verticillatis, dorsales magnitudine vix excedentibus; stramineus, capite dorsoque posteriore nigro-punctatis, dorso anteriore nigro, stramineo transversim fasciato.

1-7, Yangihissár, 8, Karghalik, south of Yárkand, both in the plains of Eastern Turkestan.

Description.—General form apparently more slender than in *Stellio caucasicus* or *S. tuberculatus*; body and base of tail depressed; tail 1.5 times to nearly twice the length of the body; the fore limb laid backward does not reach the thigh (except in very young specimens); the hind limb laid forward extends to about the ear. Head depressed, its length considerably exceeding its breadth. The largest specimen collected measures 14.75 inches, of which the head and body from the snout to the anus measure 5.4, fore limb to end of toes 2.6 inches, hind limb nearly 4, third toe of hind foot without the claw, measured from between the third and fourth toes, 0.65. In a smaller specimen the head and body measure 4.6, tail 8 inches.

The scales on the upper surface of the head are convex, those on the occiput being submucronate, those on the supra-orbital bosses are rather smaller and flat. Supra-orbital ridge and *canthus rostralis* prominent, loreal region concave, bearing small scales, some of which, like most of the scales on the side of the head, are bluntly keeled. Nostrils directed backwards, situated in the hinder part of a single shield below the *canthus*. Rostral more than twice as broad as high. Labials not much larger than the neighbouring scales. Mental the same breadth as the rostral and pointed below. Eyelids covered with small granular scales, those along the edges of the lids rather larger and pointed. Some rather large scales bluntly keeled or submucronate between the eye and the tympanum. Some spinose scales round the tympanum : groups of spinose scales are scattered over the sides and back of the neck, the former being the larger. There is no trace of a crest. Sides of the neck between the larger scales covered with very small conically mucronate scales.

Scales on the back of the neck granular, passing gradually into the bluntly keeled scales of the middle of the back; these are considerably larger than the scales of the sides, being about twice as broad. The scales on the lateral portions of the body are distinctly keeled, in tolerably well-marked transverse rows, and nearly uniform in size, but few conspicuously larger scales being scattered amongst them in general, though a few may occasionally be detected here and there, and these are patches of enlarged subspinose scales of pale colour about the shoulders. There is no patch of enlarged scales in the middle of the sides. Scales of the belly smooth, rhomboidal, about the same size as those in the middle of the back, and arranged in transverse series, containing towards the middle of the belly from fifty-eight to sixty-seven scales, tending, however, to pass into the keeled scales at the sides. I count about 150 to 160 scales round the middle of the body. The throat scales are similar to those of the abdomen, but much smaller.

In males there are two or three rows of thickened scales before the anus; in females the scales are a little larger than those adjoining, but not thickened. There is no patch of thickened scales in the middle of the abdomen, as there is in *Stellio caucasicus*, and several other species of the genus. All the limb scales are keeled, those above sharply, those below, and especially on the hind limb, faintly; those on the back of the thigh small, with a few larger and subspinose scales scattered amongst them; scales below the feet keeled, very similar to those above; toes covered beneath with transverse plates, each with several keels. Tail scales,

REPTILIA. 5

except near the base below, keeled, and ending in a short spine posteriorly; those near the base scarcely larger than the back scales, those behind very little smaller, all in verticils. There is a double fold below the neck, several at the side of the neck, and one which passes above the shoulder and down the side.

The general colour is pale yellowish, mixed with dusky black. The head above is straw-coloured, with a few black scales scattered over the upper surface and irregular vertical dusky bars on the side. Anterior portion of the back and upper part of forelimbs dusky, with transverse rows of pale spots, sometimes forming tolerably marked bars, especially on the shoulders and upper parts of the fore legs; hinder part of the back and sides straw-colour, speckled with black. Tail pale yellowish at the base, sometimes with indications of crossbands; hinder portion brown. Lower parts uniform pale yellow, except the chin and throat, which are dusky, more or less mottled, or speckled with pale yellow. The young is much paler in colour, with a pinkish tinge, and the scattered black scales on the back are few in number, and form rather irregular transverse lines.

There are twelve to fourteen maxillary teeth on each side of the upper jaw, and three pairs of conical teeth in front; the outer pair the largest. In the lower jaw there are twelve to thirteen teeth along each side, and two pairs of more elongate pointed teeth in front.

All the larger specimens are eviscerated. Dr. Stoliczka in his diary mentions that, at Karghalik, he found this species living in holes in sand, and that, on a low bush, he saw one specimen which, when pursued, took to the ground immediately. I have never seen any other *Stellio* which had similar habits, though probably, from its habitat, *S. aralensis* may resemble the present species. All the other species of the genus are, as a rule, rock lizards, living on the rocks, and taking refuge in clefts and under stones. *S. nuptus* in Persia is sometimes found on old walls of hardened mud, but with the exception of *S. aralensis*, I have never heard of any species inhabiting level ground and living in holes, as, from Dr. Stoliczka's note, is, I infer, the case with the present form.

S. stoliczkanus differs much from all known species. The arrangement of the scales on the body is quite distinct in *S. nuptus*[1] and *S. melanura*,[1] which belong, indeed, to a different section of the genus. *S. tuberculatus*,[2] *S. agrorensis*[3] and *S. dayanus*[4] are stouter forms; the first two are at once recognised by their more strongly keeled dorsal scales, the much greater difference between the dorsal and lateral scales, and the smaller size of the latter, and *S. dayanus* differs in having strongly keeled dorsal and lateral scales, in the numerous large scales scattered over the sides, and the larger scales on the limbs, besides other distinctions in each case. None of the three species have the spinose scales on the sides of the neck so developed as in *S. stoliczkanus*. *S. himalayanus* has the central dorsal scales smooth, besides other distinctions.

S. caucasicus[5] and *S. microlepis*[6] are also distinguished by stouter form and broader heads, by the presence of a large cluster of enlarged scales in the middle of each side, and of an oval patch of thickened scales in the middle of the abdomen in both sexes. The scales in

---

[1] DeFilippi: Giornale del I. R. Ist Lomb. vi, (1843);—*Eastern Persia*, ii, p. 317.
[1] *Laudakia (Plocederma) melanura*, Blyth: Jour. As. Soc. Bengal, 1854, xxiii, p. 738;—*S. melanurus*, Anderson: Proc. As. Soc. Bengal, 1871, p. 180.
[2] *Vide* ante.
[4] Stoliczka: Jour. As. Soc. Bengal, 1872, xli, Pt. 2, p. 113.
[5] Eichwald: Zool. Spec. iii, p. 187;—Fauna Casp. Cauc., p. 80;—*Eastern Persia*, ii, p. 322, Pl. xx, fig. 1.
[6] *Eastern Persia*, ii, p. 326, Pl. xix, fig. 2.

S. microlepis are smaller throughout. On the whole, the present species approaches *S. caucasicus* more nearly than any other form with which I am acquainted.

I have no specimen of *Stellio aralensis*[1] for comparison, and from its inhabiting the steppes east of the Sea of Aral, it may very possibly be nearly allied to the present species. According to Lichtenstein's description, it has the back scales strongly keeled and mucronate, and the toes fringed, the colouration is very different from that of *S. stoliczkanus*, being ash-grey, with pale wavy crossbands, the tail and limbs being also banded, and there is a large black spot at each side of the neck in the fold. The young have this spot peculiarly distinct and have long pale spots on the back on a bluish-grey ground. There can be but little doubt of the present being a distinct species. A form from Western Turkestan appears to have been named *L. lehmanni* by Strauch,[2] but I can find no description of it. In the list of Western Turkestan reptiles, "*S. himalayanus*, Strauch," is also included by Severtzoff.

5. PHRYNOCEPHALUS THEOBALDI.

P. tickellii, Günther: Proc. Zool. Soc., 1860, pp. 167, 173, *nec* Gray.
P. olivieri, Theobald: Jour. As. Soc. Bengal, 1862, xxxi, p. 518, *nec* Dum. et Bibr.
P. theobaldi, Blyth: Jour. As. Soc. Bengal, 1863, xxxii, p. 90;—W. Blanf.: Jour. As. Soc. Bengal, 1875, xliv, Pt. 2, p. 192.
P. caudivolvulus, Günther: Rept. Brit. Ind., p. 161 (1864);—Theobald: Cat. Rept. Mus. As. Soc., p. 40 (1868);—Anderson: Proc. Zool. Soc., 1872, p. 387, *nec* Pallas?
P. stoliczkæ, Steindachner: Novara Expedition, Reptilien, p. 23, Pl. i, figs. 6, 7.
P. forsythi, Anderson: Proc. Zool. Soc., 1872, p. 390, fig. 7.

1-4, between Sonamarg and Kharbu (all probably from the Indus valley and not from the Kashmir side of the Zoji-la); 5-7, Namika-la, north-east of Shargol; 8-14, above Kharbu, 14,000 feet; 15-22, Lamayuru; 23-25, Snemo; 26-40, Leh (all the above from the Indus valley in Ladák); 41-47, Mughlib, east of Tánkse 14,000 feet; 48-53, Lukung, Pankong Lake; 54, 55, Chagra, north of Pankong Lake; 56-66, between Yárkand and Karakoram (this and all the following specimens belong to the variety *P. forsythi*); 67, Sánju; 68, Yárkand; 69, Kizil; 70-72, Yangihissár; 73, 74, Kashkasu, on road from Yangihissár to Sarikol.

I feel convinced that there must be some mistake in uniting the *Phrynocephalus* of Western Tibet with *Lacerta caudivolvula* of Pallas.[3] In the first place, Pallas' description, *L. corporis squamis minutissimis lævibus, cauda longiuscula lævissima, subtus apice rubro nigroque variegata*, does not appear to agree well. The tail in the Tibet *Phrynocephalus* is not nearly so long as would be inferred from the above description and from the measurements of *L. caudivolvula* by Pallas,—whole length 3 inches 3 lines, tail 2 inches, so that the proportion of the head and body to the tail is 5 to 8. In a large number of specimens from Tibet and Eastern Turkestan I find the proportions of the head and body to the tail vary between 5 to 5·6 and 5 to 6·3, the last being exceptional. The tail, moreover, can scarcely be called very smooth; the scales towards the extremity, as a rule, are keeled. Then the colouration is different, and especially that of the tail, which is said by Pallas, in his more detailed description, to be *subtus a medio ad apicem interrupte nigra et rubra*. The colouration in *P. theobaldi* is extremely variable, as noticed by Steindachner in his description (of *P. stoliczkæ*), but I have never seen

---
[1] *Agama aralensis*, Lichtenstein, Eversmann's Reise von Orenburg nach Buchara p. 144. It is by no means clear that Lichtenstein's species was really a *Stellio*.
[2] Severtzoff: Turkistanskie Jevotnie, p. 71.
[3] Zoogr. Ros. As., iii, p. 27.

an approach to the mixture of red and black described by Pallas. It is true that these red colours are probably seasonal, and that they tend to disappear in spirits, but the colouration in the specimens before me is so well preserved, that it would be surprising if no trace of red remained in any of them, and they were collected at various seasons, some in autumn, others in spring and early summer.

I am unable to find Pallas' figure of *Lacerta caudivolvula*, but there are two figures of the species, both accompanied by descriptions, by Eversman[1] and Eichwald.[2] These figures I have compared with the Tibet *Phrynocephalus*, and I find both agree with Pallas' description of *P. caudivolvulus*, and differ from *P. theobaldi*. It is true that Eversman gives the length of the body as 1 inch 11 lines and of the tail 2 inches 1 line, but his figure shows a longer tail than this, whilst Eichwald gives the lengths of the body and tail respectively as 1 inch 8 lines and 2 inches 5 lines, a proportion of 5 : 7·5. Eichwald describes the tail as having black rings towards the end, the interspaces below being red; Eversman merely says that there are black bands on the lower surface of the tail towards the extremity, with red interspaces. A comparison of Eversman's or Eichwald's figure with Steindachner's bears out the view I have expressed of the species represented being distinct.

Dumeril and Bibron[3] appear to me to have described a different species from Pallas', under the name of *Phrynocephalus caudivolvulus*. The tail is said to be but little longer than the body, and to be black at the end, with blackish spots along the sides of the remaining portion, and the ventral scales are said to be keeled, whereas Pallas, Eversman and Eichwald describe them as smooth. Dumeril and Bibron's description agrees, except in having the ventral scales keeled, with *P. theobaldi*. Now, the specimen described by the French herpetologists came from Berlin, and was very probably identical with that compared with the Tibet *Phrynocephalus* by Professor Peters.[4].

But what especially guides me in rejecting Pallas' name for the Tibet *Phrynocephalus* is that name itself, and the circumstance from which it was derived. Pallas says "*percepto inimico caudam coloratam versus dorsum in spiram promptissime revolvit, quod in nulla alia specie vidi.*" Now, there are two groups of *Phrynocephali*, to one of which belong *P. olivieri*[5] and *P. persicus*[6]; both of these I have seen alive in large numbers, and I never yet saw one coil its tail, whilst to the other belongs *P. maculatus*[7] and a species to be described immediately, both of which have been observed, the latter by Stoliczka, the former by myself, to have a habit of coiling their tails. These last are much smoother, as a rule, than the *Phrynocephali* of the former group, and their tail is much longer, whilst in *P. maculatus* the under surface of the tail, when alive, is frequently red in part. I think it is to this group that the true *Lacerta caudivolvula* must belong, whilst *P. theobaldi* certainly belongs to the former group. In Mr. Theobald's very good account of the habits of the present species[8] which he obtained on the Tso-Morari in Rupshu, he makes no mention of having seen it coil its tail, nor does Stoliczka notice any such habit, although he especially describes it in the case of the other Turkestan species, and gives a sketch in his diary of the appearance presented.

[1] *Lacertæ Imperii Rossici*, Nouv. Mem. Soc. Imp. Nat. Moscou, iii, p. 362, Pl. xxxii, fig. 2.
[2] *Fauna Caspia Caucasia*, Nouv. Mem. Soc. Imp. Nat. Mosc., vii, p. 107, Pl. xii, figs. 6, 7, Pl. xiii, figs. 9—14.
[3] Erp. Gén., iv, p. 522.
[4] Günther: Rept. Brit. Ind., p. 161.
[5] Dum. et Bibr.: Erp. Gén. iv, p. 517;—Eastern Persia, ii, p. 327.
[6] DeFilippi: Archiv. Zool. Genova, ii, p. 387;—Eastern Persia, ii, p. 329.
[7] Anderson: Proc. Zool. Soc., 1872, p. 388;—Eastern Persia, ii, p. 331.
[8] Jour. As. Soc. Bengal, 1862, xxxi, p. 518.

I have not overlooked the fact mentioned by Dr. Günther in the "Reptiles of British India,"[1] and to which reference has already been made, that specimens from Tibet had been compared by Professor Peters of Berlin with typical examples of *L. caudivolvula*, and found specifically identical. I confess that it appears at the first glance as if the opinion of so high an authority on the Reptilia as Professor Peters must be more correct than mine, but I think there must be some mistake, as I have already indicated when noticing the description of *P. caudivolvulus* by Duméril and Bibron. The original types of Pallas can scarcely be in Berlin, and it has frequently happened that other species have been sent from Russia under Pallas' names. Under any circumstances I cannot but think, for the reasons given above, that Pallas must have described a different lizard.

Steindachner in his description of *P. stoliczkæ*, which is certainly the same lizard as *P. theobaldi*, several of the specimens examined by Steindachner being from the typical locality of the last-named species, points out that *P. stoliczkæ* differs from *P. caudivolvulus* in its shorter tail and in having smooth scales on the upper surface of the limbs. The latter character, however, is not constant. Keels may generally be detected in *P. theobaldi* on the scales of the tarsus, and not unfrequently on the thigh and forearm, and in the Turkestan variety, *P. forsythi*, they are the rule. The length of the tail is, however, a characteristic distinction, though, I believe, it is not the only one.

It is only after long and repeated comparison that I have come to the conclusion, that *P. forsythi* of Anderson cannot be separated from *P. theobaldi*.[2] At the first glance, they appear distinguished by colour and by the Turkestan form having some scattered, whitish, enlarged scales on the back, and keels on the scales covering the upper surface of the limbs. Individuals, however, vary greatly in the scales of the back; in some these are convex and granular, in others flat, smooth, and even subimbricate; in some larger in the middle of the back, in others nearly the same size throughout. The scales on the top of the head are scarcely alike in any two individuals; some have the scales large on the occiput and very small on the supra-orbital region, in others all are of about equal size; in some the enlarged superciliary scales almost reach the nasals, in others three or four small scales intervene. The keels on the limb scales and the enlarged scales on the sides of the back are no more constant than the other characters. I find specimens from Western Tibet with a few scattered enlarged scales, and with distinct keels on the limb-scales, and I find specimens from Eastern Turkestan in which the enlarged scales are wanting and the keels can scarcely be detected.

Even in colouration, I do not think the difference, although it is usually marked, is constant. *P. forsythi* has almost always a row of rather distant dark spots, arranged in pairs down each side of the back. These spots consist of rather pointed scales. *P. theobaldi* varies exceedingly in colour. Some specimens, perhaps the most, are rather irregularly spotted, others have large ocelli on the back; in others again there are no markings whatever. But there is very often a tendency to a double row of spots down the back, and in some cases a very near approach to the colouration of *P. forsythi*, and in the latter the spots

---

[1] p. 161.

[2] I may here remark that I believe Dr. Anderson was misled by his collectors into supposing that the specimens of *P. theobaldi* described by him, Proc. Zool. Soc., 1872, p. 387, under the name of *P. caudivolvulus*, were from Yárkand. Like the gecko named by him *Cyrtodactylus yarkandensis*, I think it almost certain that the *Phrynocephali* in question must have been collected in the Upper Indus valley, in Ladák. Every specimen from Yárkand and Eastern Turkestan in Dr. Stoliczka's large collection has the colouration of *P. forsythi*, whilst the specimens described by Dr. Anderson, which I have examined, are undistinguishable from some of those procured by Dr. Stoliczka in Ladák.

are often faint and small, or some of them are wanting, whilst in other specimens additional spots are found on the sides. It is evident that the colouration varies, in the case of *P. theobaldi*, to a great extent, and therefore it would be impossible to found a specific distinction upon it without stronger differences.

There is one difference which, if constant, would be of great importance. *P. theobaldi*, as Theobald has shown, is viviparous (or, more correctly ovo-viviparous), and I find fœti in the females, whereas I find only eggs in the oviducts of a female *P. forsythi*. But this may depend on the time of year, the pregnant females of the former species having been captured at a later period of the season.

Dr. Anderson omits to point out the characters which led him to suppose that *P. forsythi* was a distinct species. I may have overlooked some difference, but I have examined both forms carefully, and I do not think the two can be distinguished by any constant character.

The following is a description of *P. theobaldi* from the specimens before me:

General form as in *P. olivieri*; tail a little longer than the head and body, rather thick at the base, tapering beyond, but much thicker throughout in some specimens than in others. In some cases the tip is laterally compressed, in others round. Limbs rather short, the hind limb reaches beyond the shoulder, and often to the head; the fore limb does not reach the thigh. Scales of back small, flat, or convex, often granular. The scales in the middle of the back usually larger than those of the sides. A few rather larger scales are sometimes scattered over the sides, but they are not much larger than the others. The black scales forming spots on the back are sometimes more pointed. Scales on the head larger than those on the back; usually the largest are on the occiput. Upper labials with projecting rounded, or pointed margin along the edge of the lip, lower labials straight edged. Scales on upper surface of limbs generally faintly, sometimes more strongly keeled, often almost or quite smooth. Scales of lower parts smooth. Tail scales smooth, except towards the end, where they are usually keeled, more strongly below than above.

Usual colour above olive-grey, varying in tint, and more or less spotted with black; sometimes the ground colour is pale, almost cream-coloured, and the spots form ocelli. Sometimes, besides the black marks, there are whitish spots of various sizes. The variety *P. forsythi* has usually four or five pairs of black spots on the back, and is bluish-grey in colour. The tail has dusky spots along each side; these are never, except towards the tip, joined across the lower surface as in *P. olivieri*, but they frequently meet above. Tip of the tail not unfrequently black, and in many specimens (especially males, though it is not confined to them) the central portion of the abdomen is black; this colour sometimes extending to the whole, or nearly the whole, lower surface of the body and head.

6. PHRYNOCEPHALUS AXILLARIS. Plate I, fig. 4.

W. Blanf.: Jour. As. Soc. Bengal, 1875, xliv, Pt. 2, p. 192.

P. major, lævis, caudâ elongata, pede anteriore in adulto vix femur attingente, squamis omnibus lævibus, caudæ apicem versus exceptis; supra griseus, maculâ rubrâ utrinque post axillam notatus, membris caudâque fasciis fuscis transversis signatis, hac ad medium fusco-

annulatá, nunquam ad apicem nigrá, subtus albidus. Long. tota poll. 5—6, caudæ ⅔ totius longitudinis subæquante.

1, 2, south of Yárkand; 3-9, Yárkand; 10-16, Akrobát near Yárkand; 17-20, Kízil; 21-33, Yangihissár—all in Eastern Turkestan.

Description.—General form depressed; head not so short as in *P. theobaldi, P. olivieri,* &c., and tail longer. In adults the fore limb falls short of the thigh, or barely reaches it; the hind limb extends to the eye; the tail is one-third to one-half longer than the body with the head. The base of the tail is depressed and slightly dilated, thence the tail tapers gradually; it can be coiled upwards near the end. Toes rather long; the fourth toe on the hind foot exceeds the third by more than the length of the claw, and has both sides fringed; the outer edge of the third toe is also fringed; the fifth toe of the hind foot without the claw falls short of the point of union of the third and fourth toes. Claws yellowish, strong; the claw of the fifth toe on the hind foot longer than the rest. Nine to ten triangular teeth on each side in both jaws; six pointed anterior teeth in the upper jaw, four in the lower; the outer pair in each jaw elongate. The largest specimen measures 6·25 inches, of which the tail from the anus is 3·75, head 0·75, fore limb to the end of the toes 1·4, hind limb 2·2.

Scales of the head above convex, tubercular, not varying much in size, as a rule; each nostril in a larger scale, sometimes divided horizontally. Scales of the superciliary ridge larger; each eyelid with a fringe of about nine rather larger scales, the lower row pointed. Upper labials twenty-seven to thirty-one, with convex margins; rostral scarcely larger. Mental or lower rostral generally much larger than the lower labials, which are, as a rule, rather fewer in number than the upper labials. Scales near the lower labials rather larger than the remaining scales of the throat. Scales of the body about the same size above and below; all on the back and belly are smooth, rhomboidal, and arranged in transverse rows, those on the back subimbricate; on the sides they are smaller and more granular, especially behind the shoulder, where the red patch consists of small granular scales. I count from 95 to 103 scales round the middle of the body in adults, rather fewer in young specimens. Scales on the limbs usually rather bluntly keeled above, smooth below; those beneath the feet sharply keeled, cross-plates beneath the toes with several keels. The pointed scales forming a fringe on the outer side of the fourth toe on the hind foot are longer than those on the inside of the same toe, or on the outside of the third toe. On none of the other toes is there any distinct free margin. Tail scales all keeled, except at the extreme base; they are about the same size as the back scales, and are arranged in rings; the keels form longitudinal lines below the tail, but not above; at each side of the tail close to the base is a large patch of spinose scales.

Colour above pale slaty-grey, nearly uniform or speckled with whitish, or, in young individuals especially, with three or four dark crossbands on the body. In some cases the back is tinged in parts with pale copper red. Dusky crossbands on the limbs and tail. In the middle of the tail, about 1·5 inch from the end, two or three dusky bands pass round the under surface; these are sometimes nearly black, at other times so faint as to be barely perceptible, but they are not entirely absent in any of the specimens collected; end of the tail never black. Lower parts white throughout, except the bands round the tail. Behind the axil, so as to be partly concealed by the fore limb when laid back along the body, there is a red patch at each side; this in the living animal is said by Dr. Stoliczka to be bordered by blue. The red colour has faded greatly in spirit, but can still be detected.

This species of *Phrynocephalus* is very closely allied to the Persian *P. maculatus*,[1] and probably to the true *P. caudivolvulus*, Pallas *nec* Günther. It appears to be a much larger form than the latter. From the former it is distinguished by its limbs, when adult, being shorter, the fore limb, as a rule, not reaching the thigh, whereas it always reaches or even exceeds it in *P. maculatus*. The fifth toe of the hind foot in *P. maculatus* is longer; the fringe on the outer edge of the fourth toe less developed, and there is scarcely any fringe on the inside of the toe. But the most important distinctions are in colouration. *P. maculatus*, of which I have collected many living examples, never has a red spot behind the shoulder, and it always has a black tip to the tail, below at all events. The colouration of the tail in *Phrynocephali* is, as a rule, very constant.

P. axillaris is said by Dr. Stoliczka to be very active, to run at a great pace, and to have the habit of coiling its tail upwards at the end. It, doubtless, inhabits open plains, like its Persian ally.

From the above, it would almost appear as if I had proposed a new species on characters of no more importance than those which I have just before shown to be insufficient in the case of *P. forsythi*. But in the present case the characters appear constant, probably because the two forms *P. axillaris* and *P. maculatus* inhabit distant and isolated areas, whilst in the case of *P. forsythi* and *P. theobaldi* there is great variation, and no constant distinction can be detected even in colouration; moreover, so far as my experience of the genus goes, I should say that the red patch behind the axil in *P. axillaris* and the black tail tip in *P. maculatus* are more important than the back markings which distinguish *P. forsythi*. When *P. maculatus* exhibits bright colours, as it very often does, they are confined to the lower surface of the tail and hinder parts of the thighs.

Family—*GECKOTIDÆ*.

7. TERATOSCINCUS KEYSERLINGI.

Strauch: Bull. Acad. Sci. St. Pet., 1863, vi, p. 480;—Mel. Biol., vi, p. 554;—Zool. Record, 1864, p. 111.
Kuli-yailáng, Yárkandi (Scully).

1, Yangihissár, Eastern Turkestan.

This is a new locality for this very remarkable gecko. *Teratoscincus keyserlingi* was originally discovered by Count Keyserling in the Persian province of Khorassan, at a spot called Sar-i-cháh, and it has since apparently been found in Western Turkestan, as it is included by Severtzoff in his list of the Reptiles[2] found in that province. It thus appears to have a considerable distribution in Central Asia. The original description was copied in the "Zoological Record."

The single specimen obtained by Dr. Stoliczka is not in a very good state of preservation, but still the characters are easily distinguishable. The following is a description:

[1] Anderson: Proc. Zool. Soc., 1872, p. 389;—Eastern Persia, ii, p. 331.
[2] Turk. Jev., p. 71.

Habit stout, head and body depressed, limbs strong, toes rather short, tail shorter than the body. The hind limb reaches to the shoulder, the fore limb not quite to the end of the snout. Head covered with small granules above and below. Pupil vertical. Nostrils between the rostral, first labial and three enlarged plates behind; upper labials eleven, the hinder small, lower labials ten. Rostral nearly twice the breadth of two labials; mental also large, square behind. Some enlarged scales along the edges of the lower labials. Scales of the body all round large, smooth, imbricate, and rounded behind, those of the abdomen scarcely larger than those of the back; I count about thirty-two round the body, but they are a little irregular; scales on the limbs similar to those of the body, except behind the upper arm and thigh, where, as well as on the side of the trunk behind the shoulder, they are small and granular. Feet and toes covered with imbricate scales above, and with minute spinose tubercles below; all the toes provided with claws and fringed at the sides. Tail covered with smooth imbricate scales, those below, and near the base above, similar to those of the body; the posterior two-thirds of the tail covered above with large imbricate scutes, seventeen in number, the whole breadth of the tail. Region around the anus, before and behind, granular; two large pores, one on each side, behind, none in front. Length 5·1 inches, tail 2·1, forelimb 1·2, hind limb 1·6.

Colour grey above, with a few small blackish spots on the back, most strongly marked between the shoulders. According to Strauch, the pupil is circular, and young specimens are transversely banded, but Dr. Scully, who has seen a living specimen, assures me that the pupil is vertical, and this is borne out by the specimens I have examined. Comparing this specimen with *Teratolepis fasciata*,[1] the type of which, originally described by Blyth, is in the Indian Museum, I find that the differences pointed out by me in the "Zoology of Persia[2]" from the descriptions, hold good, and the two forms must be placed in distinct genera. *T. fasciata* has the basal portion of the toes dilated, and furnished with a double row of enlarged plates, but the toes are not fringed at the sides, and there is no external ear.

Another specimen of *Teratoscincus* has since been brought from Yárkand by Dr. Scully, who has ascertained that it is not very common, and that (according to the information given by the people) it inhabits waste ground, and is found about stones. The colouration of the back, when alive, is greenish, lower parts whitish, limbs pinkish fleshy.

8. GYMNODACTYLUS STOLICZKÆ.

Steindachner: Reptilien, Novara Expedition, p. 15, Pl. ii, fig. 2.
Cyrtodactylus yarkandensis, Anderson: Proc. Zool. Soc., 1872, p. 381, fig. 3 (figura mala).

1-5, Chiliscomo; 6-13, Kargil; 14, 15, Kharbu; 16, Lamayuru; 17, Snemo; 18-46, Leh;—all in the Indus valley, Ladák.

I have compared the specimens obtained by Dr. Stoliczka with the type of Dr. Anderson's *Cyrtodactylus yarkandensis*. They agree perfectly. *Gymnodactylus stoliczkæ* was

[1] Günther: Proc. Zool. Soc., 1869, p. 504;—*Homonota fasciata*, Blyth: Jour. As. Soc. Bengal, xxii, p. 468.
[2] Eastern Persia, ii, p. 356.

originally described by Steindachner from a specimen obtained by Dr. Stoliczka himself near Dras in 1865; and the latter mentions in his diary having found some of the specimens now obtained, those from Chiliscomo, under stones in exactly the same place in which he procured the type on his former visit. The specimens described by Dr. Anderson as *Cyrtodactylus yarkandensis* were brought, with others, by a collector, who accompanied Dr. Henderson on the mission which was sent to Yárkand in 1870; this mission traversed precisely the same route through Kashmir and Leh as the second in 1873-74, and I do not think there can be any reasonable doubt that the real locality whence *Cyrtodactylus yarkandensis* was obtained must have been Ladák, and not Yárkand, because this species appears to be replaced in Yárkand by the next, and because Dr. Anderson was, I think, similarly misinformed by his collector as to the true locality of the *Phrynocephalus* which he assigned to *P. caudivolvulus*. It is fortunate that Dr. Anderson's name does not stand, since it has, I think, been given under an erroneous idea of the locality.

The woodcut in the "Proceedings of the Zoological Society" representing this species is very poor. Steindachner's figure is much better. Dr. Anderson's specimens had lost their tails and their epidermis, and he consequently described the upper surface as smoothly granular with enlarged scales, none of which are tubercular. As this does not agree with the fresher specimens before me, and as the tail is very characteristic, I give a fresh description. Steindachner's is in German.

Description.—Form moderately stout, head and body depressed, tail usually much swollen and depressed at the base and tapering regularly. The fore foot laid forward does not quite reach the end of the snout, laid back it extends more than half-way to the thigh, the hind leg laid forwards reaches to the axil, or a little beyond it. Surface of the head covered with subequal granules, three shields behind the nostril very little larger than the other scales of the snout. Rostral large, and with a groove running down the upper part of its surface. About ten upper labials on each side, the hinder ones very small; about six lower labials. Mental large, triangular, with two (sometimes three) pairs of enlarged chin-shields behind the labials. Pupil vertical. Ear-opening round and small, but larger than the dorsal tubercles. Back granular, with scattered, enlarged, convex tubercles (these are wanting in the specimens from Kharbu). Upper surface of limbs granular; occasionally there are a few enlarged tubercles on the thigh and tarsus. Scales on the lower surface flat and hexagonal. No femoral or præanal pores. Claws very small. Tail when perfect ringed, with three enlarged blunt tubercles at each side of each ring, the uppermost the smallest; upper surface of the tail granular in the middle, lower surface covered with small smooth scales, no enlarged plates. When reproduced, the form of the tail is the same, and it is much swollen at the base, but it is uniformly granular and not ringed.

Colour grey, with numerous darker crossbands, slightly wavy and irregular on the back, limbs and tail. An adult measures 4 inches in length, tail 2·2.

The tail is very rarely perfect. Steindachner, however, appears to have been mistaken in supposing that of the specimen figured by him to have been reproduced.

This species seems hitherto to have been found only in the Indus valley in Ladák, where it appears to be abundant.

G. lawderanus[1] is closely allied, but the tail seems different.

[1] Stoliczka: Jour. As. Soc. Bengal, 1872, xli, Pt. 2, p. 105.

9. GYMNODACTYLUS ELONGATUS. Pl. II, fig. 2.

W. Blan.; Jour. As. Soc. Bengal, 1875, xliv, Pt. 2, p. 193.

G. elongatus, corpore gracili, cauda attenuata, membris exilibus, dorso tuberculis majoribus latis confertis ornato, inter tuberculas squamis rotundis parvulis induto, caudâ subtus scutis majoribus instructâ, verticillatâ, serie ultimâ verticilli cujusque ex squamis majoribus carinatis superne et ad latera omnino compositâ, poris præ-analibus ad 5; griseus, transverse fusco-fasciatus. Long. poll. 5, caudæ 2·8.

1-5, Yangihissár, Eastern Turkestan.

Description.—General form more elongate than is usual amongst geckoes, head depressed, sloping gradually down to the snout, body rather slender, tail very thin, regularly attenuate, very little, if at all, swollen at the base, exceeding the head and body in length. Limbs slender, elongate, the fore limb laid forward extends to the end of the snout, laid back it reaches more than three-quarters of the distance to the thigh; the hind limb brought forward comes some distance in front of the shoulder. Toes elongate, rounded, all with very small claws. Pupil vertical. Length of a perfect specimen 5 inches, head 0·65, tail 2·8, fore limb to end of toes 1, hind limb 1·3.

Surface of the head granular, granules nearly uniform, and about equal in size to the scales of the abdomen; nostrils between the rostral, first labial and two small shields behind, which are slightly swollen. Upper labials about twelve, the hinder very small, and passing into granules; lower labials nine or ten. Rostral rather higher than the other labials and twice as broad, with the upper portion of the anterior surface grooved. Mental the same breadth as the rostral, and pointed behind; two or three pairs of enlarged chin-shields. Back granular, with numerous broad triangular keeled tubercles, each nearly as large as the small ear-orifice; they are not arranged in regular rows, but about twelve may be counted across the back; the granular scales between the tubercles much smaller than the head granules. There are tubercles on the forearm, thigh and tarsus. I count about twenty-five larger scales across the abdomen. Tail verticillate, covered with trapezoidal or subtrapezoidal keeled scales, the posterior row of each ring larger, but without any granules or small scales between, so that there are no distinct tubercles. Lower surface of the tail, except near the base, with a row of large plates about as broad as long, two to each verticil. Præanal pores about six in a V-shaped line.

Colour in spirits pale grey, with darker transverse bands on the body, limbs, and tail.

This species belongs to the same group as *G. caspius*,[1] *G. scaber*,[2] *G. kotschyi*,[3] *G. kachhensis*,[4] *G. brevipes*,[5] &c., but is much more slender in form than any of them, and has no tubercles, with smaller scales intervening, on the tail, all the scales of the last row in each verticil being enlarged and submucronate.

Only one of the specimens obtained is in good condition.

[1] Eichwald: Fauna Casp. Cauc., p. 114, Pl. xv, figs. 1, 2.
[2] Rüpp. Atlas: Rept., p. 15, Pl. iv, fig. 2.
[3] Steindachner: Sitzungsber. K. K. Akad. Wiss. Wien., lxii, Pt. 1, p. 320, Pl. i, fig. 1.
[4] Stoliczka: Proc. As. Soc. Bengal, 1872, p. 90.
[5] W. Blanf., Eastern Persia, ii, p. 344, Pl. xxii, fig. 2.

REPTILIA. 15

10. GYMNODACTYLUS MICROTIS. Pl. II, fig. 1.

Jour. As. Soc. Bengal, 1875, xliv, Pt. 2, p. 193.

G. parum robustus, capite brevi, depresso, meatu auditorio minimo; caudâ attenuatâ, lāvi, haud verticillatæ; membris breviusculis; dorso granulato, tuberculis subcarinatis ornato; arenarius, fusco minute punctatus, subtus albescens. Long. tota 3·2 poll., caudæ 1·8.

1-20, Yárkand; 21-27, Yangihissár; 28-66, Káshghar, 67-75, no label, probably Káshghar.

Description.—General form moderately slender; head short, blunt, slightly depressed, convex towards the snout, ear-opening very small, tail stout at the base and regularly attenuate, smooth, not ringed; limbs rather short, the fore limb laid forward reaches between the eye and the snout, laid back it extends more than half-way to the thigh, the hind limb laid forward does not reach the shoulder. Toes rather short, rounded, all with minute claws; pupil vertical. Length of a perfect specimen 3·2 inches; tail 1·8.

Head granular above, granules of the occiput, region between the eyes, and sides of the head behind the eyes equal in size, those of the snout and loreal regions rather larger. Nostril in an angle between the rostral, first labial, and the points of two posterior shields, the inner of which is usually the larger.

There are generally nine or ten upper labials on each side, the first five being the largest, and in most cases there are four large lower labials followed by smaller scales, but these charac_ ters are not constant; sometimes there are six lower labials. Mental ending behind in an obtuse angle, two or three pairs of enlarged chin-shields; all the scales near the lower labials larger than the flat granules of the throat. Back granular, with bluntly keeled enlarged tubercles; about eight to ten of these may be counted across the back; they are not very regularly disposed, and all are larger than the minute ear-opening. Abdomen covered with flat hexagonal scales, which diminish in size laterally, but come farther up the sides than in most geckoes and pass into the dorsal granules. Five præanal pores in males in a transverse row in front of the anus. Limbs granular above. Tail granular throughout, granules convex above, rather larger and flat below.

Colour sandy above, whitish below, a pale line from the nose to the eye above the rather darker loreal region, and sometimes a pale line down each side of the back. Under the lens the upper parts are seen to be minutely puncticulated with brown, more closely in some parts than others, and there are also in places fine spots on the abdominal scales.

This species is probably allied to *Lacerta pipiens*[1] of Pallas, a species apparently overlooked by most herpetologists, and which, like the present species, is described as possessing a minute ear and a smooth tail; it, however, has no tubercles on the back, to judge by Pallas' description, and it is very differently coloured. On one of Dr. Stoliczka's labels the present species is said to be found under stones and about old walls, and it is evidently common.

In his list of the Western Turkestan reptiles, Sevértzoff includes *G. caspius*, Eich. *G. scaber*, Rüpp., and *G. eversmanni*, Strauch. I have not been able to find the description of the last.

[1] Pallas: Zool. Ros. As., lii, p. 27;—*Ascalabotes pipiens*, Licht., Eversman's Reise, p. 145.

Family—*LACERTIDÆ.*

11. EREMIAS YARKANDENSIS. Pl. II, fig. 3.

W. Blanf.: Jour. As. Soc. Bengal, 1875, xliv, Pt. 2, p. 194.
E. caruleo-ocellata, Anderson: Proc. Zool. Soc., 1872, p. 373, *nec* Dum. et Bibr.
? *E. multiocellata,*[1] Günther: Ann. and Mag. Nat. Hist., 1872, Ser. 4, vol. x, p. 419.

E. gracilis, supra grisea vel olivacea, nigro-maculata, ocellis albidis nigro-marginatis utrinque ad dorsum in seriem longitudinalem dispositis; subtus albida; scutis nasalibus haud tumidis, præfrontali unico, a rostrali supranasalibus atque a verticali postfrontalibus longe disjuncto; infra-orbitali ad labrum pertinente; dentibus palatalibus nullis; scutis ventralibus in series longitudinales (potius obliquas) 14-16, et in transversas ad 30 dispositis; poris femoralibus utrinque 9-14; squamis infradigitalibus vix carinatis. Long. 6 poll., caudæ 3·7.

1-4, Sánju; 5-23, Yárkand and Yangihissár; 24-28, Káshghar; 29, near Fyzabad, east of Káshghar; 30-33, Kashkasu, between Yangihissár and Sarikol; 34-44, Sarikol; 45-46, west of Sarikol.

Description.—General form rather slender, tail when perfect about one and a half times the length of the head and body, limbs rather short; the fore limb reaches to between the eye and snout, the hind foot extends to the axil. The nasal scales are not swollen, the lower eyelid is opaque and granular. Scales below the toes very faintly keeled. No palatal teeth. Usual length 5 to 6 inches. A fine specimen, in which only the tip of the tail appears renewed, measures 6·2 inches, of which the tail is 3·8, head 0·6, fore limb 0·8, hind limb 1·25.

Scales of the back rounded, arranged in transverse rows, becoming flatter and rather larger on the sides. Ventral scales in transverse and oblique rows; usually 14 to 16 in each transverse row in the middle of the abdomen (very rarely 18) and in 28 to 35 (generally 30 or 31) transverse rows. Tail scales not keeled, as a rule, on the anterior portion, though occasionally they are bluntly keeled above; on the posterior portion they are more or less distinctly keeled throughout. Præanal scales all small. From 9 to 14 femoral pores beneath each thigh. The enlarged scales below the tarsus extend about two-thirds of the distance across. Scales beneath the feet granular, not keeled. Collar free, the scales towards the middle enlarged, nearly as large as the abdominal plates, but varying in number; and often passing into small scales at the sides; usually there are ten to twelve enlarged scales.

Head shields.—Nostrils between three shields, an upper, lower, and posterior nasal, which are not swollen, but merely slightly convex, as are all the other head shields. Præfrontal single; the supranasals meet in a broad suture, and so do the postfrontals. Two large supraorbital shields, with granules outside and in front of them, but none inside. Præoccipitals each about the same size as a postfrontal; central occipital smaller, but variable in size. Postoccipitals large, each three or four times the size of a præoccipital, no azygos shield behind them. Upper labials six, in front of the large supraorbital shield which descends to the lip, its lower margin along the lip being nearly equal to that of the preceding shield. Temples covered with small granular scales. Edge of ear not denticulate.

Colour.—Olive-grey above, spotted with black, and with a more or less well marked line of whitish black-edged ocelli along each side of the back. The dark spots on the back often form longitudinal lines.

[1] This name will have priority if, as is probable, the species are the same.

This species was referred by Dr. Anderson to *Eremias cæruleo-ocellata* of Duméril and Bibron,[1] but it appears to me to differ in several characters. The nasal shields are not swollen, the dorsal scales are close together and scarcely any granules can be detected amongst them, whereas in *E. cæruleo-ocellata* they are said not to be very close, and each is surrounded by some granules. That species, moreover, has the tail scales keeled; as a rule, they are smooth in the Turkestan form, and the limbs are proportionally longer in the former, the hind legs nearly reaching the ear.

I have already[2] expressed doubts as to whether *E. cæruleo-ocellata* is the same as *E. velox*,[3] as the former has no palatal teeth, and the latter appears to possess them; but if they resemble each other at all closely, as is probable from the circumstance of most authors uniting them, I think the species now described differs much in habit, being a more slender form, and it is also distinguished by having the scales beneath the feet granular and not distinctly keeled.

The closest ally appears to be a species described by Dr. Günther from the Gobi Desert under the name of *E. multiocellata*. It is possible that this may be the same, but it is described as having an azygos shield between the postfrontals, a large central scale in the collar, and eighteen longitudinal rows of scutes across the abdomen. None of these differences is of much importance, but taking them together, they present a considerable distinction and render it possible that other differences exist. I should not think Dr. Günther would have overlooked the peculiar character of the nasal shields not being swollen, in which the present species differs from all other *Eremias* with which I am acquainted.[4]

11a. EREMIAS YARKANDENSIS *var.* SATURATA. Pl. II, fig. 4.

W. Blanf.: Jour. As. Soc. Bengal, 1875, xliv, Pt. 2., p. 194.

E. yarkandensis *magis infuscata, scuto infraorbitali horizontaliter diviso, parte superiori a labro discretâ.*

1-13, Valleys of the Kuenluen range, south of Yárkand.

This variety differs from the typical form in being much darker in colour and frequently in having much less distinct ocelli along the sides of the back. In one or two specimens the back is uniformly slaty-grey. Another difference is generally found, and it would, if constant, justify the giving a specific name to the variety. This is that the infraorbital shield is divided below the eye, and does not reach the lip, the lower divided portion forming the seventh supralabial. But in one specimen this infraorbital descends to the lip, as in the normal form.

The specimens were not labelled, and they were amongst the last collected; but Dr. Stoliczka notices this form in his diary as replacing the ordinary *Eremias* of the Yárkand plain at the commencement of the valleys leading to the Kuenluen.

[1] Erp. Gén. v, p. 295.
[2] Eastern Persia, ii, p. 374.
[3] Pallas: Reise, i, p. 718.
[4] Since the above was written, Dr. Günther has very kindly compared specimens of *E. yarkandensis* with the type of *E. multiocellata*, and informs me that they are probably the same, the only distinction of any importance, so far as can be detected, being that the fore and hind claws appear much larger in *E. multiocellata*. The type of this species is so much shrunk, that it is difficult to ascertain whether it had swollen nasals, but apparently it had not. I leave the account of the species as originally written, but I think there is every probability that *E. multiocellata* and *yarkandensis* are identical.

12. EREMIAS VERMICULATA. Pl. II, fig. 5.

Jour. As. Soc. Bengal, 1875, xliv, Pt. 2, p. 104.
? *Podarces (Eremias) pylzowi*, Strauch, Przewalski's Reptiles, p. 28, Pl. vi, fig. 1.

E. supra grisea, nigro-vermiculata, subtus albida, elongata, gracilis; dorso granuloso, scutis nasalibus tumidis, præfrontali unico a rostrali supranasalibus atque a verticali postfrontalibus longe disjuncto; supraorbitalibus convexis, omnino squamis minimis rotundis circumdatis; infraorbitali late ad labrum pertinente, dentibus palatalibus nullis; scutis ventralibus in series 16-20 longitudinales (potius obliquas), atque 36-41 transversas dispositis; poris femoralibus utrinque 19-23; squamis infradigitalibus vix carinatis. Long. 7·4 poll., caudæ 5·1.

1, 2, Yárkand; 3, Kizil, Eastern Turkestan.

Description.—General form very slender, the tail more than twice as long as the head and body. Limbs moderate, the fore limb reaches nearly to the end of the snout, the hind limb in front of the shoulder, nasal plates swollen. Scales beneath the toes but little keeled. No palatal teeth. Length of the largest specimen 7·4 inches, of which the tail measures 5·1, head 0·55, the fore limb is 0·85 long, hind limb 1·5.

Scales of the back round, granular, minute in the middle, becoming larger on the sides, all arranged in transverse rows. Ventral scales in transverse and oblique rows, 18 to 20 across the abdomen and 41 along it in the two Yarkand specimens, but only 16 across and 36 along in the Kizil individual. Tail scales all keeled, except below near the base. None of the præ-anal scales are much enlarged. Femoral pores from 19 to 23 beneath each thigh. The enlarged scales below the tarsus extend about half-way across. Scales beneath the soles of the feet granular and very small. Plates beneath the toes on the fore feet keeled, but not prominently, those on the hind feet are smooth, except towards the end of the toes, collar free, scales about the same size as those of the abdomen, rather irregular in the specimens examined, and passing gradually into the small granules of the throat.

Head shields.—The single præfrontal is large, and is separated from the rostral by the supranasals and from the vertical by the postfrontals; suture between the supranasals about equal to that between the postfrontals, and, in each case, in the specimens examined, about half the length of the præfrontal. Nasals normal. The supraorbitals are somewhat more convex than the other shields and are completely surrounded by granules, those separating them from the vertical and præoccipitals being rather larger than those towards the superciliary ridge.

Præoccipitals each about a quarter the size of a postoccipital. A small central occipital, no azygos shield behind it, five or six supralabials in front of the large infraorbital, which descends to the lip, the lower edge being equal to that of the preceding shield or longer. Lower eyelid granular. Temples covered with small granular scales. Edge of ear not toothed.

Colour.—Grey above, finely vermiculated with black lines, which tend to form longitudinal bands along the middle of the back. Upper surface of head and limbs the same; lower parts white.

This is easily distinguished from the former species by being much more elongate, with a much longer tail and hind limbs, by its having more numerous ventral scales, and swollen

REPTILIA. 19

nasal shields, by the presence of granules on the inner side of the supraorbital shields and by colouration.

From Western Turkestan Severtzoff[1] quotes, besides *E. variabilis* and *E. cæruleo-ocellata* (? *E. velox*), two species which he calls *E. intermedius*, Strauch, and *E. erythrurus*. Neither of these species, so far as I know, has been described; *E. erythrurus*, Severtzoff himself suggests, may be the young of *E. velox*. Two species of *Scapteira* and *Lacerta stirpium* are also included in the list of reptiles obtained in Western Turkestan.

E. vermiculata may be the same as *E. pylzowi* collected by Przewalski in the deserts of Alashan, 27 degrees of longitude east of Yárkand. The principal characters of the two species are similar, and so is the colouration, but, judging from the figure, the toes of the fore foot are considerably shorter in *E. pylzowi*.

Family—*SCINCIDÆ*.

13. EUMECES TÆNIOLATUS.

Eurylepis tæniolatus, Blyth : Jour. As. Soc. Bengal, xxiii (1854), p. 740.
Plestiodon sentatus, Theobald : Cat. Rept. Mus. As. Soc., p. 25.
Eumeces scutatus, Jerdon : Proc. As. Soc. Bengal, 1870, p. 73.
Mabouia tæniolata, Anderson : Proc. As. Soc. Bengal, 1871, p. 184.
Eumeces tæniolatus, Stoliczka : Proc. As. Soc. Bengal, 1872, pp. 75, 88.

1, Chakoti on the road from Mari to Srinagar, in Kashmir.

This is a very much larger specimen than the types, and so much stouter, that at first I was much inclined to consider it distinct. But the proportions are the same, and the only structural distinction I can find is, that there are twenty-three rows of scales round the body instead of twenty-one. This amount of variation is commonly found in scinques.

The length of the specimen is 13 inches ; tail, probably renewed when young, 6 ; circumference round the middle of the body, 3 ; head, 0·95 long ; fore limb, 1·35 ; hind limb, 1·75, both to the end of the claws. The colour noted by Dr. Stoliczka on the living specimen is brown above, with a dark central stripe, upper parts of sides darker and with small white spots in longitudinal rows ; the upper portion of the limbs also spotted, lower portion of sides greenish, this colour extending across the ears to the lower labials ; feet below pale fleshy, the whole of the lower surface deep waxy yellow. In spirits the middle of the back is very little darker than the lateral portions.

14. MOCOA HIMALAYANA.

Eumeces himalayanus, Günther : Rept. Brit. Ind., p. 86.
Euprepes himalayanus, Steindachner : Novara Expedition, Reptilien, p. 45.
Eumeces sikkimensis, partim, Jerdon : Proc. As. Soc. Bengal, 1870, p. 73 ;—Anderson : Proc. Zool. Soc., 1871, p. 158 ;—Blyth ?
Mocoa himalayana, Stoliczka : Jour. As. Soc. Bengal, 1872, xli, p. 127.

1-10, Mari, Punjab ; 11, 12, between Mari and Srinagar ; 13-25, Sonamurg ; 20-32, Mataian.

[1] Turk. Jev., p. 71.

Although I feel far from satisfied that the western form is really separable from the eastern (*M. sikkimensis*), most of the differences pointed out by Dr. Stoliczka appear sufficiently marked to justify the two being kept apart. The general aspect and colour of the two forms are different, and the number of scales round the body appears larger in *M. himalayana*, though this is variable. In specimens from Mari, there are almost constantly twenty-eight rows round the body, whilst in the Sonamurg examples the prevailing number is only twenty-six.

There is certainly one specimen in the Indian Museum, labelled *E. sikkimensis* from Darjiling and presented by Dr. Jerdon, which has thirty rows of scales round the body, but the colouration is so different from that of all other Sikkim specimens, that I cannot but suspect there is some mistake in the locality, for Dr. Stoliczka had large collections from Sikkim, and found no marked variation, whilst the colouration of the specimen from Dr. Jerdon is precisely that of the North-Western form, and it has a large strongly denticulated ear-opening.

The distinctions noticed by Dr. Stoliczka between the head shields of *Mocoa himalayana* and *M. sikkimensis* are not borne out by the large series before me, nor is there, so far as I can see, any constant difference in the limbs, but the ear-opening, as a rule, is decidedly larger and more denticulated in *M. himalayana*. There are more scales round the body, and there is a marked difference in colouration, Sikkim specimens being much browner and wanting the greenish white line along the lower portion of the side, which is conspicuous in *M. himalayana*. Still it is highly probable, as indeed Dr. Stoliczka suggested, that intermediate forms may connect the two.

This species appears to be common in Kashmir. The specimens labelled from Mataian were probably collected on the road from Sonamurg, for every other *Mocoa* from the Indus valley in Ladák belongs to the next species. Mataian itself is on the north side of the mountains which separate the Kashmir valley from Ladák.

15. MOCOA STOLICZKAI (?=*M. ladacensis*).

Euprepes stoliczkai, Steindachner: Novara Expedition, Reptilien, p. 45.
E. kargilensis, Steindachner: ib., p. 46.
Eumeces ladacensis, Anderson: Proc. Zool. Soc., 1872, p. 375;—*forsan* Günther: Rept. Brit. Ind., p. 88.

1-3, Mataian; 4-8, Kargil; 9, Namika-la; 10-16, Kharba; 17-19, Lamayuru—all in the Indus valley, Ladák; 20-24, no label.

It is most probable that there is really only one species of *Mocoa* in the Upper Indus valley, and that the different names above enumerated belong to it. If this be the case, and if the specimen described by Dr. Günther be really identical, the species must bear the name of *Mocoa ladacensis*. But I am unable to identify the specimens brought by Dr. Stoliczka with Günther's species, because in not one of the individuals collected does the fore foot reach the snout,[1] and because, although the three rows of scales beneath the tail are rather broader than those above, and the middle row is slightly more developed than the other, there is scarcely such a difference as I should suppose to be implied by the character of "subcaudals broad." It must be borne in mind, too, that the locality of Dr. Günther's type rests upon the authority of Messrs. von Schlagintweit, whose want of accuracy with reference to the localities assigned to their reptilian collections is notorious.

[1] This was noticed also by Dr. Anderson l. c.

It is true that in Steindachner's description of *Euprepes stoliczkai*, there is said to be a row of broader shields beneath the tail. But then the only difference stated to exist between *E. stoliczkai* and *E. kargilensis* is that in the former there are five, in the latter four supralabials in front of the infraorbital. That this character is of no specific value is proved by the circumstance that both forms occur together in the present collection, and that there are some specimens which have four shields on one side of the head and five on the other. Now, some of the specimens before me are typical *E. kargilensis* from the same locality as the original specimens procured by Dr. Stoliczka himself in 1865. The only other distinction between the descriptions of *Euprepes stoliczkai* and *E. kargilensis* is that in the former the middle denticulations on the anterior edge of the ear are larger than the others, in the latter the uppermost is largest. This is certainly of no importance.

In different individuals the number of scales round the body varies from thirty-two to thirty-eight, not depending apparently on age. In one very young specimen from Mataian there are only twenty-eight rows, but this individual is so immature, that its characters are ill marked, and it perhaps belongs to the last species. The usual number is thirty-four or thirty-six.

The colouration appears very constant; the back is brownish-olive, rather paler towards the sides, and spotted, the spots consisting of a whitish dot with a larger blackish mark behind or at the side of it. These spots sometimes, but not often, tend to form longitudinal lines. Sides with a broad band of dark olive brown broken by small pale spots and extending from the eye to the root of the tail and sometimes continued as a narrower broken line down the tail. A few dark marks forming irregular longitudinal lines on the upper surface of the tail; lower parts bluish-white.

Order OPHIDIA.

Family—*TYPHLOPIDÆ*.

16. TYPHLOPS PORRECTUS, var.

Stoliczka : Jour. As. Soc. Bengal, 1871, xl, Pt. 2, p. 426, Pl. xxv, figs. 1-4.

1, Ambor in the Jhilam valley, north-east of Mari.

The only specimen of a *Typhlops* in the collection is evidently that mentioned in Dr. Stoliczka's diary of the 18th July, and considered by him a new species. It differs in some respects from the description of *Typhlops porrectus*, but still agrees so nearly with that form, that I do not like to distinguish it on the strength of a single specimen.

The solitary example obtained is so tightly coiled towards the tail, that all the caudal portion is difficult to examine. The following is a brief description.

Scales smooth, shining, in eighteen longitudinal rows. I count (with great difficulty owing to the condition of the specimen) 393 scales along the body and eight along the tail. The body is much compressed posteriorly, but this is probably due to pressure when coiled. The diameter is nearly the same throughout, the circumference about one-twentieth of the length.

Head short and flat, rostral occupying about one-third of the upper surface, and having its lateral margins parallel above ; below it is scarcely narrower. Fronto-nasal united to the nasal above the nostril, separate below, the nasals extending a little behind the end of the rostral, but not quite touching. Nostrils rather in front. Præocular and ocular about equal, neither of them as large as the nasal, anterior margin of præocular very convex, that of ocular straight and vertical, except on the top of the head, where it is curved back. Præfrontal, postfrontal, supraocular, and interparietal scarcely exceeding the back scales in size; the parietals are considerably broader. Upper labials four, the first very small, in contact with the rostral and fronto-nasal; the second below the fronto-nasal and nasal, and just reaching the præocular; the third between the præocular and ocular, but not rising much on the side of the head; the fourth, which is considerably the largest, beneath the ocular and extending some distance back beyond it. Eyes quite invisible.

This differs from the description of *T. porrectus* in being rather less slender, in having fewer longitudinal rows of scales, and only eight instead of eleven to twelve rows round the tail, and, to judge by Dr. Stoliczka's figure, in the smaller size of the frontals, interparietals, and supraoculars.

Family—COLUBRIDÆ.

17. COMPSOSOMA HODGSONI.

Günther: Rept. Brit. Ind., p. 246;—Stoliczka: Jour. As. Soc. Bengal, 1870, xxxix, Pt. 2, p. 189.

1, Kashmir.

This specimen, which is young, being only 24·5 inches long, has the scales absolutely smooth throughout, and a second præocular, formed of a detached portion of the supralabial series, between the third and fourth labials. A similar specimen has been described by Stoliczka, loc. cit., from the North-Western Himalayas. Ventrals 227, subcaudals in 79 pairs.

18. PTYAS MUCOSUS.

1, 2, Kashmir.

These specimens do not differ from the ordinary Indian form. Kashmir must, I should think, be at the extremity of this snake's range to the north-west.

19. ZAMENIS RAVERGIERI.

Coluber ravergieri, Ménetries: Cat. Rais., p. 69 (1832).
Zamenis caudalineatus, Günther: Cat. Col. Snakes, Brit. Mus., p. 104 (1858);—Jan. Icon. Ophid. livr. 23, Pl. iii.
Zamenis ravergieri, Strauch: Schlangen des Russ. Reichs, Mem. Acad. Sci. St. Pet., xxi, No. 4, p. 127 (1873);—W. Blanford: Eastern Persia, ii, p. 417 (1875).
Z. fedtschenkoi, Strauch: Schlangen des Russ. Reichs., p. 135, Pl. iv (1873).

1, Yárkand; 2, 3, Yangihissár.

The spots on the head and back are larger than in Persian specimens, and somewhat resemble those of *Z. diadema*, whilst the dark band along the upper part of the tail has a tendency to be broken into spots, and the bands along the sides of the tail are faint or wanting. Otherwise there appears to be no constant difference.

The colouration is that of the form to which Dr. Strauch has given the name of *Z. fedtschenkoi*, and which is mainly distinguished from the typical *Z. ravergieri* by the tail being spotted instead of striped. Dr. Strauch adds that, as a rule, in *Z. fedtschenkoi* the number of longitudinal rows of scales is twenty-three, twenty-one being the exception, whilst the reverse is found in *Z. ravergieri*. He also calls attention to a slight difference in the form of the head, which is rather broader and less depressed in the first-named form. *Z. fedtschenkoi* is said to be common in Russian Turkestan.

In the three specimens from Eastern Turkestan, the rows of scales round the body are twenty-one in number, and the head is of the same form as in typical *Z. ravergieri*. I have already[1] shown that the two forms pass into each other in Persia, and the specimens from Eastern Turkestan tend to the same conclusion.

In both the specimens from Yangihissár, there are three postoculars on each side, but only two, as usual, in the Yárkand example. In the latter there are 222 ventrals and ninety-one pairs of subcaudals.

20. TROPIDONOTUS HYDRUS.

1, Káshgbar; 2-15, Yangihissár, Eastern Turkestan.

This snake is apparently as common in Eastern Turkestan as it is, according to Strauch,[2] farther to the westward. The specimen from Káshgbar was procured on the 2nd February, and is noted on the label as having been found frozen in a field; the Yangihissár specimens were collected in April.

The majority of the snakes of this species obtained in Eastern Turkestan appear to have five postoculars. They are olivaceous above, with the back spots rather indistinct as a rule, and a great portion of the ventral shields is black.

21. TROPIDONOTUS PLATYCEPS.

1, Mari; 2, 3, Kashmir.

I can see no difference between these specimens and those from other parts of the Himalayas. This species, which had previously been obtained by Dr. Jerdon in Kashmir,[3] appears to be one of the Himalayan forms, like *Compsosoma hodgsoni*, which range farther to the north-west than do most of the species characteristic of the Himalayan region.

Family—PSAMMOPHIDÆ.

22. TAPHROMETOPUM LINEOLATUM.

Coluber (Taphrometopon) lineolatum, Brandt: Bull. Ac. Sci. St. Pet., iii, p. 243 (1837) ;—Peters: Proc. Zool. Soc., 1861, p. 47.

[1] Eastern Persia, ii, p. 418. [2] Schlang. Russ. Reichs., p. 173.
[3] Stoliczka: Jour. As. Soc. Bengal, 1870, xxxix, Pt. 2, p. 192.

Psammophis doriæ, Jan.: De Fil., Viag. in Persia, p. 356.
Taphrometopon lineolatum, Strauch: Schlang. Russ. Reichs, Mem. Acad. Sci. St. Pet., xxi, No. 4, p. 185, Pl. v;—W. Blanf.: Eastern Pers., ii, p. 422.

1, Deshterek, south of Karghalik, Eastern Turkestan.

This characteristic Central Asiatic snake has been fully described and figured by Strauch. The only specimen obtained is of moderate size, being 33½ inches long, of which the tail measures 8. Ventrals 195, subcaudals about a hundred, the last three or four injured. The markings on the back are rather less distinct than in Strauch's figure, those on the belly are more developed, there being a subtriangular blackish mark in the middle on the anterior shields; this passes gradually into a trapezoidal dusky patch, with black lateral margins in the centre, and a row of black spots along the side, and this again gradually into two oblique lines on each side of the ventrals, becoming fainter posteriorly, but quite visible as far as the commencement of the tail. Similar colouration is described by Strauch as occurring in a specimen from Krasnovodsk; and another of unknown locality, loc. cit., p. 192.

Family—*VIPERIDÆ*.

23. VIPERA OBTUSA.

Dwigubsky, *teste* Strauch;—W. Blanf.: Eastern Persia, ii, p. 428.
V. euphratica, Martin: Proc. Zool. Soc., 1838, p. 82;—Strauch: Schlangen Russ. Reichs, Mem. Acad. Sci. St. Pet., xxi, No. 4, p. 221, Pl. vi.
Echidna mauritanica, Dum. and Bibr.: Erp. Gén., vii, p. 1431.

1, Kashmir.

In structure this specimen agrees fully with one which I obtained in Persia, but the colouration is very different, being almost uniform dark olive, with a little mottling of pale straw colour on the labials, chin, and ventral shields.

The discovery of this species in Kashmir adds considerably to its known range. It is found in Northern Africa, Asia Minor, and other parts of Western Asia, the Trans-Caucasian provinces of Russia, and Persia.

Family—*CROTALIDÆ*.

24. HALYS HIMALAYANUS.

Günther: Rept. Brit. Ind., p. 393, Pl. xxiv, fig. A;—Steindachner: Novara Reise, Reptilien, p. 87.

1, Mari, Panjab; 2, Kashmir? or Indus valley near Dras.

In both specimens there are twenty-one rows of scales round the body, not twenty-three. Steindachner has already pointed out that the number is variable. In two specimens in the Indian Museum, one from north-east of Simla, the other labelled from Ladák, the same number of rows of scales occurs *conf.* Anderson: Proc. Zool. Soc., 1871, p. 196. Judging from these specimens, it would appear that twenty-one is the number most frequently met with to the westward. Dr. Günther's original specimens, with twenty-three rows of scales, were from Garhwal.

AMPHIBIA.

The Amphibia are very poorly represented in Dr. Stoliczka's collections. Only four species are represented, and only one was procured from Eastern Turkestan; all are well known forms of Batrachia. No examples of *Urodela* were met with.

Order BATRACHIA.

Family—*RANIDÆ.*

1. RANA CYANOPHLYCTIS.

Schneider *apud* Günther: Rept. Brit. Ind., p. 406;—Stoliczka: Jour. As. Soc. Bengal, 1870, xxxix, Pt. 2, p. 146; Proc. As. Soc. Bengal, 1872, pp. 85, 102, 130;—W. Blanf.: Jour. As. Soc. Bengal, xxxix, Pt. 2, p. 374; Eastern Persia, ii, p. 433.

1—3, between Mari and Kashmir.

This species had previously been recorded by Dr. Stoliczka from Mari. It is common throughout the peninsula of India, and is the only abundant frog in the dry western parts of the country, Kachh (Cutch), Sind, &c., extending to the west into Baluchistan.

2. DIPLOPELMA CARNATICUM.

Engystoma carnaticum, Jerdon: Jour. As. Soc. Bengal, 1853, xxii, p. 534.
Diplopelma carnaticum, Jerdon: Proc. As. Soc. Bengal, 1870, p. 85;—Stoliczka: Jour. As. Soc. Bengal, 1870, xxxix, p. 154; Proc. As. Soc. Bengal, 1872, p. 110.
? *D. ornatum,* Dum. Bib., *apud* Günther: Rept. Brit. India, p. 417; see also Proc. Zool. Soc., 1875, p. 568.

1, Tinali, on the road from Mari to Kashmir.

The single specimen obtained agrees very well with specimens in the Indian Museum from the peninsula of India and Burma. No representative of the genus had, so far as I am aware, been previously met with so far to the north-west.

It is not without some hesitation that I retain the name *D. carnaticum* for this species, as Dr. Günther has recently repeated his opinion that both *Engystoma carnaticum* (in part at least) and *E. rubrum* of Jerdon, or rather specimens identified as such by Jerdon, are identical with *E. ornatum* of Dumeril and Bibron, but Dr. Jerdon has pointed out that *E. carnaticum* does not agree with Dumeril and Bibron's description, whilst the form inhabiting Malabar, whence the type of *E. ornatum* was obtained, is probably distinct from that found in Central and Northern India. I must say that I feel much doubt as to whether *E. carnaticum* is the species described by Dumeril and Bibron, the colouration described by those authors differing greatly from that of the present form, so far as I am acquainted with it.

3. BUFO VIRIDIS.

Laur. *apud* Steindachner: Novara Expedition, Amphibien, p. 40;—Stoliczka: Jour. As. Soc. Bengal, xxxix, 1870, p. 155; Proc. As. Soc. Bengal, 1872, pp. 113, 131.

1-3, Kashmir; 4-11, Yárkand; 12-15, Yangihissár; 16-23, Káshghar; 24, Zung, Wakhán.

The Kashmir specimens appear to differ a little from those of Turkestan. They have a shorter fourth toe on the hind foot, and the parotoid glands are somewhat more elongate. The differences, however, are not great, and specimens from Persia and from various parts of the Himalayas appear to be intermediate to some extent.

4. BUFO CALAMITA?

1, Kashmir.

A single very young toad from Kashmir probably belongs to this species. I find an older specimen, also from Kashmir, and presented by Dr. Jerdon, in the Indian Museum, and the two agree well in colouration, but I cannot find the characteristic gland on the leg in the young specimen. Its absence may, however, be due to immaturity.

SECOND YARKAND MISSION.
REPTILIA.

PLATE I.

Fig. 1. *Stellio stoliczkanus*, adult.
 ,, 2.   ,,   ,,   young.
 ,, 3. *Stellio agrorensis*, and head of the same from above.
 ,, 4. *Phrynocephalus axillaris*.

SECOND YARKAND MISSION.
REPTILIA.

PLATE II.

Fig. 1. *Gymnodactylus microtis.*
,, 2. *G. elongatus.*
,, 3. *Eremias yarkandensis*, with sketches of head from above and from the side.
,, 4. *E. yarkandensis*, var. *saturata*, sketch of head from side.
,, 5. *E. vermiculata*, and sketches of head from above and from the side.

SCIENTIFIC RESULTS

OF

THE SECOND YARKAND MISSION;

BASED UPON THE COLLECTIONS AND NOTES

OF THE LATE

FERDINAND STOLICZKA, Ph.D.

GEOLOGY.

BY

W. T. BLANFORD, F.R.S.

Published by order of the Government of India.

CALCUTTA:
OFFICE OF THE SUPERINTENDENT OF GOVERNMENT PRINTING.
1878.

SCIENTIFIC RESULTS

OF

THE SECOND YARKAND MISSION;

BASED UPON THE COLLECTIONS AND NOTES

OF THE LATE

FERDINAND STOLICZKA, Ph.D.

GEOLOGY.

BY

W. T. BLANFORD, F.R.S.

Published by order of the Government of India.

CALCUTTA:
OFFICE OF THE SUPERINTENDENT OF GOVERNMENT PRINTING.
1878.

CALCUTTA:
PRINTED BY THE SUPERINTENDENT OF GOVERNMENT PRINTING,
8, HASTINGS STREET.

SCIENTIFIC RESULTS

OF

THE SECOND YARKAND MISSION.

GEOLOGY.

BY W. T. BLANFORD.

INTRODUCTION AND GENERAL SKETCH OF THE GEOLOGY OF WESTERN TIBET.

IT is, of course, very difficult to do justice to a rough travelling diary, such as Dr. Stoliczka's. In such a diary first impressions are very often recorded, and subsequent observations do not always show how far the first notes require modification. To the writer this is a simple matter—his notes are memoranda serving to recall details to his mind; but to another, who does not possess the clue, it is very often difficult to ascertain how far the notes in the diary agree with the final conclusions of the diarist.

Of the greater portion of Dr. Stoliczka's journey the geological results have already been published by himself in the Records of the Geological Survey of India[1] and the Quarterly Journal of the Geological Society.[2] A comparison of these papers with the original notes shows that everything of interest in the latter, with the exception of an occasional section, has been extracted and condensed. These papers will, therefore, be here republished in sequence, with the addition only of such sections as can be extracted from the diary. The papers already mentioned contain the record of the geological observations from Leh, in Ladák, to Káshghar, and during two excursions from Káshghar to the northward. The notes from the Panjáb, at Mari, through Kashmir, to Leh, refer to ground which had been previously examined either by Dr. Stoliczka himself, or by other geologists; but as very little geological information has yet been published concerning Kashmir, the notes are here repeated. Of the journey from Káshghar to the Pámir nothing has hitherto appeared in print.

A brief summary of Dr. Stoliczka's previous geological observations in the North-Western Himalayas will aid the reader in understanding the notes made in his last journey. His earlier travels enabled him to classify the rocks seen in the mountain ranges of Spiti, Kulu, Lahaul, Rupshú, Záskar or Zánskar, Ladák, and the neighbouring districts south of the Indus

[1] Vol. VII, 1874, pp. 12, 49, 51, 81; and Vol. VIII, 1875, p. 13.
[2] Vol. XXX, 1874, pp. 568, 571, 574.

valley, and to show that several formations, some of which had not previously been detected, are represented in this portion of the Himalayas. In his last journey he has ascertained the extension of some of the same rocks to the northward; and as the regions lying east and west of his route are almost unknown, and those to the northward but imperfectly explored, almost the whole geological interest of his journey, with the exception of his observations on a part of the Thian Shan range north of Káshghar, depends upon the connection of the formations found by him in the Kashmir territories north of the Indus, and in the ranges known on our maps as the Mastágh (Karakoram), Kuenluen, and Bolor, with those previously explored in the country south of the Indus between Simla, Spiti, and Kashmir.

Dr. Stoliczka spent the summers of 1864 and 1865 in the North-Western Himalayas and Western Tibet, exploring the geology of the ranges. On his first journey, when he was accompanied by Mr. Mallet, he went north-east from Simla, crossing the Sutlej at the Wangtu bridge, and traversing Bissahir: he crossed the Bhabeh pass, and examined the Spiti valley, already known to be rich in fossil remains from the researches and collections of Gerard, Strachey, and others. From Spiti he marched nearly due north to the Indus, near Sangdo, by the Parang pass and the Tso-morari.[1] After two days' march up the Indus, he returned to the Spiti valley by a more eastern route, traversing Hanle, and crossing the Tagling pass. After spending some days in the examination of the important formations of Spiti, he marched back to Simla, through Lahaul and Kulu, at some distance to the west of his journey northwards.

The journey in 1865 occupied six months, from the beginning of May to the end of October. The area examined lay for the most part to the north-west of his former route, and extended to Leh, Kargil (north of Drás), and Srinagar. Starting from Simla, as before, he marched north by west, through Suket and Mandi, to Kulu, and thence, across the Rotang pass, to Kyelang in Lahaul. Thence he turned east by north, and crossed the Baralatse pass to the Tsaráp valley, and proceeded across several other passes to Korzog, in Rupshú, on the Tso-morari. Here he turned north-west, and travelled by the Taglang pass to the Indus, and to Leh. From Leh he went almost south-west, across the mountains, to Padam; thence north-west again to Suroo and Kargil, from which place he visited the Indus valley to the northward. This was his furthest point to the north-west in any of his journeys. From the Indus, north of Kargil, he marched south-west by Drás into the Kashmir valley, and, after a few days spent at Srinagar, he returned by the direct route, *viâ* Islamabad, Kishtwar, Budrawar, Chamba, and Kángra, to Simla. He suffered greatly from exposure to cold during part of this journey, especially in the mountains of Záskar, south of Leh; and although he gradually recovered from the effects of his Himalayan travels, it is probable that permanent injury to his constitution—not very strong originally—was produced by them.

The results of his explorations, and especially of his first journey, were very great. It has been already mentioned that the occurrence of fossils in the Spiti valley, and in some other parts of the trans-Himalayan region, had long been known; and considerable collections had been made by Gerard, Strachey, the brothers Schlagintweit, and others,—one having been obtained by Messrs. Theobald and Mallet, of the Geological Survey. The fossils collected had, moreover, been to a great extent described. Dr. Gerard's collection was, partly and imperfectly, illustrated by the Rev. R. Everest in the Asiatic Researches, Vol. XVIII, p. 107, plates I & II, and fully described by Mr. H. F. Blanford in 1863.[2] A large collection

---
[1] Tso = lake.
[2] Journal of the Asiatic Society, Bengal, Vol. XXXII, p. 124.

GEOLOGY.

formed by Colonel Strachey, chiefly at Niti, was described by Messrs. Salter and Blanford in 1865;[1] whilst the Schlagintweits' collections were entrusted to Professor Oppel, and descriptions and figures of them published by him.[2] Other less important notes had appeared, and several imperfect descriptions of the geology; but no thorough sections had been made, and, beyond the general fact that fossils of silurian, carboniferous, triassic, liassic, and jurassic forms were represented in the various collections, very little, indeed, had been done towards elucidating the geological structure of the country.

This work was admirably carried out by Dr. Stoliczka. In the course of a single season's work, in a most difficult country, amongst some of the highest mountains in the world, he clearly established the sequence of formations; and, from his extensive palæontological knowledge, was able to do this with an accuracy, which has stood the test of subsequent research. He, moreover, added to the list of known formations the representatives of rhætic and cretaceous rocks not previously detected, and showed that some of the other groups might be sub-divided.

The presence of this remarkable series of marine fossiliferous beds in the North-Western Himalayan region—a series in which all the principal European palæozoic and mesozoic groups, except the cambrian, devonian, permian, and neocomian, are represented—is none the less surprising, that scarcely any of the formations, except a few oolitic and cretaceous strata, are found in the peninsula of India, beyond the Indus river basin. In the hills of the Panjáb some of the formations have been detected, but they were until recently very imperfectly known.

The following is the sequence of formations, with the fossils found in them by Dr. Stoliczka:—

I. SUB-RECENT OR NEWER TERTIARY.			River and lacustrine deposits.—*Karewah* deposits of Godwin-Austen, &c.; *Mammalian bones.*
II. TERTIARY	... EOCENE	...	(Nummulitic) Indus or Shingo beds.—*Nummulites ramondi*; *N. exponens.*
III. MESOZOIC	... CRETACEOUS	(9)	Chikkim shales.
		(8)	Chikkim limestone.—*Rudistes* (fragments), *Nodosaria*, 2 sp., *Dentalina (annulata?), Rotalia*, sp., *Textilaria*, 2 sp., *Haplophragmium*, sp., *Cristellaria*, sp.
...	UPPER JURASSIC	(7)	Gieumal sandstone.—*Ostrea*, sp., near *O. gregaria*; another species near *O. sowerbii*; *Gryphæa*, sp., *Avicula echinata*, *Mytilus mytiloideus, Lima*, sp., *Amusium demissum, Pecten bifrons, Anatina spitiensis,* Stol., *A.* sp., nov., *Opis*, sp.
	MIDDLE JURASSIC	(6)	Spiti shales.—*Salenia ?* sp., *Terebratula* sp., *Rhynconella varians, Ostrea,* sp., *Pecten lens, Amusium* (conf. *Pecten stolidus*), *Aucella blanfordiana*, Stol., *A. leguminosa*, Stol., *Lima*, sp., near *L. rigida*, *Inoceramus hookeri, Macrodon egertonianum*, Stol., *Nucula*, sp., *Nucula cuneiformis, Cyprina trigonalis, Trigonia costata, Astarte unilateralis, A. major, A. spitiensis,* Stol., *A. hiemalis,* Stol., *Homomya tibetica, Pleurotomaria,* 2 sp., *Ammonites acucinctus, A. strigilis, A. macrocephalus,*[1] *A. octagonus, A. hyphasis, A. parkinsoni, A. theodorii, A. sabineanus, A. spitiensis, A. curvicosta, A. braikenridgii, A. nivalis,* Stol., *A. liparus, A. triplicatus, A. biplex, A. alatus, Aniscoceras gerardianum, Belemnites canaliculatus, B. clavatus.*
		(5)	Clayey slates.—*Belemnites*, sp., *Posidonomya ornata*.

---

[1] Palæontology of Niti, printed for private circulation, Calcutta.
[2] Palæontologische Mittheilungen, 1863, p. 267; 1865, p. 289.
[3] According to Dr, Waagen, Palæontologia Indica, Ser. IX, 3, p. 237, foot-note, this and several other species are not identical with the European fossil forms to which they were referred by Dr. Stoliczka.

III. MESOZOIC	MIDDLE LIASSIC	(4) Upper Tagling limestone.—*Terebratula sinemuriensis, Modiola*, sp. (resembling *Mytilus subreniformis*), *Neritopsis* (conf. *N. elegantissima*), *Chemnitzia undulata, Trochus latilabrus, Trochus epulus, T. attenuatus, Eucyclus (Amberleya)*, sp., *Actæonina* (conf. *A. cincta*), *Nerinea* (conf. *N. goodhalii*), *Belemnites*, sp., *Ammonites* (conf. *macrocephalus*).
	LOWER LIASSIC	(3) Lower Tagling limestone.—*Terebratula gregaria, T. pyriformis, T. punctata, T. (Waldheimia) schafhautli, Rhynconella obtusifrons, R. pedata, R. flucicostata, R. austriaca, R. variabilis, R. ringens, Ostrea* (conf. *O. acuminata*), *O.* (conf. *O. anomala*), *Amusium*, sp., *Pecten* (conf. *P. palosus*), *P. moniliger, P. sabal, P. bifrons, P. valoniensis, Lima densicostata, Avicula inæquivalvis, A. punctata, Gervillia*, sp.(near *G. oliflex*), *Arca (Macrodon)*, sp. (apparently *A. lycetti*), *Dentalium*, sp. (near *D. giganteum*), *Nerita*, sp. nov., *Natica* (conf. *N. pelops*), *Chemnitzia* (conf. *C. coarctata*), *C.*, sp. (near *C. phidias*), *Nerinea*, sp. (near *N. goodhalii*), *Ammonites* (conf. *A. germanii*, *A.*, sp. (conf. *A. macrocephalus*), *Belemnites budhaicus*, Stol., *B. bisulcatus*, Stol., *B. tibeticus*, Stol.
	RHÆTIC	(2) Para limestone.—*Dicerocardium himalayense*, Stol., *Megalodon triqueter*.
	TRIASSIC	(1) Lilang series.—*Encrinus cassianus, Spirifer*, sp. n., *S. (Spiriferina)*, (conf. *S. fragilis*), *S. (Spiriferina) stracheyi, S. (Spiriferina) lilangensis*, Stol., *S. spitiensis*, Stol., *Rhynconella mutabilis*, Stol., *R. theobaldiana*, Stol., *R. salteriana*, Stol., *R. retrocita* var. *angusta*, Stol., *Athyris strohmeyeri, A. deslongchampsi, Waldheimia stoppanii, Halobia lommeli, Monotis salinaria*,[1] *Lima* (conf. *L. ramsaueri*), *L.*, sp. nov., *Myoconcha lombardica, Discohelix*, sp., *Pleurotomaria* (conf. *P. buchi*), *P. sterilis*, Stol., *Orthoceras*, sp., *O. salinarium, O. latiseptum, O. dubium, Nautilus spitiensis*, Stol., *Clydonites oldhamianus*, Stol., *C. hauerinus*, Stol., *Ammonites floridus, A. jollyanus, A. khanikofi, A. gaytani, A. diffissus, A. aussæanus, A. gerardi, A. medleyanus*, Stol., *A. studeri, A. thuillieri, A. malletianus*, Stol., *A. batteni*, Stol.
IV. PALÆOZOIC	CARBONIFEROUS	Kuling series.—*Spirifer moosakhailensis, S. keilhavii, S. tibeticus*, Stol., *S. altivagus*, Stol., *Productus purdoni, P. semireticulatus, P. longispinus, Avicula*, sp., *Cardiomorpha*, sp., *Aviculopecten*, sp., *Orthoceras*, sp.
	SILURIAN ?	Muth series.—*Syringopora*, sp., *Cyathophyllum*, 2 sp., Crinoid stems, *Orthis* sp. (near *O. thakil*, var. *striato-costata* and var. *convexa*), *O.* (near *O. compta*), *O.* (near *O. tibetica*), *O.* (conf. *O. resupinata*), *Strophomena*, sp., *Tentaculites*, sp.
	SILURIAN	Bhabeh series.—*Orthis*, sp. ? *Chætetes yak*.
V.	METAMORPHIC	Central gneiss.

But, although the general sequence of the beds was established, the observations made were insufficient to enable a map to be prepared showing the distribution of the different strata. Further examination was necessary for this purpose; and Dr. Stoliczka always hoped to return to the Himalayas and complete the work he had so well begun. The severe and long-continued labour necessary for the preparation of his great work on the cretaceous fossils of Southern India engrossed the whole of his time; and, as has already been mentioned, his health was seriously affected by the exposure he underwent in his second Himalayan journey, so that, for a year or two at least, he was unfitted for work involving severe exertion. Thus the sketch he made—for such it was—has never been filled up; no geological map of the Western Himalayas has ever been published, and the idea which can be formed of the distribution of the known strata is, at the best, fragmentary.

It is as well, before proceeding further, to point out, in such a manner as to render it easily recognised on the map, the area to which Dr. Stoliczka's observations were chiefly con-

GEOLOGY.   5

fined. This area has somewhat the form of an oblong, with the longer axis north-west and south-east. Its north-eastern boundary is formed by the Indus, whilst the south-western boundary is far less regular, and, bulging out near the southern corner, includes a considerable tract of country about Spiti, Kulu, and Lahaul. The south-eastern limit of the area examined is formed by a line drawn north-north-east from Simla to the Indus, the north-western extremity being near Kargil and Drás. The south-western boundary is formed first by the range which separates the Kashmir valley from that of the Indus, and the continuation of the same in the Záskar range as far as the Baralatse or Baralacha pass, whence the boundary turns southward and embraces the country between the Baralatse range and the snowy ridges north of the Sutlej valley, near Simla.[1]

The general formation of the mountains near Simla is too irregular for any definite range of great length to be distinguished. The ridges throughout the North-Western Himalayas and Western Tibet have a general north-west and south-east direction, shown by the main course of both the mountains and river valleys; and this direction is, of course, due in a great measure to the strike of the various rocks, and the outcrop of softer or harder strata. Commencing at the south,[2] the range north of the Sutlej, opposite Simla, usually considered the true Himalaya, and well known to all visitors to Simla as the snowy range, is chiefly composed of the rock called by Dr. Stoliczka "central gneiss."[3] The mineralogical character by which this rock is distinguished is the presence of albite in large quantities, with quartz, orthoclase, and biotite, and a still more marked peculiarity in the constant occurrence of veins of albite granite, which traverse the mass in every direction.

To the south of the central gneiss various metamorphic rocks are found : to the north or north-east of it commences the sedimentary area of Tibet. It is palpable that this central gneiss is not only pre-silurian in age, but that it must, in all probability, have been metamorphosed before the deposition of the silurian strata. Hence its importance: for whilst other metamorphic formations of the Himalayas and Tibet are, probably, represented by fossiliferous sedimentary deposits in other parts of the range, the central gneiss appears to belong to an older period altogether.

To the north-west this gneissic formation extends but a short distance. The natural continuation of the range formed by it would be the Pir Panjál, south-west of Kashmir; but this consists of newer formations. Dr. Stoliczka was inclined to consider the Záskar ridge as the probable continuation of the central axis, as he considered it, and to look upon the gneiss of which that range consists as the representative of the central gneiss. It, however, wants the albite granite.

The highest peaks of the snowy Himalayan range consist of silurian rocks dipping northward, and followed in ascending order by carboniferous, triassic, and jurassic strata.

[1] For convenience sake, it may be as well to point out that the principal ranges of the North-Western Himalaya and Western Tibet, all running nearly north-west and south-east, are, commencing on the north, the Kuenluen range on the edge of the Yárkand plain; the Mastágh range traversed by the Karakoram pass, and forming the main ridge, separating the Indus watershed from that of the Yárkand plain; the Ladák range running along the northern (or north-eastern) bank of the Indus, and separating its valley from that of the Shayok; the Záskar range, which forms the south-western limit of the Indus drainage, extending along the north-eastern boundary of Kashmir, and the continuation of which to the south-west is sometimes known as the Baralatse range, and the Himalaya proper, the north-western continuation of which is the Pir Panjál.

[2] The account which follows is derived in great part from Mr. H. B. Medlicott's sketch of the Geology of the Panjáb and its dependencies in the Panjáb Gazetteer.

[3] Some important additional information concerning this rock has recently been furnished by Colonel C. A. McMahon, who has determined by microscopical examination that this gneiss possesses the characters of an igneous rock, in parts at all events, and that it must probably have been in a more or less plastic or fluid state.—Records, Geol. Surv. Ind., X, p. 322.

b

The cretaceous rocks have only been found at a few localities in Spiti and Rupshú; but the jurassic and liassic strata upon which they rest occupy a large area, constantly spoken of by Dr. Stoliczka as the jurassic ellipse, and having an elliptical form, with the long axis in the normal north-west, south-east direction. These beds were traced from Spiti and Southern Rupshú to Záskar, where they end out against the great granite and syenitic mass of Little Tibet. To the south-west the same jurassic rocks are known to exist in Northern Kumaon. Except close to the Karakoram pass, where liassic beds occur, and a little farther east by south in the Lokzhung range, capped by cretaceous rocks,[1] none of these middle and upper mesozoic rocks have hitherto been found in Western Tibet beyond the limit of this basin; nor have they hitherto been found in Kashmir proper, although some of them recur in the hills near Mari (Murree).

The silurian, carboniferous, and triassic (including the rhætic[2]) formation have a far wider range, and it is probable that their altered representatives form no inconsiderable proportion of the metamorphic rocks, which occupy so large an area in the Indus valley and its neighbourhood.

The silurian rocks on the south of the jurassic area have been traced at intervals from the Bhabeh pass, through Northern Lahaul and Záskar, to the neighbourhood of Drás, and they are probably, in Dr. Stoliczka's opinion, represented by some of the lower beds seen in the Indus valley below Leh, and in the Marka valley to the south. North-west of the jurassic area they have not been detected, and they may be represented by some of the metamorphic rocks.

The carboniferous series is distinctly developed both to the south-west and north-east of the jurassic area in the Spiti country, and it becomes even more prominent to the north-west. It occupies large areas in the Indus valley south-west and west of Leh, and reappears in the Kashmir valley. The triassic rocks appear everywhere to overlie the carboniferous, and to have nearly an equal extension.

Northern and Eastern Rupshú, to the north-east of the Spiti area, consists mainly of gneiss and other metamorphic rocks. The same crystalline formations form the whole of the range north of the Indus, from the sharp bend made by the river to the southward, north of Hanle, to Leh.

In the Indus valley itself, apart from all the secondary series of the Spiti basin, sandstones, shales, and clays are found, which have been proved to be of eocene age by the discovery in them of nummulites and other fossils. Where these were first observed by Dr. Stoliczka in Northern Rupshú, they were unfossiliferous, and their old and altered appearance made him suspect that they might be palæozoic. But near Leh they are much newer in appearance, and contain fossils which prove their age. Similar beds are seen west of Leh, as far as Kargil.

Lastly, eruptive rocks, containing serpentine, diallage, and epidote, occupy a considerable area around Hanle, east of Rupshú, and extend for many miles to the north-west, towards the Indus. Syenite is largely developed near Leh, and extends westward, towards Drás, occupying a considerable area about Kargil. Serpentine is associated with it.

If we look upon the snowy range north of Simla and the Záskar range as identical, and as forming the axis of the Himalayas, we may consider the palæozoic and mesozoic rocks of the Indus and Spiti valleys as lying between two great metamorphic ranges—that just mentioned and the Ladák range north of the Indus. To the north of Kashmir, however, the

[1] See note, p. 47.
[2] This formation was kept distinct by Dr. Stoliczka in his first paper, but subsequently he was disposed to unite it with the triassic group.

GEOLOGY.   7

carboniferous and triassic beds completely lap round and replace the older metamorphies. In his last journey Dr. Stoliczka has shown that another great sedimentary region in the Karakoram area lies between the crystalline Ladák ridge and the gneissic rocks forming the Kuenluen. But in this region no oolitic or cretaceous beds have hitherto been found, the highest fossiliferous rocks observed being liassic.[1] North of the Kuenluen, however, the presence of a cretaceous formation was detected.

As occasional reference must be made in the ensuing pages to the names given by Mr. Medlicott to particular formations on the southern slopes of the Himalayas, a list of these groups, with their supposed trans-Himalayan equivalents, is appended. It must be remembered that the identifications are little more than surmises,[2] and were only suggested as probable by Dr. Stoliczka, no fossils having been found in the cis-Himalayan rocks below the nummulitics.

Age.		Cis-Himalayan.	Trans-Himalayan.
PLIOCENE and MIOCENE	Siwalik . . . . . Náhan		Mammaliferous. Deposits of Tibet (? Karowah, in part).
EOCENE (*Nummulitic*)	Sirmúr	Kasauli (purple and grey sandstones) Dagshai (red clays, purple and grey sandstones). Sabáthu (brown and grey clays and limestones).	Indus or Shingo beds.
TRIAS . . .	.	Krol (limestone)	Lalang series.
CARBONIFEROUS	.	Infra-Krol (sandstone and carbonaceous shales)	Kuling series.
SILURIAN	.	Blini (limestone and conglomerate) . . . Infra-Blini (slates and sandstone) . . .	Muth. Dbabch.

On the other hand, there is some slight possibility of the Krol limestone being nummulitic, and Mr. Medlicott at one time, and before the trans-Himalayan rocks had been classified by Dr. Stoliczka, was rather inclined to this view,[3] but he never considered the evidence in its favour of much importance.

In the following pages the order preserved is that of the journey: first, the notes taken from the diary of the route from the Panjáb to Leh, then the (previously printed) geological descriptions of the journey from Leh to Sháh-i-dula, and from Sháh-i-dula to Káshghar; next, the excursions from Káshghar to the Chadyr-kul and to Altyn Artysh; and finally, the notes from the diary of the journey to the Pámir, and of the return march from Yárkand to the Karakoram pass. The sections illustrative of the geology of the country are from sketches in Dr. Stoliczka's note-book; they are introduced, as they serve greatly to explain the relations of the rocks, but it should be remembered that the original drawings are frequently rough, and they may not, in some instances, have been quite correctly interpreted. Should subsequent research show the sections to require modification, the circumstances under which they were prepared should be remembered.

Dr. Stoliczka himself spoke of his geological results as meagre. This is, probably, the first impression of most travellers: either they have traversed enormous areas composed of

---

[1] Some obscure unfossiliferous sandstones near Kium, in Changchenmo, and at Aktágh, north of the Karakoram pass, were referred with doubt to the tertiary epoch.

[2] Mr. Lydekker's surveys, made since the above was written, have indicated that some modification is probably necessary in the above list of correlated strata. It appears now more probable that the Krol limestone is carboniferous,—Records, Geol. Survey of India, XI, p. 63.

[3] Memoirs, Geological Survey, Vol. III, p. 170.

one or two rock groups, and the geology appears to them monotonous in the extreme, or they have been compelled to leave behind sections only half examined, in which the various formations succeed each other too rapidly for their sequence to be determined in a hurried journey. But in all cases, as with all discoveries in science, the observations require record and comparison for their value to appear. However useless they may seem at the time, no one can tell when the information may prove of the last importance.

For details as to the route, the map and diary should be consulted. All the explanatory notes in brackets and foot-notes in the subsequent pages are by myself, with the exception of the foot-notes marked (S) on page 18 and 20.

NOTE.—Since the above was in type, Dr. Waagen has kindly sent to me a paper, which he has just published in the Denkschrift Kais. Acad. Wiss. Wien. (Math. Naturwiss. Classe) for 1878, entitled "Ueber die Geographische Vertheilung der fossilen Organismen in Indien." In this paper he points out that, although the classification of the Spiti shales is still imperfect, and further subdivision may be necessary, it is clear that the great mass of these strata must be classed as Upper Jurassic (Kimmeridge and Tithonian), several of the Cephalopoda having been at first wrongly identified with European forms, and being of later age than was supposed. Dr. Waagen also notices that further to the westward in the Alpine Panjáb, near Mari (Murree), the Gieumal sandstone or its equivalent contains the *Trigoniæ* (*T. ventricosa*, &c.) characteristic of the Umia (Portlandian) group in Cutch.

According to Dr. Waagen, also, only the upper Tagling limestone, the representative of the 'Hierlatz beds' of the Alps, should be classed as Lias, the lower Tagling limestone, the equivalent of the Alpine 'Kössen beds,' being of Rhætic age. The Para limestone should be classed as upper triassic, and the Lilang series in part as middle triassic (Muschelkalk). Most of these relations had been pointed out by Dr. Stoliczka himself.

PART I.

NOTES ON THE GEOLOGY FROM MARI (MURREE) IN THE PUNJAB TO LEH IN LADÁK.

[THE following notes, it should be remembered, commence in the Panjáb, at Mari (Murree), the sanatarium lying a short distance north of Rawal Pindi. A "rough section showing the relation of the rocks near Mari," by Dr. Waagen, was published in the Records of the Geological Survey of India.[1] He showed that Mari is built on red slates and sandstones, newer than the nummulitics, but unfossiliferous, and that these beds are succeeded (the formations are too much crushed and contorted for anything like order in descent to be made out) by nummulitic limestone, jurassic and triassic beds; the jurassic beds being identified with the " Spiti shales." Dr. Waagen gave a section round Chamba Peak, from Kairagali to Changligali. Dr. Stoliczka describes that seen on the road round the other side of the mountain.

An account of the geology of the neighbourhood of Mari hill station in the Panjáb has also been given by Mr. A. B. Wynne,[2] of the Geological Survey of India, accompanied by a map and section. In this paper many additional details of the geology are given, and the same section is described which is here extracted from Dr. Stoliczka's note-book. Dr. Stoliczka's notes were made before Mr. Wynne's paper was published, although the latter had been written long before. Within the last two years the systematic geological survey of Kashmir has at length been commenced, and a large amount of information as to the distribution and relations of the different beds has been added by Mr. Lydekker.[3] In a few cases, as at the Zoji-la, slight changes have been shown to be necessary in the views formed by Dr. Stoliczka on his hurried journeys, but as a general rule his opinions have proved correct.]

July 3rd to 6th.—The Mari hill consists of sandstone and shales, the former full of *fucoids*, but I could not find a trace of other fossils.[4] The geological section from Mari

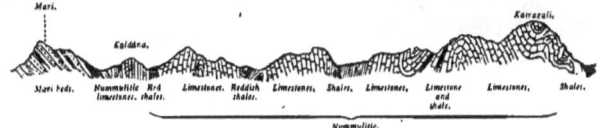

Section from Mari to Kairagali, distance 8 miles.

to Kairagali is rather simple, but thence along the road to Changligali it is rather complicated, and on the whole similar to that made by Waagen on the corresponding road passing

---
[1] Vol. V, 1872, p. 15.
[2] Records, Geological Survey of India, Vol. VII, p. 64.
[3] Rec. G. S. I., IX, p. 155; XI, p. 30.
[4] The Murree Beds of Mr. Wynne, see Quarterly Journal, Geological Society, 1874, p. 71, &c., and Rec. G. S. I., Vol. VII, p. 66.

round the other side of Chamba Peak.[1] The section from Changligali to Dangagali is a little more simple.

On the saddle at Kaldána the Mari beds dip towards the nummulitic shales, but at Sunnybank they are turned up sharply against the latter. There must have been a tremendous slip along this boundary. After some shales and crumbling sandstones, the southern side of the Kaldána hill consists chiefly of limestone, and then follow reddish shales and sandstones, very like those of the Mari group in general character. The shales are seen on the next saddle, succeeded chiefly by limestone and grey shale and carbonaceous sandstone, often very impure. These beds, the calcareous especially, are often full of nummulites, with an occasional pelecypod or gastropod.

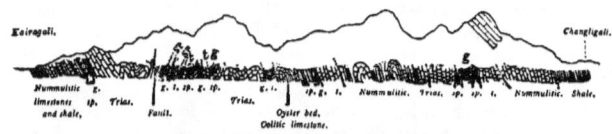

Section from Kairagali to Changligali, distance a little above 2 miles.

The section on the western side of the Chamba Peak is even more contorted than that made by Waagen on the other (eastern) side. The general dip of the rock is towards the north-west, and the consequence is, that the rocks are dreadfully twisted in every stream: on the whole, the section is much more contorted than in the sketch.

The triassic limestone in contact with the Spiti shales is semi-oolitic, just like the Krol limestone in some places. Its thickness is generally from 10 to 30 feet, and then follows more compact grey limestone, sometimes full of small oysters. About half a mile from Kairagali, I got a good *Rhychonella* in it. Changligali lies on shales, but the next

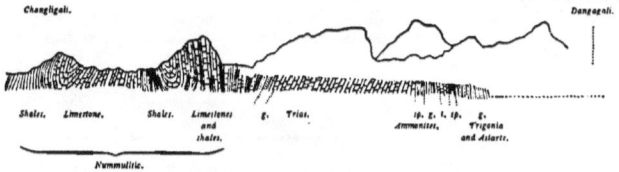

Section from Changligali to Dangagali, distance about 6½ miles.

hill is limestone, mostly vertical, and dreadfully old-looking. If I had not occasionally got a nummulite out of the intermediate calcareous shales, I should certainly have taken the limestone for triassic. But, as a rule, the nummulitic limestone is highly bituminous, while the compact triassic limestone is apparently never bituminous, and the semi-oolitic (triassic) limestone is occasionally slightly bituminous, but generally not. Nummulitic beds continue about half-way to Dangagali. There is a great thickness of triassic limestone, and then

[1] Rec. G. S. I., Vol. V, 1872, p. 16.

at the last corner, before the road turns towards Dangagali, there is a repeated alternation of Gieumal, Spiti, and triassic beds. In the sandy beds of the Spiti shales I found a fragment of an Ammonite; and in the Gieumal sandstone, which occupies the whole corner, I got an *Astarte*, which is apparently the same as that I got at Lunari in the lower Umia beds,[1] and a *Trigonia*, but this is difficult to make out. The saddle on which Dangagali lies is again nummulitic shales.

[The most interesting point in the preceding sections is the identification of the Gieumal sandstone (upper jurassic). Dr. Waagen had previously recognised the Spiti shales, and had suggested that the sandstone represented the upper jurassic beds of Spiti[2]—a suggestion which Dr. Stoliczka confirmed. The red Mari beds are called Náhan (newer tertiary) by Dr. Stoliczka in his notes; but Mr. H. B. Medlicott, who is by far the best authority on the subject, considers that this is due to a mistake in the identification of the Náhan beds themselves near Simla, as proved by some notes in Dr. Stoliczka's diary, and that the rocks with which Dr. Stoliczka really identified the Mari beds belong to Mr. Medlicott's Dagshai division (older tertiary). Under these circumstances, I have ventured in the notes to substitute Mr. Wynne's name "Mari beds" for "Náhan," leaving the question of identification undecided.]

July 15th, Mari to Koháia.—Mari sandstone and shale are seen all the way dipping in various directions: near the Jhelum the dip is about north or north by east. The older rocks are seen on the left bank of the river, at the base of the Dangagali hill. The boundary between nummulitic and Mari beds runs along the stream coming from Kaldána: on the right bank are Mari sandstones and shales, dipping at about 40° or 50° towards north-east or east.

16th, Chatarkelas.—All the way I saw nothing but the same Mari sandstone and shale, mostly dipping to north-east or north-east by east.

17th to 23rd, Chatarkelas to Uri.—The Mari beds prevailed throughout the whole distance, and no others were seen on the left bank of the Jhelum, along which river the road lay for a great part of the distance. On the opposite bank dark shales, either Spiti or Sabáthu, were noticed between Raru and Tinali, and limestones opposite Uri. From Tinali to Hatian the general dip of the Mari beds is south-east; near Uri they are much contorted.

24th, Urambu.—Uri is on a high river plateau. After crossing a stream, very red shales are seen, and blocks of limestone, looking exactly like Krol limestone, which it probably is. I am not sure whether the shales are nummulitic: more probably they belong to the Krol series. Further on more are chloritic and quartzose schists, which continue to Urumbu. The Urumbu bungalow is built at the foot of some very fine cliffs of a metamorphic quartz and schist.[3]

25th, Baramula.—The same metamorphic quartzose rock, with bacillary structure, continues a long way until the road opens into a portion of the old lake: this portion is separated by a ridge from 200 to 300 feet (high?) of lake clay and gravel deposit. The same form the low hills to the south for several miles. The lake must formerly have been much larger and wider than it now is, its water extending far up the Sind valley.

July 26th to August 6th.—Baramula to Srinagar and thence to Gandarbal.—[No description is given in the diary of the rocks about Srinagar, although reference is made to

[1] Of Cutch.
[2] Records, Geological Survey of India, V, p. 15.
[3] Lydekker, Rec. G. S. I., IX, p. 158, describes this section more fully. The limestone (Kiol) appears to be identical with Krol, as Dr. Stoliczka suggested. See also Rec., G. S. I., XI, p. 62.

them subsequently.] Passing the village Malshabagh (near Gandarbal), I saw a sub-recent conglomerate, which was deposited fully 50 feet above the present level of the lake, and in places it was overlain by terraces of clay (level), which seemed to reach about 30 to 50 feet higher.

7th, Kangan.—The rocks on both sides of the road are the same as about Srinagar—the green plutonic rock, often with zeolite cavities, and sometimes not to be distinguished from greenstone. In other places it is distinctly stratified, and it is probably a metamorphic silurian or devonian rock.

8th to 12th, Kangan to Sonamarg.—[No mention of any geology on the road.] The triassic limestones come almost down to the valley about three miles before reaching Sonamarg. At Sonamarg they are in some parts rather slaty and thin-bedded: I got no fossils in them. They dip north and south on the right and left bank of the valley respectively.

13th, Baltal.—About four miles east of Sonamarg, schists below the limestones occupy the greater heights, particularly on the north, and they extend in a north-easterly direction along these heights. At Baltal all the rocks are these schists, which are probably carboniferous. They often contain carbonaceous bands full of crystals of iron pyrites.

14th, Mataian.—[Crossing the Zoji-la,[1] 11,800 feet.] The schistose beds, which are in places almost mica schist, are followed, a couple of miles north of the Zoji-la, by more carbonaceous beds, which are probably true carboniferous, and then, about a mile south of Mataian, they are overlain on the right and left bank by the usual thin-bedded triassic limestones. These are sometimes quite white and dolomitic, alternating with black and earthy beds. I saw several *Rhynconellæ* and sections of large bivalves, like *Megalodon* and *Dicerocardium*, and small oysters; but nothing sufficiently determinable. [Further examination of the beds near the Zoji-la has shown that there is inversion, and that the rocks at the crest of the pass are of later age than the triassic limestones seen on each side.—Lydekker, Rec. G. S. I., XI, p. 45.]

15th, Mataian.—I looked over the limestones near the village, but found no determinable fossils.

16th, Drás.—About three miles after we left Mataian the green rocks cut off the limestone on the left bank, and for a few miles the boundary between the two rocks runs in the valley. After about the seventh or eighth mile, the base of the valley is all of green rock, which is generally quite massive, like greenstone; only occasionally it is thinly bedded with bacillary structure. To all appearance they are the same rocks as about Srinagar. About two or three miles before reaching Drás, the green rocks cross over entirely on to the right bank, and extend in a north-easterly direction, the trias limestones keeping to the heights. At their contact with the green rocks the limestones are more slaty. North by west of Drás the green rocks decompose very readily, and weather out reddish, as greenstones often do. About the camping ground numbers of syenite rocks are strewn about. The whole plain about Drás is filled with a deposit of shingle to about a hundred feet above the level of the river.

17th, Tashgaon.—For some distance from Drás the rugged, barren hillsides consist of greenstone. This rock gradually passes into a greenish syenite, with large quantities of schorl; but on both sides of the valley there is still the green rock *in situ*: higher up on the left bank is syenite.

[1] *La*, a pass Tibetan.

GEOLOGY.

18th, Chiliscomo.—The green rock becomes rather schistose about half-way between the last camp and this, and nearly opposite Kharbu the syenite comes down to the river, and cuts off the green rock: the former about here is light coloured and of the ordinary type.

19th, Kargil.—Syenite rocks seen the whole way.

20th, Shargol.—The tertiaries on the Kargil plain are much covered by diluvial conglomerate. The Pashkyumkur is built on serpentine rock; and from this spot to near Shargol all the rocks are serpentine, sometimes rather slaty and splintery, in other places much purer and solid, so that it could be worked for ordinary cups, &c. All along the river the diluvial conglomerate forms an almost continuous strip, particularly along the left bank of the stream.

Wherever the valley widens a little, as at Lotsun, the conglomerate is found on both sides, the horizontal banks rising up to 500 or 600 feet above the stream. About a mile from Shargol, grey and greenish and reddish shales come in from the hills to south-west and west, and are greatly developed north of Shargol. These shales appear to belong to the Sabáthu group, although they look rather metamorphic in some places, but in others they are more recent looking and micaceous. All about Shargol lumps of serpentine are sticking out of them, and the whole are covered along the left bank of the stream with a conglomerate rising to 600 and more feet above the river. Beyond this, south and south-east of Shargol, the higher hills all consist of triassic limestone, alternating near the base with rather highly metamorphic and sometimes strongly carbonaceous shales, which it is very difficult to distinguish from the tertiary beds. I found no trace of fossils in the tertiaries, but the determination of the triassic limestone is tolerably certain. It is the same as above Drás, and has often the peculiar pseudo-foraminiferous or semi-oolitic structure.

21st, Kharbu.—A good long march of 18 miles: we went by the Namika-la, and then turned almost south up the stream for about four miles to Kharbu. The diluvial conglomerate extends all the way along the river, mostly developed on the left bank, until we turned up the stream almost north and then north-east and east towards the Namika-la. A couple of miles from Shargol the monastery is built upon triassic limestone, and there are lumps and patches of it very often sticking out of the so-called tertiary shales. The great figure of Buddha a little further on is also cut in a single block of triassic limestone. When we left the conglomerates at the Wakha river, we turned almost north. There was nothing but very soft and crumbling grey and greenish (tertiary) shales as far as the Namika pass, and for some distance on the other side, extending more to north about two miles east of the pass; and the high hills to the north consisted of serpentine, while south of the Namika-la was a high solitary rock of trias limestone. The diluvial conglomerates were again seen in the little stream from the Namika-la, and are very highly developed in the Kharbu stream. Approaching this, we had up to Kharbu, along the right bank, all trias limestone, underlain by highly carbonaceous and metamorphic-looking shales and slates, which are always distinctly silky and micaceous on the planes of bedding, and often very much contorted.

22nd, Kharbu.—I went out in a north-easterly direction across the stream, and found the ground composed of various kinds of shales for several miles. First, the shales were rather carbonaceous; then they became more slaty, gray, greenish, and red, but all rather highly metamorphic. It is clear they cannot be tertiary; for they all lay under the trias. The top of the high hills appears to consist partly of serpentine. Among the higher slates there are often beds of the same green rock that I saw south of Drás.

d

23rd, Lamayuru, crossing the Fotu-la.—Leaving Kharbu, the triassic limestones pass over to the right bank of the stream after the second or third mile, where the stream makes a bend; but further on the carboniferous shales occupy the whole of the right and the base of the left bank, the limestones keeping to the greater heights. The diluvial conglomerate is locally of great extent; and in ascending the Fotu-la, it reaches to within about 200 feet of the top of the pass, that is, up to about 13,200. On the Fotu-la the southern hills are trias limestone. The pass itself is formed of carboniferous shales; and these shales extend down to Lamayuru. Unfortunately I could not find any fossils in them.

24th, Snurla on the Indus.—For more than a mile after leaving Lamayuru there are extensive shaly deposits, some of them well stratified; they reach to about 300 feet high on the slopes. The shales are at first in places very carbonaceous, and when decomposed they are covered with a white efflorescence of soda and alum. About two miles or a little more further on, these carbonaceous shales overlie nearly vertically bedded green and red shales; the latter alternate with beds of strong green sandstone, very similar to the "green-rock," and the whole group evidently represents the Bhabeh series, just as the former does the Muth series. In one place only I saw, in the Bhabeh slates, a bit of an impression, something like a portion of a *Trilobite;* and in another place I got a few traces of worms. These Bhabeh slates, shales, and sandstones are variously contorted, but for the most part approach the vertical position, dipping highly towards south or south-west. Towards the Indus the Bhabeh series is cut off by serpentines, which reach down to the valley. Only in one place, I think, there is a portion of syenite left, the ground about a mile from the Indus being strewn with boulders of syenite. The opposite bank of the Indus is occupied by greenish and reddish slates and sandstones—evidently the treacherous tertiary rocks, like in North Rupshú and Záskar. The bridge across the Indus to Khalchi is built over serpentine, and there are a good many patches of serpentine also on the right bank, and near these the sandstones and shales appear to be almost metamorphic. There is also, about half-way between Khalchi and Snurla, a lump or two of a grey or bluish limestone, full of bivalves. It looks triassic; still I do not know how it could be that. Fragments of it were locally full of large pelecypods and indistinct gastropod traces, and in some round rolled fragments I thought I saw nummulites, but I cannot be sure of it. Similar lumps of the same limestone I saw in the serpentine region before reaching the Indus, and it is just possible that some of the slates and sandstones here are really tertiary. I rather think this very probable. At Snurla the tertiary slates and shales, greenish and reddish beds alternating with each other, occupy both banks of the Indus, mostly dipping at high angles towards the south. Conglomerates are locally to be found reaching to a couple of hundred feet or less along the whole road.

25th, Saspúl.—All the way we passed through the tertiary red and greenish shales and sandstones, mostly along the strike of the rocks, which dip at a high angle of between 60° and 80° to south-west or south by west. The crystalline rocks appear to occupy the hills above Himis. Diluvial conglomerate is extensively developed along the river, and particularly about Saspúl.

25th and 26th, Saspúl to Leh.—The same rocks for the greatest part of the distance; the gneiss and hornblendic gneiss do not touch the river till just before Pittuk, beyond the village of Phayang. The diluvial deposits are very extensive, and are very thick just east of Snemo.

GEOLOGY. 15

PART II.

THE HILL RANGES BETWEEN THE INDUS VALLEY IN LADÁK AND SHÁH-I-DULA ON THE FRONTIER OF YÁRKAND TERRITORY.

[This section is copied, with a few verbal alterations, from the Records of the Geological Survey of India, V^{ol}. VII, p. 12.]

THE following brief notes on the general geological structure of the hill ranges alluded to are based upon observations made on a tour from Leh, *viâ* Changchenmo, the high plains of Lingzí-thung, Karatágh, Aktágh to Sháh-i-dula, and upon corresponding observations made by Dr. H. W. Bellew, accompanying His Excellency Mr. Forsyth's camp along the Karakoram route to this place.

Before proceeding with my account, I will only notice that our journey from Leh (or Ladák) was undertaken during the second half of September and in October, and that we found the greater portion of the country north of the Changchenmo valley covered with snow—the greatest obstacle a geologist can meet on his survey. While on our journey the thermometer very rarely rose during the day above the freezing point, and hammer operations were not easily carried out. At night the thermometer sank, as a rule, to zero, or even to 8° below zero, in our tents, and to 26° below zero in the open air. Adding to this the natural difficulties of the ground we had to pass through, it was occasionally not an easy matter to keep the health up to the required standard of working power.

Near Leh, and for a few miles east and west of it, the Indus flows on the boundary between crystalline rocks on the north and eocene rocks on the south. The latter consist chiefly of grey and reddish sandstones and shales, and more or less coarse conglomerates, containing an occasional *Nummulite* and casts of *Pelecypoda*. These tertiary rocks extend from eastward south of the Pankong lake, following the Indus either along one or both banks of the river, as far west as Kargil, where they terminate with a kind of brackish and fresh-water deposit, containing *Melaniæ*.

Nearly the entire ridge north of the Indus, separating this river from the Shayok, and continuing in a south-easterly direction to the mouth of the Hanle river (and crossing here the Indus, extending to my knowledge as far as Demchok), consists of syenitic gneiss, an extremely variable rock as regards its mineralogical composition. The typical rock is a moderately fine-grained syenite, crossed by veins which are somewhat richer in hornblende, while other portions contain a large quantity of schorl. Both about Leh and further eastward extensive beds of dark, almost black, fine-grained syenite occur in the other rock. The felspar often almost entirely disappears from this fine-grained variety, and quartz remains very sparingly disseminated, so that gradually the rock passes into a hornblendic schist ; and when schorl replaces hornblende, the same rock changes into layers which are almost entirely composed of needles of schorl. Again, the syenite loses in places all its hornblende, the crystals of felspar increase in size, biotite (or sometimes chlorite) becomes more or less abundant, and with the addition of quartz we have before us a typical gneiss (or protogine gneiss), without being able to draw a boundary between it and typical syenite. However, the gneissic portions, many of which appear to be regularly bedded, are decidedly subordinate to the syenitic ones. As already mentioned, the rock often has a porphyritic structure, and the felspar becomes pink, instead of white,—as, for instance, on the top of the Khardung pass and on the southern slope of the Chāng-la, where large fragments are often met without the slightest trace of hornblende. To the north of the last-mentioned pass the syenitic gneiss gradually passes into thick beds of syenite-schist, and this again into chloritic schist, by the hornblende becoming replaced by chlorite, while the other mineral constituents are gradually almost entirely suppressed. The syenitic and chloritic beds alternate with quartzose schists of great thickness. The schistose series of rocks continues from north of the Chang-la to the western end of the Pankong lake, and northwards to the Lankar-la, generally called the Marsemik pass. On the western route Dr. Bellew met similar rocks north of the Khardung pass at the village Khardung, and traced them northwards across the Shayok up the Nubra valley to near the foot of the Sasser pass.

Intimately connected with the metamorphic schistose series just noticed is a greenish chloritic, partly thin-bedded, partly more massive rock, which very closely resembles a similar rock found about Srinagar. Only in this case certain layers, or portions of it, become often distinctly or even coarsely crystalline, sometimes containing bronzite sparingly disseminated, and thus passing into diallage. This chloritic rock forms the greater part of the left side of the Changchenmo valley, and also occurs south of the Sasser pass. I think we have to look upon this whole series of schistose and chloritic rocks as the representatives of the *silurian formation*.

After crossing the Changchenmo valley to Gogra, we met with a different set of rocks. They are dark, often quite black, shales, alternating with sandstones. Many beds of the latter have a comparatively recent aspect, and are rather micaceous, without the least metamorphic structure, while the shales accompanying them very often exhibit a silky, sub-metamorphic appearance on the planes of fracture. I observed occasionally traces of *fucoids* and other plants in these shales, but no animal fossils. On the Changchenmo route these shaly rocks form the ridge of the Chang-lung pass, as well as the whole of the western portion of the Lingzi-thung; and they are met again after crossing these high plains and entering the Karakūsh valley, as far as Shinglung (or Dunglung). On the Karakoram route Dr. Bellew brought specimens of similar rocks from the Mastāgh (Karakoram) range itself. There can be but little doubt,—judging from similar rocks which I saw in Spiti, and from their geological relation to certain limestones, of which I shall presently speak,—that we have in the shaly series the *carboniferous formation* represented.

In many localities along the right bank of the Changchenmo river, then at the hot springs north of Gogra, and on the southern side of the Chang-lung pass, we find the carboni-

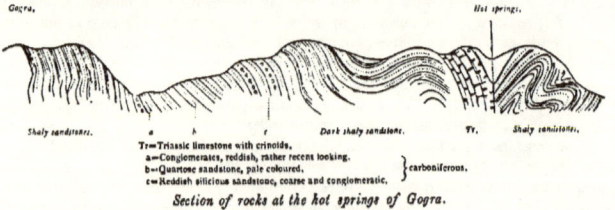

Section of rocks at the hot springs of Gogra.

ferous beds overlain by *triassic limestone*, which often has the characteristic semi-oolitic structure of the Krol limestone, south of Simla. At Gogra and several other places dolomitic beds occur; and, in these, sections of *Dicerocardium Himalayense* are not uncommon. In other places beds are met with full of Crinoid stems. North of the Lingzi-thung plain—to the west of which the hills are mostly composed of the same triassic limestone—a red brecciated, calcareous conglomerate is seen at the foot of the Compass-la, but this conglomerate gradually passes into the ordinary grey limestone, which forms the ridge, and undoubtedly belongs to the same group of triassic rocks. The last place where I saw the triassic limestone was just before reaching the camping ground Shinglung: here it is an almost white or light grey compact rock, containing very perfect sections of *Megalodon triqueter*, the most characteristic triassic fossil. On Mr. Forsyth's route Dr. Bellew met with similar triassic limestones on the northern declivities of the Sasser pass, and also on the Karakoram pass, overlying the carboniferous shales and sandstones previously noticed. On the Karakoram the triassic limestone contains spherical corals, very similar to those which were a few years ago described by Professor Ritter von Reuss from the Hallstadt beds in the Alps, and which are here known to travellers as Karakoram stones.[1]

Returning to our Lingzi-thung route, we leave, as already mentioned, the last traces of triassic limestone at Shinglung, in the Upper Karakásh valley. Here the limestone rests upon some shales, and then follow immediately the same chloritic rock which we noticed on the Lankar-la, alternating with quartzose schists, both of which must be regarded as of upper palæozoic age.

At Kizil-jilga regular sub-metamorphic slates appear, alternating with red conglomerate and red sandstones; and further on dark slate is the only rock to be seen the whole way down the Karakásh, until the river assumes a north-easterly course, some fourteen miles east of the Karatágh pass. From here my route lay in a north-westerly direction towards Aktágh, and the same slaty rock was met with along the whole of this route up to the last-mentioned place. Dr. Bellew also traced these slates from the northern side of the Karakoram to Aktágh. They further continue northwards across the Súget-lá, a few miles north of the pass, as well as in single patches down the Súget river to its junction with the Karakásh. The irregular range of hills to the south of the portion of the Karakásh river, which flows almost east and west from Sháh-i-dula, on its southern side entirely consists of these slates, while on the northern side it is composed of a fine-grained syenite, which also forms the whole of the Kuenluen range along the right bank of the Karakásh river, and also is the sole rock composing the hills about the camping ground at Sháh-i-dula. The slates of which I spoke are, on account of the close cleavage, mostly fine, crumbling, not metamorphic, and must, I think, be referred to the silurian group. They correspond to the metamorphic schists on the southern side of the Karakoram ranges.

Thus we have the whole system of mountain ranges between the Indus and the borders of Turkistan bounded on the north and south by syenitic rocks, including between them the silurian, carboniferous, and triassic formations.[2] This fact is rather remarkable, for, south of the Indus, we have nearly all the principal sedimentary formations represented, from the silurian up to the eocene, and most of the beds abound in fossils.

The only exception to which I can allude on the Changchenmo route is near Kium, in the Changchenmo valley. Here there are on the left bank of the river some remarkably

---

[1] We are still somewhat in the dark as to the true nature of those curious fossils. Dr. Waagen considered them allied to some sponges (*Astylospongia*) described by Professor Ferd. Römer from Tennessee and from the Silurian pebbles in the drift of Silesia, and certainly the resemblance externally and on cut sections is very great, but hitherto no spicules have been detected in the Karakoram stones. The specimens have now been sent to Europe for identification.

[2] On his subsequent journey from Yarkand, Dr. Stoliczka found that the highest portions of the Karakoram pass consist of liassic rocks (Tagling). See concluding portion of Geology, p. 45.

recent-looking sandstones and conglomerates, dipping at an angle of about 45° to north by east, and at the foot of these beds rise the hot springs[1] of Kium. I think it probable that this conglomerate has eastward a connection with the eocene deposits, which occur at the western end of the Pankong lake[2] and in the Indus valley south of it.

In the previous notes I have scarcely alluded to the dip of the rocks at the different localities. The reason is, that there is, indeed, very great difficulty in directly observing both the dip and the strike. At the western end of the Pankong lake the dip of the metamorphic schists is mostly south-westerly, but further on nearly all the rocks dip at a moderate angle to north-east, north by east, or to north. On the Lingzi-thung, just after crossing the Chang-lung, the shales are mostly highly inclined, but further on the limestones lie unconformably on them and dip to north-east. Wherever the hills consist merely of shales and slates, their sides are generally so thickly covered with debris and detritus, that it becomes almost an exception to observe a rock *in situ*.

The debris is brought down in large quantities by the melting snow into the valleys, and high banks of it are everywhere observable along the water-courses. At a somewhat remote—say diluvial—period this state of things has operated on a far greater scale. Not only were the lakes, like the Pankong, much more extensive, but valleys, like the Chang-chenmo, or the Tánkse valley, sometimes became temporarily blocked up by glaciers, or great landslips, and the shingle and clay deposits were often accumulated in them to a thickness of two or more hundred feet. Near Aktágh similar deposits of stratified clay exist of about 100 feet in thickness, and extend over an area of more than 100 square miles.[3] There can be but little doubt that when these large sheets of water were in existence, the climate of these now cold and arid regions was both milder and moister, and naturally more favourable to animal and vegetable life than it is now. A proof of this is given, for instance, by the occurrence of subfossil *Succineæ*, *Helices*, and *Pupæ* in the clay deposits of the Pankong lake, while scarcely any land mollusk could exist at the present time in the same place.

Note regarding the occurrence of jade in the Karakásh valley on the southern borders of Turkistan.

[From Records of the Geological Survey of India, Vol. VII, p. 51; and Quart. Jour. Geol. Soc., 1873, XXX, p. 508.]

The portion of the Kuenluen range which extends from Sháh-i-dula eastward towards Khotan appears to consist entirely of gneiss, syenitic gneiss, and metamorphic rocks, these being quartzose, micaceous, or hornblendic schists. On the southern declivity of this range, which runs along the right bank of the Karakásh river, are situated the old jade mines, or rather quarries, formerly worked by the Chinese. They are about 7 miles distant from the Kirghíz encampment Balakchi, which itself is about 12 miles south-east of Sháh-i-dula. I had the pleasure of visiting the mines in company with Dr. Bellew and Captain Biddulph, with a Yárkandi official as our guide.

[1] The temperature of these hot springs varies from 60° to 125°. They form no deposit of gypsum, like the springs north of Gogra, but there is a good deal of soda deposit round them. (B).

[2] I can find no mention of any eocene deposits at the western end of the Pankong lake in the diary. Some deposits are noticed which contain fresh-water shales, but are evidently much more recent. Some recent-looking yellow conglomerate or coarse sandstone is mentioned in the Rimdi valley, north of the Pankong lake. There may be some mistake in the wording of the text here, due to its having been printed in Dr. Stoliczka's absence.

[3] For a description of the alluvial deposits of Ladák and the Upper Indus basin, see Drew, Quart. Jour. Geol. Soc., 1873, XXIX, p. 441.

GEOLOGY.

We found the principal jade locality to be about 1¼ miles distant from the river, and at a height of about 500 feet above the level of the same. Just in this portion of the range a few short spurs abut from the higher hills, all of which are, however, as usual, thickly covered with debris and sand—the result of disintegration of the original rock. The whole has the appearance of being produced by an extensive slip of the mountain-side. Viewing the mines from a little distance, the place seems to resemble a number of pigeon-holes worked in the side of the mountain, except that they are rather irregularly distributed. On closer inspection we saw a number of pits and holes dug out in the slopes, extending over a height of nearly a couple of hundred feet, and over a length of about a quarter of a mile. Each of these excavations has a heap of fragments of jade and rock at its entrance. Most of them are only from 10 to 20 feet high and broad, and their depth rarely exceeds 20 or 30 feet; only a few show some approach to low galleries of moderate length, and one or two are said to have a length of 80 or 100 feet. Looking on this mining operation as a whole, it is no doubt a very inferior specimen of the miners' skill; nor could the workmen have been provided with any superior instruments. I estimated the number of holes at about a hundred and twenty; but several had been opened only experimentally—an operation which had often to be resorted to on account of the superficial sand concealing the underlying rock. Several pits, also, which were probably exhausted at a moderate depth, had been again filled in; their great number, however, clearly indicates that the people had been working singly, or in small parties.

The rock, of which the low spurs at the base of the range are composed, is partly a thin-bedded, rather sandy, syenitic gneiss, partly mica and hornblendic schist. The felspar gradually disappears entirely in the schistose beds, which on weathered planes often have the appearance of a laminated sandstone. They include the principal jade-yielding rocks, being traversed by veins of a pure white, crystalline mineral, varying in thickness from a few feet to about forty, and perhaps even more. The strike of the veins is from north by west to south by east, or sometimes almost due east and west; and their dip is either very high towards north, or they run vertically. I have at present no sufficient means to ascertain the true nature of this vein rock, as it may rather be called, being an aggregate of single crystals.[1] The mineral has the appearance of albite, but the lustre is more silky, or perhaps rather glassy, and it is not in any way altered before the blowpipe, either by itself or with borax or soda. The texture is somewhat coarsely crystalline, rhombohedric faces being on a fresh fracture clearly traceable. It sometimes contains iron pyrites in very small particles, and a few flakes of biotite are also occasionally observed. This white rock is again traversed by veins of nephrite, commonly called jade; which, however, also occurs in nests. There appear to be two varieties of it, if the one, of which I shall presently speak, really deserves the name of jade. It is a white tough mineral, having an indistinct cleavage in two different directions, while in the other directions the fracture is finely granular or splintery, as in true nephrite. Portions of this mineral, which is apparently the same as that usually called white jade, have sometimes a fibrous structure. This white jade rarely occupies the whole thickness of a vein; it usually only occurs along the sides in immediate contact with the white vein rock, with which it sometimes appears to be very closely connected. The middle part of some of the veins, and the greater portion of others, consist entirely of the common

[1] The only specimen in the collection made by Dr. Stoliczka at this place which agrees with his description proves to be dolomite.

green jade, which is characterised by a thorough absence of cleavage, great toughness, and rather dull vitreous lustre. The hardness is always below 7, generally only equal to that of common felspar, or very little higher, though the polished surface of the stone appears to attain a greater hardness after long exposure to the air. The colour is very variable, from pale to somewhat darker green, approaching that of pure serpentine. The pale-green variety is by far the most common, and is in general use for cups, mouth-pieces for pipes, rings and other articles used as charms and ornaments. I saw veins of the pale green jade amounting in thickness to fully 10 feet; but it is by no means easy to obtain large pieces of it, the mineral being generally fractured in all directions. Like the crystalline vein-mineral, neither the white nor the green variety of jade is affected by the blowpipe heat, with or without addition of borax or soda. Green jade of a brighter colour and higher translucency is comparatively rare, and, on that account, no doubt much more valuable. It is usually only found in thin veins of one or a few inches; and even then it is generally full of flaws.

Since the expulsion of the Chinese from Yárkand in 1864, the jade quarries in the Karakásh valley have become entirely deserted. They must have yielded a considerable portion of the jade of commerce; no doubt the workmen made a good selection on the spot, taking away only the best coloured and largest pieces; for even now a great number of fair fragments, measuring 12 to 15 inches in diameter, form part of the rubbish thrown away as useless.

The Balakchi locality is, however, not the only one which yielded jade to the Chinese. There is no reason to doubt the existence of jade along the whole of the Kuenluen range, as far as the mica and hornblendic schists extend. The great obstacle in tracing out the veins, and following them when once discovered, is the large amount of superficial debris and shifting sand which conceal the original rock *in situ*. However, fragments of jade may be seen among the boulders of almost every stream which comes down from the range. We also observed large fragments of jade near the top of the Sanju pass, which, on its southern side at least, mostly consists of thin-bedded gneiss and hornblendic schist.

Another rich locality for jade appears to exist somewhere south of Khotan, from whence the largest and best coloured pieces are said to come; most of them are stated to be obtained as boulders in a river bed, though this seems rather doubtful. Very likely the Chinese worked several quarries south of Khotan, similar to those in the Karakásh valley, and most of the jade from this last locality was no doubt brought into Khotan, this being the nearest manufacturing town. A great number of the better polished ornaments, such as rings, &c., sold in the bazar of Yárkand, have the credit of coming from Khotan; possibly they are made there by Chinese workmen, but the art of carving seems to have entirely died away, and indeed it is not to be expected that such strict Mahomedans as the Yárkandees mostly are would eagerly cultivate it. If the Turkistan people will not take the opportunity of profiting by the export of jade, or if no new locality of that mineral is discovered within Chinese territory, the celestial people will feel greatly the want of the article, and good carved specimens of jade will become great rarities. The Chinese seem to have been acquainted with the jade of the Kuenluen mountains for the last two thousand years, for Khotan jade is stated[1] to be mentioned "by Chinese authors in the time of the dynasty under Wuti (B. C. 148—86)."

[1] Yule's Marco Polo, Vol. I, p. 177. (S.)

PART III.

FROM SHÁH-I-DULA TO YÁRKAND AND KÁSHGHAR.

[From Records of the Geological Survey of India, Vol. VII, p. 49 ; and Quart. Jour. Geol. Soc., 1874, Vol. XXX, p. 571.]

IN a former communication I had already occasion to notice, that the rocks composing the Kuenluen range near Sháh-i-dula chiefly consist of syenitic gneiss, often interbedded, and alternating with various metamorphic and quartzose schists. Similar rocks continue the whole way down the Karnkásh river for about 24 miles. After this the road follows, in a somewhat north-westerly direction, a small stream leading to the Sanju (or Grim) pass. Here the rocks are chiefly true mica schist, in places full of garnets. Near the summit, and on the pass itself, chloritic and quartzose schists prevail, in which veins of pale-green jade occur, numerous blocks containing this mineral having been observed near the top of the pass. All the strata are very highly inclined, often vertical, the slopes of the hills, and in fact of the entire range, being on that account rather precipitous, and the crests of the ridges themselves very narrow.

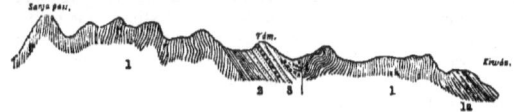

1. Metamorphic rocks.   1a. Submetamorphic schists.   2. Slates.   3. Sandstones and conglomerates.

Section from the Sanju Pass to Kiwáz.

To the north of the Sanju pass we again meet with metamorphic, mostly chloritic schists, until we approach the camping place Tám, where distinctly bedded sedimentary rocks cap the hills of both sides of the valley. They are dark, almost black, silky slates, resting unconformably on the schists, and are overlain by a grey, partly quartzitic sandstone, passing into conglomerate. The last rock contains particles of the black slates, and is, therefore, clearly of younger age. Some of the conglomeratic beds have a remarkably recent aspect, but others are almost metamorphic. In none of the groups, the slates or sandstones or conglomerates, have any fossils been observed ; but they appear to belong to some palæozoic formation. They all dip at from 40° to 50° towards north-east, extending for about 1¼ miles down the Sanju valley. Here they are suddenly cut off by metamorphic schists, but the exact place of contact on the slopes of the hills is entirely concealed by debris. The schists are only in one or two places interrupted by massive beds of a beautiful porphyritic gneiss, containing splendid crystals of orthoclase and biotite ; they continue for about 18 miles to the camp Kiwáz. On the road, which often passes through very narrow portions of the valley, we frequently met with old river deposits, consisting of beds of gravel and very fine clay, which is easily carried off by only a moderate breeze, and fills the atmosphere with clouds of dust. These old river deposits reach in many places up to about 150 feet

f

above the present level of the river, which has to be waded across at least once in every mile.

At the camp, Kiwáz, the hills on both sides of the valley are low, composed of a comparatively recent-looking conglomerate, which in a few places alternates with beds of reddish, sandy clay, the thickness of the latter varying from 2 to 5 feet only. These rocks strikingly resemble those of the supra-nummulitic group, so extensively represented in the neighbourhood of Mari. They decompose very readily, covering the slopes of the mountains with loose boulders and sand, under which very little of the original rock can be seen. Near the camp the beds dip at about 40° to north-east, but about one mile and a half further on a low gap runs parallel to the strike, and on the other side of it the beds rise again, dipping with a similar angle to south-west, thus forming a synclinal at the gap. Below the conglomerate there crops out a grey, often semi-crystalline limestone,[1] containing in some of its thick layers large numbers of Crinoid stems, a *Spirifer*, very like *S. striatus*, and two species of *Fenestellæ*. Following the river to north by east, this carboniferous limestone again rests on chloritic schist, which, after a mile or two, is overlain by red sandstone, either in horizontal or very slightly inclined strata. Both these last-named rocks are very friable, easily crumbling between the fingers, particularly the latter, from which the calcareous cement has been almost entirely dissolved out. At Sanju the red sandstones underlie coarse grey calcareous sandstones and chloritic marls, some beds of which are nearly exclusively composed of *Gryphæa vesiculosa*,[2] many specimens of this most characteristic middle cretaceous fossil being of enormous size. The *Gryphæa* beds and the red sandstones are conformable to each other; and although I have nowhere seen them interstratified near their contact, there is strong evidence of their being both of cretaceous age. Both decompose equally easily, and the *Gryphæa* beds have indeed in many places been entirely denuded. They have supplied the greater portion of the gravel and beds of shifting sand, which stretch in a north-easterly direction towards the unknown desert land.

1. Chloritic schist. 2. Carboniferous limestone. 3. Red sandstone. 4. Sandstones and marls with *Gryphæa vesiculosa*. 5. Conglomerate with reddish clay. (? tertiary).

Section from Kiwáz to Sanju, distance about 2 miles.

On the road from Sanju to Yárkand, which first passes almost due west, and after some distance to north-west, we crossed extensive tracts of these gravel beds, and of low hills almost entirely composed of clay and sand, though we only skirted the true desert country. Locally, as, for instance, near Oi-toghrak and Bora, pale reddish sandstones crop out from under the more recent deposits, but they appear to be newer than the cretaceous red sandstones, underlying the *Gryphæa* beds: the former most probably belong to some upper

---

[1] This carboniferous limestone had been previously noticed by Dr. Henderson, who gave a sketch of the section: "Lahore to Yárkand," p. 107.

[2] *G. vesicularis* in the original; but as this is an upper cretaceous species, and the specimens resemble *G. vesiculosa*, I think the latter is the name which Dr. Stoliczka intended to use.

GEOLOGY. 23

tertiary group.[1] Among the sandy and clayey deposits I was not a little surprised to find true *Loess*, as typical as it can anywhere be seen in the valleys of the Rhine or of the Danube. I might even speak of "Berg" and "Thal-Löss," but I shall not enter into details on this occasion, for I may have a much better opportunity of studying this remarkable deposit. At present I will only notice that commonly we meet with extensive deposits of *Loess* only in the valleys. Its thickness varies in places from 10 to 80 and more feet—a fine yellowish *unstratified* clay, occasionally with calcareous concretions and plant fragments. In Europe the origin of this extensive deposit was, and is up to the present date, a disputed question. Naturally, if a geologist is not so fortunate as to travel beyond the "Rhein-" or "Donau - thal," and is accustomed to be surrounded with the verdant beauty of these valleys, he might propose half a dozen theories; and, as he advances in his experience, disprove the probability of one after the other, until his troubled mind is wearied of prosecuting the object further. Here in the desert countries, where clouds of fertile dust replace those of beneficial vapour, where the atmosphere is hardly ever clear and free from sand, nay occasionally saturated with it,—the explanation that the *Loess is a subaërial deposit* is almost involuntarily pressed upon one's mind. I do not think that by this I am advancing a new idea; for, unless I am very much mistaken, it was my friend Baron Richthofen who came to a similar conclusion during his recent sojourn in Southern China.

Yárkand lies about 5 miles from the river, far away from the hills, in the midst of a well cultivated land, intersected by numerous canals of irrigation; a land full of interest for the agriculturist, but where the geological mind soon involuntarily falls into repose. And what shall I say of our road from Yárkand to Káshghar? Little of geological interest, I am afraid.

Leaving Yárkand, we passed for the first few miles through cultivated land, which, however, soon gave way to the usual aspect of the desert, or something very little better. A few miles south-west of Kokrabát a low ridge runs from south-east to north-west. If we are allowed to judge from the numerous boulders of red sandstone and *Gryphæa* marl, some of considerable size and scarcely river-worn, we might consider the ridge as being composed of cretaceous rocks. But one hardly feels consoled with the idea that in wading through the sand he is only crossing a former cretaceous basin, and that the whole of this country has remained free from the encroachment of any of the cænozoic seas. It is very dangerous to jump to conclusions regarding the nature of ground untouched by the geological hammer. The answer to any doubt must for the present remain a desideratum.

On the fourth day of our march, approaching Yangihissár, we also crossed a few very low ridges; but these consisted entirely of gravel and marly clay beds, most of them dipping with a very high angle to south by east, the strike being nearly due east and west. South of Yangihissár the ridge bent towards south-west, and there was also a distant low ridge traceable in a north-easterly direction, the whole having the appearance of representing the shore of some large inland water-sheet. From Yangihissár to Káshghar we traversed only low land, usually more or less thickly covered with a saline efflorescence, but still to a considerable extent cultivated.

[1] From a note in the diary of May 31st, made on the return journey from Yárkand, it appears that Dr. Stoliczka ultimately considered these rocks the equivalents of some examined north of Káshghar, which he termed Artysh beds.

PART IV.

GEOLOGICAL OBSERVATIONS MADE ON A VISIT TO THE CHADYR-KUL, THIAN SHAN RANGE.

[From Records of the Geological Survey of India, Vol. VII, p. 18; and Quart. Jour. Geol. Soc., 1874, Vol. XXX, p. 174.]

AFTER a stay of nearly a month in our embassy quarters at Yangishahr, near Káshghar, the diplomacy of our envoy secured us the Amir's permission for a trip to the Chadyr-kul, a lake situated close on the Russian frontier, about 112 miles north by west of Káshghar, among the southern branches of the Thian Shan range. Under the leadership of Colonel Gordon, we—Captain Trotter and myself—left Yangishahr about noon on the last day of 1873, receiving the greeting of the new year in one of the villages of the Artysh valley, some 35 miles north-west from our last quarters. On the 1st of January 1874 we marched up the Toyanda river for about 20 miles to a small encampment of the Kirghiz, called Chung-terek; and following the Toyanda, and passing the forts Mirza-terek and Chakmák, we camped on the fifth day at Turgat-bela, about 11 miles south of the Turgat pass, beyond which, 5 miles further on, lies the Chadyr-kul. On the sixth we visited the lake, and on the day following retraced our steps, by the same route we came, towards Káshghar, which we reached on the 11th January.

Having had a day's shooting at Turgat-bela, and one day's halt with the King's obliging officers at the Chakmák fort, we were actually only nine days on the march, during which we accomplished a distance of about 224 miles. It will be readily understood that, while thus marching, there was not much time to search for favourable sections in out-of-the-way places, but merely to note what was at hand on the road. I can therefore only introduce my geological observations as passing remarks.

Leaving the extensive loess deposits of the valley of the Káshghar Daria, the plain rises very gradually towards a low ridge, of which I shall speak as the Artysh range. It is remarkably uniform in its elevation, averaging about 400 feet, somewhat increasing in height towards the west and diminishing towards the east, which direction is its general strike. This range separates the Káshghar plain from the valley of the Artysh river, which cuts through the ridge about 8 miles nearly due north of the city. Viewed from this, the entire ridge appears very regularly furrowed and weather-worn on its slope, indicating the softness of the material of which it is composed. One would, however, hardly have fancied that it merely consists of bedded clay and sand, mostly yellowish white, occasionally reddish, and sometimes with interstratified layers of greater consistency, hardened by a calcareous or silicious cement. On the left bank, in the passage of the river through the ridge, the beds appear in dome shape, gently dipping towards the Káshghar plain on one side, and with a considerably higher angle into the Artysh valley on the other. On the right bank at the gap all the exposed beds dip southward, those on the reverse of the anticlinal having been washed away by the Artysh river up to the longitudinal axis, and thus exposing almost vertical faces. These remarkably homogeneous clayey and sandy beds may appropriately be called *Artysh beds*; and although I could nowhere find a trace of a fossil in them, it seems to me very probable that they are of marine origin and of neogene age.

GEOLOGY. 25

The southern slopes of the ridge are on their basal half entirely covered with gravel, which in places even extends to the top, assuming here a thickness of from 10 to 15 feet. Locally the gravel beds are separated from the main range by a shallow depression, forming a low ridge which runs along the base of the higher one, and from which it is, even in the distance, clearly discernible by its dark tint. The pebbles in the gravel are mostly of small size and well river-worn; they are derived to a very large extent from grey or greenish sandstones and shales, black or white limestone, more rarely of trap, basalt, and of gneiss. With the exception of the last-named rock, all the others had been met with *in situ* in the Upper Toyanda valley. The pieces of gneiss belong to a group of metamorphic rock which is usually called *Protogine*. It is mainly composed of quartz and white or reddish orthoclase, with a comparatively small proportion of a green chloritic substance. The white felspar variety generally contains as an accessory mineral schorl, in short, rather thick, crystals. I shall subsequently allude to the probable source from which the protogine pebbles might have been derived.

From Artysh we marched, as already stated, northwards, up the Toyanda river, and for the next 22 miles one was surprised to find nothing but the same Artysh and gravel deposits, the former constantly dipping at a high angle to north by west, and the latter resting on them in slightly inclined or horizontal strata; while among the recent river deposits in the bed of the valley itself the order of things appeared reversed. The gravels, having first yielded to denudation, here underly the clays derived from the Artysh beds, thus preparing an arable ground for the agriculturist, whenever a favourable opportunity offers itself. A few miles south of Chung-terek, the laminated Artysh beds entirely disappear under the gravel, which from its greater consistency assumes here the form of a rather tough, coarse conglomerate. In the bend of the river the latter has a thickness of fully 200 feet, and is eroded by lateral rivulets into remarkably regular Gothic pillars and turrets. It is rare to meet with a more perfect imitation of human art by nature. The general surface of the gravel deposits is comparatively low, from 400 to 500 feet above the level of the river; it is much denuded and intersected by minor streams and old water-courses.

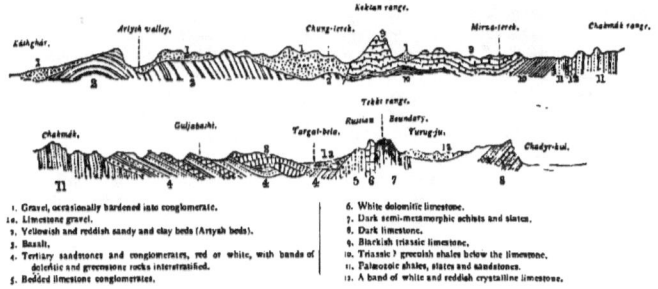

1. Gravel, occasionally hardened into conglomerate.
1a. Limestone gravel.
2. Yellowish and reddish sandy and clay beds (Artysh beds).
3. Basalt.
4. Tertiary sandstones and conglomerates, red or white, with bands of doleritic and greenstone rocks interstratified.
5. Bedded limestone conglomerates.
6. White dolomitic limestone.
7. Dark semi-metamorphic schists and slates.
8. Dark limestone.
9. Blackish triassic limestone.
10. Triassic ? greenish shales below the limestone.
11. Palæozoic shales, slates and sandstones.
12. A band of white and reddish crystalline limestone.

Section from Káshghar to the Chadyr Lake.

At a couple of miles north of Chung-terek the Koktan range begins with rather abrupt limestone cliffs, rising to about 3,000 feet above the level of the Toyanda. Nearly in the middle of it are situated the forts Mirza-terek and Chakmák, some ten miles distant from each other. The southern portion of this range consists at its base of undulating layers of greenish or purplish shales, overlain by dark-coloured, mostly black, limestone in thick and thin strata, the latter being generally earthy. The limestone occupies all the higher elevations, and, as is generally the case, greatly adds to the ruggedness of the mountains. About 5 miles north of Chung-terek, I found in a thick bed of limestone an abundance of *Megalodon triqueter*, a large *Pinna*, a *Spiriferina* of the type of *S. stracheyi*, blocks full of *Lithodendron* corals, and numerous sections of various small *Gastropoda*. Thinner layers of the same limestone were full of fragments of Crinoid stems, and of a branching *Ceriopora*, the rock itself bearing a strong resemblance to the typical St. Cassian beds. In this place the shales, underlying the limestone, were partly interstratified with it, in layers of from 5 to 10 feet; and from this fact it seems to me probable that they also are of triassic age, representing a lower series of the same formation.

Proceeding in a north-westerly direction, the *Megalodon* limestones are last seen near Mirza-terek. From this place the greenish shales continue for a few miles further on, much disturbed and contorted; and at last disappear under a variety of dark-coloured shales, slates, and sandstones, with occasional interstratified layers of black, earthy limestone. The strike of the beds is from east by north to west by south, and the dip either very high to north or vertical. At Chakmák the river has cut a very narrow passage through these almost vertical strata, which rise precipitously to about 3,000 feet, and to the south of the fort appear to be overlain by a lighter-coloured rock. It is very difficult to say what the age of these slaty beds may be, as they seem entirely unfossiliferous, and we can at present only regard them as representing, in all probability, one of the palæozoic formations.

About 5 miles north-west of Chakmák a sensible decrease in the height of the range takes place, and with it a change in the geological formation. The palæozoic beds, although still crossing the valley in almost vertical strata, become very much contorted; while, unconformably on them, rest reddish and white sandstones and conglomerates, regularly bedded, and dipping to north-west with a steady slope of about 40 degrees. The rocks, though evidently belonging to a comparatively recent (cænozoic) epoch, appear to be much altered by heat, some layers having been changed into a coarse grit, in which the cement has almost entirely disappeared. I have not, however, observed any kind of organic remains in them. A little distance further on, they several times alternate with successive, conformably bedded, doleritic trap. The rock is either hard and compact, being an intimate, rather fine-grained mixture of felspar and augite in small thin crystals, or it decomposes into masses of various greenish and purplish hues, like some of the basic greenstones.

After leaving the junction of the Suyok and Toyanda (or Chakmák) rivers, and turning northwards into the valley of the latter, the panorama is really magnificent. Shades of white, red, purple, and black compete with each other in distinctness and brilliancy, until the whole series of formations appears in the distance capped by a dark-bedded rock.

Although, judging from the greater frequency of basaltic boulders, we already knew that this rock must be found further north, we hardly realised the pleasant sight which awaited us on the march of the 4th January, after having left our camp at Gulja, or Bokumbashi. The doleritic beds increased step by step in thickness, and after a few miles we passed through what appeared to be the centre of an extensive volcanic eruption. Along the banks of the river columnar and massive basalt was noticed several times, with occasional small heaps of slags and scoriæ, among a few outcrops of very much altered and disturbed.

GEOLOGY.

strata of red or white sandstone, thus adding to the remarkable contrast of the scene. In front of us, and to the right, stretched in a semicircle a regular old Somma; the almost perpendicular walls rising to about 1,500 feet above the river, and clearly exposing the stratification of the basaltic flows, which were successively dipping to north-east, east, and south-east. On our left, as well as in an almost due western direction, portions of a similar Somma were visible above the sedimentary rocks, all dipping in the opposite way from those ahead of us. The cone itself has in reality entirely disappeared by subsidence, and the cavity was filled with the rubbish of the neighbouring rocks.

Passing further north we crossed a comparatively low country, studded with small rounded hills and intercepted by short ridges, with easy slopes; the average height was between 12,000 and 13,000 feet. This undulating high plateau proved to be one of the head-quarters of the *Kulja* (*Ovis karelini*), chiefly on account of the very rich grass vegetation which exists here. For this the character of the soil fully accounts. The entire ground was shown to consist of limestone gravel and pebbles of rather easily decomposing rocks, mixed with the ashes and detritus, evidently derived from the proximity of the volcanic eruption. Only rarely was an isolated basaltic dyke seen, or the tertiary sandstone cropping out from under the more recent deposits.

Viewing the country from an elevated position near our camp at Turgat-bela, the conglomerate and gravel beds, well clad with grass vegetation, were seen to stretch far away eastwards, and in a north-easterly direction across the Turgat pass; while on the south they were bounded by a continuation of the somewhat higher basaltic hills. Towards the west I traced them for about 7 miles, across a low pass at which a tributary of the Toyanda rises in two branches; while on the other side two similar streams flow west by south to join the Suyok river. To the north the proximity of a rather precipitously rising range shut the rest of the world out of view. For this ridge the name Terak-tagh of Humboldt's map may be retained; its average height ranges between about 16,000 and 17,000 feet. In its western extension it runs almost due east and west, composed at base of a tough limestone conglomerate of younger tertiary origin, followed by white dolomitic limestones, and then by a succession of slaty and dark limestone rocks, the former occasionally showing distinct signs of metamorphism, and changing into schist. All the beds are nearly vertical or very highly inclined, dipping to north by west, the older apparently resting on the younger. North of Turgat-bela the range makes a sudden bend in an almost northerly direction, and continues to the Chadyr-kul, where it forms the southern boundary of the lake plateau. By this time the white dolomitic, and afterwards the slaty beds, have entirely disappeared, and with them the height has also diminished. A comparatively low and narrow branch of the range which we visited consists here entirely of dark limestone, which in single fragments is not distinguishable from the trias limestone of the Koktan mountains, but here it does not contain any fossils. The ridge itself, after a short stretch in a north-east by north direction, gradually disappears under the much newer conglomeratic beds.

Across the Chadyr-kul plain the true Thian Shan range was visible, a regular forest of peaks seemingly of moderate and tolerably uniform elevation. The rocks all exhibited dark tints, but most of them, as well as the hills to the west of the Chadyr-kul, near the sources of the Arpa, were clad in snow. The lake itself was frozen, and the surrounding plain covered with a white sheet of saline efflorescence.

Brief sketch of the geological history of the hill ranges traversed.—In order that the preceding remarks may be more easily understood, I add a few words regarding the changes which appear to have taken place at the close of the cænozoic epoch within the southern offshoots of the Thian Shan which we visited.

Short as our sojourn in the mountains was, it proved to be very interesting and equally instructive. Humboldt's account of the volcanicity of the Thian Shan, chiefly taken from Chinese sources, receives great support; but we must not speculate further beyond confiding in the expectation that both meso- and cænozoic rocks will be found amply represented in it.

As far as our present researches on the physical aspect of the country extend, we may speak of three geologically different ranges: the *Terek range*, which is the northernmost, the *Koktan* in the middle, followed by the *Artysh range*, below which begins the Káshghar plain. All three decrease in the same order in their absolute height, the last very much more so than the middle one. The first consists of old sedimentary rocks, the second of similar rocks in its southern parts, while younger tertiary and basaltic rocks occupy the northern portions; the third is entirely composed of young tertiary deposits. The general direction of all the ranges is from west to east, or nearly so: this direction evidently dating from the time when the whole of the Thian Shan chain was elevated. The undulating high plateau between the Terek and the Koktan is, near Turgat-bela, about 8 miles wide, the distance between the two ranges diminishing westward, while in the opposite direction it must soon more than double. Judging from the arrangement of the pebbles, which, as already noticed, are half derived from limestone, the direction of the old drainage must have been from west to east, and must have formed the head-waters of the Aksai river, which on the maps is recorded as rising a short distance east of the Chadyr-kul. Similarly, the gravel valley between the Koktan and Artysh ranges indicates a west to east drainage, and its width appears to have approximately averaged 20 miles. About 3 miles north of Chung-terek a secondary old valley exists, also extending from west to east, and is diametrically cut across by the Toyanda river. In this valley, which was formerly tributary to the one lying more southward, the gravel beds accumulated to a thickness of fully 100 feet. As the Artysh range did not offer a sufficiently high barrier, masses of the gravel passed locally over it or through its gaps into the Káshghar plain, which itself at that time formed a third large broad valley.

Thus, at the close of the volcanic eruptions in the hills north of Chakmák, we find three river systems all flowing eastward, and made more or less independent of each other by mountain ranges, about which it would, however, not be fair to theorise (in the present state of our knowledge) on the causes of their assumed relative position. It must have been at that time that the pebbles of protogine were brought down from some portion of the hills lying to the west; and it would be interesting to ascertain whether or not this rock is anywhere in that direction to be met with *in situ*.[1] When the turbulent times of Vulcan's reign became exhausted and tranquillity was restored, the whole country south of the axis of the

---

[1] In Severtzof's journey to the western portion of the Thian Shan (Jour. Roy. Geogl. Soc., 1870, pp. 352, &c.) metamorphic rocks are stated to be largely developed in the ranges further to the north-west. A large tract of geologically unexplored mountains intervenes, however, between the southern limits of Severtzof's examination and the Chadyr-kul. Baron Osten-Sacken's journey *vid* the Chadyr-kul, from Vernoye to the neighbourhood of Káshghar (Jour. Roy. Geogl. Soc., 1870, p. 250), contains scarcely any information as to the geology of the countries traversed. He does not even notice the volcanic rocks south of the Chadyr-kul. See remarks at the end of Part V, p. 33.

It is perhaps as well to point out here, what will probably have occurred to many geologists who have read thus far. The geological school to which Dr. Stoliczka belonged has not, I believe, accepted the views prevalent amongst most English geologists as to the extent of subaërial denudation. It is far from improbable that some of the geological phenomena attributed by Dr. Stoliczka to subsidence might by other observers be considered as a simple effect of disintegration and removal by rain-water.

GEOLOGY.

Thian Shan must have greatly subsided, and the wider the valleys, the more effectively was the extent of subsidence felt. To support this idea, by an observation, I may notice that north of Chung-terek, at the base of the Koktan range, the Artysh beds have entirely disappeared in the depth, and the gravel beds overlying them dip partially under the Trias limestone,—a state of things which cannot be explained by denudation, but only by subsidence and consequent overturning of the older beds above the younger ones. A similar state of things is to be observed on the Terek range, where the young tertiary limestone conglomerate is in some places of contact overlain by the much older dolomite. Now, if the broad valley of the Káshghar plain sank first, and gradually lowest, as it in all probability did, we find a more ready explanation of the large quantities of loose gravel pouring into it and accumulating at the base of the Artysh range.

The sinking in of the volcanic centre north-west of Chakmák first appears to have drained off the former head of the Aksai river, making it the head of the Toyanda instead; and to the north of the Terek ridge it was most probably the cause of the origin of the Chadyr-kul. The subsidence of the country followed in the south, making it possible for the united Suyok and Toyanda rivers to force their passage right across the Koktan range, strengthen the Artysh river, cut with facility through the Artysh range, and join the Káshghar Daria. While thus indicating the course of the comparatively recent geological history of the ground, it must be, however, kept in mind that this change in the system of drainage had no essential effect upon the direction of the hill ranges. This, dating from much older times, was mainly an east-westerly one, following the strike of the rocks which compose the whole mountain system.

PART V.

ALTYN-ARTYSH.

[From the Records of the Geological Survey of India, Vol. VIII, p. 13.]

UNDER the personal guidance of the Envoy, we—Dr. Bellew, Captain Chapman, Captain Trotter, and myself—left Yangishahr on the 14th of February, reaching Altyn-Artysh at a late hour the same day. A halt of two days was desirable to enable us to make all necessary arrangements for our further movements. However, before I proceed, I shall endeavour to give the reader an idea of the geographical position and limits of the country, of which I shall speak in the subsequent lines.

The data are derived from a general survey by Captain Trotter, and from information given by the Hakim Mahomed Khoja.

Altyn-Artysh, which is the chief place of the province, lies approximately about 23 miles north by east of Yangishahr. It is situated in the western part of the Yilak[1] on the Bogoz, here called Artysh river, and north of a low ridge which separates the Artysh valley from the plains. The southern boundary runs along this ridge for about 10 miles west of Altyn-Artysh, and from there almost due north to the crest of the Koktan range; then along this range eastwards of the Belauti pass, and from thence in a south-western direction to the village of Kushtignak, some 15 miles north of Faizabád. From here the southern boundary runs close to the right bank of the Káshghar river, until almost opposite to where the Artysh river runs into the plains.

During the first four days we all marched in company up the valley of the Bogoz river to the fort Tongitár, about 23 miles to the north by west; then to a Kirghiz camp, Bashsogon, in a north-easterly direction; Tughamati almost eastern, and Ayok-sogon in a south-eastern direction; the directions being from the last camps respectively.

At Ayok-sogon Captain Trotter and I separated from the rest of the party, and marched northwards along the Ushturfan road to Jaitapa, and from thence across the Jigda Jilga in a north-east by east (?) direction to the camp at Uibulák, crossing the Uibulák pass, passing a second jilga, and turning then for almost 9 miles more northwards to the Belauti pass, beyond which lies the valley of the Kakshal or Aksai river. On our return we passed Ayok-sogon, Karáwal, about a mile from our former camp of the same name, and visited Kulti-ailak and Faizabád, returning to Yangishahr on the 3rd of March.

It was not a very favourable time for travelling in these regions, not so much on account of the cold, as in consequence of the heavy falls of snow which appear to occur over the whole of the Thian Shan during the second half of February and first half of March. During the last few days of February we were almost constantly wading in fresh-fallen snow, though on the saline plains it melted very rapidly. The snow naturally interfered seriously with our observations.

[1] Yilak, or Ailak, is the summer, Kishlak the winter, residence. Amongst the pastoral wandering tribes of Central Asia, it is the practice to drive all the animals to higher elevations for pasture in summer, and to bring them to lower ground when the upland pastures are covered with snow. The terms mentioned are used by the Túrk tribes.

GEOLOGY.

From a geological point of view the trip proved in many respects to be of considerable interest, particularly as supplementing some former observations made more to the west. Although there is not much variety in the rock formations, we may distinguish three successive series.

1. The most southern part of the province, along the foot of the hills, is formed of alluvial gravels and sand, in whose unfathomable depths are swallowed both the Artysh and Sogon rivers before they can reach the Káshghar Daria.

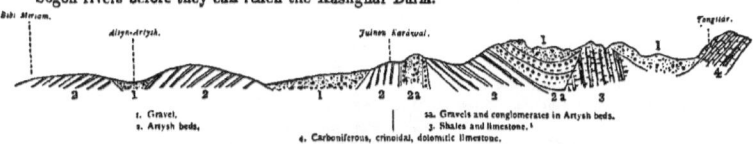

1. Gravel.
1. Artysh beds.
2a. Gravels and conglomerates in Artysh beds.
3. Shales and limestone.[1]
4. Carboniferous, crinoidal, dolomitic limestone.

Section from the Káshghar plain to Tongitár, about 25 miles.

2. The second series includes the low hills which extend diametrically from north to south over about 30 miles, while the prevalent strike is from north-east by east to south-west by west. All these lower hills are occupied by Artysh beds, of which I spoke in a former communication.[2] They are separated into two groups. The lower beds consist of greenish or reddish clays or sandstones, and the upper of coarse conglomerates, which on a hill south of Tongitár have a thickness of about 1,000 feet. At their contact both groups generally alternate in several layers. An anticlinal runs almost through the middle of their superficial extent. At the fort Ayok-sogon it is caused by a low ridge of old dolomitic limestones on which the Artysh clays and sandstones found a firm support. To the south of it the beds dip at angles of about 40° and 50° towards the Káshghar plain, in remarkably regular and successive layers. North of the ridge, which has no doubt a considerable subterranean extent in an east to west direction, all the beds dip towards north by west at a similar angle. Approaching the higher range, more recent diluvial gravels cover most of the slopes. The geological puzzle of finding strata of young beds as a rule dipping *towards* a higher range composed of comparatively much older rocks seems to me to be due, at least in this special case, to the phenomenon that the atmospheric waters which, descending on the crest, flow down the slopes of the high ridge, gradually soften them, and if a subterranean outlet facilitate it, the softened beds are worn away. While this process is going on, the more distant beds simply subside in order to fill the vacant spaces. In some cases a sinking or rising of the main range, or even an overturn of high and precipitous cliffs, seem to go hand in hand with the action of erosion, but it is not always the case. I hope to illustrate this idea by a few diagrams, partly derived from actual observations, on some future occasion.

3. A third series of entirely different rocks forms the main range of hills, which are a continuation of the Koktan range, and in which, more to the westward, are situated the Terek and Chakmák forts. The average height of the range above the plain of Káshghar is here between 1,200 and 1,300 feet, single peaks rising to about 1,500 feet. The whole of the southern portion consists, as far as I could see, of carboniferous rocks, in which, however, there is a great variety of structure. The lowest beds are very often a peculiar breccia-limestone passing

---

[1] In his field books Dr. Stoliczka speaks of these as probably triassic, but he may have changed his opinion subsequently, for in his published notes he classes them with the dolomitic limestone, and refers all to the carboniferous period.

[2] *Ante*, p. 24.

into regular limestone conglomerate. Above this are beds of solid grey dolomitic limestone, partly massive, partly stratified; the former possessing the character of reef limestone, and portions of it are indeed full of reef-building corals, crinoid stems, and a large *Spirifer*, the sections of which, when seen on the surface, have a striking resemblance to those of *Megalodon*.

North of Tongitár and about Básh-sogon I met in several places great numbers of fossils, but they were so firmly cemented in a calcareous matrix that only a few could be extracted. Among these I could recognise a small *Bellerophon*, *Productus semireticulatus*, and an *Athyris*. A new *Terebratula* was also very common. Here, about Básh-sogon and Tughamati, greenish shales occurred often interstratified with the limestones, beds of which were highly carbonaceous; the shales appeared to be unfossiliferous.

The limestone hills, which, as already stated, are a continuation of the Koktan range, extend in a north-easterly direction the whole way to south of the Belauti pass, where they are overlaid by a particularly well-bedded dark limestone very similar to that containing *Megalodon* north of Chung-terek. On this limestone rest greenish and purplish sandstones and shales which occupy the pass and the adjoining hills to the north-west of it; mineralogically these last rocks are quite identical with what we understand under the name of "*Bunter sandstein*," and it is by no means improbable that the Belauti beds are also of triassic age, as they succeed in regular layers those of the carboniferous formation.

A peculiar feature in this part of the hills consists in the occurrence of extensive plains to which the name *jilga* is generally applied. It means originally, I think, merely a watercourse, and, on a large scale, these plains may be looked upon as water-courses of former water-sheets. They occur at the base of the high range, and in some respects resemble the *dúns* of the southern slopes of the Himalayas. North of Tongitár one of these large plains occurs within the limestone rocks, being surrounded by them on all sides. It must be about 30 miles long from east to west, and about 10 from north to south. Several isolated limestone hills and ridges occur in it, and it is drained off by the Bogoz and Sogon rivers, the former rising in the south-west, the latter in the south-east corner. The average elevation is about 5,000 feet. The greater portion is covered with a low scrubby vegetation, and, near the rivers, with high grass. The principal camping grounds are Básh-sogon and Tughamati. The whole plain, which affords good pasturage ground, is occupied by about 120 tents of Kirghiz during the summer.

The next jilga is the Jigda Jilga. It differs considerably both in its physical situation and in its general character from the former. It stretches from west by south to east by north for about 35 miles, while the diameter of the eastern half is about 20 and that of the western about 12 miles. Save for a few low hillocks it is almost a level plain throughout. On the north-western, northern, and north-eastern side it is bounded by the Koktan range, from which several water-courses lead into it, one about the middle from the north, and one from north-east of considerable size, this containing a large quantity of crystalline pebbles; the rock from which they are derived must be *in situ* near the axis of the ridge. A third big stream comes from the east, leading from the Uibulák pass. None of these streams had any water in them. On the south, east, and south-east the plain is bounded by the much lower hills composed of Artysh beds; their slopes covered with gravel.

An elevated gap or saddle situated in the south-west corner appears to connect this jilga with that of Tughamati. There is no drainage from this jilga; all the water is absorbed by the enormous thickness of sand and mud which fills the entire basin.

GEOLOGY. 33

The southern part of the jilga, particularly south-east of Jaitupa, is lowest, and here a large quantity of pure salt, in small cubical crystals, is collected. The fact that there is such a large quantity of saline matter together with salt swamps in the southern part, seems to prove that this jilga at least, and probably most of the others, had been washed out by the sea, and that, while others had gradually, though only partially, drained off the saline matter, this one retained it, because it has at present no outlet. It is in fact a dried-up saline lake, which at some remote time was cut off from the sea, of which it was a fiord.

A third jilga is south of the Belauti pass and north-east of the Uibulák pass. It is about 8 miles in breadth and the same in length. There are two large water-courses leading to it from the range. On the southern side it is enclosed by Artysh and gravel beds but whether an outlet exists is not known. A southerly outlet very likely exists.

[Some little information as to the geology of the Thian Shan may be gained from Russian travellers, although, so far as I am aware, no general description of the range has been hitherto attempted by them; nor, indeed, have the mountains been sufficiently explored to enable its geology to be thoroughly understood.

With the exception of publications in the Russian language, the only original papers in which the geology of the Thian Shan is treated, so far as I know, are those by Semenoff and Severtzoff,[1] Osten-Sacken's interesting journey across the mountains, from Vernoye to the neighbourhood of Káshghar,[2] affording very little geological information. A very good general résumé of the section across the Thian Shan is given by Professor Suess[3] in a work which has recently appeared on the "Origin of the Alps," in which the geology of various mountain chains is discussed. The following translation will probably serve to give a better idea of the constitution of these mountain ranges than any which I could compile from the same materials.

After describing Dr. Stoliczka's discoveries, Professor Suess says, referring to the Russian explorers,—

"From these works it appears that these mountains are solely composed of old rocks, stratified and unstratified. To granite, syenite, and diorite succeed old slates, and then palæozoic limestones, amongst which the existence of mountain limestone is proved by fossils. The newest formation is Permian (Rothiegende) in the form of red sandstone and conglomerate, locally containing salt and gypsum. A band of red porphyry runs along the northern foot of the most northerly of these chains, the Trans-Ili-Alatau.[4]

"No mesozoic or tertiary beds are known to occur; consequently the succession of strata is nearly the same as in the Kuenluen, and as, according to Richthofen, in a great portion of the Chinese empire. The mountains are composed of great folds, the strike of which occasionally corresponds with that of the separate chains.

"The main chain of the Thian Shan consists, according to Semenoff, of two parallel axes of granite and syenite, the southern of which forms the principal ridge of the mountains,[5] the northern the ridge of a

---

[1] Semenoff; Erforschungsreise im Innern Asiens im Jahre 1857, Pet. Mit., 1858, p 350; Narrative of an exploring expedition from Fort Vernoye to the western shore of Issik-kul Lake, Eastern Turkestan.—Jour. Roy. Geogl. Soc., 1869, p. 311.
Severtzoff: A journey to the western part of the celestial range (Thian Shan), Jour. Roy. Geogl. Soc., 1870, p. 343 (translated from the Russian).—Erforschung des Thian Schan Gebirgssystems, &c., Ergänzungsheft No. 42, 43, Pet. Mit., 1875.
[2] Jour. Roy. Geogl. Soc., 1870, p. 250.
[3] Entstehung der Alpen, 1875, pp. 135, 142.
[4] The names adopted for these various mountain chains by Russian and German geographers are cumbrous, and might be simplified with advantage. The Trans-Ili-Alatau is the range just south of Fort Vernoye, and is the more northern of two parallel chains north of Lake Issik (Issik-kul).
[5] The main range is considered to be that lying south of Lake Issik. The highest and best marked portion of this main range lies further to the eastward than the meridian of the lake.

parallel secondary chain. Between the two the palæozoic rocks rise to a considerable elevation, forming synclinal and longitudinal valleys. We shall follow the section to the north-east, from the foot of the principal ridge, according to Severtzoff's latest accounts, and begin at the Naryn River, the valley of which is bounded on the south by an outer range of the Thian Shan, the Chakir-tau. This consists of granite and mica schist, the opposite slope of the valley being entirely composed of contorted clay-slate, which locally, overlaid by dark violet porphyry conglomerate, extends to the north-west to the top of the Sari-tau, in which, at the pass of Barskoum, syenite is exposed.

"Proceeding from this pass towards Lake Issik diorite and serpentine are first seen; then mountain limestone, which forms a synclinal. This synclinal coincides with the longitudinal valley separating the Sari-tau from the next range to the north, the Terskei-Alatau, and this latter corresponds to the Sari-tau ridge precisely, so that, on the north side of the intervening valley, first mountain limestone with the slope reversed, then diorite, and finally syenite, are met with. Below, on the shores of Issik-kul, sandstone is found, which may be compared with the carboniferous strata of the Kara-tau.[1] At the eastern end of Issik-kul the little range of Kisil-kija[2] consists of red argillaceous sandstone; this range lies nearly in the direction of the greatest (longitudinal) diameter of the lake itself, and in the line of strike of the Rothliegende at the western end of the lake, in the gorge of the Boam stream and on the northern slopes of the Khighiz Alatau.[3] Proceeding over the Santash pass into the region of the Trans-Ili-Alatau, this is found to consist of granite intersected by two or more bands of limestone standing at high angles or bent into trough-shaped curves; one of these bands forming the ridge between the rivers Chilik and Chanishk.

"Finally, the granite northern slopes of the Trans-Ili-Alatau, as already stated, are terminated, towards the north, by a long but rather low chain of hills which consist of porphyry."]

[1] North of Chemkend and Tashkend.
[2] Tasma mountains on some maps.
[3] Now called on many maps Alexandrovski range.

PART VI.

From Yangihissár, Káshghar, to Panjah, in Wakhán, by the Little Pámir, and return journey by the Great Pámir.

[This section, like that describing the country between Mari and Leh, is simply compiled from Dr. Stoliczka's diary. It commences from Yangihissár, two marches, or about 40 miles, from Káshghar, on the road to Yárkand. Thence the route followed led in a south-west direction through the district of Sarikol (Sirikol) to the frontier of Wakhán, at or near Aktúsh, a distance of about 150 miles in a direct line, and thence in a west-south-west direction for 120 miles more across the Pámir steppe to Panjah or Kila Panjah in Wakhán. The road from the Yárkand frontier to Panjah traverses a district known as the Little Pámir, and follows the more southern of the two streams which unite near Panjah to form the head of the Amu or Oxus; the return route to Aktásh was by the northern stream (that followed by Wood) and the Victoria lake. From Aktásh the party with which Dr. Stoliczka was associated returned by the same route as before to Yangihissár. The geological notes made on this portion of the return eastward journey have been incorporated with those made in the same localities on the westward route. The former largely supplement the latter, which were made when the ground was much concealed by snow.]

March 21st, Yangihissár to Ighiz Yar.—Started for Sarikol under Gordon, with Biddulph and Trotter. March of about 18 miles almost due south. A mile from Yangihissár we crossed several low ridges, extending for about a couple of miles, of what appeared to be upper Artysh beds, consisting of sand, clay, and conglomeratic beds. The dip was at first north by east, then the beds were horizontal, and further on they dipped to south by west. Crossed the Yangihissár stream, and traversed, first, a saline plain, and then one of gravel. The ascent throughout was very gradual, but must have amounted altogether to more than 1,000 feet.

March 22nd, Aktala.—A march of about 18 miles, chiefly in a south-western direction. The low hills west of Ighiz Yar are composed of lower Artysh beds, hardened greenish sandstones much contorted.[1] Leaving Ighiz Yar, we crossed the plain for about 3 miles, and then entered the narrow valley of the Rin or Ring river.

The rocks at the entrance are lower Artysh sandstones, much contorted and disturbed. They continue for fully a mile, and are then succeeded by greenish sandstones and shales of a much older appearance. These rocks are again at first very much disturbed, but further on they dip regularly at a low angle to west by south, or even to west. The general dip, however, appears to be north-east. Nearer to the camp Aktala, the sandstones and slates alternate with highly carbonaceous shales and slates, and some highly ferruginous or hæma-

---

[1] On the return route from the Pámir and Wakhán on this march from Aktala to Ighiz Yar the following remark occurs: "The same slates and thin-bedded sandstones continue all the way. Towards the plain they alternate with coarser and conglomeratic beds; but they all appear to belong to the same old series." I infer from this that Dr. Stoliczka was finally inclined to believe that the rocks near Ighiz Yar, which he at first assigned to the Artysh beds (tertiary), were really older.

titic beds. These older beds very much resemble those we saw about Chakmák,[1] which also may turn out to be the same we saw north of Tám.[2] The sides of the hills are more or less thickly covered with *loess* dust, which much obscured the bedding of the rocks. I found no fossils.

Among the river boulders I noticed boulders of the red sandstone we saw south of Sanju, and a greenish syenitic rock.

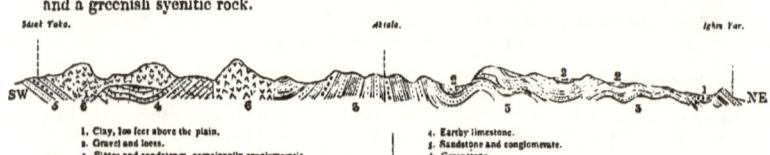

1. Clay, 100 feet above the plain.
2. Gravel and loess.
3. Slates and sandstones, occasionally conglomeratic.
4. Earthy limestone.
5. Sandstone and conglomerate.
6. Greenstone.

Section from Sásak Taka to Ighis Yar.

March 23rd, Sásak Taka, 13½ *miles.*—The dark slates, shales, and sandstone continued for a couple of miles, then followed greenish chloritic and felspathic rocks, very much like those south of Sanju, but more massive, being in fact a form of greenstone. These cap the whole series, and in one or two places come down to the bed of the river. Next follow earthy limestones, whitish or dark in colour, without any fossils, and then shales, carbonaceous slates, &c., with occasional conglomeratic beds and coarse sandstones. The whole of this series appears to be the same we saw on the road from Tám to Sanju. Some of the strata very highly carbonaceous, but not a trace of a fossil anywhere.

March 24th, Kaskasu.—Fourteen miles up the river Kaskasu. Nothing but the same carbonaceous slates and shales which are probably palæozoic, or occasional beds of grey more or less coarse sandstone, or even conglomerate. Not a trace of a fossil anywhere. The beds are mostly much disturbed and contorted, but where traces of regularity occur, they are seen dipping to south-west at an angle of about 50°. About half-way the old rocks were overlain by an old alluvial deposit, mostly consisting of boulders of the red sandstone, somewhat sparingly intermixed with boulders of gneiss. I have, however, not seen anywhere *in situ* the red sandstone; the greater portion seems to have come from a valley leading into the Kaskasu from the west about 4 miles east of our camp at Kaskasu. In several of the streams coming from the north, pebbles of white dolomitic limestone are seen containing a fossil like *Bellerophon*. These are probably from the white limestone, which is seen further on from the pass, and which is probably carboniferous. There were also blocks of a black earthy limestone, full of crinoid stems; this last is probably Silurian and interbedded with the black slates. A very similar limestone was seen on the road, but it contained no crinoids.

March 25th, Chehil Gombáz.—A short march of 11 miles across the Kaskasu pass. The bed of the Kaskasu river was strewn with boulders of gneiss, which must have come from the head of the stream. East of the pass the rocks are the same as before ; palæozoic slates, sandstones, and conglomerates striking north and south, nearly vertical, much contorted, but sometimes dipping to the westward. On the pass the beds apparently dip north-east, but the strike is very indistinct, the surface being covered with fine clay, partly derived from the

[1] 'North-north-west of Káshghar, p. 26.' | [2] 'Near the Sanju pass, south of Yárkand, p. 21.'

decomposition of the slates, but principally, in all probability, a subaërial deposit, like the loess. In some places this clay covering is thin, and on a sharp incline parts of it are often carried away, so that some of the slopes have a rather rugged appearance. Looking north from the pass, I saw what was evidently limestone on one of the hills; it was probably the same as the carboniferous limestone seen south-west of Sanju, but there was no possibility of getting near the hill. In a north by east direction I saw red thin-bedded sandstones capping one or two hills, the beds apparently dipping to north-east. This red rock was very probably identical with the cretaceous red sandstone north-west of Sanju, thus remarkably indicating that this portion of the hills is a continuation of the Kuenluen.

From the pass to Chchil Gombáz the rocks are palæozoic carbonaceous slates, very variable in strike and dip. Near the pass the strike is indistinct: in the valley north of Chchil Gombáz, it is nearly east and west, the beds being vertical and much contorted.

March 26th, Pasrobát (across the Torat pass).—The whole way nothing but the same carbonaceous slates and shales, and partly sandstone, were seen. They were dipping at a very high angle to north by east or north-east by east. In some places they were interbedded with crystalline limestone, and with white quartzite, in strata of about 40 to 50 feet in thickness. At the junction of the two streams, the Pasrobát and the Tongitár, and much higher up, I noticed old diluvial gravel, in some places up to the thickness of 300 feet, the boulders mostly consisting of crystalline gneissic rock: some of the boulders are of huge dimensions, and all are well-rounded. These boulder deposits must have been formed by enormous rivers and large quantities of snow. The gneiss is either fine-grained, with biotite mica, sometimes almost schistose, or it is porphyritic with rosy quartz, white felspar and a greenish mica. There is little schorl to be observed in any of the pieces.

March 27th, Tárbáshi, about eight miles in a western direction.—The carbonaceous slates and sandstone continued for about a mile from camp, seeming, however, more micaceous. Then they gradually changed into dark carbonaceous mica schists with garnets; this again gradually into light-coloured mica schist, with more white quartz and less garnets, and this after about two and a half miles from camp into gneiss. All the strata were dipping at about 50° to north-east and north-east by east. In many places gravels conceal the rocks to a height of 150 feet above the river. On the greater heights dark-coloured schistose rocks are seen; they are mostly hornblendic.

March 28th, Balghun.—A march of about 20 miles across the Chichiklik plain and the Kokmainák pass. All the rocks around are gneiss, which gets gradually schistose, but it is cleaved in all directions and breaks up easily; the irregular cleavage entirely obliterates the bedding.

March 29th and 30th, Balghun to Chushman, and thence to Tashkúrgán (Sarikol).—Two marches of rather more than 20 miles altogether. The rocks are all metamorphic schists, rarely micaceous, but chiefly chloritic, quartzose, and hornblendic. North-west of the camp the dip is west by north; previously it was east by south. On the western side of the valley are thick gravel deposits, the boulders mostly of gneiss and syenite.

April 2nd, Kanshubar, 16 miles.—The whole way nothing but gneiss, in different variations, was to be observed. At first where we entered the Tongitár (valley), the fine-grained pale-whitish gneiss was interstratified with dark gneiss and syenitic gneiss, full of schorl; further on, syenitic gneiss prevailed, then bands of beautiful reddish gneiss occurred in it, with reddish-brown quartz, reddish glassy felspar in large crystals, and bits of schorl. Further on, the gneiss became more ordinary, both coarse and fine grained.

April 3rd, Kogachak, near Aktásh.—[Frontier of Sarikol belonging to Káshghar, and Wakhán under the rule of Kabul.] Followed up the valley for about a mile, when the gneiss was apparently underlain by black palæozoic slates, strike almost from east to west, and the dip very little towards the gneiss—or, rather, the beds were vertical. I could not find a trace of fossils. The slate is brittle, and very much cleaved in different directions : it would not do for roofing purposes, unless large quarries were opened. The slates continued for more than a mile, then they gradually became calcareous, and a series of thin-bedded whitish limestones followed—first, again, almost vertical, but, a little further on, distinctly dipping at an angle of about 50° towards the slates, though evidently younger. The limestone was dolomitic and highly bituminous, but unfossiliferous. After about a mile it changed to grey limestone, and became slaty. Then followed a band of greenstone for about half a mile, overlain by brownish-black shales, apparently carboniferous; and these shales were overlain by greenish dolomitic crinoidal limestones, lithologically the same as those which I found to be carboniferous in the Artysh district. I dare say this limestone is also carboniferous. However, the upper beds of this limestone series are paler, and apparently less dolomitic; and in them I found a cordiform pelecypod, like *Megalodon*, very common. Possibly the whole of the limestones, but certainly those on the western side of the range, are triassic. They rest here on purple and greenish shales and slates, which are afterwards traversed by greenstone. (See also diary of May 6th.)

April 4th, Onkul.—A march of about 24 miles. Crossed a spur over an old gravel deposit, and traversed a valley, the rocks on both sides of which were whitish triassic limestone, resting on reddish shaly rock, which, again, overlaid black slates, evidently palæozoic. Before we reached camp the slates rested on gneiss.

April 5th, Oi-kul or Kul-i-Pámir Khurd (Little Pámir Lake).—Marched about 24 miles along the valley of Pámir Khurd, or Little Pámir. The rocks composing the hills to the left of the valley are all gneiss to an elevation of 2,000 or 2,500 feet above the valley; those to the right are higher and more sharply ridged, but their composition could not be ascertained.

April 6th, Langar.—Marched about 24 miles. After 6 miles, in a west by south direction, the hills to the north became black slates, resting on gneiss. These same slates were seen dipping at an angle of about 60° to north-east by north at the entrance into the valley, which was here very narrow. They were overlain higher up by reddish slates and conglomerates, and the whole of the series has bands of quartzite, often intercalated : one of these quartzite bands seems to have passed right across the stratification of the slaty rocks at the entrance of the narrow part of the valley from the Pámir, which here terminates. The gneiss on the Pámir appears to have had only a very slight dip to north. The black slaty rock continued all the way to camp.

April 7th, Daraz-diwán, 15 miles.—Black slates, dipping north by east, were seen on both sides of the valley, and on the right the purplish or reddish slates and conglomerates rested on them. The conglomerates consisted of angular boulders of white quartzite in a reddish or purplish matrix. I saw fragments of similar conglomerate in the Sanju river.

April 8th, Sarhada.—March of 11 miles. For the first 2 miles black slates were seen along the road, which was above the level of the river; further on, the slates rested on the same finegrained gneiss which we had seen at Pámir Khurd, until within half a mile of Sarhada, where the slate again came down into the valley.

Throughout the valley, from the spot where it was entered from Pámir Khurd, old banks of bedded clay and gravel are seen up to 1,200 and 1,500 feet above the present level of the river. They are generally seen at the turns of the river, and can be traced all the way down, but are nowhere more extensive. Before the river cut its present deep bed, its course was probably often interrupted, and small lakes formed, or, at least, its course was retarded, so as to form these deposits.

April 9th, 10th, and 11th, Sarhada to Patir, halting at Patuch and Yúr.—Three marches of 4½, 15, and 12 miles. Black slates alone were seen till 9 miles beyond Patuch, thence gneiss (fine-grained) and metamorphic rocks for the remainder of the way. The gneiss is sandy, and disintegrates easily.

April 12th and 13th, Patir to Panjah, or Kila Panj, halting at Zang; 20 miles from the former, only 3 from the latter.

[No special description of the geology is given. The beds seen were probably all metamorphic, the same as before. A hot spring opposite Patir is said to rise in black metamorphic slates.]

All the hills at Panjah consist of a metamorphic quartzose schist, which composes the hills on the left bank of the valley. The rocks dip to south or south by east into the valley: a few miles west they are overlain by dark hornblendic schist.

[After a halt of 12 days in Panjah, the party marched back to Káshghar territory by the Great Pámir, re-entering their former line of march at Kanshubar, east of Aktásh.]

April 26th and 27th, Panjah to Langerkish, 6 miles only.—Visited the hot spring near Zang: the water is 120°. The rocks are quartz, hornblendic, and mica schist, with garnets, dipping to the south-east.

April 27th, Yumkhana, 16 miles.—Old clay deposits reach to about 2,000 feet above the present level of the river. The metamorphic schists are very variable, but highly micaceous throughout (containing biotite); they still dip to the south-east, and include beds of white marble. On the left bank of the river they seem to dip under the gneiss, which is not distinctly stratified.

April 28th, Yolmazár, 12 miles.—Rocks same as before—all fine-grained gneiss, with biotite,—very much resembling the Himalayan central gneiss, with biotite mica, traversed mostly by thin veins of albite granite, with muscovite. It really seems that this is the continuation of the central gneiss, in which the Spiti and Záskar secondary rocks may form a bay, extending from south-east towards north-west. About Drás the secondary rocks go over a saddle into Kashmir, but the gneiss continues northward. Hornblendic beds often occur in the gneiss; they consist of dark, rather homogeneous rocks, which include hornblende and staurolite crystals.

April 29th to May 1st, Yolmazár to Lake Victoria (Wood's Lake).—Three marches, altogether about 37 miles.

[Rocks throughout described as gneiss; that on the first march described as containing a little green mica or chlorite; on the second but little rock was seen in place, the valley being largely occupied by beds of pebbles and boulders, which form terraces along the sides, whilst the hills were covered with snow. The gneiss seen was "remarkably altered, craggy, conglomeratic, split in all directions, and as if it had been burnt," but no trace of an eruptive rock was seen.]

The shingle boulders were mostly rounded; some of very large size only slightly so, and mixed with sand. The whole mass must have been accumulated more by the agency of snow and ice than running water.

[The hills around the lake are described as entirely of gneiss, and rather sharply pointed.] The lake is about two miles in width, and surrounded by terraces of rounded worn boulders, mixed with sand. These terraces rise to at least 100 feet above the lake, and show that the lake was formerly much more extensive than it now is. [The details will be found in the diary.]

May 2nd, Shashtupa, 18 miles.—For the first 6 or 7 miles the rocks are apparently gneiss; further, black slates and shales overlie the metamorphic rocks, and the hills on both sides become more rounded. Immediately above the gneiss the slates look rather metamorphic, but, further on, they are of the usual type, and reddish beds overlie them near the camp. The dip is low to north by east.

[The whole march nearly was over what Dr. Stoliczka terms "shingle beds," and the watershed was formed by a mixture of boulders and sand. See diary.[1]]

May 3rd, Isligh, 18 miles.—About three miles north of camp the upper reddish slates of the silicious group are overlaid by darkish grey limestone, dipping to north by east. I found no fossils in it. This limestone (*a*) is about 1,000 feet thick, and extends for about a quarter of a mile. Then follows a very indistinctly stratified white or light grey limestone (*β*), which must be at least 2,000 feet thick, and extends for about one mile. I saw *Crinoid* stems in it, but nothing else. After this follows, again, a darker grey limestone, evidently belonging to a different series, being unconformable on the former. This series of limestones forms the highest ridge, some of the rugged mountains rising to fully 20,000 feet; and the thickness of the rocks must be from 3,000 to 4,000 feet. The general strike is west by north to east by south, and the dip to north by east, or almost north, with angles ranging from 80° to 90°. The thickness of this limestone series must be about 3,000 feet. The whole of these limestones appear to be of palæozoic age—probably for the most part carboniferous.

After this follows a great series of dark shales, with beds of limestone. The shales themselves (*δ*) are highly carbonaceous, and the limestones are earthy, mostly thin-bedded, but greatly contorted, rising in more or less vertical ridges.

May 4th, Aktásh, 36 miles.—After four miles over the plain, the road led for two miles through a narrow gorge between limestone (*ε*), on which, further on, rest brownish, rather silicious sandstone, and grey, then black, crumbling shales. The road crosses a low pass, and then follows through these shales, in almost a due eastern direction, to the junction with the Isligh. The whole road passes through these shales, with a little sandstone, but more of the earthy limestone. The series extended north, as far as I could see, the shale hills being rounded, and the limestone ridges sharp. Greenstone appears to pierce through it in the distance, and the elevations of the hills appears to decrease. South of the road runs the high limestone range in a west by north to east by south direction towards Aktásh. The shales (*δ*) and limestones (*ε*) appear to be triassic. Near Isligh I saw a lot of *Rhynchonellæ* in one of the earthy limestone beds, but could not extract any thing very recognisable.

May 5th, halted at Aktásh.—Going about a mile north of camp, and then tunring in a western direction up a gorge, I found myself north of the great limestone hill, and here, resting on the limestone, were the dark crumbling shales, exactly like the Spiti shales in mineralogical character. The shales (*δ*) contained a few beds of the brown sandstone, but both appeared entirely unfossiliferous. In the interbedded limestone (*ε*) I found, however, a great number of *Rhynchonellæ*, which decidedly appear to be triassic, if not younger. In the more compact limestone I could only see crinoids, no other distinguishable fossil; not a trace of a *Cephalopod*. In a block of more earthy grey limestone loose in the stream bed I got several *Rhynchonellæ*; but I am not sure whether that limestone is (*ε*); it seems more probably (γ)

---

[1] It is not quite clear from the diary what Dr. Stoliczka's views were on the subject of these accumulation. He repeatedly says they must have been brought down by snow, or snow and ice. He never mentions glaciers or moraines, and never notices the presence or absence of striation on the rocks.

Afterwards I went south of the camp, where on our road westward[1] I got a section like that of a *Megalodon*. The limestone is mostly dolomitic, white or light grey, and less bituminous than (₁). I got crinoid stems in it, and a small *Pecten;* I could not say whether lower trias or carboniferous.

May 6th, Kanshubar (same camp as on April 2nd).—Two and a half miles from Aktásh, at a spot where the stream from the Nezatásh pass is joined by another flowing from the south-east, there is a mass of greenstone in the shales, and east of that mass the shales are very much altered, evidently indicating that the outburst of the greenstone must have taken place after the deposition of the triassic shales. Looking north, the shales continue for about a couple of miles, composing the hills, which rise to about 3,000 feet above the valley. To the north-west is a great mass of greenstone again, while a sharp ridge of limestone runs through the shales, coming from the west, and disappearing and broken up towards the east. Further on, the shales are seen to be overlain by reddish sandstones and shales, towards the top much alternating with greenish-grey beds; and this series is again capped by a light-brownish rock of inconsiderable thickness. These last rocks and the limestones dip north by east, but the crumbling shales are very much contorted, mostly by the greenstone.

The section from Aktásh to the north is something like this:—

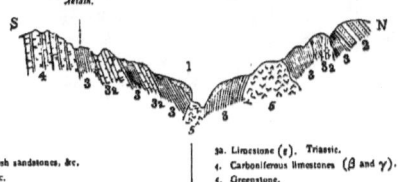

1. Gravel.
2. Reddish and greenish sandstones, &c.
3. Shales (δ) Triassic.
3a. Limestone (ε). Triassic.
4. Carboniferous limestones (β and γ).
5. Greenstone.

Sketch section of the rocks north of Aktash.

Proceeding towards the Nezatásh pass, I found in the limestone (ε) dark beds full of *Halobia Lomelli?*, and I also noticed the *Rhynchonella* limestone, which is very earthy and brown, *in situ* in the shales. In the limestone (ε) *Rhynchonellæ* are very rare, or, at least, very difficult to observe. The limestone (ε) is, however, always very much less bituminous than (γ), and usually darker, and weathers off in flakes, which peel off the surface, while (γ) is usually massive. Limestone (ε) forms the Nezatásh pass in a ridge crossing the pass, but the passage itself is in shales, which are also seen in a kind of basin east of the pass, the basin being quite encircled by very high cliffs of limestone (ε). Crossing into the stream, which comes from the south, and combines with that flowing eastward from the pass, I observed a number of pelecypod sections in the limestone, which appear to belong to *Megalodon*. They were rather large, but otherwise not distinguishable.

Further on, the shales were several times crossed by greenstone, and then followed the bedded grey rock. The carboniferous limestone ridge runs from Aktásh almost due eastward, and about 5 miles before reaching Kanshubar it turns gradually to south-east, still retaining its great height.

[From Kanshubar the return route to Yárkand *via* Ighiz Yar was over ground previously traversed, and the geological notes have already been incorporated with those of the journey westward.]

[1] See notes for April 3rd, p. 38.

PART VII.

FROM YÁRKAND TO BURTSI, SOUTH OF THE KARAKORAM PASS, viá KUGIÁR, THE UPPER VALLEY OF THE YÁRKAND RIVER, AKTÁGH, AND THE KARAKORAM PASS.

[THIS route lies in general considerably to the west of that traversed by Dr. Stoliczka in the preceding autumn. For two marches from Yárkand to Karghalik the road is the same as before; thence it leads a little west of south across the Kuenluen to the upper valley of the Yárkand river; it turns eastward up the valley of that stream as far as Aktágh, where it meets the former route, but it then turns southward across the Karakoram pass. The following notes commence from Karghalik and are copied, like those in the preceding section, from the diary.]

May 31st, Karghalik to Beshterek, 20 miles.—The first 10 miles over gravelly desert; thence the road lies up the Kugiár stream, a broad desert valley, nearly a couple of miles wide. Gravel beds, as much as 150 feet thick in places, extend up to the village: they are evidently alluvial, and not Artysh beds, though the reddish sandstones at Bora [1] belong to the latter. Loess rests on the gravel, and in places has been re-deposited by the river and stratified. There is a good deal of this stratified loess in the valley itself, but it is chiefly sand.

June 1st, Kugiár, 17 miles.—For 14 miles the road lay across desert, over somewhat elevated terrace land of sand and gravel. About 4 miles north of Kugiár, Artysh beds, clayey sandstone, and fine conglomerate are seen below horizontal beds of diluvial gravel. Further on, they again entirely disappear under the diluvial terraces, which rise about 200 feet above the elevated ground. The amount of sand, clay, and gravel brought from the hills is something enormous. The Artysh beds evidently form the axis of the low ridge, which runs from east to west, about 4 or 5 miles north of Kugiár; but they are covered with diluvial gravel.

June 2nd, Ak Masjid, about 27 miles.—The first half of the road is entirely over gravel beds, then a grey dolomite begins to crop out. The beds undulate, but the general dip is north : not a trace of a fossil could be detected. Further on, close to camp, a reddish, somewhat silicious sandstone, and thin-bedded streaked limestone of the same colour, with a high northerly dip, underlies the grey dolomite, and rests upon other grey and whitish dolomitic limestone, less distinctly stratified. As a rule, dust covers all the slopes of the hills so thickly that, except on a precipitous cliff, not a trace of solid rock can be seen. In the valley, loess attains a thickness of fully 30 feet ; it is partly stratified, but the accumulation appears mostly due to moisture.

June 3rd, Chiklik, 13 miles.—Up to the foot of the pass the grey limestone rock continues, gradually becoming in places thinner bedded, streaked, and metamorphic. Near the foot of the pass it changes to a stratified chloritic rock, while the grey limestone occupies the

---

[1] These were noticed in Part III, and were observed on the road between Sanju and Yárkand; *ante*, p. 22.

GEOLOGY.

greater height. The green rock alternates with thick beds of a white quartzose and calcareous schist, and beyond the pass the green rock becomes more solid, loses its stratification, and becomes a regular greenstone, exactly like that I met with east of Sastekke, on the Sarikol road. Black slate I only saw in one or two places, and then in mere fragments or blocks; but it is evident that the whole series of rocks is the same as that south-west of Sanju.

June 4th and 5th, Chiklik to camp, about 2 miles west of Mazarkhoja.—Two short marches, together about 16 miles. Nearly all the way nothing was seen but greenstone, similar to that near Sasúk Taka : towards the end of the second march this unstratified greenstone is overlain by chloritic schists and other bedded metamorphic rocks, resembling those to the north of the Sanju pass.

June 6th and 7th, Mazarkhoja to Grinjikalik.—Two marches, together 18 or 19 miles. A mixture of metamorphic rocks was met with, like those north of the Sanju pass, dipping at a rather high angle to north-west, west, and south-west. The whole series seems much disturbed. The prevalent rock is a quartzitic and highly hornblendic schist, traversed in all directions by ramifying veins of white quartz, with some schorl, and by other darker veins, containing hornblende.

June 8th, Jiraksheldi, 10 miles.—The same metamorphic rocks continue for about a mile beyond yesterday's camp, and rest here on light-coloured, rather fine-grained gneiss, which is indistinctly stratified, and dips to the north-west. It is traversed by dark hornblendic veins. This greyish white gneiss continues for a couple of miles, and rests on an unstratified mass of fine gneiss porphyry,[1] similar to that I saw west of Sarikol. This feldspathic gneiss seems to form the axis of the whole metamorphic mass; for, further to south by east from this camp, within about a mile, it is again overlain by the same somewhat fine-grained greyish-white gneiss, dipping to the south. This gneiss is, again, overlain at the camp by almost vertical and much-contorted beds of black shale, grey sandstone, and conglomerate, the same as I saw north of Tám. The coarse conglomerate has a comparatively recent aspect, but the whole series of rocks must be upper palæozoic, although one cannot help doubting the fact.

June 9th, Kulunaldi, 12 miles.—[This march led across the main ridge of the Kuenluen by the Yangi pass (16,000 feet), and down again into the upper valley of the Yárkand river. The corresponding pass to the eastward crossed on the journey to Yárkand is that of Suget.]

From yesterday's camp, the sandstones, conglomerates, and interbedded shales continued up the pass, where the conglomerates were of great thickness, evidently occupying the top of the series, and dipping with a slight angle to west. On the other or western (southern) side of the pass, the conglomerates and sandstones all continue for about 2½ miles highly inclined, and dipping towards east by north; they rest at about the third mile from the pass on black slates, which soon pass into dark grey and greenish metamorphic schist, sometimes with small garnets.

1, Conglomerate; 2, Sandstone; 3, Shales; 4, Black slates; 5, Metamorphic rocks, dark-coloured, with quartzite; 6, Fine-grained gneiss; 7, Unstratified granitoid porphyritic gneiss.

Section across the Yangi Pass, north of Yárkand River.

The metamorphic series is often traversed by veins of a solid greenstone-like rock, and towards the Yárkand valley there is a considerable thickness of a white quartzitic schist,

---

[1] Evidently, from the description, a granitoid rock.

more or less massive: under it lies a brownish sub-metamorphic schist, which is also found on the other side of the Yárkand river at the camp.

June 10th, Kirghiz Jangal, 16 miles.—The sub-metamorphic schists near Kulunaldi are overlain by a reddish, very coarse conglomerate, and from beneath this, further along the road, a series of grey and pink metamorphic schists crops out, occasionally with graphitic layers interstratified. The last continue up to camp. The beds dip first to west 30° north, and afterwards almost west. The coarse, reddish conglomerates are the same which I saw in going from the Pámir-kul to the camp Langar, in Wakhán.

June 11th, Kashmir Jilga, about 24¼ miles.—The rocks are all grey silky, or brownish mica schist. For the first 14 miles it is difficult to see any stratification, the schists having distinct bacillary cleavage; but further on, the schists dip to the north-east on the right bank, and for the last 6 or 7 miles the valley runs along an anticlinal, the beds dipping on the right bank to north by east, and on the left bank to south by west, at an angle of about 50°. The schists decompose easily, and cover the slopes with fine debris. Almost all along the bank of the river there are extensive deposits of detritus, some of them containing beds of clay and sand, left by the river. At the openings of the lateral ravines there are extensive fans of debris, some of them more than 100 feet thick.

June 12th, Kufelang, 11 miles.—Rocks the same all the way; greenish metamorphic schists, often alternating with graphitic layers. The schists decompose and break up very readily, and the hill-sides, in some cases up to the top, are covered with debris, loose or cemented together. The dip is very variable, usually at a high angle to south-west or south.

June 13th, Aktágh, about 20 miles.—[Here the road joins that followed on the journey northward, but it immediately diverges again.]

1, Argillaceous beds (? tertiary); 2, Shaly sub-metamorphic beds (trias); 3, Grey limestones (trias); 4, Red limestones, with *Ammonites batteni*, &c. (trias); 5, ? Trias; 6, Dark triassic limestone; 7, Limestone (? carboniferous); 8, Grey limestones (? carboniferous); 9, Red calcareous sandstone; 10, Sub-metamorphic schists.

Section near Aktágh.

The schists, greenish and metamorphic in general, but blackish and sub-metamorphic in parts, continue for about a couple of miles along the river; they are mostly almost vertical. Then some of the beds incline to the south, and are more regularly bedded; but there does not seem to be any distinct break between these latter and the vertical beds. After the second mile the greenish silky schists are overlain *unconformably* by reddish earthy and calcareous sandstones of about 150 feet in thickness, dipping regularly to south by east at an angle of about 30°. These reddish beds pass into distinctly bedded grey limestone and whitish marl of some 500 feet at least, the dip being to the south, but the angle gradually decreasing until the beds, after some 8 miles, become almost horizontal. Further on, they again dip to the southward, and the top beds have a reddish colour. There are greenstones in these rocks, like those which I saw about Aktásh on the Pámir; and the limestones must be carboniferous or triassic, but I could not find a trace of a fossil. The higher beds are often brownish and sandy; some beds almost a calcareous sandstone, alternating with conglomeratic beds.

Near Aktágh the series is overlain by much more recent looking earthy and conglomeratic beds, readily yielding to decomposition. The hill Aktágh at camp consists of these (? tertiary) beds, dipping at about 45° or 50° to the south.

GEOLOGY. 45

There must be greenstones somewhere in this southern direction among the dark crumbling rock.

The light-coloured bedded limestone strikes over to Karatágh lake, and the hills to the west, east, and south-east appear to consist of it. I noticed, when I marched last year, that their steepness indicates in part limestone cliffs, and some of them at least were of a light colour.

This is also the pale limestone seen north of our camp, some miles north of Khush Maidan, and no doubt these limestones extend to the south of Aktásh. [That is to say, that this pale limestone, which is probably of carboniferous age, appears to stretch across from the high ground between the Mastágh and Kuenluen ranges to the eastern edge of the Pámir.]

June 14th, Woabjilga, 12 miles.—The hills all covered with detritus.

A little way south of Aktágh the grey limestones, which appear to be carboniferous, are overlain by dark crumbling dolomitic limestone and sub-metamorphic shales, in several places in contact with greenstone, which is again either typical, like that near Aktásh, or it is dark, and very homogeneous in texture, and at first strikingly resembles basalt. Further on, the grey dolomitic limestones again crop out from under the detritus of the valley; and near the camp the sub-metamorphic schists are overlain by more compact grey dolomitic limestone, which rises high upon a hill a little south by east of our camp. These grey dolomitic limestones regularly bend over at the top, and in the centre are exposed what may be called *Hallstädt* or *St. Cassian* beds—a red, somewhat earthy, marble, with *Arcestes? johannis austriæ, Ammonites batteni, Aulacoceras,* and Crinoids. I shall speak of this red marble as the *A. batteni* bed.

The *A. batteni* bed is seen exposed far towards the west, overlain by the grey limestone, and is mostly highly inclined towards the north. I must see more of the whole triassic series to-morrow.

June 15th, Karakoram-brángsa, 14 miles.—Starting from Woabjilga, the grey triassic limestones were met with, afterwards the red limestones succeeded them, and continued to camp, often interrupted by patches of greenstone, which is greatly developed at the camp north of the pass.

June 16th, Daulatbeg Uldi (crossing the Karakoram pass), *about 22 miles.*—Leaving camp, the greenstones are underlain by black crumbling shale, in mineralogical character like the Spiti shales, but very likely triassic, like that near Aktásh. Then follows an alternation of grey or whitish limestones and shales and the triassic red limestone; and on these rest blackish and grey marly shales, which are overlain by almost horizontal strata of brown limestone, very much like the lower Taglang limestone, and which contains fragments of *Belemnites.* These *liassic rocks* form the Karakoram range proper, and extend far eastward. The hills to the west are much higher, and do not allow a distant view.

After crossing the pass, the road skirts the base of the centre ridge in a south-east direction; and here the liassic limestones come down several times, and about four miles from the pass grey marly shale, or almost marly limestone, crops out from under the brown limestone: both are evidently liassic. On the right bank of 'the stream more massive limestones occur, dipping to north-east, but very indistinctly. I should think that these are triassic limestones. They very readily crumble to pieces, being highly dolomitic; and these often contain reddish beds interstratified.

June 17th, Burtsi, 24 miles.—First we crossed the Dipsang plain, with solitary low hills, probably still belonging to the Taglang series. Then we ascended towards the watershed.

The low worn-down hills to the west were thickly strewed with round pieces of whitish or reddish compact limestone, intermingled with boulders, large and small, of fine-grained syenitic gneiss. This rock must be *in situ* somewhere near the head of the watershed. Further on were many greenstone boulders coming down from the west, and this rock must also be found in that direction. At last we descended into a narrow gorge, the sides of which for fully a mile consisted of a limestone conglomerate, the boulders of white, grey, or black limestone being well rounded and worn and cemented together by a stiff bright red clay. Upon this followed dolomitic limestone, rather indifferently bedded, massive and white, and this was overlain by bluish shales and well-bedded limestone, extending from about 6 miles north of Burtsi to the camp. These limestones appear to be triassic: they are compact, with layers full of small gasteropods, amongst which I recognised a *Nerinea*. The so-called Karakoram stones, *i.e.*, corals, occur in dark shales below the limestones, which are capped by a yellowish-brown limestone, well bedded, but of unascertained age. The whole series dips south-west, at a moderate angle. [The last paragraph closes the diary, and is here repeated, as it is entirely geological.]

GEOLOGY.

Concluding Summary.

As this collection of Dr. Stoliczka's geological notes on the countries traversed during his journey was introduced by a brief account of his previous geological work in the Himalayas and Western Tibet, it may most fitly be concluded by a general sketch of the additional information which he has obtained in the countries north of those explored in earlier years.

His explorations in his last journey extended over portions of Northern Ladák, of the Mastágh or Karakoram, Kuenluen, Pámir, and Karatágh ranges, the last being a part of the Thian Shan. He also examined the plains of Yárkand and Káshghar, and the upper valleys of the streams which form the source of the Oxus or Amu. The notes on Kashmir, and on the Indus valley west of Leh, although interesting and affording some addition to our previous knowledge of the geology, do not touch on fresh ground, or add more than details to what was known before. Each of the other areas demands a few notes separately.

The Ladák range, north of the Indus, proved, so far as it was examined, to consist entirely of metamorphic rocks, principally syenitic gneiss. The same formations extend to the northward to the western end of the Pankong lake, and, so far as is known, throughout the greater portion of the Changchenmo, Shayok, and Nubra valleys, passing in places into a greenish chloritic rock, more or less schistose. These metamorphic rocks are believed by Dr. Stoliczka to be of silurian age. In the northern portion of the valleys named beds of dark shales and sandstones are met with, probably belonging to the carboniferous series: they are unfossiliferous, but agree with rocks of that age in Spiti and elsewhere, and they are succeeded, in ascending order, by fossiliferous triassic limestones, red and grey in colour, with dark shales; whilst the crest of the Karakoram pass, and some of the smaller hills immediately south of it, are composed of liassic rocks, containing fragmentary *Belemnites*. At one spot alone near Kium, in the Changchenmo valley, sandstones and conglomerates of comparatively recent aspect were observed, which are perhaps tertiary, and may belong to the same eocene formation as the rocks in the Indus valley near Leh.[1]

The valley of the Upper Yárkand river between the Mastágh (Karakoram) and Kuenluen ranges consists of metamorphic and sub-metamorphic schists and slates, reddish calcareous sandstone, and grey limestones, all unfossiliferous. The schists and slates are considered by Dr. Stoliczka as probably silurian; the other rocks, carboniferous. Some triassic limestones are found on the northern slopes of the Karakoram pass; and at Aktágh some recent-looking argillaceous beds were noticed, perhaps tertiary.

Two sections across the Kuenluen were examined—one, on the Karakásh river, the Suget and Sanju passes; the other, further west by the Yangi Diwán. On the former route the greater portion of the range consists of syenitic gneiss, associated with various forms of schists, with some of which pale-green jade is associated. On the more western route the same metamorphic rocks are found, but the syenitic gneiss is less developed, and there is a great quantity of greenstone.

---

[1] Drew (Jummoo and Kashmir, p. 343) has noticed the occurrence of hippuritic limestone (cretaceous) resting unconformably on older encrinital limestone (? palæozoic) in the Lokzhung range, north of the Lingzi-thung plain and east by south of the Karakoram pass. In the same work there is an excellent account of the extraordinary high plateaus of northern Ladák, west of which appear to be of lacustrine origin.

North of the metamorphic axis of the Kuenluen range, the hills sloping down to the plain of Yárkand consist principally of various forms of schistose rock, slates, and limestone. In the latter, north of Sanju, carboniferous fossils were found in some places, but the rocks are, as a rule, destitute of organic remains. On the western route the only limestone seen was dolomitic and unfossiliferous. Towards the edge of the plain, formations of later date crop out; and near Sanju red sandstones, capped by grey calcareous sandstones and chloritic marls, are found, the latter containing cretaceous fossils; and upon these, again, rest gravels and clays of still later date. The cretaceous rocks were not observed further west.

The ranges lying west of the Yárkand plain, and intervening between it and the Pámir watershed, appear to be composed chiefly of the same rocks as the Kuenluen, south of Yárkand. Only one section was examined, and this was traversed twice. Near the plain the prevailing beds are carbonaceous slates, sandstones, and conglomerates, probably palæozoic, with which greenstone is associated. A few limestones were seen, and traces of the red cretaceous sandstones of Sanju: the latter, however, was not examined *in situ*. No fossiliferous beds were observed, but the slates, sandstones, and conglomerates are probably palæozoic, like the corresponding rocks in the Kuenluen. Further from the plain, in the district of Sarikol, the slates and their associated beds become metamorphosed, and pass into schist and gneiss, upon which, close to the frontier of Wakhán, near Aktásh, rest black slates, and limestones of apparently carboniferous age; and above these, again, other limestones with triassic fossils, and sandstones.

The Pámir itself between the Yárkand frontier at Aktásh and Panjah, the principal village of Wakhán, was twice crossed, the return route lying a little north of the other, and each following one of the two streams, which unite to form the head of the southern or main source of the Oxus. The geology throughout is of the very simplest description. The carboniferous and triassic limestones were only found for a very short distance west of the Yárkand frontier; and thence to Panjah the whole country consisted of black slates, occasionally capped by reddish slates and conglomerates, and resting upon gneiss, which forms the great mass of the plateau. The slates are, doubtless, palæozoic; but no evidence of their precise age was obtained. The gneiss is fine-grained; it contains biotite, and is, in places, traversed by veins of albite granite, and it altogether so much resembles the "central gneiss" of the Himalayas north of Simla, that it may be a continuation of the same rock. Immense accumulations of boulders and sand were observed on the Pámir, in all the river valleys and around the lakes.

The two journeys made to the mountains north of Káshghar, which are a continuation of the Thian Shan range, and unite it to the Pamir or Bolor, scarcely extended beyond the southern skirts of the range, the greater portion of which lies within the Russian territory. The first of these journeys extended nearly 100 miles in a direction north by west, from Káshghar to a lake called the Chadyr-kul; the second, to a distance of about 120 miles north-east to the Belauti pass. After passing the gravel slopes on the edge of the Káshghar plain, and some ridges of sand and clays, which appear to be of tertiary date, and which Dr. Stoliczka calls the Artysh beds, the first range met with to the westward consists of dark triassic limestones, resting on greenish shales, and the next range of old shales, slates, and sandstones, with crystalline limestone. More to the eastward all the fossiliferous rocks are of carboniferous age: they consist of grey dolomitic limestone, resting on a limestone breccia, passing into conglomerate, and locally interstratified with greenish shales. This series, probably, represents the old slates and their associates seen further to the west. On this eastern route the carboniferous limestones extend to the Belauti pass, where they are capped by darker limestones, on which

GEOLOGY.

rest greenish and purplish sandstones and shales,—all which rocks are possibly triassic. North of the old palæozoic formations to the westward volcanic outbursts of very recent date are found, and the remains of old craters are conspicuous; and beyond these, again, are limestones and slates of undetermined age, the latter occasionally showing signs of metamorphism. Some of the limestones resemble the triassic rocks in character, but no fossils were detected in them. The presence of metamorphic rocks in the ranges north of Káshghar is proved by the occurrence of gneiss pebbles in the gravels derived from the hills.

It is probable that coal occurs in places in the carboniferous formation, as specimens brought from the mountains were examined and roughly analysed by Dr. Stoliczka when in Káshghar.[1] Rocks of the carboniferous period are largely developed in Western Turkestan, and coal has been found in several places.

The plains of Yárkand and Káshghar consist of recent deposits of clay and sand, with occasional ridges of gravel and marly clay. They, doubtless, resemble closely the other great plains of Central Asia, all of which, having no exit, are basins of deposit, and are being gradually raised by the alluvium brought from the surrounding hills by rivers and streams, which dry up and lose themselves on the plains. Towards the edge of all such plains there are immense gravel accumulations,[2] which greatly conceal all the rocks. Below these gravels, all round the edge of the Káshghar plain, there is found a series of clays, sandstones, and conglomerates, often much disturbed, but evidently not of old date, called by Dr. Stoliczka Artysh beds, from the Artysh valley north of Káshghar, where they are extensively exposed. No fossils were found in them, but their discoverer was inclined to consider them marine. They present a marked resemblance, both in composition and in their position at the base of higher ranges, to the Sub-Himalayan rocks of Northern India, and the *molasse* of the Alps.

All of these deposits, and the rocks on the slopes of the hills for some distance from the great plain, are much concealed by an extremely fine unstratified accumulation, precisely similar in character to the *loess* of the Rhine and Danube, and which is evidently composed of fine dust, deposited by the atmosphere. The air in Eastern Turkestan, as in parts of China, is constantly, during the day, thick from the fine sand raised by the wind; so much so, that objects at a comparatively short distance are rendered invisible.

It is evident that there is great similarity in the geology of all the mountains surrounding the Yárkand basin. So far as they were examined, the prevalent formations were palæozoic, resting upon gneiss and other metamorphic rocks; and carboniferous limestones were constantly found largely developed. The only lower mesozoic rocks recognised were of triassic age, but traces of cretaceous beds were found to the south and west, whilst in the mountains north of Káshghar evidence of comparatively recent volcanic eruptions was met with. No representatives of the jurassic formations of the Himalayas and Western Tibet have hitherto been recognised in this part of Central Asia north of the Karakoram.

---

[1] Diary for 1st to 13th February. See also Severtzoff: Journal, Royal Geological Society, 1870, Vol. XL, pp. 410, &c. I am also indebted to Mr. Hume for a copy of a report by a Russian Engineer officer named Ramanoffsky, in which the occurrence of coal in Western Turkestan is described.

[2] I have described similar deposits in Persia: Quarterly Journal, Geological Society, 1873, Vol. XXIX, p. 493.

SCIENTIFIC RESULTS

OF

THE SECOND YARKAND MISSION;

BASED UPON THE COLLECTIONS AND NOTES

OF THE LATE

FERDINAND STOLICZKA, Ph.D.

SYRINGOSPHÆRIDÆ.

BY

PROFESSOR P. MARTIN DUNCAN, M.B. Lond., F.R.S.,

VICE-PRESIDENT OF THE GEOLOGICAL SOCIETY, CORRESPONDENT OF THE ACADEMY OF NATURAL SCIENCE OF PHILADELPHIA.

Published by order of the Government of India.

CALCUTTA:
OFFICE OF THE SUPERINTENDENT OF GOVERNMENT PRINTING
1879.

SCIENTIFIC RESULTS

OF

THE SECOND YARKAND MISSION;

BASED UPON THE COLLECTIONS AND NOTES

OF THE LATE

FERDINAND STOLICZKA, Ph.D.

SYRINGOSPHÆRIDÆ.

BY

PROFESSOR P. MARTIN DUNCAN, M.B. Lond., F.R.S.,

VICE-PRESIDENT OF THE GEOLOGICAL SOCIETY, CORRESPONDENT OF THE ACADEMY OF NATURAL SCIENCE OF PHILADELPHIA.

Published by order of the Government of India.

CALCUTTA:
OFFICE OF THE SUPERINTENDENT OF GOVERNMENT PRINTING
1879.

CALCUTTA:
PRINTED BY THE SUPERINTENDENT OF GOVERNMENT PRINTING,
8, HASTINGS STREET.

KARAKORAM STONES,

oʀ

SYRINGOSPHÆRIDÆ.

ʙʏ

PROFESSOR P. MARTIN DUNCAN, M.B. Lond., F.R.S., &c.

CONTENTS.

	PAGE
I.—The history of the discovery of the new order of Rhizopoda, the *Syringosphæridæ*, and the literature of the subject	1
II.—The general morphology of the Fossils, their histology, and their position in the classificatory scale. List of the genera and species	3
III.—A description of the genera *Syringosphæria* and *Stoliczkaria*	10
IV.—A description of the species *Syringosphæria verrucosa, Syringosphæria monticularia, Syringosphæria tuberculata, Syringosphæria plana,* and *Syringosphæria porosa*	11
V.—A description of *Stoliczkaria granulata*	16
VI.—Description of the plates	18

SCIENTIFIC RESULTS

OF

THE SECOND YARKAND MISSION.

KARAKORAM STONES,

OR

SYRINGOSPHÆRIDÆ.

BY P. MARTIN DUNCAN, M.B. LOND., F.R.S.

I.—THE HISTORY OF THE DISCOVERY OF THE *SYRINGOSPHÆRIDÆ* AND THE LITERATURE OF THE SUBJECT.

A number of spheroidal and of spherical stones, ornamented naturally on the surface, and which give no indications of ever having been attached to other bodies, could not but attract the attention of those geologists who years since travelled in Kashmir. Measuring in some instances two or three inches in diameter and in others not half an inch, and resembling stone balls in shape, these fossils, from the Karakoram range, became known to the curious as "Karakoram stones." But that they were not simple mineral productions was evident from the first to the educated collector; nevertheless, the nature of their external anatomy was singularly mistaken by those palæontologists into whose hands they first came. Dr. Verchère, when writing on the geology of Kashmir in the *Journal of the Asiatic Society of Bengal* in 1867, had the benefit of the palæontological skill of M. de Verneuil, and two plates of figures accompanied the descriptions of these remarkable forms.[1]

The description given of one species was that the bodies are "perfectly globular, covered with small rounded warts, sharply defined. The whole shell, between the warts, is pierced with minute pores. No traces of plates; no mouth nor stalk scar visible." The locality whence the specimens were derived was the rocky plains at the foot of the Masha Brum, Karakoram chain. The generic position was stated to be that of *Sphæronites*.

Another species had the name *Sphæronites ryallii*, Verch., given to it; and the diagnosis is as follows:—"Globular, large warts well set apart and not very sharply defined. The whole shell is covered with pores. No mouth. A stalk stem very conspicuous." A third specimen, also classed as a *Sphæronites*, is thus noticed :—"Depressed, no warts or spines: no plates or traces of plates, no stalk scar. The whole surface pierced by minute pores." These two specimens were derived from the same locality as the first.

[1] Journal, Asiatic Society of Bengal, 1867, Pt. 2, No. 3, Appendix p. 208, Plate VIII, Figs. 5 and 6, and Plate IX, Fig. 1.

The illustrations of this essay of Dr. Verchère do not assist the comprehension of the subject, and they were evidently drawn with a crinoidally disposed pencil. The so-called stalk stem is evidently an adventitious and accidentally adherent body.

The only other notice of the Karakoram stones previously to that of Stoliczka was attached to a specimen of one which was presented to the Geological Society by Major, now Colonel Godwin-Austen, and collected by him. This specimen closely resembles a *Parkeria*, and this did not escape the accomplished palæontologist, who, at the time of the reception of the fossil, had charge of the Museum of the Society. Professor Rupert Jones, F.R.S., wrote on the label of the specimen "*Parkeria*."

The next and the most important notice of the Karakoram stones was the last effort of Stoliczka, whose lamented death occurred soon after he concluded his short description of their geological position.

The following extract from Stoliczka's last diary places the subject at the point whence the present attempt to explain the morphological characters and the classificatory position of the Karakoram stones may be said to commence :—[1]

Extract from Stoliczka's last diary.

"*June 15th, Karakoram-brángsa, 14 miles.*—Starting from Woabjilga, the grey triassic limestones were met with, afterwards the red limestones succeeded them, and continued to camp, often interrupted by patches of greenstone, which is greatly developed at the camp north of the pass.

"*16th, Daulatbeg Uldi* (crossing the Karakoram pass), *about 22 miles.*—Leaving camp, the green stones are underlain by black crumbling shale, in mineralogical character like the Spiti shales, but are very likely triassic, like that near Aktásh. Then follows an alternation of grey or whitish limestones and shales, and the triassic red limestones; and on these rest blackish and grey marly shales, which are overlain by almost horizontal strata of brown limestone, very much like the lower Taglang limestone, and which contains fragments of *Belemnites*. These *liassic rocks* form the Karakoram range proper, and extend far eastward. The hills to the west are much higher, and do not allow a distant view.

After crossing the pass, the road skirts the base of the centre ridge in a south-east direction ; and here the liassic limestones come down several times, and about four miles from the pass grey marly shale, or almost marly limestone, crops out from under the brown limestones; both are evidently liassic. On the right bank of the stream more massive limestones occur, dipping to north-east, but very indistinctly. I should think that these are triassic limestones. They very readily crumble to pieces, being dolomitic; and these often contain reddish beds interstratified.

"*17th, Burtsi, 24 miles.*—First we crossed the Dipsang plain, with solitary low hills, probably still belonging to the Taglang series. Then we ascended towards the watershed. Upon this followed dolomitic limestone, rather indifferently bedded, massive and white, and this was overlain by bluish shales and well bedded limestone, extending from about six miles north of Burtsi to the camp. These limestones appear to be triassic; they are compact, with layers full of small gasteropods, among which I recognised a *Nerinæa*. The so-called Karakoram stones, *i.e.*, corals, occur in dark shales below the limestones, which are capped by a yellowish-brown limestone, well bedded, but of unascertained age. The whole series dips south-west at a moderate angle. [The last paragraph closes the diary.]"

[1] See the portion of the present work relating to Geology, by W. T. Blanford, page 45.

KARAKORAM STONES, OR SYRINGOSPHÆRIDÆ.   3

The late distinguished Palæontologist to the Geological Survey of India had traced these remarkable spheroids to their time and place in the succession of rocks, and he expressed an opinion regarding their zoological position. They were found in shales beneath limestones which were certainly lower than the Lias, and which were probably triassic in age. The term "coral" was singularly justified, for some of the superficial markings on the stones resemble, in their radiate appearance and regularity, the casts of the calices of minute *Madreporaria* of the genera *Astrocœnia* and *Stylocœnia*. But it is only necessary to remark that Stoliczka's great knowledge of the *Anthozoa* would have led him to the expression of a different opinion had his specimens been prepared for microscopic examination.

The so-called Karakoram stones collected during the second Yarkand Expedition by my lamented friend were placed in my hands by Mr. W. T. Blanford in 1878.

The specimens are numerous and in very perfect condition; the weathering to which some have been subjected rendering the outside details all the more visible. Their surfaces are free from other fossils, and a broken serpula tube is the only one to be recognised.

Fossilization has occurred by the introduction of calcite, and this is usually somewhat dark in colour, but is transparent in thin sections. The original structure of the body now consist of carbonate of lime of a different and lighter colour to the infiltrated calcite, and it appears that on the outside of the fossils the original structure has usually disappeared and the intermediate or infiltrated mineral has lasted.

Carefully made radial and tangential sections of the fossils, assisted by biting out with dilute acids, and the use of low and high powers of the microscope, assisted by the polarising apparatus, rendered their remarkable construction evident, and also that it was necessary to include all the Karakoram stones in a new order of *Rhizopoda* called the *Syringosphæridæ*. A notice of this new order was published in the Annals and Magazine of Natural History for October 1878, Ser. 5, Vol. II, page 297.

II.—THE GENERAL MORPHOLOGY OF THE FOSSILS, THEIR HISTOLOGY, AND THEIR POSITION
IN THE CLASSIFICATORY SCALE.

The Karakoram stones are either nearly perfectly spherical, or more or less spheroidal or ellipsoidal in shape. They may be of small size, and some are more than three inches in their greatest diameter; but they are always symmetrical, and there is no trace of a stalk or of any former attachment by the surface to other bodies. Some forms are nearly smooth, others are minutely granular, each granule having a definite construction, and the most numerous types have tubercules, wart-like growths, and large eminences crowded, more or less, with papillæ and little warts upon them. There is one group of forms with a very verrucose surface, and, on the other hand, another type is covered with a finely granulate surface; nevertheless this external structure does not interfere with the general curvature of the mass, the tops of the highest and lowest eminences never exceeding their symmetrical position.

The more rugose and mammilated surfaces of the fossils have small circular or deformed shallow pits scattered here and there; they are very numerous in some of the types with rounded surface tubercles, and are but scantily distributed in others, and whilst they crowd the surface of one form with a granular surface, they do not exist on another. These pits become elongate on the equatorial part of some of the spheroidal fossils, and are found on the sides and on the edge of the bases of some of the papillæ, tubercles and warts of other types. Their resemblance to minute oscula of sponges is superficially evident; but it is to be shown that one great group of the fossils under consideration does not possess them, that they differ in their number in different parts of the same fossil and in different individuals of the same species. I have called them "pores," and their absence in one of the groups of the fossils has led me to divide these Karakoram *Syringosphæridæ* into two genera—one with pores on the outer surface is termed *Syringosphæria*, and that without pores I have dedicated to Stoliczka's memory, terming it *Stoliczkaria*.

The method of examination of the fossils is necessarily a simple one. Their surfaces are usually well preserved and not over-weathered, and the insides, in the majority of instances, yield good sections, both radial and tangential. Careful washing adds to the details of the surface, and biting with hydrochloric acid and water is necessary to distinguish tube structure from the intertubular calcite of fossilization which sometimes simulates it.

The sections, on account of the brilliant opacity and white or white-brown colour of the tubes, can be well studied by reflected light, and indeed it is advisable to do this preparatory to the examination by transmitted rays. The dilute acid is very useful in some confused sections, for it dissolves the infiltrated calcite which exists between the tubes, and leaves their granular wall to a certain extent untouched. The paths of tubes can then be seen by reflected light very well. If the acid is allowed to act too strongly, all structure disappears.

The tubes, both radial and interradial, are easy to see in the majority of instances, but in one particular case polarized light and the selenite plate determined the visibility of the structures, which were hidden amongst a confused mass of calcite. The calcite which was introduced during fossilization fills the tubes as well as their interspaces, and it has taken on definite or indefinite cleavage planes. These must be studied under polarized light, for the dark lines they produce to ordinary transmitted light, and which simulate cœnenchymal structure, can then be decided to be only divisions between crystals or parts of different polarizing influence on the ray.

Low powers of the microscope suffice for most of the examination, but a good ⅛-inch object glass is required to distinguish the granules and granule-spiculate elements of the tubes.

No other form of fossilization but that by calcite has been noticed, and silica does not enter into the composition of the bodies.

On examining the surface of a rugged or tuberculate specimen of either of these genera with a hand lens, a reticulate appearance is seen between the projections. In very good specimens, on the ordinary level of the surface, after biting with dilute acid, or sometimes without this proceeding, this reticulation resolves itself into a gyrose tubulation; the tubes coming to the surface, running along it in close proximity, dipping down again suddenly and re-appearing, and sometimes bifurcating. Between the tubes is a more or less linear interspace filled with dark calcite. Weathering sometimes has destroyed the tubulation and left the thin interspace to look like a mesh, or the interspace has been left void, the tubules remaining.

Besides this reticulation, there are in some types numerous, and in others but a few minute openings from $\frac{1}{1000}$ to $\frac{1}{500}$ inch in diameter, and they have a margin or tube layer. They are sometimes separate, and at others they are clearly the outside opening of one of the superficial tubes just mentioned. Usually the caliber of the tubes is filled with brownish coloured calcite, or with granular carbonate of lime, but in some instances the presence of a very delicate tube wall, unattached by its outside to any structure, is evident.

On the projections, whether mammilated, wart-like, papillate, tuberculate or granular, there are markings to be seen which are of two kinds. On the top or centrally are circular markings, few or many, which on careful examination turn out to be the openings of tubes. They are often very minute, and their caliber is smaller than that of the tubes seen in the interspaces just alluded to. On the sides, and converging to the margins of the top of the eminences, are numerous close, straight lines, usually continuous, but sometimes wavy, broken and bifurcate. They are, according to the condition of the fossil, either the preserved calcite of converging tube interspaces, or they may be the walls of the tubes themselves, or both. These tubes may be traced on the surface to be continuous with some of those of the spaces between the projections, to appear from within the fossil and to run up outside the eminences. In many instances they open, finally, at the surface around those smaller ones which appear in the centre of the top.

In some forms, especially where the eminences are broad and low, these converging tubes open all over the projection.

It is evident that the projections, whether they are simple or compound, are made up of the outsides of tubes, tube openings, and of calcite which fills up the interspaces between them; there being much bifurcation and side inosculation of the tubes also. The projections, mammilation or granulate tube openings and convergings belong to a *radial tube series*, and the tubulation between these eminences to an *interradial series*. No cœnenchyma or skeleton exists.

The pores are spaces in the superficial interradial tubulations, but in rare instances they are found elsewhere. They are surrounded and limited at their margin by tubes bounded within by others, and their shallow floor has the outward openings of deeply-seated tubes on it. The distinction between the interradial tube reticulation and the radial tube series is best seen in the genus *Stoliczkaria*, on account of the definite intervals, without pores, which exist between the granules containing the end of the radial series. It is well seen in the pore bearing *Syringosphæriæ*, which have distinct eminences, and it is the least apparent in some spheroidal kinds, where there is as much space occupied by pores as by eminences.

The relative positions of the radial and interradial series of tubes, and the close and converging character of the one and the reticulate appearance of the other, must be kept in mind as this description proceeds, for they have the same definite relation within the fossil. In some species, moreover, the radial tubes are readily distinguished, because they are smaller than those of the interradial series.

This persistence of the radial series of tubes, and the environing interradial and reticulate tubulation, can be well seen in tangential sections of those types in which the structure is close; for instance, in *Stoliczkaria granulata*, especially if the thin slice is taken rather close to the surface of the body. Then a number of star-shaped masses are seen, separated from one another by a denser structure. The centre of the star contains small tubes cut across, and giving off small branches to the outside and separating structures, which consist of sections of larger tubes made in different directions; such as oblique, transverse, and longitudinal. The small tubes of the centre of the star are well separated from each other, except where they bifurcate, but the surrounding tube reticulation is close, the tubes being nearly in contact. Clear calcite fills the spaces between the small tube ends of the star, and there is less of it amongst the large tubes around. The opacity of the calcareous structure of the walls is evident, and they are usually brilliantly white or brown under reflected light. Here and there the lumen of a tube may be seen filled with calcite. (Plate III, Fig. 5.)

In other types the limitation and surrounding of the radial series of tubes by the larger and more extensive series of reticulating ones is readily seen in tangential sections close to the surface, but it becomes rather confused at some distance within, on account of the obliquity of the radial series in relation to the surface. For they start as it were from a central point in the fossil, and radiate in all directions, increasing in width and in their number of tubes. The distinction between the two series is readily made, however, for the interradial is usually the largest in extent and its tubes are eminently bifurcate and form close reticulations, bending often suddenly and showing geniculate outlines. Tangential sections further in, even although they are less distinct, show that the breadth of the radial and interradial series diminishes centrally.

In one group of the fossils, forming the genus *Stoliczkaria*, no trace of the surface-pores exists and no vestige of any of them can be distinguished in sections. But in the other group forming the genus *Syringosphæria*, the pores can usually be distinguished in some parts of the fossil, besides the surface, especially in tangential sections, as circumscribed structureless spaces filled with clear or opaque calcite. It does not appear that the tubes which pass out of the pores, at the surface, are restricted to one particular series, and they may belong to the radial group, or more frequently to the reticulate or interradial set.

Radial sections of the fossils show structures which correspond to those seen in tangential sections, and the morphology of the forms is divisible into two categories. In one, the structures consist of numerous conical congeries of bifurcating radial tubes, the apex being central and the base at the surface of the body; and of a reticulate tubulation separating the cones, joining their external tubes and arising from them. The cones and the intermedial reticulation increase in size towards and at the surface of the body, on which are widely or closely-placed tubes passing radially, tangentially, and obliquely. Hence the surface of the body presents the ends of the radial tubes and those of the reticulation, and it is mainly composed of the tubes which are placed tangentially over the circumference.

In those types of the *Syringosphæridæ*, where there are eminences with radial tubes surrounded by much space occupied with tubo-reticulation, the radial sections illustrate the structure of the whole admirably. One of these sections may be considered in three parts in order to explain the morphology. Near the centre fossilization confuses the structure, but it appears that a simple tubular structure arises from around a foreign body, such as a many-chambered Foraminifer, or that one tube branches suddenly in every direction. The tubes radiate in separate groups, each tube bifurcating frequently as it recedes outwards, and there are frequent lateral tubes connecting them together. Hence the mass of tubes increases in the number of its tubes, and forms in section a more or less triangular outline, the apex being towards the centre of the body. At the same time the tubes of the outside of the triangle or longitudinal section of the cone give off others which form in part the reticulation of the interradial part. This is small at first, but increases in section in the middle of the body.

In the middle of the body, in sections, the radial series is seen to be broader and the interradial to form large meshes. Close to the surface of the body, in sections, the radial series of tubes is seen to bifurcate to the last, and to open directly on an eminence so far as its central tubes are concerned, and many of the outer tubes pass obliquely on the flanks and open at the top. The interradial series also opens by its radial tubes at the surface and by its oblique tubes, but those parallel with the circumference pass over it. (Plate III, Figs. 1, 4, 6).

The relative size of the radial and interradial series is apparently of specific importance.

In the radial sections the pores are seen to be spaces surrounded by interradial tubes, some of which open on the floor.

But in the radial sections of those types which have a great number of radial series and a very scanty surrounding reticulate tubulation, the appearances under the microscope are not so striking as in the other instances. In these the radial cone is very long, and bifurcation occurs comparatively scantily, so that it is narrow, and the sides of the series often appear to be parallel. The tubes of the radial series, moreover, are smaller than the surrounding series; they are not so close together side by side, and their course is almost invariably straight. The interradial surrounding tubes are closer and larger than the others, and they bend so as to present oval or geniculate knots, the continuity of the tube being often lost to sight, a cross line denoting the upward or downward bend. They bend laterally also, and touch here and there and bifurcate. The size of this series is usually larger than the other, so that in these radial sections a radiating series of light lines is separated by broader dark ones.

This close structure is best seen in the group without pores, but it exists in the other, in some species. (Plate III, Fig. 6).

In one type of the *Syringosphæridæ* the pores are very developed, especially equatorially.[1] In radial sections their presence is evident in the body or from the surface. They extend in long rectangles one outside the other, and evidently bound radial series, but they are situated just within the interradial. On either side of them are elongate tubes, offshoots of the environing series, and separating one space from another; that is to say, from within outwards is a bridge of cross and reticulate tubes parallel with the circumference, like a tabula of a hydrocoral. Several of these bridges exist, and the last one is incomplete, often quite at the surface where a pore is about to be occluded.

In tangential sections, the circular outline of the pores may be seen surrounded with tubes.

In other specimens, this absence of tube-structure along definite lines, that is to say, the presence of pores, is not so visible, but they can be detected as vacant pits or circular spaces filled up with extraneous material.

No special tubes enter the pores.

The tubes forming both series are continuous, bifurcating, and inosculating; and, as has been already noticed, some are in the main straight and others are curved and form the edges or sides of greater or less meshes or vacant spaces.

The tubes are much larger in some types than in others, and they range from $\frac{1}{1000}$ inch to $\frac{1}{500}$ inch in diameter; they usually retain the same caliber for some distance or altogether, but frequently in some types they swell out, become varicose, flat, and again return to their original cylindrical condition. (Plate III, Figs. 6, 8.) The union of tubes is by small offshoots usually, but the bifurcation, often at an acute angle, gives origin to two tubes of equal size to the parent, or nearly so.

The tubes have a wall and a lumen, and the thickness of the wall varies; moreover, some of the constituents of it pass irregularly into the caliber, as well as occasionally surround the tubes like a furry investment.

[1] *Syringosphæria porosa*, Duncan. Plate III, Fig. 3.

There are no diaphragms in the tubes. In some types a part of the tube-wall is so homogeneous as to render the possibility of the former existence of a membrane well worthy of consideration; but in the majority of instances, the construction of the wall is evidently of close and semi-spiculate granules and of shapeless granules, and was probably not quite impervious. The tubes are filled with calcite. They are often perfectly transparent, and at other times impervious to light. Under high powers the structural element of the tube is shown to be mainly spiculo-granular and molecular; the grains usually being $\frac{1}{10000}$, $\frac{1}{15000}$ inch, or less in breadth. But in some instances there are elongate pieces with spiny processes on them, all being however excessively small. The structure of the tube-wall was organic in its origin, and not the result of simple adhesion of foreign or arenaceous particles.

The question whether there is an intertubular cœnenchyma of fibres, or a reticulate skeleton, which supports the tubes, separates them, and allows the symmetry and ornamentation of the surface to be kept up, is by no means readily answered. The examination of the forms of *Syringosphæridæ*, with the radial series of tubes separated by much tube reticulation, leaves this question not satisfactorily solved. The fossilization is by calcite, and the cleavage planes, commencing cleavage planes, irregular crystals, and cracks show dark lines by transmitted light, which in many instances resemble sponge structure, and even in one instance a hexactinellid spicule was suggested to the eye. Polarized light, with or without the selenite plate, resolves these markings into the limiting lines of different crystals, and, although one or two evidently extraneous organic bodies have been seen amongst the tubes, no continuous or partial interskeleton can be determined to exist now. In the centre of the masses, the confusion of tube radiations, cleavage planes, and the presence of some foreign body, which formed in some instances the nucleus, or rather the starting point of the *Syringosphæridæ*, renders it impossible to decide dogmatically whether there is a cœnenchyma or not. On the other hand, in those forms where the tubes are close, even in the interradial series, the absence of cœnenchyma is evident enough. Under correction, and relying on the specimens examined, I do not think that there ever was a structure in them external to the tubes and which supported and separated them after the manner of a cœnenchyma.

The position of these spherical and spheroidal masses of radiating and interradiating tubes in the classificatory scale must be low. The minute size of the tubes, their bifurcating so frequently, and inosculating, and giving off others from small offshoots, and the structure of the wall, do not render the *Syringosphæridæ* polyzoan in their nature. The analogy with the tubular or more or less globular masses of *Fascicularia* found in the English Crag is of the slightest in degree. It is tempting to theorize, so as to place a Gastrozooid in each pore, supplying it by the radial tubulation, and to decide that the tubes of the interradial series opening at the surface were those of Dactylozooids, the whole being a hydroid. But the absence of pores in some forms, the evidence that there are places where growth is not proceeding in others, and the deficiency of surrounding open tube mouths in most, prevents this idea from having any value. There are moreover no tabulæ in the tubes.

That these great and small spherical and spheroidal masses are corals is, of course, out of the question, and the evidence of their sponge nature is small.

Had there been a cœnenchyma between the tubes, the bodies would have resembled foraminifera, with gigantic canal systems, but its absence and the peculiar nature of the tube-wall remove these forms from that polymorphic group. The absence of labyrinthic spaces, and the fact that the tubes are not formed by arenaceous particles, separate the *Syringosphæ-ridæ* from the arenaceous foraminifera of the *Parkeria* group.

It is evident that the calcareous granules and spicules were not collected by these tube-makers mechanically, and their occasional presence in the tubes themselves, and their extending beyond them, but still clinging to the furry outside in other instances, show that the tube structure is organic in origin and that it resembles that of some *Rhizopoda*. The symmetry of the bodies could only have been maintained by a common sarcode, enveloping the whole; food could only have been obtained by pseudopodia from the tubes, and these soft external substances would not be unfavourable to the shape of the mass, and to its never being found worn by resting or attrition.

That these fossils are rhizopodous is almost a necessary belief, but it is evident that they cannot be brought within the order *Radiolaria* any more than they can within any group of the foraminifera. It remains, therefore, to establish a new order, the *Syringosphæ-ridæ*, amongst the class *Rhizopoda*, and to include these triassic or lower liassic fossils within it.

Class: RHIZOPODA.

Order: *SYRINGOSPHÆRIDÆ.*

Genus: **SYRINGOSPHÆRIA.**

Species SYRINGOSPHÆRIA VERRUCOSA.
,, S. MONTICULARIA.
,, S. TUBERCULATA.
,, S. POROSA.
,, S. PLANA.
Variety S. MONTICULARIA var. ASPERA.

Genus: **STOLICZKARIA.**

Species STOLICZKARIA GRANULATA.

III.—A DESCRIPTION OF THE GENERA *SYRINGOSPHÆRIA* AND *STOLICZKARIA* OF THE ORDER *SYRINGOSPHÆRIDÆ.*

Order: SYRINGOSPHÆRIDÆ.

Body free, spherical or spheroidal in shape, consisting of numbers of limited, more or less conical, radiating congeries of minute, continuous, long, bifurcating and inosculating tubes; also of an interradial close or open tube reticulation arising from and surrounding the radial congeries. Tubes opening at the surface on eminences and in pores, and ramifying over it. Tubes minute, consisting of a wall of granular and granulospiculate carbonate of lime. Cœnenchyma absent.

The presence of pores on the surface of some forms of the order, and their absence in others, and the very close nature of the interradial reticulation in the poreless kinds, necessitates its division into two genera.

Genus: **SYRINGOSPHÆRIA.**

Body large, symmetrical, nearly spherical or oblately spheroidal, covered with large compound wart-like prominences with intermediate verrucosities, or with compound monticules having rounded summits, with solitary eminences between them, or with close broadly rounded tubercles, or with minute granulations. Rounded, or oblique, or linear depressions occur on the surface usually between the eminences, but sometimes upon them; they are shallow and are bounded by tubes, some of which open on their floor. The surface has tubes opening on it from the internal radial series, and also from the interradial tube reticulation; also masses of tubes running over it, converging on the eminences, and more or less reticulate elsewhere. Radial congeries of tubes numerous and defined, and the interradial tubulation is open or close and varicose.

Genus: **STOLICZKARIA.**

Body very large, symmetrical, oblately spheroidal, covered with a great number of minute distinct granulations, which are circular at the base, short and rather flat where free, and which are separated by an amount of surface about equal to their breadth. No pores exist. Tube openings occur on the granulations, and tubes, with or without openings, converge to their base and cover the intermediate surface. The tubes opening on to the granulations are terminations of the very numerous radial series, and are small; and the others, which are larger, belong to the closely-packed varicose and much contorted interradial series. The body within consists of a vast number of small, not very conical, but rather straight, radial series, whose rather distant tubes give off minute offshoots to the surrounding large tubes of the close interradial series. No cœnenchyma can be discovered.

I have named the most remarkable of all these fossils, those which belong to the poreless division of the order, after the distinguished Palæontologist, whose loss, whilst in the performance of his duty and whilst studying these very forms, is greatly and justly regretted.

IV.—A DESCRIPTION OF THE SPECIES OF THE GENUS *SYRINGOSPHÆRIA*.

There is nothing more unsatisfactory than the endeavour to separate and define rhizopodal forms into species, and the attempt would not have been made in this instance were there not five well-characterised types of the first, and one of the second genus.

As the presence and absence of pores have been held to be of generic value in classifying the order, so the paucity or abundance of them can enter into the specific diagnosis; moreover, the surface ornamentation, although of doubtful value, becomes more important to the specialist when it is accompanied, or not, by an open or close condition of the interradial tube series.

There is one group of the genus *Syringosphæria* in which the pores are in excess, and occupy as much of the surface as the eminences do. This forms a specific distinction and is all the more important, because the presence of former pores can be detected within the body, and the interradial tube reticulation is rather close. These, then, are the specific characters of *Syringosphæria porosa*, Plate II, Figs. 3 and 4.

The kinds with compound verrucose elevations have a moderate number of pores and a very open tube reticulation in the interradial series; they form, with the group possessing compound and simple monticules, a tolerably well-defined set, divisible into two species by the surface growths. They are *Syringosphæria verrucosa* and *Syringosphæria monticularia;* Plate I, Figs. 1 to 12; Plate III, Figs. 1 to 4, 8 and 9. The species *Syringosphæria monticularia* is, however, subject to variation, and the monticules may be very flat, the whole surface being nearly level, or the eminences may be sharply defined. The forms classified under the last head constitute the variety *aspera;* Plate II, Figs. 6 and 7. A form with granular and minute processes with pores leads to the next genus. It is *Syringosphæria plana.* All these are well defined and readily recognised species.

There is but one species of the genus *Stoliczkaria*, the granulate, poreless surface of which distinguishes it from all other forms of the order.

SYRINGOSPHÆRIA VERRUCOSA, Duncan. Plate I, Figs. 1 to 3.

The body is spheroidal in shape, and the surface has numerous large compound wart-like or rounded or conical mammiliform eminences on it, and also solitary mammiliform projections, as well as small, distant, sharp granules. Numerous minute, shallow, circular pores exist, especially on the bases of the verrucose and mammiliform projections, and there are some on the surface between them. The largest of these eminences are on the equatorial region. The surface between the great and small verrucosities and mammiliform eminences supports the majority of the small granulations, and is covered with closely-packed tubes and many tube openings. The tubes run short courses, bend and dip down, and are from $\frac{1}{200}$ to $\frac{1}{100}$ inch in diameter. They are separated by linear, low projections of dark coloured calcite, and very frequently the tube has disappeared and left these limiting products of fossilization only. The openings of the tubes at the surface are surrounded by circular rims of the dark calcite.

The top of every mammiliform, conical or verruciform eminence is smooth, and many tubes open on the summit and resemble circular patches of a slightly different colour to the brownish calcite which environs them. On the sides of the eminences, and reaching around and more or less on to the summit (Plate I, Fig. 3), are converging, wavy, linear projections of calcite, separated by long broad spaces. The spaces are the remains of tubes, and amongst them are wavy tube openings, limited by calcite rims. The pores have tubes around them and opening on their shallow floor, and they appear to be parts where the upward growth of some radial systems has not been as rapid as the interradial. The height of the body is $1\frac{1}{2}$ inch, and the breadth is $1\frac{3}{4}$ inch. The diameter of the base of a large compound verrucose prominence is $\frac{3}{10}$ inch. In the fossilization of this form the tube-wall is light brown and the calcite, which has been infiltrated, is darker brown and smooth.

SYRINGOSPHÆRIA MONTICULARIA, Duncan. Plate I, Figs. 4 to 12; Plate III, Figs. 1, 2, 3, 4, 8 and 9.

The body is oblately spheroidal in shape, and the surface has wide-apart, low, rounded, compound mammillæ on it, consisting of one large rounded eminence surrounded by many smaller; also solitary, short, flatly rounded mammillæ, and very small blunt granules of two or three sizes may exist. The pores are very numerous and are small, being found everywhere on the surface, and opening directly or obliquely.

The intermammillate surface is marked mainly with the openings of tubes, and by a few sides of tubes passing for a short distance on the surface and converging on the eminences. Most of the tubes are $\frac{1}{100}$ inch in diameter. The mammillæ are crowded with tube openings which are circular, and often the lighter colour of the substance within the tube is seen surrounded by infiltrated calcite. In some specimens the tubes are excessively bent and geniculate, and they dip down or end suddenly. They surround the pores and open into them. The tubes are crowded, close, and the linear dark calcite often alone remains, indicating the lateral limits of former tubulation.

KARAKORAM STONES, OR SYRINGOSPHÆRIDÆ. 13

Radial sections show the radial series of tubes to bifurcate or inosculate frequently, and to increase in size in varicosities. These tubes mainly go to the surface and open there directly; and some of them give off branches on all sides to form the interradial tube reticulation. As much of this reticulation consists of radiating tubes, the last series of them opens at the surface. The tubes of the outer meshes are also represented at the surface by flat or bent tubes. The interradial series thus formed separates, very distinctly, the wide conical radial congeries from each other. Almost every mammilla has its radial congeries of tubes. The diameter of the smallest lateral tubes given off is $\frac{1}{1000}$ inch, but the average size of the tubes is $\frac{1}{500}$ inch in diameter. Near the surface there are occasionally great differences in the size of the tubes, many of which become flat, and the same spreading out is seen further in, where the granular element of the tube-wall has been formed in excess.

The typical specimen is $\frac{1}{15}$ inch high and 1 inch broad. The diameter of the pores is $\frac{1}{80}$ inch to $\frac{1}{70}$ inch. (Plate I, Figs. 4, 5, 6).

A young specimen has the compound mammillæ hardly formed, but the single ones and the pores are abundant. It is more spheroidal than the type (Plate I, Figs. 7, 8, 9). The magnified radial sections (Plate III, Figs. 1, 8, 9) were taken from this form.

A variety of the species has a larger body than the type (Plate I, Figs. 10, 11, 12), but the mammillæ are low and insignificant. The magnified oblique section, showing the divergence of the very open tube series (Plate III, Fig. 4), is from this form, as is also the top of a monticule showing tubes and tube openings (Plate III, Fig. 3).

SYRINGOSPHÆRIA MONTICULARIA, variety ASPERA, Duncan. Plate II, Figs. 6, 7.

This transitional variety has very few compound mammillæ, but a great number of single ones and pores. It is a large form, and is oblately spheroidal, about 1 inch in height and 2 inches in breadth. It was collected by Colonel Godwin-Austen, and is introduced here in exemplification of the series.

The radial section shows that the radial congeries are very widely separated by reticulate tubulation; that the tubes are large, usually $\frac{1}{300}$ inch, that they have a very delicate wall, are often varicose, and that they pass in great multitudes to the surface close together. Further in, the intertubular space equals the diameter or the tubes, and gives rise to much confusion, and it is difficult to know, except by reflected light, which is tube and what is calcite infiltration.

In some parts the tube reticulation is close, and the tubes crowded together, and in this there is an approximation to the next species.

SYRINGOSPHÆRIA TUBERCULATA, Duncan. Plate II, Figs. 1, 2.

The body is spherical and symmetrical in shape, and is covered with numerous low, rounded, broad elevations, separated by indistinct interspaces. There are minute pores scattered over the whole surface. The eminences about $\frac{1}{16}$ inch across at their base, are not $\frac{1}{2}$ of that measurement in height; they are sometimes irregularly shaped. In some parts the interspaces are as broad as the bases of the eminences, but usually the slope of one eminence merges into that of another, the interspaces being confined to the concavity. The interspaces are covered with a very crowded and close arrangement of the tubes; many

D

tubes pass out radially on them, and the orifices are only seen; others come up to the surface and bend down again suddenly, leaving a geniculate swelling visible; and others enlarge and diminish in their caliber. Some of these pass along the surface for a very short distance, and all very close together laterally, and others pass up the flanks of the eminences converging close to the summit and opening on them with their orifices, or more frequently on the centre of the tubercular elevations.

The pores are numerous, small, shallow, and universal; they are limited by lateral tubes, and some open on their floor. The fossilization is by calcite, and in many places the interspace between the surface tubes infiltrated with calcite has been preserved, the tubes having weathered away. The tubes are so close together that the infiltrated calcite is difficult to distinguish from tube; but its breadth is usually much the smaller.

In radial sections the radial series of tubes are numerous and large, but the interradial systems are not very distinct from them, there being no wide tube reticulation.

The tubes of the radial series are rather close, large, bifurcate, varicose, geniculate often, suddenly diminishing in size where joining others; they join much with each other, side by side, are usually distinctly radial in their direction, which, however, is locally irregular, and they have thin walls and a large caliber.

The interradial tubes, very radial in their course, however, are often seen passing for short distances, parallel with the circumference, in all parts of the body. They are more varied in their courses than the radial series, and are usually close together and crowded, the distance between them being small. They unite with the radial systems by offshoots of tubes, and it is evident that at the surface of the body most of the interradial tubes open directly outwards.

There is no very definite relation between the outward opening of the tubes within and the eminences and interspaces; moreover, the pores are situated without order.

The majority of the tubes are nearly $\frac{1}{500}$ inch in diameter, some being $\frac{1}{500}$ inch, but very small tubes are rare.

The fossilization of the interior of the body has led to radiating portions being infiltrated with a denser semi-granular calcite which hides much structure, and especially centrally. In some places the tubes are filled with opaque matter, and the intertubular spaces are readily distinguished, whilst in others the intertubular spaces are large, and the tube has either disappeared or remains in very transparent calcite. Under this condition, it is difficult to distinguish tube from continuous infiltrated calcite in section. Relics of the pores, as clear spaces, are to be seen in radial sections. The height of the body is $2\frac{3}{10}$ inches, and the whole resembles a *Parkeria*.

SYRINGOSPHÆRIA PLANA, Duncan.

The body is oblately spheroidal, almost smooth, on the surface, with many minute granules on it, and numerous small shallow scattered pores. The granules are flat, with rounded, or elongate, or irregular bases, and are about the same size as the pores. Many tubes open on them, forming circles on their periphery, and also into the pores, and there is considerable variation in their caliber. No tube reticulation exists on the surface, but the massing of the tubes is closer in some places than in others.

In radial sections of the body a very marked tube arrangement is to be seen. A very considerable number of long, narrow, radial series pass on all sides to the surface, bounded and environed by broader interradial series, with slightly larger, closer, and very bent tubes. The tubes of the radial series are wider apart than the others, although their course is usually radial and straight; they often bend much here and there, are irregular, and are often geniculate at the sides. They unite by means of very small offshoots, and bifurcate, but rarely increase in number sufficiently to present the aspect of a cone in the mass. They rather form linear radial lines.

The larger and closer interradial series bend, unite, bifurcate, and are singularly gyrose, varicose, and irregular in their course in many places. They are often so close together that they resemble knots of tubes, and then the section having cut across many, exhibits the more or less circular incision in the tube-wall and the lumen.

The tubes are usually $\frac{1}{500}$ inch in diameter, those of the interradial series being the largest. Throughout the number of tubes in the interradial series is very great.

In some spots calcite has filled up a vacant space which was evidently once a surface pore, and in one or two places the tubes end at one of these places. New tubes were formed distally to the space by the arching over of side ones, and the branches taking a radial direction. In some parts the radial tubes are smaller than in others, and then there is manifest difference between them and those of the adjoining interradial series, which branch give off offshoots from one side, and twist in a close and remarkable manner.

The interspaces between the radial tubes are the largest, and those of the interradials are very minute.

Towards the centre of the section a confused mass of convoluted tubes exists, and the radial and interradial series appear to start from it. The tubes are thin at the wall, and the structural element, granular, molecular and thinly set, is minute in the extreme.

At the surface of the body every granule with its circlet of pores is the outlet of a radial series, and the space between the granules, pores included, represents the interradial structure within.

The greatest breadth of the spheroidal body is one and a half inch.

Syringosphæria porosa, Duncan. Plate II, Figs 3, 4.

The body is very oblately spheroidal in shape and symmetrical. The surface is covered with minute low, rounded granules. The granules vary much in size, the pores are exceedingly numerous and unequally distributed, and the space between many of them is in ridges, giving a boldly reticulate appearance, especially equatorially. No large amount of tube reticulation is visible on the surface; on the contrary, it appears, except at the pores, to be made up of tubes opening directly with circular or oblique outlines, and of wide intertubular interspaces filled with dark calcite. Where there is much space between the pores, the irregularity of this calcite indicates the former existence of peripheral tubes which have weathered out; but where the granules show any structure, it is that of tubes on their sides, converging upwards and opening at the top, and of tubes opening on the centre of the top. The pores are clearly spaces where tube-growth has not progressed equally with that of the surrounding parts. The sides of the pores present tubes passing radially, and tubes open on their floor.

Tangential sections, under low powers, exhibit localised and more or less circular groups of tubes which correspond to granules. In some the tubulation is reticulate, and in others, so radial that only the cut ends of tubes are seen. There are spots where the reticulation is very diffused, the tubes being very irregular in size, shape, and position in the section. In some places the tubes are very close, bifurcate, as in the other instances, and are more or less around the circular groups. There is not much difference in the size of the tubes, which vary from $\frac{1}{500}$ to $\frac{1}{150}$ inch. There are spots without any tube structure, and those are circumscribed and are the relics of old pores, passed by during the radial growth of the body.

In radial sections there is in many places such an exact relation in shape between the tube-structure, whether reticulate or radial, and the interspaces, that it is very difficult to distinguish interspaces filled with clear calcite from very transparent tubes. So many circular spaces exist, $\frac{1}{500}$ inch in diameter, in these parts of the section, that they may be taken for tube sections, surrounded by a whitish and rather opaque calcite. But they are really interspaces, the true tubes having the translucent walls. The radial series is not, on the whole, very distinguishable from the interradial, but the pores exist as vacant elongate spaces bounded by tubes all around, and bridged over tangentially by tube reticulation. They are not lined by any special structure.

The minute structure of the tubes is a finely granular substance (carbonate of lime), lightish red to transmitted light, and there are dark granules like minute dendrites. There is no trace of a cœnenchyma, and the fossilization simulates many structures, which are, however, readily resolved by even low powers of the microscope.

The height of the body is $1\frac{1}{10}$ inch, and the breadth 2 inches.

V.—THE SPECIES OF *STOLICZKARIA*.

One species of this genus is amongst the collection, and its forms are readily known by their great size, minutely, but not sharply, granular appearance, and the absence of pores.

STOLICZKARIA GRANULATA, Duncan. Plate II, Fig. 5; Plate III, Figs. 5, 6, 7.

The body is large, spheroidal, and symmetrical; it is covered with a vast number of minute eminences and interspaces. The eminences are separated by about their own breadth, or they may be closer, touching at their bases; they are usually circular in outline, low, flat or rounded at the free extremity, and are about as tall as their base is broad. There are usually five, and the corresponding interspaces, in $\frac{1}{10}$ inch. In some places the bases are continuous so as to form long narrow gyrose ridges, and in others they are absent, the circular base existing only. Here and there are some larger ones, and minute granules are interspersed.

Rather large tubes are on the outside and flanks of the eminences, and they open around and close within the circular top edge. They pass on to the spaces between the eminences, and are closely crowded, very bent, and form a dense reticulation, some opening there outwards.

The inner or central part of the upper surface of the eminences has a few, rather wide-apart tubes opening there; they are radial and small, and are readily distinguished from the interradial series around. Where an eminence is rudimentary, the central radial tubes may be seen separated by a little interspace from the dense reticulation of larger and closer interradial tubes.

Sections of the body tangentially show a vast number of small circular radial systems, surrounded by encircling interradial tube-structures (Plate III, Fig. 5). The tubes are for the most part seen cut across, and the radial are very small, few in number, and are wide apart. The surrounding mass of tubes consists of those of large caliber, often with minute offshoots to the radial series, and usually very varied in shape and size on account of their gyrose, varicose, rapidly bending course, of their inosculating and bifurcating, and of the necessary obliquity of their section. They are close and crowded. Both series have the tube-wall developed and thin, and the radial tubes are usually $\frac{1}{147}$ inch in diameter, the others measuring usually not much less than $\frac{1}{168}$ inch. The section gives the appearance of a multitude of stars by transmitted light, the centre of each being most distinct and occupied by the radial tubes. These combined series do not increase much in their size from within outwards, and they are $\frac{1}{70}$ inch across. The interradial tubes of one system communicate with those of the neighbours, and with the surrounding radial series sometimes. The sections of some of the interradial tubes present a flask-shaped outline, and this arises from the radial tubes or the interradial now and then giving off very delicate tubes of connection.

The sections made radially present a totally different appearance to those just described.

A little way below the surface a series of nearly equal parallel systems of tubes is seen; one set of tubes is closely crowded, and they are close, large, swell out here and there, bend, bifurcate, and give off minute offshoots. The other consists of a few wide-apart, narrow, not over-straight, tubes which give off tubes of their own size or a little smaller to each other and to the larger tubes of the set at their side. The larger set is the interradial system, seen, longitudinally or radially, and the smaller by its side is a radial system. Next comes another interradial system, about as broad as the radial one thus included, or perhaps a little broader; (Plate III, Fig. 6).

When the radial section is examined, close below and at the surface, the large tubes of the interradial systems are seen in lines, with the smaller radial ones parallel with them.

The height of the body is $2\frac{1}{14}$ inches, and the breadth nearly 3 inches.

VI.—DESCRIPTION OF THE PLATES.

PLATE I.

Fig. 1. The body of *Syringosphæria verrucosa*, Duncan. Natural size.
,, 2. A portion of the surface of the same specimen magnified to show the superficial projections, pores, and tubulation.
,, 3. The top of a large eminence, with pores on its sides; the tubes are seen crowding the surface, and many round markings at the apex denote the openings of internal tubes. The specimen is the same as the last, and is more highly magnified.
,, 4. The body of *Syringosphæria monticularia*, Duncan. Natural size.
,, 5. The same specimen magnified in part to show the monticules, pores and openings of tubes, with many ramifying and superficial tubes on the surface.
,, 6. A monticule more highly magnified to show canal openings, canals and spaces between them, also some small monticules.
,, 7. The body of a smaller and less mature specimen of *Syringosphæria monticularia*.
,, 8. A portion magnified, the radiating canals and the canal openings being shown on the monticules.
,, 9. A portion more highly magnified, showing a large monticule and smaller ones, with superficial tubulation and the exit of internal tubes. Pores are also shown.
10. A part of the body of a large specimen of a mature *Syringosphæria monticularia*.
11. A portion considerably magnified, showing a minute monticule and two pores. The tubulation is between the dark lines, and the dots on the monticule and elsewhere are the openings of internal radiating tubes.
12. A portion less highly magnified, showing numerous minute pores and larger monticules.

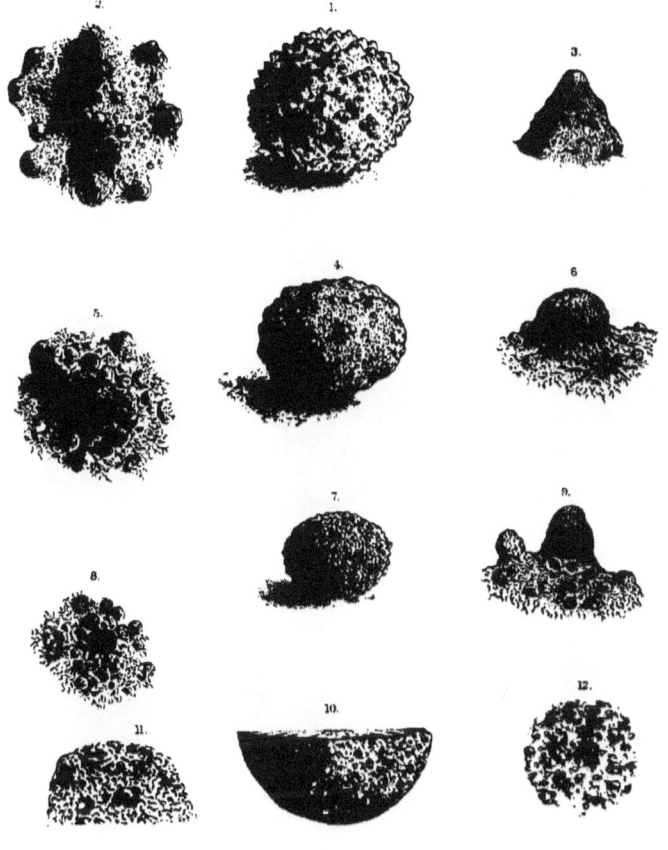

SYRINGOSPHÆRIDÆ.

PLATE II.

Fig. 1. The body of *Syringosphaeria tuberculata*, Duncan. Natural size.
,, 2. A portion magnified, showing the tubulation on the surface of the body and monticules, and a few pores.
,, 3. The body of *Syringosphaeria porosa*, Duncan, shown in outline, with a portion indicating the numerous pores. Natural size.
,, 4. A portion magnified, showing numerous round pores with canal openings and the intermediate surface with indistinct tubulation.
,, 5. The body of *Stoliczkaria granulata*, Duncan, shown in outline. The upper portion of details is of the size of nature, and indicates the numerous irregularly disposed granulations. The lower portion is in part magnified to show the numerous granulations, the tube openings on their top and their radiating tubulation on their sides and in the intervening space.
,, 6. The body of *Syringosphaeria monticularia*, Duncan, variety *aspera*. Natural size.
,, 7. A portion magnified, showing the openings of tubes on the monticules and the other tubulation, the black lines being interspaces between stout, crooked tubes.

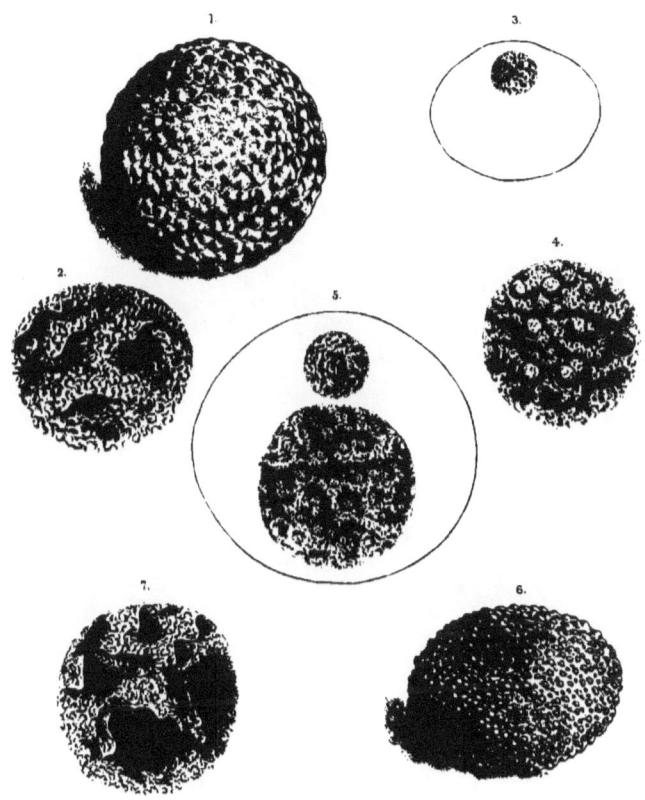

SYRINGOSPHÆRIDÆ.

PLATE III.

Fig. 1. A section taken from *Syringosphæria monticularia*, Duncan, the specimen being figured on Plate I in figure 7. The section is radial, and the top represents a small monticule at the surface; the lower part is towards the centre of the body. Many tubes are seen reaching to the surface and opening, some on the faintly rounded monticule, and others in the depressed part. The tubes in the centre of the section are essentially part of a radial congeries. At the sides there is tube reticulation and some of the endings of those tubes are seen at the surface. Swellings of the tubes are seen in some places. Magnified, half-inch object glass.

" 2. The surface of a specimen of *Syringosphæria monticularia* magnified, showing on the right a small pore with one large tube opening and two smaller. The dark, straight, bent, and branching dark lines elsewhere are the calcite intertubular infiltration, and the white or shaded spaces between them are tubes, some running, as on the left, a short course and opening on the surface, others bounding the pore, and some only showing geniculate portions of their track.

" 3. The top and sides of a small monticule of the same specimen, less highly magnified. There are tube openings of the radial series in the centre, and portions of tubes, partly of the radial and partly of the interradial series, covering the sides of the monticule, and opening externally around the top.

" 4. An oblique section near the centre of a specimen of *Syringosphæria monticularia* magnified. In the centre is what may be called a parent tube which gives off others that in turn bifurcate and radiate. Those on the sides of the section are becoming interradial reticulations, and are here and there irregularly swollen. Many small tubes cut across are seen disconnected. The central tubes are two radial sets, and the bifurcating is very characteristic.

" 5. A tangential section of one of the granules of *Stoliczkaria granulata* magnified. The small radial tubes open in the midst directly, and the large interradial tubes, most irregular in their outline of section, are, some of them, provided with neck-like prolongations. These are connected with the small radial series.

" 6. A longitudinal section of the same specimen and through a granule. In the centre are a few inosculating and bifurcating small tubes, and three of them open at the surface on the top of a granule, being equivalent to the central openings in figure 5. On either side are large interradial tubes, two uniting with the radial series by small short neck-like tubes. Magnified under quarter-inch object glass.

" 7. The surface of a rugged part of the same *Stoliczkaria* slightly magnified. The granules show tube openings on them and some large tube reticulation, the dark lines being intertubular weathering.

" 8. A longitudinal or radial section of *Syringosphæria monticularia* magnified. The depression is a pore, and the relation of some tubes to it is shown. Other tubes are opening out at the surface close by.

" 9. A radial section of the same specimen showing an interradial tube reticulation opening at the surface and running over it, forming there a tubular series. Elsewhere the irregular size of the tubes is shown and their general reticulation.

Plate III

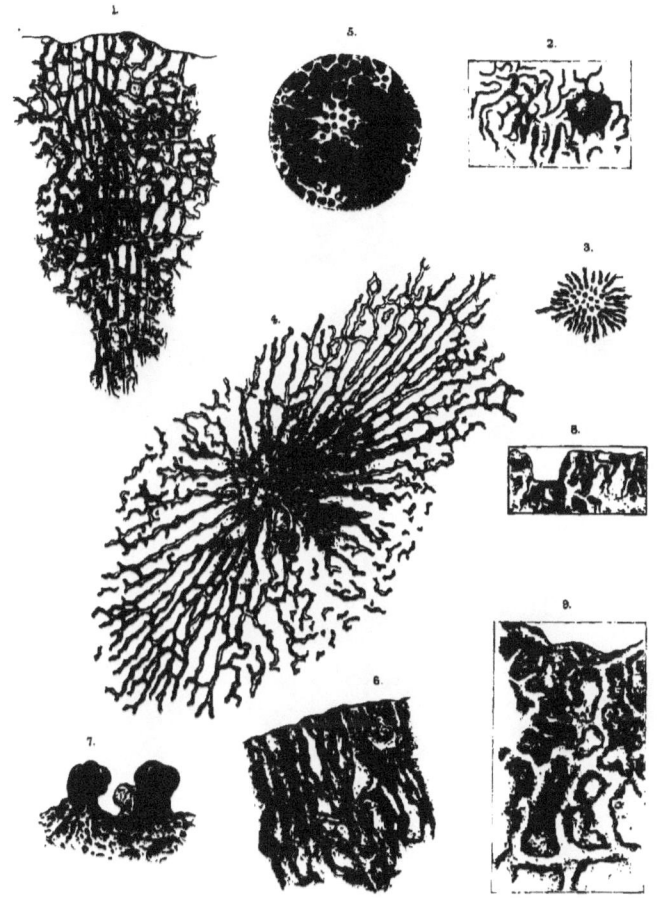

SYRINGOSPHÆRIDÆ.

www.ingramcontent.com/pod-product-compliance
Lightning Source LLC
Chambersburg PA
CBHW051235300426
44114CB00011B/747